second edition

Cytogenetics

The Chromosome in Division, Inheritance and Evolution

Carl P. Swanson
University of Massachusetts

Timothy Merz
Medical College of Virginia

William J. Young
University of Vermont

Prentice-Hall, Inc., Englewood Cliffs, NJ 07632

Library of Congress Cataloging in Publication Data

Swanson, Carl P
  Cytogenetics.

  Bibliography: p.
  Includes index.
  1. Cytogenetics. 2. Chromosomes. I. Merz, Timothy, joint author. II. Young, William
Johnson, date   joint author. III. Title.
QH430.S98   1981   574.87'3223   79-27393
ISBN 0-13-196618-9

Previously published as *Cytogenetics*.

Interior design and editorial/production supervision by Steven Bobker.
Manufacturing buyers: Edmund Leone and John Hall

Printed in the United States of America

10 9 8 7 6 5 4 3

Prentice-Hall International, Inc., *London*
Prentice-Hall of Australia Pty. Limited, *Sydney*
Prentice-Hall of Canada, Ltd., *Toronto*
Prentice-Hall of India Private Limited, *New Delhi*
Prentice-Hall of Japan, Inc., *Tokyo*
Prentice-Hall of Southeast Asia Pte. Ltd., *Singapore*
Whitehall Books Limited, *Wellington, New Zealand*

To: Dot, Phyllis, and Joan

Contents

Preface to the Second Edition

During the thirteen years that have elapsed since the appearance of the first edition of this book, the science of cytogenetics has undergone a phenomenal expansion, acquiring in the process a much more varied and dynamic aspect than that which has characterized its immediate past, and including within its scope a far wider range of species from every taxonomic group. The chromosome, the principal object of cytogenetic attention, has been moved front and center on the biological stage. Once studied primarily by the light microscopist for its own sake, or because of parallel interests in inheritance or evolution, the chromosome, from viruses and bacteria to man, has now become the research object of choice for a broad spectrum of biological scientists seeking answers to an equally broad range of questions. The chromosome crosses disciplinary lines in all directions. Virologists, bacteriologists, protozoologists, immunologists, biochemists, human geneticists, genetic evaluators, molecular biologists and electron microscopists are now generating information that can readily be intermeshed with the more classical data and ideas of the past, creating thereby a subject matter far richer in approaches, techniques, concepts and promise, and one that, potentially, embraces all living organisms instead of only those selected few species— *Drosophila melanogaster,* corn, *Neurospora crassa, E. coli,* Datura, Oenothera, wheat, Tradescantia, and grasshoppers—knowledge of which provided the early basis of cytogenetics. This poses a problem for the beginning student of cytogenetics who must acquire the philosophy, and learn to use the jargon, equipment, and techniques, of a number of biological subdisciplines, but the difficulty is also a challenge, for it is at the interfaces of these subdisciplines where the excitement of discovery and synthesis is most likely to occur.

The basis of this explosive growth stems from the fact that the molecules that form, or are determined by, the chromosome—that is, the macromolecular triad of DNA, RNA, and protein—are manageable and manipulable in a myriad of new and exciting ways, while the chromosome, as a complex whole, has exhibited unexpected and revealing variations when exposed to techniques of specific staining and labelling. The arsenal of the cytogeneticists of today—using the term "cytogeneticist" in its broadest sense to include all who contribute to our understanding of the structure and behavior of chromosomes—is, therefore, a varied and powerful one: techniques of differential staining with Giemsa and fluorescent dyes; density gradient centrifugation for the isolation of fragments of DNA or RNA of specific sizes and base composition; DNA/DNA and DNA/RNA hybridization *in vitro* and *in vivo;* the sequencing of DNA and RNA, aided by electrophoresis and the discovery of restriction enzymes which can cleave nucleic acids at highly specific sites; the cloning of genes or fragments of DNA by the use of plasmids and the recombinant DNA techniques; the production of somatic cell hybrids across wide taxonomic boundaries; the isolation of whole chromosomes of eukaryotes as well as prokaryotes; and the laboratory synthesis of a functional gene. Out of these techniques and approaches, coupled with an environment in which new and different questions can legitimately be asked, has come the discovery of repetitive as well as of unique, or single-copy, sequences of DNA; the uncovering of noncoding spacer DNA, both within and between genes; the establishment of detailed cytological and genetical maps of the human species; an understanding of the post-transcriptional modification of RNAs to form appropriate messenger molecules; a clarification of the nature of the gene in chromosomal and molecular terms as well as in a genetic context; the sequencing of entire genomes such as those of the ϕX174 and SV40 viruses; the discovery of the existence and structure of the nucleosome as a DNA-histone complex, and as a first step in the contraction of DNA; the discovery of gene amplification and of chromosome diminution as phenomena of growth and differentiation; a partial understanding of the processes of gene regulation in eukaryotes as well as in prokaryotes; and, most unexpectedly, the discovery of overlapping, included, and interrupted gene sequences.

Few of these techniques and discoveries were to be found in the first edition of this volume. To convey the content and excitement of this new information, and to intergrate it with the earlier and still valid substance of cytogenetics, required the preparation of a volume much larger than its predecessor. The number of pages, illustrations, and tables have been more than doubled, and an extensive bibliography now terminates each chapter. Two new chapters have been added: to deal with the recent discoveries in molecular cytogenetics and to explore our current understanding of the evolution of the karyotype. However, this is still a textbook, rather than a reference volume. An attempt has been made to keep the text as readable as possible for the student; it is not fragmented, therefore, by the insertion of names and dates, but the bibliography, selected for its relevancy, informative titles, and availability in recently issued and commonly found journals, should permit the student and the instructor to pursue any topic mentioned beyond the level presented in the text. In addition, and since any textbook, by its very nature, is past history, an appendix lists those journals and review series which are the sources of current articles of cytogenetic interest, thus enabling student or instructor to keep up to date in any area.

The increased size of the volume means that, to an even greater degree than in the first edition, we are indebted to others for the source and use of ideas, illustrations, tables and citations; to all of these individuals and publishers we acknowledge our debt and express our thanks; we trust that we have justified their generosity by the appropriate and accurate use of their materials, data, or concepts. We would also like to acknowledge, with equal gratitude, the critical review efforts of Jim and Marta Walters of the Department of Biological Sciences, University of California, Santa Barbara, and Allyn Bregman of the Department of Biology, College of New Paltz, New Paltz, New York. They have called our attention a number of errors and omissions, and have been a source of strength, stylistically as well as scientifically. If errors or misrepresentations remain, the fault is ours.

The passage of time has not altered all things, however; so it is that, with renewed pleasure, we re-dedicate this volume to our wives.

C.P.S.   T.M.   W.J.Y.

Preface to the First Edition

There are very few good reasons for a preface, especially to a text-book. One would hope that such a book describes its own goals and demonstrates its own reason for being. We would, however, like to place this book in its most useful setting for the students who read it.

We have attempted here to present as meaningful a picture of cytogenetics as is possible. A textbook, by definition, deals with past history; but the past has a way of changing, and each generation of students must reinterpret and make use of the past as time and they move into the future. We would like to believe, therefore, that we have prepared an open-ended text; and we have done this by indicating some of the important questions still to be answered in the future at the same time that we have assessed and recorded the accomplishments of the past.

This text is part of a series of related volumes, and as such it does not contain material that is both peripheral to its central theme and is also carefully described in its companion volumes. We have, however, included any material we thought relevant and pertinent, regardless of any overlapping descriptions in the other texts. A certain amount of duplication inevitably results, but since none of the authors of the series views things in the same light, perhaps this is just as well.

For the student who wishes to pursue further readings in this field, a selected and partially annotated bibliography is included at the end of each chapter. We have also appended a list of journals that regularly include reports and reviews relating to cytogenetics. As a supplement to the text, these are invaluable in keeping the student abreast of a field that has been, and continues to be, rapidly changed by new techniques, instrumentation, discoveries, and ideas. All of these have provided us with an increasingly intimate knowledge of the structure and behavior of chromosomes, which provide the physical basis of most genetic phenomena.

As authors, we are indebted to many others in the field of cytogenetics: for discussions, ideas, and the use of illustrative material. To all of these individuals and to the cited sources, we express our grateful thanks. If we have used their materials unwisely or inaccurately, the fault rests with us. We also take a good deal of pleasure in dedicating this book to our wives; only they and we know how well they have earned this small recognition.

C.P.S.   T.M.   W.J.Y.

I

Introduction

The science of cytogenetics is based on the fact that the hereditary material of an organism, be it virus, bacterium, or mammal, is ordered into one or more chromosomes. By means of a wide variety of physical, chemical, and biological techniques, it has been possible to examine the structure and function of these organelles, and to correlate chromosomal characteristics with patterns of genetic function and of phenotypic inheritance and distribution.

Such an ordered arrangement of heritable material possesses obvious advantages. As greater and greater degrees of complexity of structure and behavior were introduced into the biological world through evolutionary change, and as increased multicellularity was accompanied by cell and organ differentiation, an increasing number of genetic units, both structural and regulatory, were required to provide the information necessary to specify and control that complexity during the processes of growth and development. In a very

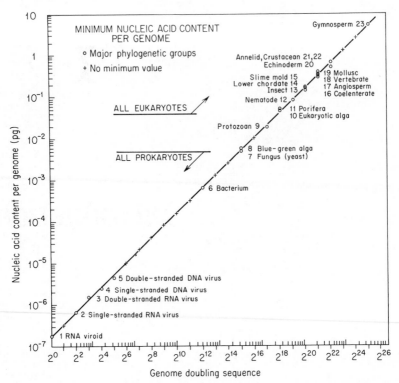

Fig. 1.1. The nucleic acid content, in picograms, of the major groups of prokaryotes and eukaryotes plotted on a log scale as a theoretical sequence of genome doublings. The RNA viroid value of 2^0 is equal to about 100,000 daltons, or about 300 nucleotides, a figure assumed to be the basic repetitive sequence found in the eukaryotes. Each exponent of 2 along the x-axis indicates the number of doublings assumed to have taken place. In locating the position for the several groups of organisms, only the minimum nucleic acid values in each group were used. The relationships suggest that the origin of the several groups in evolutionary time was accompanied by a doubling phenomenon, but the derivative nature of the viruses and the uncertain relation of the prokaryotes to the eukaryotes casts doubt on the meaning and significance of these data (Fig. 4 of Sparrow and Nauman, 1976, copyright 1976 by the Amer. Assoc. Adv. Science).

general way, and making use of minimum genome size in the major taxonomic groups as one variable, such increases have been found to follow a DNA doubling pattern (Fig. 1.1), although it is equally evident that within each major group wide genomic size differences are also encountered. By having these genetic units grouped into chromosomes, an economy in numbers of segregating units in both somatic and reproductive cell divisions becomes possible, and the physiological dangers of unbalanced gain or loss of essential genetic information is greatly minimized. In addition, the grouping into one or more chromosomes makes possible a differentiation of function between chromosomes and among the segments of individual chromosomes, and it permits the evolution of degrees of control and interaction not possible among a group of genes, each acting independently of other similar units. The chromosome, therefore, is a complex and highly ordered organelle whose activity as a whole transcends the function of any single part; as such it possesses many of the advantages that accrue to any organized structure at the same time that the malfunction of any single unit through mutation could lead to serious disruption of the system. It follows, consequently, that the spectrum of detectable mutations, from the advantageous and neutral to the lethal, would broaden progressively as biological complexity increased, but the data to prove this statement would be difficult to collect.

The above statement does not imply that the most efficient genetic system is one having the lowest number of segregating units. Our knowledge of the meaning of chromosome numbers, other than their more general evolutionary relationships, is too meager to make such an unequivocal statement. Rather, the idea being conveyed is that the chromosomes of a species as ordered instead of random aggregates of genetic units are the products of evolutionary pressures, and because no organism is without one or more chromosomes, they must therefore have had a high selective value from the time of their inception.

As far as is known, no gene exists as a solitary unit. One of the smallest, if not the smallest, chromosomes known is that found in R17 or MS2, single-stranded viruses that infect the bacterium, *E. coli* (Fig. 1.2). The chromosome is an RNA molecule about 1.1 μm in length, and it consists of about 3,400 nucleotides divided among

Fig. 1.2. The genome of R17 or MS2, single-stranded DNA viruses which infect cells of *E. coli*. The approximately 3400 nucleotides include only three genes: an A protein necessary for penetration into the bacterial cell is derived from the first gene on the left; a coat or capsid protein from the middle gene; and a replicase needed for making more viral DNA from the third gene. The significance of the two untranslated ends is not known; the spacer DNA separating the three genes does not code for protein, but presumably serves some purpose during the life cycle of the virus.

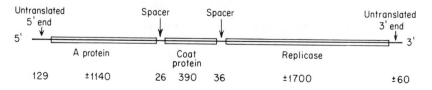

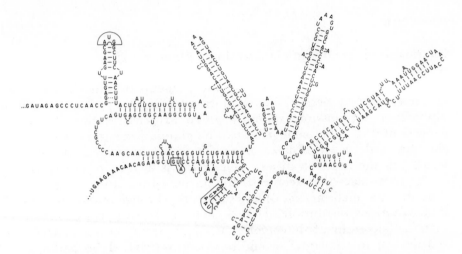

Fig. 1.3. The nucleotide sequence of the coat protein gene of the MS2 virus. The initiating codon—AUG—is enclosed at the top of the first hairpin loop on the left of the diagram, the reading frame is from left to right, and the initiating codon is preceded by a recognition sequence to which the replicase first attaches itself. The double terminating codons—UAA and UAG—are enclosed in one of the hairpin loops at the bottom; these are followed by a stretch of spacer nucleotides, and then the initiating codon—AUG—of the viral replicase gene which follows. A recognition sequence probably precedes the replicase gene unless the whole viral genome is read by one replicase, in which case only a single recognition sequence is needed. The entire coat protein gene is about 390 nucleotides long. The significance of the hairpin loops and the elaborate "flower" structure remains to be determined, although stability resulting from the paired regions is a possibility; the structure, however, is too regular to be without meaning (modified from Min Jou, et al., *Nature* 237: 82, 1972).

three genes which code for three different proteins: an A protein necessary for absorption to, and penetration into, the bacterium, a coat protein which provides an outer covering for the viral particle, and an enzyme, a replicase, which, together with host proteins, is responsible for RNA replication. The complexity of the small coat gene is indicated by the flower-like structure brought on by the pairing of complementary nucleotides and the formation of the many hairpin loops. (Fig. 1.3). All other materials and energy needed for its continued existence are derived from the host cell. By way of contrast, the stretched lengths of chromosomes of many higher organisms such as some amphibians are measured in meters instead of in micrometers. When used in its generic sense, therefore, the term "chromosome" includes the compound heritable units of viruses and bacteria as well as those of all higher plants and animals, despite the fact that, during the course of evolution and in various groups of organisms, these units have come to differ quite widely in mode of transmission, size, molecular complexity, patterns of internal control, and, most particularly, genetic constitution. Wherever found, however, all chromosomes share common attributes. They are capable of regular transmission to viral progeny or to daughter cells, they can be rendered visible by one technique or another (Figs. 1.4, 1.5, and 1.6), and they contain nucleic acids as their heritable component: DNA in the great majority of organisms, RNA in some viruses. Consequently, the cytogenetic properties of a viral or bacterial chromosome are as relevant to our understanding of the physical basis of inheritance as are those of the more familiar chromosomes of maize, *Drosophila,* or human beings.

The universal occurrence of chromosomes suggests, as indicated above, that they made their appearance very early in the history of life, with their adaptive value rapidly asserted. Whether the earliest chromosomes bore a resemblance to those now in existence, and whether they competed with and displaced other ways of packaging genetic units, we do not know. Their adaptive value can probably be stated in a variety of terms: precision of replication, coupled with

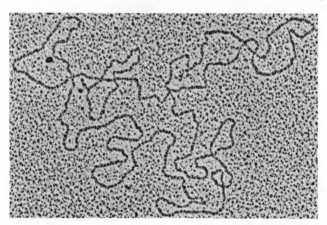

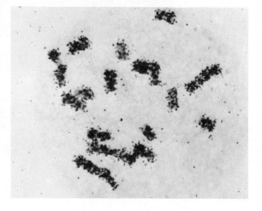

Fig. 1.4. Top, electron micrograph of the circular chromosome of the P22 bacteriophage, which infects the bacterium *Salmonella typhimurium.* The chromosome is spread on a protein film on the surface of water and then picked up on a grid before being photographed. Courtesy of Dr. L. McHattie. Right, metaphase chromosomes of the Chinese hamster grown in tissue culture and labeled with tritiated thymidine in the preceding S period of interphase. Courtesy of Dr. T. C. Hsu.

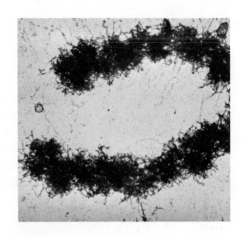

Fig. 1.5. Top, electron micrograph of a single metaphase chromosome from a bovine kidney cell grown in culture; the cell was burst, and the chromosomes were isolated through centrifugation, floated on a protein film in a Langmuir trough, transferred to a grid, and dried prior to being photographed. Courtesy of Dr. S. Wolfe. Bottom left, polytene chromosomes from the salivary glands of *Sciara*. Courtesy of Dr. Helen Crouse. Bottom right, metaphase chromosomes from a haploid microspore of *Trillium erectum*. Courtesy of Dr. A. H. Sparrow.

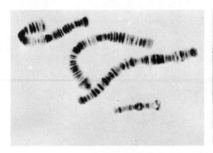

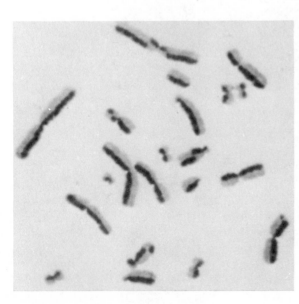

Fig. 1.6. Chromosomes from a somatic cell of the Chinese hamster that had replicated in the presence of 5-bromodeoxyuridine, and then been stained with a combination of a fluorescent dye and a Giemsa stain. The two chromatids of each "harlequin" chromosome stain differentially because of different amounts of 5-bromodeoxyuridine; the dark ones would have only one of the two polynucleotide strands of the double helix containing 5-bromodeoxyridine, while the light chromatids would have both strands so labelled. Sister chromatid exchange occurs during the cell cycle to bring about the reciprocal exchange of light and dark material (courtesy of Dr. D. Stetka).

sufficient built-in copy-error, or mutation, to provide for a supply of diversity through segregation and recombination; efficient informational packaging of the genetic code and its decoding through the processes of transcription and translation; selective control of gene activity through a variety of feedback mechanisms; minimizing of damage to the genetic system through the presence of repair systems; uniqueness (single-copy) or redundancy of genes to meet, in ways not yet fully understood, the needs of the organism at all times. However, neither human nor viral chromosomes are capable of self-replication; in isolation they may contain all of the information necessary for their own and the organism's needs, but this information can be expressed only within the framework of a still higher order of complexity, the cell. Chromosomes, then, are dynamic structures, and like all other parts of the cell or an organism subject to the varied pressures of continued existence, they continue to evolve. Comparative studies of chromosomes are, of course, possible, and yield some information, but difficulty is encountered in trying to account historically for all of the past and present attributes of chromosomes; so our concern, therefore, is with the chromosome of today.

Chromosomes are replicated time and again with a remarkably low margin of error. Errors of nucleotide insertion into a replicating chromosome have been judged to occur at a rate of one wrong nucleotide for every 10,000 to 50,000 incorporated, but the repair mechanisms of the cell, which can detect and correct these mistakes, keep the ultimate errors to an extraordinarily low level of occurrence. Chromosomes are distributed to daughter cells or viral progeny with a comparable degree of exactitude. Continuing irregularity in chromosome distribution would, of course, result in extensive genetic repetition or loss among daughter cells, leading to genetic imbalance and possibly to cell or viral impairment or death. Gain or loss of chromatin, as well as extensive redundancy of some chromosomal regions, are well-documented phenomena. It was once thought that most genetic phenomena could be explained by assuming a chromosome of rather simple structure and behavior, i.e., genes strung end to end as beads on a string, while the cytologist sought explanations for a chromosome which underwent intricate maneuvering during the cell cycle and which seemed to be enormously complex in structural features. Today, discoveries at the molecular level force us to view the chromosome as a most intricate and integrated structure, exquisite in the precision of its parts and in the control of its behavior.

As biological complexity has evolved, a division of labor, and presumably efficiency of operation, are achieved by a specialization among the parts of the chromosome. All parts, to be sure, are capable of being replicated, but a differentiation of function has arisen even though the molecules of inheritance are uniformly sequences of nucleotides, singly or in pairs, throughout their length. Those which, through the processes of transcription and translation, code for proteins are *structural genes;* some of these proteins are enzymatic in function, some, such as the histones and the acidic proteins of the nucleus, are intimately concerned with the character and behavior of eukaryotic chromosomes, while others enter into the formation of membranes, flagella, or microtubules. Other genes participate only in the process of transcription, producing the RNA molecules which are involved in ribosome formation or, which as tRNAs, play a role in the translational process. Still others have a *regulatory* function. These would include *families of repetitive genes* in the eukaryotic as well as the *promoter* and *operator* genes of prokaryotic systems. Other parts of the chromosome are concerned with movement during the cell cycle; these are the *centromeres* or *kinetochores,* conceivably compound genes to judge from their relatively large size and complex structure. But despite the fact that many parts of the chromosome can be identified and mapped because of their morphological and genetic uniqueness, there are, nevertheless, other regions of the chromosome which appear to be genetically "silent" even though recognizable by cytological techniques. These regions, consisting of segments of varying length and represented in repetitious fashion, constitute *heterochromatin* as opposed to the more genetically active *euchromatin,* but the role of heterochromatin remains to be completely understood.

Cytogenetics, therefore, focuses its major attention on the chromosome. It is concerned with its chemical and genetic organization and with the processes through which this organization determines the transmission of hereditary characteristics from one generation to the next; with the behavior of chromosomes in dividing as well as nondividing cells; with the relation of the parts of the chromosome to metabolic function and its timing during the life cycle of an organism; and with chromosomally determined patterns of differentiation at cellular, organ, and organismic levels. In another sense, cytogenetics is concerned with both the structures and processes which deal with the conservative aspects of heredity, i.e., with the preservation of type, as well as with those which provide for the introduction of diversity into the systems of inheritance. It is the former which perpetuate the species as a taxonomic entity; it is the latter which confer uniqueness of each individual and provide a material basis for the operation of natural selection.

Historical Background. Cytogenetics, as its name implies, is a hybrid science, compounded of the subject matter and methods of genetics and cytology. Like many newly created hybrids, it exhibited hybrid vigor at its outset, a characteristic which it has maintained as the result of the continued infusion of new ideas and techniques arising out of electron, and fluorescent, as well as light, microscopy, bacteriology, virology, biochemistry of nucleic and amino acids, tissue and cell culture, molecular hybridization *in vitro* and *in situ,* molecular discrimination through density gradient centrifugation and electrophoresis, and somatic cell fusions. Originating with the realization of the relation of the chromosome to particulate inheritance, which developed around the turn of this century, there has been recognized an increasingly more intimate correspondence between the physical structure of the chromosome at various levels of organization to the behavior of the chromosome in inheritance on the one hand and its determination of phenotypic expression on the other. In this regard, the structure, behavior, and expression of the naked genome of a virus fits as easily and as validly under the banner of cytogenetics as does that of the more complex genome of the human being.

The science of cytology had its beginnings in the 17th century when the first simple microscope was invented (Table 1.1). For some time these instruments were few in number, and because of the structure and philosophy of science at that time, they were more of a scientific oddity used for social amusement than instruments for the systematic study of the structure of living things. Indeed, Samuel Pepys, the English diarist, once complained about the high cost of the instrument he kept for the entertainment of himself and his friends. Nevertheless, the few observations made by Hooke (1635-1703) and Grew (1641-1712) in England, Leeuwenhoek (1632-1723) in Holland, and Malpighi in Italy (1628-1694) led to an understanding of the fact that tissues and organs of both plants and animals had a common structural basis. Brown would point out in 1831 that the nucleus was a normal constituent of all living cells while at the same time a number of others tried to deal with the structure and function of either the cytoplasm or the general cell contents. These observations would later culminate and be formally expressed in the cell doctrine of Schleiden and Schwann in 1838-1839, a doctrine stating that the cell was the basic unit of structure and function in all living organisms. Like the atomic theory of John Dalton, the statement is simple, direct, and elementary: modern biology is predicated on its universality. Supported as it was by prominent 19th-century biologists, the cell doctrine focused attention on the cell at a time when compound microscopes, adequate for cytological research, had become generally available. Progress in cytology was rapid and inevitable, and it led to the opening of an immense and varied field of research.

The third quarter of the 19th century saw the development of those concepts which gave concrete expression to the emerging ideas of genetic continuity: the conservation of species through reproduction and the origin of the diversity expressed in individuals. The first of these concepts, the idea that all cells arose from the division of preexisting cells, was developed in 1858 by Remak and Virchow. We know, at the present time, that cells do not arise spontaneously any more than does life itself, even though we must suppose that the first cell, or cells, had a spontaneous, if laboriously slow, origin and evolution. Carried to its logical extreme, therefore, the idea of cell lineage relates every organism to every other organism, living or dead, for unless we interject the idea of a continuing spontaneous origin of cells, an unbroken chain of cells must extend from the present time throughout the entire past history of cellular life.

The interrelatedness of all life was given further expression through the theory of evolution put forth by Darwin and Wallace in 1858. It is in the operation of evolution that we encounter the principle of genetic continuity enriched by variation, for only when heritable diversity exists is evolution possible. A fundamental feature of life, therefore, is that it both reproduces to conserve the species and it changes to produce the diversity upon which natural selection acts. Because the cell is the fundamental unit of structure, these two crucial features of evolution must have a cellular basis.

At the end of that quarter century (1875), Pasteur, working with microorganisms, demonstrated the spurious nature of any hypothesis that was based upon the immediate spontaneous generation of life. The cell doctrine, in addition to expressing a structural and functional principle, also stated that cells give rise to cells; the evolution theory, that species may become extinct, perpetuate themselves through time, or give rise to other species through directed change; and Pasteur's biogenetic law, that life arises only from preexisting life. The principle of genetic continuity can consequently be expressed at many levels of organization, from the molecular to that of populations. Still to be accounted for were the origin, nature, mode of transmission, and retention of variation in individuals and in populations.

Inheritance of Variation. The Darwinian concept of evolution through natural selection postulated that some of the variations observed among individuals in a population were heritable. He visualized each of these variations as having only a slight effect on an individual, or on a population, but a significant effect when viewed cumulatively through time. Being heritable and, potentially at least, having an influence on the survival and reproductive capabilities of the individual, these variations were acted upon by the forces of natural selection. Late 19th-century biologists, including Darwin, had difficulty, however, in dealing with these postulated variations and in incorporating them into the general biological thought of the period. The blending hypothesis of inheritance, current at that time, could not embrace small variations as being of any significance since it was assumed that their effects would tend to be diluted to greater and greater degrees with each new generation. It would be the foresight and experimental astuteness of Gregor Mendel (1822–1884), the Augustinian monk, that would eventually prove Darwin correct, and show that variations, singly or collectively, were heritable realities which could be dealt with in a precise, quantitative, and predictable manner.

The Mendelian laws of inheritance, formulated in 1865, still form the basis of our understanding of the transmission of heritable variation, and the inheritance test is still the basic technique for following phenotypic characters through the several generations. It is the genius of Mendel that a few, but quantitatively predictable, results enabled him to postulate that an abstract entity—the element or factor, as he called it, the gene as we now designate it—was responsible for the phenotypic character he was following in his breeding experiments, that this entity was singly represented in gametes and doubly so in zygotes and somatic tissues, and that factors, when together in pairs, could exhibit dominance or recessiveness in expression. Since the biology of that period had no theoretical structure into which Mendel's data and concepts could be comfortably incorporated, it is not surprising that he failed to get a hearing from his scientific peers even though he published in a journal that was generally read throughout most of Europe.

The Chromosomal Theory of Inheritance. It would remain for others to give Mendel's abstract factor a physical character and location. Mendel's genetic discoveries were ignored, or given no credence, from 1865 to 1900, when several scientists recognized that their breeding results had already been anticipated and previously published. By this time, however, the scientific world was in a more receptive mood, and a number of cytological studies permitted the union of cellular facts with patterns of inheritance, to give rise through this union to the science of cytogenetics.

The egg and sperm of animals, despite obvious differences in size and shape, had earlier been shown to be cells; the physical continuity between generations was, therefore, a slender cellular bridge. Fertilization was shown to be the union of gametes, with the fusion of parental nuclei in the cytoplasm of the egg being the crucial event, and with each nucleus providing equal, or nearly equal, numbers of chromosomes to the newly formed zygote and hence to the developing embryo. As the embryo grew by adding to its cell numbers, the critical acts of cell division were the longitudinal replication of each chromosome, and the segregation of these longitudinal halves to the two daughter cells, thus providing both a physical basis for the qualitative and quantitative chromosomal equality of each daughter nucleus and a functional basis for the conservation of the genotype. Meiosis, a kind of cell division leading to the formation of eggs or sperm in animals and spores in most plant species, was revealed as a mechanism for halving the chromosome number; thus providing a means for counteracting the chromosome doubling that took place earlier in the fusion of parental nuclei in fertilization and making clear the patterns of haploidy and diploidy in the life cycle. Chromosomes had been demonstrated to have a physical continuity from one cell to another, and from one generation to another, and to be qualitatively different from each other insofar as they affected developmental processes. By 1902, Garrod, through his study of the metabolic diseases of humans—"inborn errors of metabolism," as he called them—would forge the first link that would couple the biochemistry of disease syndromes to patterns of inheritance. With the further observation that species were characterized by a constant number of chromosomes, although the two sexes of a species may differ slightly from each other, it was becoming more and more obvious that the patterns of inheritance were somehow linked to, or mirrored in, the behavior of chromosomes.

In retrospect, therefore, the discoveries made during the last quarter century of the 1900s and the first few years of the 20th century made it increasingly clear that the chromosomes were key elements whose behavior in division and fertilization revealed a regularity that could, on the one hand, account for the transmission of hereditary factors from one generation to another, whether of cells or of individual organisms, and on the other for the conservatism of species reproduction, i.e., the preservation of type, generation after generation. The stage was consequently set for the coupling of two previously unrelated phenomena that had been developing in somewhat parallel fashion: the transmission and behavior of the abstract factors of Mendel and the physical behavior of chromosomes in mitosis, meiosis, and fertilization. The Sutton-Boveri chromosomal theory of inheritance, advanced in 1902–1903, brought about the union, just as earlier, the concept of cell lineage of Virchow had coupled the cell doctrine with the Darwinian theory of evolution. The science of cytogenetics was thus launched with a brilliant correlation of factor (gene) and chromosome behavior.

The basis of the theory, as stated by Sutton, is as follows:

1. In somatic cells, arising from a fertilized egg, or zygote, the chromosomes consist of similar groups, one of maternal origin inherited through the egg, the other of paternal origin and inherited through the sperm. Each somatic nucleus, therefore, contains pairs of like chromosomes, or homologues, the number of pairs being the same as the haploid set of chromosomes in a gamete. Hence, the chromosomes, like Mendel's factors, are doubly represented in the somatic cells of an organism, and singly, or pure, in the gametes.

2. The chromosomes retain their structural individuality and their continuity throughout the life cycle of an organism. Again, the abstract factors of Mendel retain their individuality and continuity even though the character they determine might not be expressed. The basis of genetic homogeneity and heterogeneity, dominance and recessiveness, was suggested by chromosomal behavior.

3. In meiosis, synapsis brings together pairs of homologous chromosomes and leads to their subsequent segregation into different cells, establishing thereby a quantitative basis for both the segregation of abstract factors and the independent assortment found when two pairs of contrasting factors are considered together.

4. Each chromosome, or chromosome pair, plays a definite role in the development of the individual. This conclusion derived not from any specific facet of Mendel's investigations but rather from studies of Boveri of abnormal larvae of sea urchins which lacked certain normal chromosomes, as well as Sutton's own studies of the size differences among the chromosomes of the insect *Brachystola.*

Sutton, consequently, visualized the chromosomes as the physical carriers of Mendelian factors and the segregation of a pair of homologous chromosomes and the independent assortment of non-homologous chromosomes as the physical basis for the qualitative and quantitative aspects of Mendelian segregation. Sutton, together with Roux and Boveri, also anticipated the phenomenon of linkage when he stated that all the factors in any one chromosome must be inherited together. It would take studies during the next quarter century to prove beyond reasonable doubt that the chromosome theory of inheritance could take its place alongside the cell doctrine and the theory of evolution as one of the unifying concepts of biology, much as the notion of stratigraphical succession underlies the science of geology and the notion of atomic theory underlies the sciences of physics and chemistry. As the theory of the gene would come to replace the concept of Mendelian factors, the hereditary constitution of an organism, and eventually of a species, would come to be visualized not as an indivisible whole, as the blending hypothesis would imply, but as an aggregation of units which would be disassembled and reassembled in many and varied ways, much as molecules are not to be viewed as indivisible but as units that can be taken apart and put together into different configurations of atoms. Such notions would eventually lead to a further refinement, namely, the molecular basis of hereditary phenomena.

Molecular Cytogenetics. In a modern context, cytogenetics extends beyond just an understanding of the mechanics of the transmission and continuity of genes and chromosomes. Recognizing that life in all of its manifestations is a phenomenon of ordered form and behavior, and that the organization which gives to life its varied attributes is in conformity with the laws of physics and chemistry, the cytogeneticist today is equally concerned with the molecular architecture of the chromosome and the relation of that architecture of genetic function and evolutionary changes. The Sutton-Boveri theory made evident that there was, and is, a striking parallel between genes and chromosomes in the transmission of hereditary potentialities, and that each chromosome plays a particular role(s) in development. An extension of this to molecular levels indicates that a comparable parallelism must also exist between gene action within and between cells and the chemical organization and activity of the chromosome and its parts. This position has been fully borne out by studies during the past several decades.

The nucleic acids and proteins of the cell are the key molecules of hereditary uniqueness, and the interactions of nucleic acids and proteins the key to genetic behavior. Deoxyribonucleic acid (DNA) is the molecule in which all precoded genetic information of an organism is stored (ribonucleic acid, or RNA, performs the same role and function in some viruses), and as befits its role, it is the only more or less permanently conserved molecular species in the chromosomes. The RNA and proteins, which are also a part of, or associated with, the chromosomes of higher organisms are more transitory in nature, but they serve important structural and regulatory functions in that they assist in selective translation of the coded information into phenotypic expression.

DNA, together with its associated proteins, was discovered about 1890 by the German chemist, Miescher, as a nuclear component of cells. In 1924 Feulgen demonstrated through specific staining techniques that DNA was localized in the chromosomes. It was not until 1944, however, that DNA was shown to be the molecular basis of hereditary uniqueness by Avery, McCarty, and McLeod, thus focusing attention on the crucial molecule that had to be known structurally and functionally if its central role in inheritance was to be similarly understood. Even so, it was difficult for the biologists of that period to believe that a molecule consisting of only four different nucleotides could provide the internal diversity of coded information needed to account for all of the varied manifestations required of an organism for its inherited characteristics. The proteins, with their 20 different amino acids, which could be arranged in an almost inifinite array of permutations and combinations, seemed a much more likely candidate for so critical a role. But in 1953 Watson, a virologist by training, Crick, a physical chemist, and Franklin and Wilkins, X-ray diffraction crystallographers, made clear the structure of DNA, a discovery that ranks in importance with the cell doctrine, the theory of evolution, and the theory of the gene. As a result of the interest of George Gamow, a theoretical physicist and cosmologist, it was soon apparent that this structure could serve as a bank of coded information, an inference that would be fully borne out by subsequent studies. The abstract inherited factors of Mendel could now be visualized as molecular units, manipulable by the tools and techniques of the biochemist and physicist as well as being made visible to the cytologist by a variety of optical means and staining procedures.

A wide spectrum of newer tools and techniques, coupled with microscopic methods and breeding procedures, has enabled the experimenter to determine how DNA, together with the associated molecules of proteins and the several kinds of RNAs, functions as a precoded source of information, translated into expression during ontogeny and activities of an organism. These molecules, enormous as they are in length and molecular weight, and complex as they are in their configurational shapes, can now be handled with comparative ease and confidence, and *in vivo* activities of these molecules within the cell can often be duplicated in the test tube. Electron micros-copy; radioactive tagging of molecules; spectroscopy; chromato-graphy; electrophoresis of several sorts; differential and specific staining; density gradient centrifugation; the isolation and use of enzymes with highly specific action; techniques for the "fingerprint-ing" of proteins and nucleic acids; the transfer of DNA from one cell to another among the prokaryotes; cell culturing; and the fusion of cells both intraspecifically and interspecifically and even intergeneri-cally—all of these techniques, singly and in combination, coupled with appropriate instrumentation, more and more sophisticated and often exquisite experimentation, and most importantly a rapidly changing body of ideas, have given the science of cytogenetics a vitality and a scope that was undreamed of a generation ago. The viruses and bacteria take their place with the higher organisms as objects of legitimate cytogenetic interest, and the discovery in the 1960s of DNA in plastids and mitochondria makes possible the un-derstanding of cytoplasmic, or maternal, inheritance and the begin-nings of what might be called cytoplasmic cytogenetics.

When the element of time is taken into account, and when an attempt is made to visualize cytogenetic systems in the context of evolution, molecular techniques have also proven to be of inestima-ble value in detecting the kinds and amounts of genetic change that have taken place at the molecular level and that parallel those mani-fest at grosser levels of morphology and behavior. The melting, or denaturing, of the DNA double helix into separate polynucleotide strands, and their reannealing at lower temperatures, has permitted a comparison of genomes that parallels the comparative cytogenetics of microscopically visible chromosomes, while the shearing and reannealing of DNA fragments, together with reannealing back to intact chromosomes, have permitted a dissection of chromosomes in a way that has revealed an unanticipated mine of linear chromo-somal diversity, not all of it yet understood in terms of ontogenetical or phylogenetical significance. As Table 1.1 shows, a whole succes-sion of events and discoveries has continued to maintain the hybrid vigor that first characterized cytogenetics at its birth in 1902-1903 with the formulation of the Sutton-Boveri chromosomal theory of inheritance. The increased breadth and depth of cytogenetic under-standing, resulting from the recent technical developments in DNA, RNA, and protein structure and function, and the improved possibilities of relating these data to developmental phenomena, would seem, to the uninformed, to sever this knowledge from its historical and classical base, and to relegate pre-1953 information to the domain of the historians of science. It is our hope, however, that we can demonstrate, as we move from one topic to another, that microscopic and molecular cytogenetics complement each other and any attempt to understand one without the other is to gain only a partial picture of that amazing evolutionary structure, the chromosome.

Table 1.1. Significant Dates in the Development of Cytogenetics

Pre-cytogenetic era

1666 – Discovery of cells. Hooke
1831 – Discovery of the nucleus as a central feature of cells. Brown
1838-1839 – Formal statement of the cell doctrine. Schleiden and Schwann
1852-1853 – Theory of cell lineage. Remak and Virchow
1858 – Theory of evolution through natural selection. Darwin and Wallace
1859 – Publication of *Origin of Species*. Darwin
1865 – Discovery of the laws of segregation and of independent assortment, and that inheritance was particulate and not blending. Mendel
1875-1876 – Fertilization as the fusion of nuceli of egg and sperm. Hertwig in animals, Strasburger in plants
1882-1884 – Chromosomal aspects of cell division described. Flemming, Van Beneden
1884 – Chromosomal contributions of egg and sperm equal; constancy of chromosome numbers. van Beneden
1885 – Chromatin as the physical basis of inheritance. Weismann, Strasburger, and von Kölliker
1887-1888 – Reduction of chromosome numbers to offset fertilization. Weismann on theoretical grounds, observed by Strasburger in plants
1888 – Individuality of chromosomes in development. Boveri
1896 – Publication of *The Cell in Development and Heredity* (1st ed.). Wilson

Classical Period

1900 – Discovery of Mendel's laws and recognition of their significance. Tschermak, De Vries, and Correns
1901 – Correlation of X chromosome to sex determination. McClung
1902-1903 – Chromosomal theory of inheritance. Sutton and Boveri
1905-1908 – Discovery of gene linkage. Correns, Bateson, and Punnett
1906 – Distinction made between heterochromosomes (X and Y) and autosomes. Montgomery
1909 – Chiasmatypy theory. Janssens
1910 – Sex inheritance in *Drosophila*. T. H. Morgan*
1912 – Linkage maps of *Drosophila*. Morgan*
1913 – Development of three-point cross for determining map distances. Sturtevant
1913, 1916 – Nondisjunction as proof of the chromosome theory of inheritance. Bridges

*Nobel laureates in Medicine and Physiology except Sanger who is a Nobel laureate in Chemistry.

Table 1.1. Continued

1913, 1917 – Correlation of Mendelian segregation (independent assortment) and the
               distribution of heteromorphic homologues in meiosis. Carothers
1915 – Publication of *Mechanism of Mendelian Heredity*. Morgan*, Sturtevant, Muller*,
               and Bridges
1917, 1919, 1921 – Recognition of deficiencies, duplications, and translocations by
               genetic tests. Bridges
1921 – Recognition of an inversion by genetic test. Sturtevant
1922 – Attached-X chromosomes in *Drosophila*. L. V. Morgan
1924 – Specific stain for DNA. Feulgen and Rossenbeck
1925 – Publication of *The Theory of the Gene*. Morgan*
1927 – B-chromosomes in maize. Longley
1927 – Induction of mutations by artificial means (X rays). Muller* and Stadler
1928 – Distinction drawn between euchromatin and heterochromatin. Heitz
1928 – Discovery of a gene that affects meiosis. Gowen
1931 – Proof that crossing over involves an exchange of chromatin between homologues.
               Stern; Creighton and McClintock
1932 – Publication of *Recent Advances in Cytology* (1st ed.). Darlington
1933 – Use of polytene chromosomes for cytogenetic purposes. Painter
1933 – Use of pachytene configurations for study of chromosomal aberrations. McClintock
1934 – Nucleolar organizers in maize. McClintock
1936 – Discovery of mitotic crossing over. Stern
1936 – Preferential segregation in *Drosophila*. Sturtevant
1936 – Relation of inversions to crossing over and disjunction. Sturtevant and Beadle*
1937 – Beginnings of X-ray cytology and cytogenetics. Sax
1937 – Use of colchicine as a mitotic inhibitor. Blakeslee and Eigsti
1939 – Induction of mutations by nitrous acid. Thom and Steinberg
1941 – Induction of mutations by ultraviolet light. Hollaender and Emmons; Stadler
1941–1942 – Preferential segregation in maize. Randolph and Rhoades
1947 – Induction of mutations by sulfur and nitrogen mustards. Auerbach and Robson
1952 – Technique of osmotic shock for the study of mammalian chromosomes. Hsu
1956 – Accurate determination of human chromosome number. Tjio and Levan
1959 – Chromosomal aberrations in humans. Down's syndrome, Lejeune; Turner's syn-
               drome, Ford
1960 – Culture of human leucocytes with phytohemagglutinin. Nowell
1962 – Chloroplast chromosomes discovered. Ris and Plaut
1963 – Mitochondrial chromosomes discovered. Nass and Nass

Modern Period

1928 – Bacterial transformation. Griffiths
1941 – One gene–one enzyme concept. Beadle* and Tatum*
1944 – DNA as the physical basis of inheritance. Avery, McCarty, and McLeod
1946 – Genetic exchange in bacteria. Lederberg* and Tatum*
1949 – Genetic exchange in bacteriophage. Delbruck* and Bailey
1949 – Discovery of photoreactivation. Kelner
1952 – Discovery of A-T and G-C ratios in DNA. Chargaff
1952 – Phenomenon of transduction. Zinder and Lederberg*
1952 – DNA as basis of inheritance in bacteriophages. Hershey and Chase

---

*Nobel laureates in Medicine and Physiology except Sanger who is a Nobel laureate in Chemistry.

1952 – Sex factor (F) as a transmissible factor. Lederberg* et al.
1953 – Structure of DNA. Watson*, Crick*, and Wilkins*
1954 – Theory of the triplet code. Gamow
1955 – Ciston concept. Benzer
1957–1961 – Concept of the operon and of messenger RNA. Jacob* and Monod*
1957 – Radioactive tracers to demonstrate semi-conservative replication of eukaryotic chromosomes. Taylor, Hughes, and Wood
1957 – Discovery of resistance (R) plasmids. Ochiai; Watanabe
1958 – Semi-conservative replication of prokaryotic genome. Meselsen and Weigle
1958 – Concept of an adaptor molecule (tRNA) with an anticode. Crick*
1960 – Somatic cell fusions. Barski
1960 – Proof of the triplet code by means of reading-frame mutations. Crick* and others.
1961 – Discovery that F is a DNA plasmid. Driskell and Adelberg
1961 – Breaking of the genetic code. Nirenberg* and Matthaei
1962 – Colicinogenic (Col) factor as a DNA plasmid. Silver and Ozeki
1963 – DNA/DNA hybridization *in vitro*. Marmur
1964 – Discovery of repetitious DNA in eukaryotes. Britten and Waring
1968 – Restriction and modification enzymes. Arber*, Smith*, Nathan*
1970 – Fluorescent and Giemsa banding techniques of eukaryotic chromosomes. Caspersson et al.
1970 – DNA/DNA and DNA/RNA hybridization *in situ*. Pardue and Gall
1975 – Synthesis of an artificial gene. Khorana*
1975–1977 – Rapid sequencing of DNA. Sanger* and Coulson; Maxam and Gilbert
1977–1978 – Discovery of overlapping, included and interrupted genes. Various investigators, including Sanger*

---

*Nobel laureates in Medicine and Physiology except Sanger who is a Nobel laureate in Chemistry.

BIBLIOGRAPHY

Allfrey, V. G., et al., *Organization and Expression of Chromosomes,* Life Sciences Research Report 4 (Dahlem Konferenzen). Berlin: Abakon Verlagsgesellschaft, 1976.

Bartalos, M., and T. A. Baramki, *Medical Cytogenetics.* Baltimore: Williams & Wilkins, 1967.

Bostock, C. J., and A. T. Summer, *The Eukaryotic Chromosome.* New York: Elsevier, 1978.

Brown, W. V., *Textbook of Cytogenetics.* St. Louis: C. V. Mosby, 1972.

Carlson, E. A., *The Gene: A Critical History.* Philadelphia: Saunders, 1966.

Cold Spring Harbor Symposium on Quantitative Biology, XXXVIII: Chromosome Structure and Function. Coldspring Harbor, N.Y.: Coldspring Harbor Laboratory, 1974.

——, XLII (#1 and 2): Chromatin. Coldspring Harbor, N.Y.: Coldspring Harbor Laboratory, 1977. *(Note:* To be able to appreciate the great changes that have taken place over the last several decades, compare the contents of these two volumes with Vol. IX: *Genes and Chromosomes. Structure and Organization,* 1941).

Darlington, C. D., *Recent Advances in Cytology,* 2nd ed. Philadelphia: Blakiston, 1937.

——, and L. F. LaCour, *The Handling of Chromosomes,* 6th ed. Darien, Conn.: Hafner Press, 1970.

Dupraw, E. J., *DNA and Chromosomes.* New York: Holt, Rinehart & Winston, 1970.

Garber, E. D., Cytogenetics in *Cell Biology: A Comprehensive Treatise,* Vol. 1 (L. Goldstein and D. M. Prescott, eds.), New York: Academic Press, 1978.

Goodenough, U., *Genetics.* New York: Holt, Rinehart & Winston, 1978.

Hamkalo, B. A., and J. Papaconstantinou (eds.), *Molecular Cytogenetics.* New York: Plenum, 1973.

Hayes, W., *The Genetics of Bacteria and Their Viruses. Studies in Basic Genetics and Molecular Biology,* 2nd ed. New York: John Wiley, 1976.

Hsu, T. C. *Human and Mammalian Cytogenetics; An Historical Perspective.* New York: Springer-Verlag, 1979. (Note: Hsu has divided the history of vertebrate cytogenetics in a manner different from that in Table 1.1. The student would profit from a reading of this warmly humanistic as well as soundly scientific book.)

Hughes, A. A., *A History of Cytology.* London: Abelard-Shuman, 1959.

John, B., and K. B. Lewis, *Chromosome Hierarchy: An Introduction to the Biology of the Chromosome.* Oxford: Oxford University Press, 1975.

King, R. C. (ed.), *Handbook of Genetics,* Vols. I–V, New York: Plenum. Vol. I. *Bacteria, Bacteriophages and Fungi,* 1974. Vol. II. *Plants, Plant Viruses and Protists,* 1974. Vol. III. *Invertebrates of Genetic Interest,* 1975. Vol. IV. *Vertebrates of Genetic Interest,* 1975. Vol. V. *Molecular Genetics,* 1976.

Koller, P. C., *Chromosomes and Genes: The Biological Basis of Heredity.* New York: W. W. Norton & Co., Inc., 1971.

Lewin, B. M., *The Molecular Basis of Gene Expression.* New York: John Wiley, 1970.

——, *Gene Expression—2: Eucaryotic chromosomes.* New York: John Wiley, 1974.

Lewis, K. R., and B. John, *Chromosome Marker.* Boston: Little, Brown, 1963.

Lima-de-Faria, A. (ed.), *Handbook of Molecular Cytology.* New York: John Wiley, 1969.

Peacock, W. J., and R. D. Brock (eds.), *The Eukaryotic Chromosome.* Canberra: Australian Acad. Science, 1975.

Pearson, P. L., and K. R. Lewis, *Chromosomes Today,* Vol 5. New York: John Wiley, 1976.

Peters, J. A. (ed.), *Classic Papers in Genetics.* Englewood Cliffs, N.J.: Prentice-Hall, 1959.

Phillips, R. L., and C. R. Burnham (eds.), *Cytogenetics.* New York: Academic Press, 1977.

Rees, H., and R. N. Jones, *Chromosome Genetics.* Baltimore: University Park Press, 1977.

Roberts, P. A. (ed.), *Chromosomes: From Simple to Complex.* Corvallis: Oregon State University Press, 1974.

Sparrow, A. H., and A. F. Nauman, "Evolution of Genome Size by DNA Doublings," *Science, 192* May 7, (1976), 524-29.

Stickberger, M. W., *Genetics,* 2nd ed. New York: Macmillan, 1976.

Swanson, C. P., *Cytology and Cytogenetics.* Englewood Cliffs, N.J.: Prentice-Hall, 1957.

White, M. J. D., *Animal Cytology and Evolution,* 3rd ed. New York: Cambridge University Press, 1973.

Wilson, E. B., *The Cell in Development and Heredity,* 3rd ed. New York: Macmillan, 1925.

2

Architecture of the Chromosome: I. Prokaryotic and Viral Chromosomes

P roceeding from the principle that one examines and tries to understand simple systems before turning to those more complex, we will examine first the chromosomes of a number of prokaryotic species—that is, the bacteria and, by inference, the blue-green algae, the latter often referred to by botanists as the *Cyanophyceae,* but more correctly called the *Cyanobacteria.* The viruses are also included here because of chromosomal similarity, although technically they are neither prokaryotes, nor are they cellular. These differ from eukaryotic species in a number of ways, but the major cytogenetic variations of a structural nature are that the prokaryotic chromosome consists of a naked strand of DNA lying free in the cytoplasm, and that the entire genetic system of each species is encoded in a single chromosome. The viral chromosome also consists of a naked strand of DNA or RNA, but when outside of a bacterial or eukaryotic cell it is encased in a protein shell or coat. By way of contrast, the eukaryotic genome is encased and partially isolated within a double-layered nuclear envelope, the haploid chromosome number is always more than one, and the chromosomes are complexes of nucleoproteins even though the molecular basis of inheritance still resides in the sequences of base pairs of DNA.

Some cytogeneticists would object to "chromosome" as a term used to designate the genetic apparatus of the prokaryotes, preferring "genophore" in its place. This is more a matter of semantics than it is of cytogenetics since the prokaryotic chromosome not only has the same fundamental structure based on nucleic acid sequences, but it also carries out the same three basic functions of its eukaryotic counterpart. That is, it is the organelle which stores, replicates, and transmits the coded information of biological inheritance; which is so regulated in its action that its primary products appear at appropriate times and in appropriate amounts during the life cycle of the organism; and which is capable of recombining its genetic content with other similar individuals in such a manner that the segregating progeny of a hybrid union can have different genetic properties. The term "chromosome" will, therefore, be applied to the genetic units of both prokaryotic and eukaryotic species, with distinctions in structure being made when appropriate.

The Bacterial Chromosome

The chromosomes of *Escherichia coli* and its close relative, *Salmonella typhimurium,* are probably the most thoroughly mapped genomes of any cellular organisms, and the hundreds of identifiable genes in each species can be located in a single, circular linkage group. A single cell may have more than one chromosome at a time, each in its own nuclear area (Fig. 2.1); these are replicas of each other and result from the fact that the division of the chromosome and the division of the cell are not necessarily coordinated with each other.

The chromosome of *E. coli* is a double helix of DNA about 1,100 μm in length, with a diameter of 20 Å to 40 Å (Fig. 2.2). Such a molecule would contain more than 3.235×10^6 nucleotide pairs (each pair is 3.4 Å long, and there are 2,941 base pairs per micrometer), and has a molecular weight of 2.5×10^9 daltons (one dalton equals the weight of one hydrogen atom). This is contained in the nuclear area of the bacterial cell, and since this area is no more than a micrometer or so in diameter and lacks a nuclear membrane as a confining device, the chromosome must be enormously folded, or compacted, to fit within it. When isolated by gentle extraction

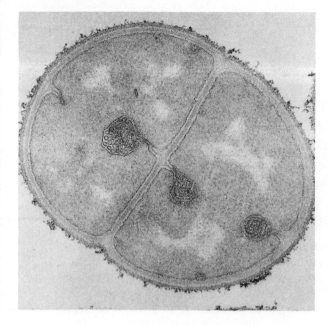

Fig. 2.1. An electron micrograph of a bacterial cell in the process of dividing. The rounded structures are *mesosomes*, formed by an in-folding of the cell membrane; the lighter areas contain the naked chromosomes which are not visible at this magnification. There is some indication that the chromosome is attached to the mesosome, and that during division the mesosome aids in separating the replicated halves of the chromosome. The mesosome also seems to be involved in the process of cell division, and a second division place can be seen forming at the outer edges of the cell (courtesy of Dr. S. C. Holt).

Fig. 2.2. The circular chromosome of *E. coli* in its relaxed and open form. In this state it would measure 350 μm in diameter, and about 1100 μm in circumference. When spread in the presence of ethidium bromide, the chromosome assumes a supercoiled and twisted form (see Delius and Worcel, 1973), much like that assumed by the supercoiled chloroplast genome in Fig. 2.19. As Fig. 2.3 indicates, RNA is involved in bringing about the supercoiled state (courtesy of Dr. S. C. Holt).

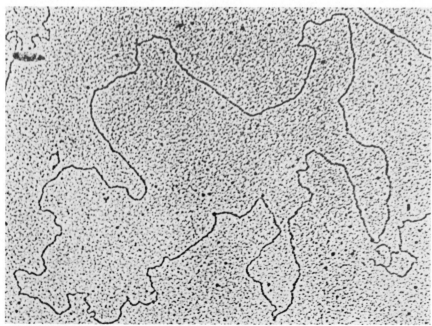

methods and prepared for electron microscopy, the chromosome takes on one to two basic appearances. Occasionally the open circular nature of the genome is evident, and the chromosome reveals itself, through the use of various techniques, to be a naked DNA molecule, that is, such a genome is not altered in appearance by the action of an RNase or a protease, indicating that neither RNA nor protein are involved in maintaining the basic circular structure or integrity of the chromosome. More usually, however, the genome is released from the physically disrupted cell as a highly folded, compact structure, which is revealed to possess a high degree of regularity based on several orders of coils which progressively reduce the general diameter of the genomic mass (Fig. 2.3). The isolated genomes, on the other hand, do contain about 30% by weight of RNA and 10% of protein. The RNA is largely transcribed messenger, transfer, or ribosomal RNA which has not, as yet, been released into the cellular milieu, together with a small amount of RNA which is responsible, in part, for the folded character of the genome; that is, RNase activity causes the chromosome partially to unfold, but proteases do not. Most of the protein has been identified as RNA polymerase engaged in transcriptional activities, an understandable contamination since a metabolically active bacterial cell would be transcribing at virtually all stages of its life cycle. A smaller proportion of the protein represents pieces of the cell membrane, a special segment called the *mesosome,* to which the chromosome is attached in the cell. Such an attachment is not typical of eukaryotic chromosomes, although the latter may come to lie against the cell membrane at certain times during the cell cycle.

The thermophilic bacterium *Thermoplasma acidophilum,* possesses an unusual chromosome for a prokaryote. Associated with its DNA is a histone-like protein that, at any one time, can complex with about 20% of the genome. Its suggested purpose is that of stabilizing the DNA during periods of denaturing conditions, and thus permitting a rapid return to normal once the environmental conditions are normal.

The folded chromosome contains about 50 (the range is 12 to 80) loops (Fig. 2.3), with the loops formed by RNA molecules crosslinking the chromosome at the base of each loop. Each loop is further thrown into supercoils, with each supercoil containing about 400 nucleotide pairs. The remarkable thing about these highly compacted chromosomes is that they can presumably replicate and transcribe, although it is probable that DNases (endonucleases) and/or RNases open up the coils or the loops for greater maneuverability.

Bacillus subtilis possesses a circular chromosome of approximately the same dimension and molecular weight as that of *E. coli,*

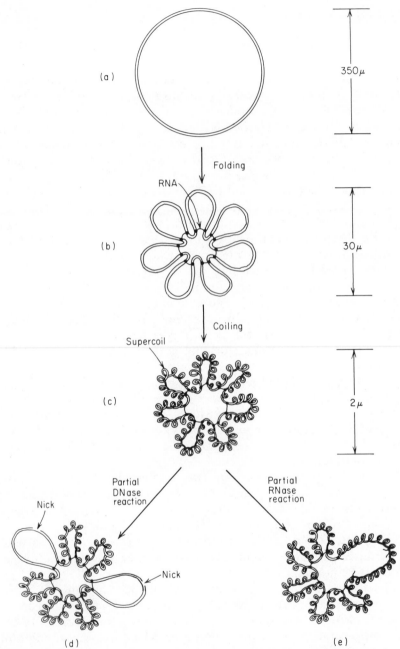

Fig. 2.3. Diagrammatic representation of the coiled and supercoiled forms of the *E. coli* chromosome, and the role played by RNA as a crosslinking molecule. The shift from a coiled to a supercoiled state remains to be explained, but RNase, which causes the larger coils to open up, demonstrates the nature of the RNA crosslinking. The coiling and super-coiling can reduce the chromosome seen in Fig. 2.2 to a compact mass about 2.0 μm in diameter (Pettijohn and Hecht, 1973, copyright 1973, by the Cold Spring Harbor Laboratory).

but there is some suggestion that it may replicate as a linear rather than a circular molecule. *Streptomyces coelicolor* and *Salmonella typhimurium* also have circular chromosomes, with the latter having at least 323 known genes. *Mycoplasma sp.* have a small circular chromosome of about 265 μm in length, but *Hemophilus influenzae* might be an exception to the rule. Its 865 μm long chromosome appears at times to have a linear form with two identifiable ends (Fig. 2.4). Whether these ends represent ruptures of a circle during extraction, and are ordinarily covalently bonded to form a closed circle, is an open question, but this seems likely.

Fig. 2.4. An electron micrograph of the chromosome of the bacterium *Hemophilus influenzae*. The cell was burst osmotically and the contents spread on a protein film on a water surface, picked up on a grid, dried and then photographed. The chromosome is about 869 micrometers long, and seems to possess only two ends, one of which is visible at the upper right; this may be an artifact of preparation, however, since other bacterial chromosomes such as *E. coli* are circular. The dark structure in the center is the remains of the bacterial wall (courtesy of Dr. L. McHattie).

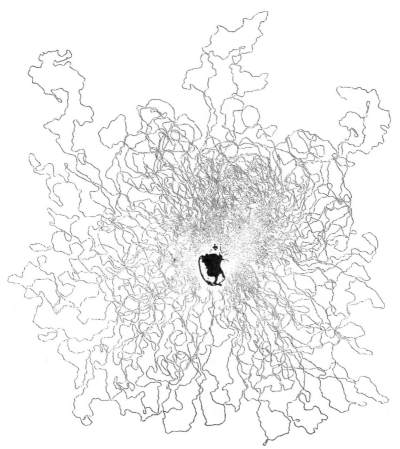

Although it is customary to think of *E. coli* and related bacteria as possessing only a single chromosome per nuclear area, the cells may in fact contain additional double-stranded DNA in the form of small circles which can persist in the cytoplasm. Referred to as *plasmids* or *episomes,* these are of several kinds, and they exercise a variety of influences on the bacteria that harbor them. The usual distinction, but not a sharp one, made between a plasmid and an episome rests on the fact that the episome is capable of being integrated into the host chromosome, and hence of simultaneous replication, while the plasmid is not integrated and can replicate independently of the host. By this definition, the F^+ agent and lambda bacteriophage are episomes while the *col* and *R* factors remain plasmids. In one sense, they seem analogous to the supernumerary or B chromosomes of higher forms in that they are dispensable; in other ways, however, their influence and behavior are such that they more closely resemble temperate (i.e., nonvirulent) bacteriophages, leading to the suggestion that they may have had their origin from defective viruses which now provide the host cells with additional, but not absolutely necessary, genes.

Through the passage of time the plasmids have acquired a structural and genetic identity of their own. They are extragenomic circular elements of DNA which can replicate autonomously within the cytoplasm of the bacterial cell, and pass from cell to cell through slender projecting appendages called *fimbriae* or pili, and independently of cell division. Although many copies of a plasmid may be present in a single cell, drawing on the metabolic machinery of the cell for their replication, they are not lethal to the cells which possess them; they may, in fact, be highly beneficial, as shall be pointed out later. They can be eliminated by growing the bacteria at high temperatures $(42°C)$, or in the absence of thymine, or by exposing the cells to acridine dyes.

Some episomes such as the *F*, or sex, factor, can under appropriate circumstances, become integrated into the bacterial chromosome (Fig. 2.5), after which it will replicate along with the host chromosome and be passed from cell to cell at the time of division. In this form it behaves like some viruses such as lambda (λ) bacteriophage, but plasmids and some of the viruses are capable of such integration. A further distinction between a virus and a plasmid is that if a virus multiplies in the cytoplasm of the cell it will eventually kill it by utilizing the cell's substance, but a plasmid will not.

The best known of the episomes are the sex factors, variously

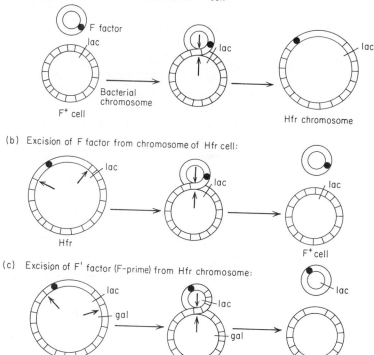

(a) Insertion of F factor into chromsome of F⁺ cell:

F factor

lac

Bacterial chromosome

F⁺ cell

lac

lac

Hfr chromosome

(b) Excision of F factor from chromosome of Hfr cell:

lac

Hfr

lac

lac

F⁺cell

(c) Excision of F' factor (F-prime) from Hfr chromosome:

lac

gal

lac

gal

lac

lac

Fig. 2.5. The insertion (top line) and excision (middle and bottom lines) of the *F* factor from a bacterial chromosome. Insertion is preceded by homologous pairing much as occurs during the insertion of lambda bacteriophage (see Fig. 2.14), and involves nucleases which can cleave the polynucleotide strands in precise ways. Insertion occurs at a precise point on the host chromosome, after which *F* becomes *Hfr*. Excision is accomplished by the reverse of the insertion events. The bottom line indicates that *Hfr* can carry a portion of the host chromosome with it to another cell where it may become integrated again in the process of transduction.

called fertility (F^+) or transfer agents. The F^+ agent of *E. coli* is a circular piece of DNA about one-fortieth of the length of the host chromosome, or about 30 μm in length, and having a molecular weight of 35×10^6 daltons. Only one F^+ agent as a rule is present per cell, and it replicates coordinately with the host chromosome. The F^+ agent has about a dozen genes in it, some of which control the formation and development of *pili*, long tube-like appendages extending from the surface of the bacterial cell. These tubes, plus other surface components which establish prior contact, provide a physical bond between cells, permitting the F^+ agents, other plasmids, and occasionally pieces of host DNA to pass from one cell to another, and at the same time to serve as attachment and entry points for some kinds of infectious DNA and RNA viruses.

During transfer from cell to cell, the F^+ agent is believed to enter as a linear molecule beginning at a specific point O, with the rest of the DNA following, as replication proceeds by the rolling circle mode of replication (Fig. 2.6). After entry it recircularizes. The transfer of F^+ is always to F^- cells, and hence is one-way. An $F^+ \times F^-$ cross can consequently yield only F^+ cells.

The F^+ agent can shift from an independent existence in the host cell cytoplasm to one of integration into the host chromosome. Integration occurs at a specific point in the circular host chromo-

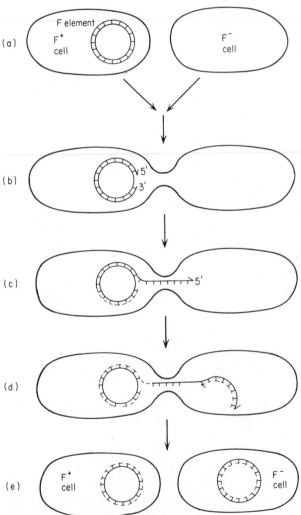

Fig. 2.6. *F* factor entering a bacterial cell: a) F⁺ and F⁻ cells approach each other, and b) a *pilus* or contact tube becomes established between them; one polynucleotide strand is nicked by a nuclease, and c) the 5′ end enters the F⁻ cell and begins to replicate itself via the rolling circle method (d); at the same time the F⁺ circle is also replicating itself, after which the two cells separate (e), with both now being F⁺.

some, indicating that integration is dependent upon a degree of shared homology of F^+ and host chromosome (see Fig. 2.14). As a result of integration, the F^+ cell is converted into an Hfr cell (high frequency of recombination), a designation related to the fact that in transferring to an F^- cell, the F^+ factor in an integrated state may, when released, carry the host chromosome, or a portion of it, with it, and recombination between donor and recipient chromosomes can occur readily. The integrated F factor can also disengage itself from the host chromosome to become an F^+ agent again. However, if in the process of excision or disengagement it carries a portion of the host chromosome with it, it becomes an F' agent (F-prime) which exerts a somewhat different genetic effect on the recipient cell than does an F^+ agent in that the genes it carries will create a condition of partial diploidy as well as an F^+ condition.

Two types of plasmids, of which there are many derivative forms, are the *colicinogenic* (*col*) and *resistance* (*R*) factors (Table 2.1). They are similar to the F^+ agent in possessing circular forms of DNA. The former, which seem to be dependent upon the F^+ agent for transmission, since they cannot pass independently from one F^- cell to another F^- cell, can cause the death of other bacterial strains lacking the *col* plasmid through the release of specific proteins called *colicins* which have a lethal effect by interfering with certain meta-

Table 2.1. Commonly used plasmids and their physical and chemical characteristics

Plasmid	Mol. Wt. in 10^6 Daltons	Number of Copies per Cell	Selective Factor*
ColE1	4.2	10–15	E1 imm.
pCR1	8.7	10–15	Kn^R or E1 imm.
pTK16	2.8	10–15	Tc^R or Kn^R
pMB9	3.5	10–15	Tc^R or E1 imm.
pBR322	2.6	10–15	Ap^R or Tc^R
RSF1030	5.6	20–40	Ap^R
CloDF13	6.0	10	Cloacin
R6K	25.0	13–38	Ap^R or Sm^R
F	35.0	1	Sex factor
R1	65.0	1–3	$Ap^R, Cm^R, Sn^R, Sm^R, Kn^R$
R6	65.0	1–3	$Tc^R, Cm^R, Sn^R, Sm^R, Kn^R$
EntP307	65.0	1–3	Enterotoxin

*Abbreviations of selective factors are as follows: E1 imm., immunity to colicin E1; *Kn*, kanamycin; superscript *R*, resistance; *Tc*, tetracycline; *Ap*, ampicillin; *Sm*, streptomycin; *Sn*, sulfonamide; *Cm*, chloramphenicol. Those plasmids beginning with *p*, and indented under ColE1, are derivatives of ColE1, and together with ColE1, they can, in the presence of chloramphenicol, increase the number of plasmids per cell to 1,000 to 2,000 (see Helinski, 1977, and Beers and Bassett, 1977 for further details).

bolic processes such as ATP production, DNA degradation or synthesis, and protein synthesis by means of ribosomal alteration. The colicins formed by some col^+ strains are strikingly similar in molecular weight and amino acid sequence to the tail proteins of bacteriophages, lending credence to the belief that they may have originated from some one or more defective viruses.

The R factors confer antibiotic resistance on the strains of bacteria possessing them. Resistance to as many as four antibiotics—chloramphenicol, tetracycline, streptomycin, and sulfonamide—may be present in a single R plasmid, thus providing the host cell with an important set of genes not normally a part of its own genome. Such degrees of resistance can be passed from cell to cell much as F^+ can be acquired by F^- cells, but the F^+ factors are not involved in this transaction since the R plasmids produce their own pili for entry into cells.

The col and R factors differ from the F^+ agent in being capable of stable integration into the host chromosome, but as will be dealt with later (p. 327), both of these plasmids, as well as others like them, are most useful vehicles for recombinant DNA experiments.

The Chromosomes of Viruses

The viruses are a widely varying group of host-specific infectious agents which are distinguished from cellular organisms by their relative simplicity of structure and by their dependence upon a host cell for continued increase in numbers. Outside a living cell, the virus exists as an inert, but potentially infectious, particle called a *virion,* which possesses a core of nucleic acid, either DNA or RNA, surrounded by a protein coat, or *capsid.* Some of the animal viruses may have this nucleoprotein particle further enclosed in a membranous envelope. The nucleic acid may be single- or double-stranded, and each virion contains a single molecule which in some may be linear in form, in others circular as in some bacteria such as *E. coli.* The length of the nucleic acid molecule varies from one virus to another (Table 2.2), being little more than a single micrometer long in R17 to over 50 μm in T7. The proteins of the viral head, or capsid, are coded for by the viral genes and may consist of a single kind of protein as in the helical outer covering of the tobacco mosaic virus (*TMV*) (Fig. 2.7), or several kinds of proteins as in bacterial viruses (bacteriophages) of the T-series (Fig. 2.8). During the intracellular phase of the viral life cycle, the viral chromosome either exists free in the cytoplasm of the infected cell or is integrated into the chromo-

Table 2.2. Characteristics of viruses that have been studied extensively

Virus	Host	Mol. Wt. in 10^6 Daltons	Length in Micrometers	Number of Genes (Approx.)	Form and Composition*
Double-stranded DNA					
Mu	*E. coli,* strain K12	25	13	21–25	L; U
SV40	Monkey	3.0	1.5–1.7	8	L
T2	*E. coli*	120	52–58	200	L; CPS; TR
T4	*E. coli*	130	—	160	L; CPS; TR
T5	*E. coli*	75	38.8	100	L; U (segmented)
T7	*E. coli*	25	12.0	30	L; U; TR
P22	Salmonella	33	17.0	40–50	L or C; U; CE
Adenovirus 2	Humans	28	14.0	40–50	L; CPS; TR
		23	12.0	—	L
Single-stranded DNA					
φX174	*E. coli*	1.7	1.8	9	C
Double-stranded RNA					
Reovirus	Mammals	17	8.5	—	L
Dwarf virus	Rice	10	5.0	—	L
Single-stranded RNA					
R17	*E. coli*	1.05–1.12	1.0	3	L; U
Polio	Humans	2.0	2.6	—	L
TMV	Tobacco	2.1	2.1	—	L
Measles	Humans	3.3	—	—	L

*Abbreviations of topology: L, linear; C, circular; U, unique (all chromosomes similar in nucleotide sequence); CPS, circularly permuted sequence (each chromosome a permutation of a given sequence of genes); TR, terminally redundant; CE, cohesive ends.

Fig. 2.7. A representation of the structure of the tobacco mosaic virus (TMV). The protein subunits, each having a molecular weight of 17,000 daltons, are helically wound about a central RNA molecule, to form a hollow rod about 3000 Å long and with a diameter of about 170 Å. The RNA is infective by itself, the protein is not. If separated from each other, and then placed in solution together, they can spontaneously reform the viral particle.

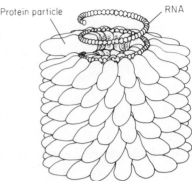

Protein particle RNA

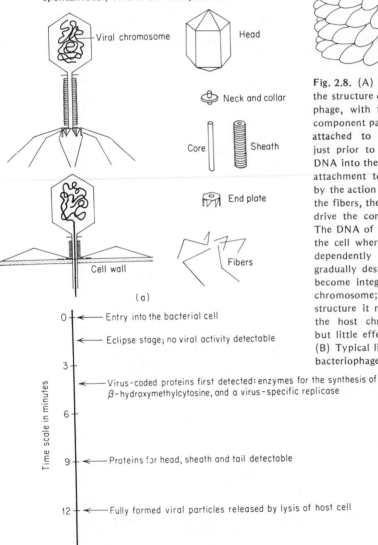

Viral chromosome

Head

Neck and collar

Core

Sheath

End plate

Fibers

Cell wall

(a)

Fig. 2.8. (A) A representation of the structure of a T-even bacteriophage, with the intact virus, its component parts, and as it appears attached to the bacterial wall, just prior to the injection of its DNA into the host cell. Following attachment to the bacterial wall by the action of the end plate and the fibers, the sheath contracts to drive the core through the wall. The DNA of the virus then enters the cell where it can replicate independently of the host cell, gradually destroying it; or it can become integrated into the host chromosome; as an integrated structure it replicates along with the host chromosome, and has but little effect on the host cell. (B) Typical life cycle of a T-even bacteriophage in a lytic state.

Time scale in minutes

0 ← Entry into the bacterial cell

← Eclipse stage; no viral activity detectable

3

← Virus-coded proteins first detected: enzymes for the synthesis of β-hydroxymethylcytosine, and a virus-specific replicase

6

9 ← Proteins for head, sheath and tail detectable

12 ← Fully formed viral particles released by lysis of host cell

(b)

some of the host. Only when the viral units are ready to be freed from the host cell through lysis are the nucleic acids of the core and the proteins of the capsid assembled to form the mature virions.

Table 2.2 lists a number of the viruses that have been studied extensively.

Double-stranded DNA Viruses. The T-series of bacteriophages, along with bacteriophage lambda (λ), are among the better understood of this group. The T-even viruses generally have larger amounts of DNA than do the T-odd viruses and they differ still further by having the nucleotide, 5-hydroxylmethylcytosine, replace the normal cytosine in the double helix. Figure 2.8 provides a general view of their overall structure; all have a life cycle which follows a typical sequence.

Contact between the bacterium and a T-even bacteriophage is necessary for infection. The fibers of the tail unwind on contact, and the base plate, together with the fibers, affixes the virus firmly to the cell wall. The sheath contracts, and in doing so, drives the core through the cell wall and cell membrane after which the DNA of the head, plus a small amount of protein, is injected into the host cell in syringe-like fashion. The entry of the DNA is facilitated by a viral lysozyme which enables the core more easily to penetrate the wall and cell membrane of the bacterium. Following injection of the DNA, the remainder of the bacteriophage plays no further role.

During the first 3 or 4 mins. after infection, the virus cannot readily be detected in the bacterial cell. This is the vegetative or eclipse period of the life cycle. Virus-coded proteins then make their appearance, in particular, the enzymes involved in the biosynthesis of 5-hydroxymethylcytosine of the T-even phages, and a virus-specific DNA polymerase for purposes of replication. Eight minutes after infection viral chromosomes are being formed; proteins that will go to form the head, sheath, and tail become identifiable after 9 mins. During this period degradation of the host chromosome is taking place. Fully assembled viral particles appear in about 12 mins., after which a lysozyme is synthesized which lyses the bacterium and frees the viruses. About 40 genes are involved in the formation of these proteins, as determined by mutations of those genes which affect the shape and character of the head, sheath, and/or tail.

The chromosomes of T_2 and T_4 are linear structures with terminally redundant ends; that is, the sequence of nucleotides at the left end of the chromosome is repeated at the right end, giving a chromosome that is longer than the basic genome (Fig. 2.9). In these two bacteriophages the amount of terminal redundancy is between

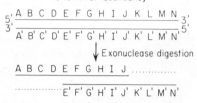

No terminal redundancy

```
5' | A B C D E F G H I J K L M N | 3'
3' |_____| 5'
     A' B' C' D' E' F' G' H' I' J' K' L' M' N'
```

↓ Exonuclease digestion

```
A B C D E F G H I J .............

............ E' F' G' H' I' J' K' L' M' N'
```

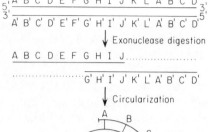

Terminal redundancy

```
5' | A B C D E F G H I J K L A B C D | 3'
3' |_____| 5'
     A' B' C' D' E' F' G' H' I' J' K' L' A' B' C' D'
```

↓ Exonuclease digestion

```
A B C D E F G H I J .....................

............ G' H' I' J' K' L' A' B' C' D'
```

↓ Circularization

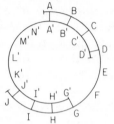

Fig. 2.9. The results of the action of exonuclease digestion in a 3' ⟶ 5' direction, followed by reanneling. When terminal redundancy is present, circularization will occur under conditions which will permit reannealing; If not present, no circularization will take place.

Unique collection of chromosome

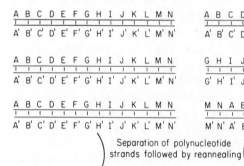

```
A B C D E F G H I J K L M N
| | | | | | | | | | | | | |
A' B' C' D' E' F' G' H' I' J' K' L' M' N'

A B C D E F G H I J K L M N
| | | | | | | | | | | | | |
A' B' C' D' E' F' G' H' I' J' K' L' M' N'

A B C D E F G H I J K L M N
| | | | | | | | | | | | | |
A' B' C' D' E' F' G' H' I' J' K' L' M' N'
```

Permuted collection of chromosomes

```
A B C D E F G H I J K L M N
| | | | | | | | | | | | | |
A' B' C' D' E' F' G' H' I' J' K' L' M' N'

G H I J K L M N A B C D E F
| | | | | | | | | | | | | |
G' H' I' J' K' L' M' N' A' B' C' D' E' F'

M N A B C D E F G H I J K L
| | | | | | | | | | | | | |
M' N' A' B' C' D' E' F' G' H' I' J' K' L'
```

Separation of polynucleotide strands followed by reannealing

```
A B C D E F G H I J K L M N
| | | | | | | | | | | | | |
A' B' C' D' E' F' G' H' I' J' K' L' M' N'
```

```
A B C D E F G H I J K L M N
| | | | | | | | | | | | | |
G' H' I' J' K' L' M' N' A' B' C' D' E' F'
```

↓ Circularization

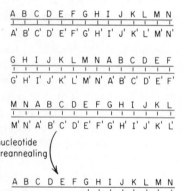

Fig. 2.10. Unique and permuted chromosomes differ in the sequence of nucleotides in each one, and differ still further if separation of the polynucleotide strands takes place, followed by reannealing of the strands. If the chromosomes are all unique, only linear double-stranded sequences can form; if permuted, double-stranded circles can form from any two complementary single strands.

2,000 and 6,000 nucleotide pairs, or about 1% to 3% of the basic chromosome. The presence of terminal redundancy can be detected by the use of exonuclease III, an enzyme that systematically removes nucleotides from the $3'$ end of the polynucleotide strands of the double helix. Following a brief period of enzymatic digestion, the chromosome can form circles (Fig. 2.9), a phenomenon that would not occur if terminal redundancy were absent.

A collection of T_2 and T_4 chromosomes has also been shown to be a group of circularly permuted molecules of DNA; that is, each chromosome is a different permutation of a common sequence of nucleotides (Fig. 2.10). Two quite different sets of observations support this concept of the chromosome. In the first place, the genetic map of T_2 or T_4 is circular even though the chromosome is linear in structure. Because of this, four linked genes may be *a-b-c-d* in one chromosome, *d-a-b-c* in another, or *b-c-d-a* or *c-d-a-b* in still other chromosomes. In the second place, if a circularly permuted collection of chromosomes is denatured into single polynucleotide strands and then permitted to reanneal, circles with gaps will form without the action of exonuclease (Fig. 2.10).

By ways of contrast, T_7 (and presumably T_3 as well) chromosomes are unique but also terminally redundant; that is, all chromosomes are identical in nucleotide sequence from one end to the other. The T_7 genetic map is linear as would be expected from such an arrangement. In addition, a T_7 chromosome is only about one-quarter the length of a T_2 or T_4 chromosome, and its terminal redundancy amounts to about 260 nucleotide pairs. Through the use of exonuclease III, a progressive removal of nucleotides from the $3'$ end, followed by reannealing conditions, will cause the chromosomes to form circles (Fig. 2.11). If denatured into single nucleotide strands

Fig. 2.11. The behavior of unique, but terminally redundant, chromosomes such as those of T7 which have been exposed to the action of an exonuclease and then allowed to reanneal. If separation of the strands followed the exonuclease digestion, circles consisting of several chromosomes can form as well linear aggregates of more than two chromosomes (concatenates).

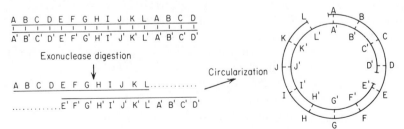

and then reannealed, circles will be more likely to form at low concentrations and concatenates (linear aggregates of more than two chromosomes with varying degrees of pairing) at high concentrations.

The uniqueness of the T_7 chromosome indicates that it has a distinct beginning and an equally distinct end, a circumstance not applicable to the circularly permuted T_2 or T_4 chromosomes. It is possible to produce deletions in bacteriophage chromosomes by radiation or by the use of chemicals. A deletion of a portion of a T_7 chromosome simply produces a shorter chromosome; if it is in a nonessential portion of the genome, the bacteriophage will still be viable and infectious. A comparably sized deletion in a T_2 chromosome would not, however, shorten the chromosome, but since a deletion removes part of the genome without a shortening of chromosome length, there must be a compensatory addition in the amount of terminal redundancy. This suggests that the chromosomes of these two phages are formed by somewhat different replicative processes and that the fixed length of the T_2 chromosome is determined by what fits inside the head of the virus, not by the basic nature of the genome, whereas that of T_7 is determined by its unique ends.

About 30 proteins as gene products have been identified through mutations in T_7, and the genes responsible can be classified into three groups, placed in serial order from left to right (Table 2.3). The Class I genes occupying about 20% of the left end of the chromosome, and terminated by a signal to the right of the gene specifying the ligase, are transcribed by an RNA polymerase of host cell origin, beginning at a point about 1.3% from the left end of the chromosome. The far left end of the chromosome has, as yet, no known genetic function, although this region probably includes the recognition sites for polymerase attachment as well as the ribosome recognition site. Among the Class I enzymes is a different RNA polymerase which subsequently will be involved in the transcription of Class II and III genes. The serial order of the genes seems to be a reasonable reflection of the serial sequence of actions of the proteins which are formed; that is, the Class I and II genes seem concerned mainly with DNA metabolism such as the destruction of host DNA and the synthesis of T_7 DNA, while Class III genes are concerned with the formation of the proteins involved in the assembly of the head and tail portions of the bacteriophage, plus the processes of maturation of the virus for eventual release from the host cell.

The T_5 bacteriophage differs from all other related viruses in the character of its chromosome. It is unique in its structure and reveals a single linear genetic map. However, when the unbroken but

Table 2.3. The genes of T7 bacteriophage with their map position (beginning at the left end of the chromosome), the proteins they specify, and the molecular weights of these proteins. See text for explanation (from Table 1, Studier, 1972, copyright, 1972, American Association for the Advancement of Science).

	Gene Position	Protein or Function	Estimated Mol. Wt. in Daltons
	0.3	Nonessential	8,700
	0.5	Nonessential	40,000
	0.7	Nonessential	42,000
	1.0	RNA polymerase	100,000
I	1.3	Ligase	40,000
	1.7	Nonessential	17,000
	2.0	Reduced DNA synthesis	—
	3.0	Endonuclease	13,500
	3.5	Lysozyme	13,000
	4.0	Reduced DNA synthesis	67,000
	5.0	DNA polymerase	81,000
	6.0	Exonuclease	31,000
	7.0	Found in phage action unknown	14,700
II	8.0	Head protein	62,000
	9.0	Head assembly protein	40,000
	10.0	Major head protein	38,000
	11.0	Tail protein	21,000
	12.0	Tail protein	86,000
III	13.0	Found in phage action unknown	14,000
	14.0	Head protein	18,000
	15.0	Head protein	83,000
	16.0	Head protein	150,000
	17.0	Tail protein	76,000
	18.0	DNA maturation	—
	19.0	DNA maturation	73,000

denatured chromosomes of other viruses, such as T_2 or T_7, are sedimented in a sucrose density gradient, they come to equilibrium in a single band or zone, indicating that the molecules are all alike in physical length and conformation and are without breaks in either of the polynucleotide strands. When T_5 chromosomes are treated similarly, four different sedimentation zones appear, leading to the interpretation that gaps or weak spots exist normally in the double helix which cause the chromosome to separate readily into smaller units. The intact chromosome is 38.8 μ in length, and it appears to have one gap in one of the polynucleotide strands and three in the other, each gap being a weak area subject to separation or fragmentation. When the chromosome is denatured, five strands are recovered,

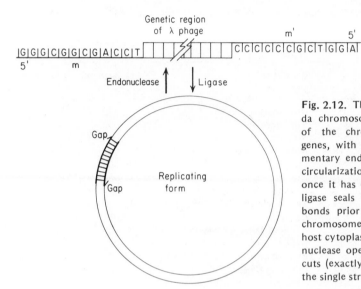

Fig. 2.12. The structure of the lambda chromosome. The central region of the chromosome contains the genes, with single-stranded, complementary ends which can bring about circularization of the chromosome once it has entered a host cell. The ligase seals the gaps with covalent bonds prior to the time when the chromosome replicates within the host cytoplasm, while a specific endonuclease opens the circle by making cuts (exactly 12 nucleotides apart) in the single strands.

two being about 4 μm long and the others being 8, 16, and 24 μm long. The uninterrupted 4-μm-long piece at one end of the T_5 chromosome is that which enters first on adsorption to the host cell. If only that 4-μm end is permitted entry (the remainder being easily sheared off by agitation in a blender), that piece can cause degradation of the host DNA, indicating that the gene or genes having that function are terminally located.

The lambda (λ) bacteriophage is another double-stranded DNA virus that has been extensively studied, particularly from the points of view of gene control and regulation (see p. 338) and of transcriptional behavior (see p. 304). Here we will consider only the general structure of its chromosome and how this structure influences its behavior. The λ chromosome has a length of about 17 μm, or about 47,000 nucleotide pairs, with each of its polynucleotide strands extended from the 5' end by the addition of 12 nucleotides (Fig. 2.12). The extensions are complementary to each other and are necessary for penetration into an *E. coli* cell. Being complementary, the ends are cohesive and the chromosome circularizes immediately upon entrance into the host cell, with the gaps being closed by host ligases. The chromosome thereby is protected from degradation by the host exonucleases. The circular form of the chromosome is also that form which is retained during replication, but despite its circularity, the genetic map is a linear one, a situation quite the reverse of that encountered in the T_2 or T_4 bacteriophages. The reason is that the

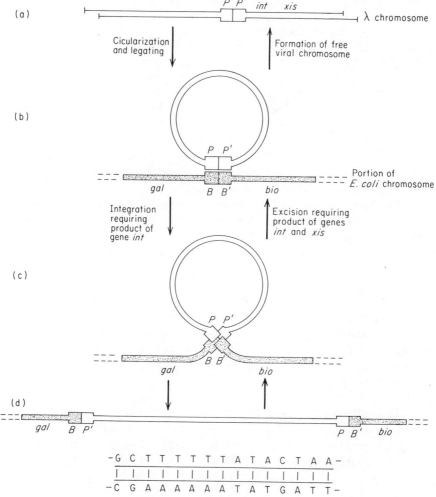

(a)

P P′ int xis

λ chromosome

Cicularization
and legating

Formation of free
viral chromosome

(b)

P P′

Portion of
E. coli chromosome

gal   B B′   bio

Integration
requiring
product of
gene *int*

Excision requiring
product of genes
int and *xis*

(c)

P P′

gal   B B′   bio

(d)

gal   B P′

P B′   bio

```
 -G  C  T  T  T  T  T  T  A  T  A  C  T  A  A-
    |  |  |  |  |  |  |  |  |  |  |  |  |  |  |
 -C  G  A  A  A  A  A  A  T  A  T  G  A  T  T-
```

The sequence of 15 nucleotide pairs in the *att* regions of phage (POP′) and bacterium (BOB′).

Fig. 2.13. Integration and excision of lambda bacteriophage from the chromosome of *E. coli*. There are several attachment (*att*) sites on the bacterial chromosome, and the POP′ site of the phage and the BOB′ site of *E. coli* consist of identical sequences of 15 nucleotide pairs for homologous pairing for integration. Integration requires the action of the product of gene *int* which maps to the right of P′ while excision requires the action of the products of both genes *int* and *xis*. See also Fig. 2.14.

lambda chromosomes are a unique group, all being similar to each other in nucleotide sequence and having distinct ends. The shift from circularity to linearity depends upon a particular region in each of the projected ends, called m or m', which can be nicked, and opened up, by a specific endonuclease. As Fig. 2.13 indicates, the lambda chromosome also possesses an additional specialized site, $P-P^1$, centrally located, which enables the circular form to recombine with, and thereby become integrated into, the *E. coli* host chromosome. When integrated, it is replicated along with the host chromosome. The recombinational site on the host chromosome, $B-B'$, which is identical in nucleotide sequence to the $P-P^1$ site, lies between the genes coding for the two enzymes which govern galactose and biotin metabolism, and it is obvious, therefore, that the map distance between the two genes is determined by the integrated presence or the absence of the viral genome.

The lambda genome can disengage itself from the host chromosome by reversing the recombinational events, at which time two things can happen: (1) It can detach itself as an intact and unencumbered genome, replicate itself in the cytoplasm of the host, become encapsulated in the head proteins, and then be released by lysis from the disintegrated host cell, after which it can reinfect other *E. coli* cells; or (2) it can detach in such a manner that an adjacent piece of the host chromosome accompanies it (Fig. 2.14). This piece must be either the *gal* or the *bio* locus, in whole or in part, and since the virus can reinfect another *E. coli* cell, this locus can subsequently be integrated into the second host chromosome, a process known as *transduction*.

Single-Stranded DNA Viruses. The ϕX174 bacteriophage, which infects the bacterium *Shigella* as well as *E. coli,* is one of a small group of single-stranded DNA viruses. Like all of the others, ϕX174 exists as a circle; there are, in fact, no known single-stranded linear forms, and circularity must be maintained if it is to be infectious.

ϕX174 is a small virus of 5,375 nucleotides, the first to be completely sequenced. As a virion it is encapsulated in icosahedral form (20-sided polygon), without a tail but with protein spikes at the junctures (vertices) of its many faces. Its small genome codes for ten proteins. The genetic map (Fig. 2.15) reveals regions A-D to be concerned with DNA replication; protein A, together with the host enzymes, is involved in the process of replication, while proteins B, C, and D are involved in the packaging of the single-stranded viral strands into the capsids prior to release from the host cell. Region E produces a protein which is involved in lysis of the host cell at the

Fig. 2.14. Formation of a transducing lambda bacteriophage. Ordinarily, when the lambda phage is to be excised from the bacterial chromosome, the process is exactly the reverse of integration (Fig. 2.13), and takes place within the *att* site. The phage, however, may carry a piece of the host chromosome with it by an "illegitimate" kind of excision with the breaks outside of the *att* site. In this figure this has occurred in a manner which involves the *gal* locus of the host chromosome, and the result is the formation of a *gal-transducing* phage; it could occur also on the other side of the *att* site, and produce thereby a *bio-transducing* phage. This piece of the host chromosome can be introduced into another cell, and integrated into that host chromosome.

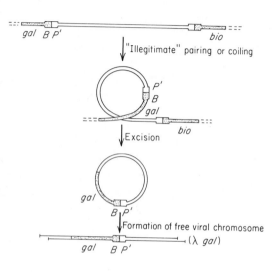

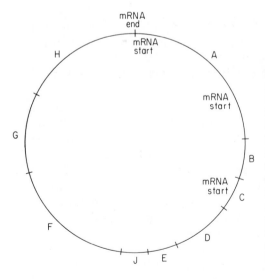

Fig. 2.15. The genetic map of φX174 phage, determined by standard bacterial techniques, showing the relative length and position of the nine genes. A tenth gene, K, is located within the limits of gene C. The function of these genes is as follows: A, double-stranded DNA synthesis; B, C, D, single-stranded DNA synthesis and packaging; E, lysis; J, structural; F, major capsid component; G, major spike component; H, minor spike component. Compare this map with that of the same phage determined by molecular techniques (Fig. 5.35).

time of viral release; region *J* is concerned somehow with a structural component of the virion; region *F* is concerned with the formation of the capsid proteins; and regions *G* and *H* are concerned with the proteins that form the spikes projecting from the head. As will be pointed out in Chapter 5, ϕX174 represents an extraordinary example of evolutionary economy in that genes B, E, and K are included within other genes, while genes A and C, C and D, and D and J overlap each other by one nucleotide apiece. Only four short regions of 39 (between *J* and *F*), 111 (between *F* and *G*), 11 (between *G* and *H*), and 66 nucleotides (between *G* and *A*)—or 227 out of 5,375 nucleotides—do not code for protein formation, but, as will be pointed out later, they serve other important functions.

The single-strandedness of ϕX174 was discovered as a result of the kinetics of lethality induced by incorporated radioactive phosphorus (^{32}P). The ability of a linear double-stranded helix to replicate and form infectious and viable virus particles could be inhibited only if two radioactive disintegrations occurred very close to each other with one in one strand and one in the other, thus breaking the genome into two distinct pieces. In T_2 the ratio of disintegrations to lethality was about 10:1; in ϕX174, however, every disintegration was lethal, suggesting that only one polynucleotide strand was present. Subsequent chemical analysis also revealed that the nucleotide ratios were nonequivalent, that is, instead of A = T and G = C, the ratios were 0.77A:1.02T:0.70G:0.58C. In addition, no thermal transition was evident; that is, the absorption characteristics of double- vs. single-stranded DNA are different, and a shift in absorption occurs when double-stranded DNA is raised to a melting, or denaturing, temperature and single-stranded molecules are formed. The only solution to these seeming ambiguities was that ϕX174 was single-stranded.

Immediately on entry into a host cell, ϕX174 becomes double-stranded by action of the host enzymes. In this form it is replicative, producing a dozen or more daughter genomes of identical character, after which these daughter duplex circles produce single-stranded versions identical in base sequence to the infecting molecule (Fig. 2.16). The infecting chromosome is referred to as the *positive* or *plus* strand which becomes double-stranded by the formation of a complementary *negative* or *minus* strand. At the time of lysis, only plus strands are found in the mature virions, indicating that only the minus strand of the double-stranded circle serves as a template to be transcribed. Interestingly enough, ϕX174 can be induced to transcribe *in vitro* both as a circle and as a linear molecule. When a circle, only the minus strand transcribes; when linear, both transcribe.

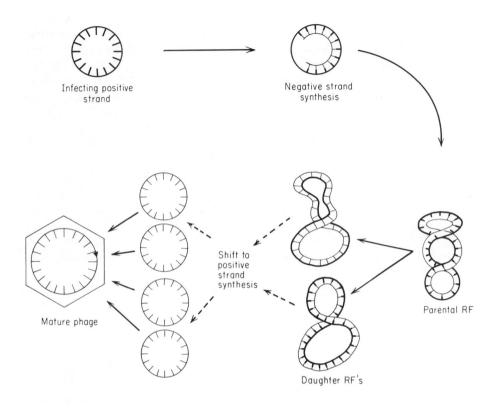

Fig. 2.16. Production of φX174 phage particles within a host cell. RF stands for replicative form. Gene A would be involved in forming both parental and daughter RFs; genes B, C and D in the production of positive single strands; genes F, G and H with the formation of the capsid and spike proteins; and gene E with lysis of the host cell.

RNA Viruses. All RNA viral genomes are linear in form, and they vary considerably in molecular weight, from 1.05 to 1.2×10^6 daltons in the R17 bacteriophage of *E. coli* to 17.0×10^6 daltons in the reoviruses of mammals (Table 2.2). The lack of circularity among the RNA viruses suggests a lack of terminal redundancy. A majority of these viruses has single-stranded instead of double-stranded genomes, and several of the double-stranded forms have their genomes represented by more than one piece of RNA rather than as a single filament of nucleic acid.

The RNA bacteriophages (R17, MS2, f2, Fr) show a considerable overlap with each other in antigenic properties, nucleotide sequences, and nucleotide ratios, indicating a close relationship to each other. They are among the smallest of the group in addition to being single-stranded, and each possesses a genome of about 3,500 nucleotides. While they are capable of infecting a number of bacterial forms other than *E. coli* and thus show a lack of species-specificity, they are similar in that they infect only male (F^+) cells, attaching themselves to, and entering by means of, the sex pili whose formation is determined by the sex factor. In overall structure, the individual virion is polyhedral in shape, with a single protein species forming the head from a number of identical subunits, each of which has a molecular weight of 15,000. The life cycle is not very different from that of other bacteriophages. The release of mature virions occurs from 25 to 30 mins after infection, with up to 20,000 particles released when lysis of the host cell occurs. This may seem to be a very large number compared to the far fewer numbers of T_2 or T_4 virions produced under similar circumstances, but the actual bulk of viral material formed and released is basically the same in each instance since the bacterial cell is limited in the amount of its content that can be contributed to viral reproduction.

R17 is illustrative of a single-stranded RNA bacteriophage. Its genome is 3,482 nucleotides long, distributed between three structural genes, two internal spacer regions, and two termini which are not transcribed (Fig. 1.2). No recombination occurs in RNA phages so that mapping of the genes has to be done biochemically rather than genetically. The structural genes code for an A protein involved in the assembly or maturation of the virion, a coat protein of 129 amino acids, and a portion of the subunits of an enzyme with dual functions; the latter can act either as an RNA-directed RNA replicase to make more viral genomes or as a transcriptase to make mRNA. As a result, the virus genome can be both a messenger and the source of a message.

The nature of the nucleotide sequences of both the coat protein and 5' terminal end sections of the genome (and possibly the remainder of the genome as well) results in a substantial amount of base pairing, leading to the flower arrangements in Fig. 1.3. A predetermined secondary structure results, therefore, which probably contributes significantly to packing within the head protein and to a molecularly stable configuration. The significance of such complementary nucleotide sequences and base pairing is not understood, however, although it is hardly likely to be an accidental, or random, arrangement of nucleotides.

The infecting single-stranded genome of R17 is the *plus* strand. Upon infection, this strand attaches itself to bacterial ribosomes where, acting as an mRNA, it specifies the formation of an enzyme which promptly acts as a replicase which reads the plus strand to form a complementary *minus* strand. The latter, together with the original plus strand, forms a double-stranded replicative unit. Many rounds of replication, with the replicase forming complementary copies of both plus and minus strands, build up the number of viral RNA units within the bacterial cell. The plus strands dominate in numbers since some will serve as mRNAs to specify more replicase, while other plus strands will become encapsulated in head proteins to form mature virions for release. Any minus strands remaining at the time of lysis are degraded or discarded.

Other RNA viruses, many of which are infectious in eukaryotic cells, can be grouped into two major classes, based on whether they reproduce independently within the host cell, and hence, like R17 or MS2, can be replicated into more viral RNA or translated as an mRNA into proteins, or whether they integrate into the host chromosome, in which case a double-stranded DNA intermediate is formed in completing the life cycle (Fig. 2.17). Among the former group, the RNA → RNA class, are such single-stranded forms as those responsible for influenza, measles, poliomyelitis, and mumps in humans, Newcastle disease in fowl, and the tobacco mosaic virus of *Nicotiana tabacum.* The last named is, by comparison, a huge, rod-shaped virus which has several thousand subunits of protein making

Fig. 2.17. Two different life cycles of the RNA viruses. Left, the infecting RNA virus can replicate to make more copies of itself, or it can act as an mRNA and produce protein by translation. The viruses that cause the common cold, influenza and poliomyelitis are of this type. Right, the other type of RNA viruses do not transfer coded information from RNA to RNA, but rather by a process of reverse transcription, produce a DNA intermediate, which, in turn, by more or less normal transcription, produces more RNA copies which, acting as mRNAs, produce protein by transcription. Viruses of this type, which include the Rous chicken sarcoma virus, are capable of inducing tumors, and in their DNA intermediate form can be integrated into the chromosomes of eukaryotes. For further details refer to Fig. 5.14.

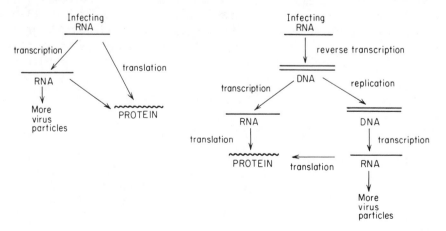

up the outer coat (Fig. 2.7). Also among the RNA → RNA class are the double-stranded reoviruses of mammals and the dwarf virus of rice. Of those RNA viruses which employ a double-stranded DNA intermediate during their life cycles, some are associated with, and therefore implicated as a causal agent in, the induction of cancer in animals. These intermediates are formed by a reverse transcriptase of viral origin which is introduced into the host cell at the time of infection. This DNA behaves similarly to that of the DNA tumor viruses such as the polyoma virus and simian virus 40 (SV40) in that the DNA becomes integrated into the host genome at a unique site, where it is referred to as a *provirus*. In its integrated, or episomal, state, it can be replicated along with the host genome without inducing transformation of the host cell to a malignant state, but such integrated genomes contain a transformation gene which, on activation, gives rise to a product capable of transforming the host cell from a normal to a malignant condition. At the same time, progeny viruses are also formed. The oncogenic (cancer-inducing) viruses carry out their activities in the nucleus of the host cell, and they differ from similar, but nononcogenic, viruses by the fact that they code for not only the proteins of their own structure but also for proteins that are not involved in the structure or maturation of virions. It is these proteins which can lead to cellular transformation.

The Rous sarcoma virus of fowl is an example of an oncogenic RNA virus. It can, unlike many viruses, be expressed in a number of differentiated cell types rather than be dependent upon a particular kind of host cell.

It is of considerable interest that the DNA of normal eukaryotic cells contains nucleotide sequences which, by annealing techniques, can be shown to be complementary to the RNA of certain viruses. Thus, normal chicken cells possess nucleotide sequences complementary to avian leukosis virus and mouse cells possess nucleotide sequences complementary to mouse leukemia virus. Therefore, there seems little doubt that these viral genomes have had their origin in the past from host genomes and that evolution has given rise to a variety of viral types with different structures, modes of existence, and pathological manifestations.

The Genomes of Mitochondria and Plastids

Chapter 3 will deal with the architecture of the eukaryotic chromosome, but within the eukaryotic cell are found sources of extranuclear, or cytoplasmic, DNA which correspond structurally in significant ways with the DNAs of prokaryotic organisms. Because examples of cytoplasmic, or maternal, inheritance have been known for some time, it has been suspected that plastids contain an autonomous means of preserving their identity from one generation to the next and that this feature must relate to an inheritance based on nucleic acid. Confirmation of this idea through various extraction methods, coupled with electron microscopy and the incorporation of radioactive nucleotides, has now been accomplished, and so it is established that plastids as well as mitochondria contain their own DNA, which often differs from nuclear DNA in base composition as well as structure. This discovery has led to the resurrection of an old theory that the eukaryotic cell is of composite origin, with the mitochondria being descendants of an invasive, aerobic bacterial form which sometime in the distant past took up symbiotic residence in a host cell, while the plastid derives from a similarly invasive, and now symbiotic, blue-green bacterial cell. The circular character of the respective genomes of mitochondria and plastids, the mode of replication of the organelles, the size of their ribosomes, and the inhibition of protein synthesis by chloramphenicol, but not by cycloheximide, are strikingly parallel to the structure and behavior of the genomes of bacteria and blue-green algae. Protein synthesis on the ribosomes of eukaryotic cells is inhibited by cycloheximide, but not by chloramphenicol, suggesting again the possible relation of the DNA of mitochondria and plastids to that of the prokaryotes and also suggesting that the eukaryotic cell has a polyphyletic origin.

Mitochondrial DNA (mtDNA). Mitochondrial DNA is found in all aerobically respiring eukaryotes as a naked double-stranded molecule varying quite widely in length (Fig. 2.18). Multicellular animals possess mtDNA of a relatively narrow range of lengths, from 4.4 μm to 5.6 μm in anuran amphibia and some reptiles, somewhat larger in other species. Such mtDNA molecules have a molecular weight of from 9 to 11 $\times$ 10^6 daltons, a weight more characteristic of bacterial plasmids than of the bacterial genome itself. Molecules of varying lengths, but as short as 1.0 μm, have been isolated from *Euglena*, but some uncertainty exists as to the relation of these mtDNA pieces to the total mtDNA.

Protozoa such as the *Amoeba* and *Paramecium* possess mtDNA molecules averaging around 16 μm, while intact yeast mtDNA is about 25 μm. The mtDNA from higher plants such as the red bean average around 19 μm, but molecules as long as 62 μm, with a molecular weight of 10^8 daltons, have been isolated. Whatever the true weight and length of plant mtDNAs, these seem substantially greater

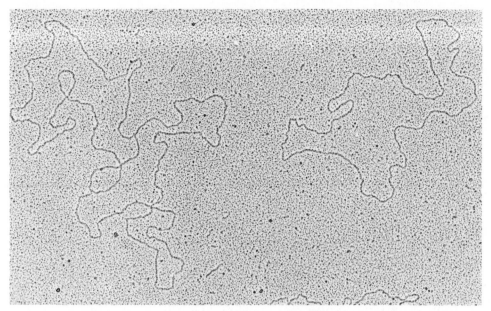

Fig. 2.18. Circular mitochondrial genomes of *Drosophila melanogaster.* The genome at the right has a contour length of 6.4 μm., equal to 19,500 base pairs and a molecular weight of 12.9 × 10⁶ daltons. The genome at the left is caught in the process of replication, with about 90% of the genome replicated. Mag. 40,500 (courtesy of Dr. D. R. Wolstenholme).

than comparable figures from animal mitochondria. Such differences are probably paralleled by comparable differences in the number of structural mitochondrial genes in plants and animals.

The animal mtDNA molecules are circular in form while the larger molecules from the bean seem to be linear. The base composition, as judged by nucleotide ratios, is generally distinguishable from that of nuclear DNA, but since mtDNA in any species may be lighter or heavier per unit length than nuclear DNA, and to demonstrate no parallel phylogenetic trend, the evolution of the two kinds of DNA seems either to have proceeded independently of each other or to have been fixed prior to any symbiotic relationship.

Whatever the form, the mtDNA is located in the matrix of the mitochondrion, but it is attached at a particular point in the genome to the inner membrane, reminiscent of the manner by which the bacterial chromosome is attached to the cell membrane by way of a mesosome (Fig. 2.1). Following replication the mtDNA is distributed to daughter mitochondria through a process of fission. The amount of mtDNA per cell and its relation to nuclear DNA is, of course, a function of the number of mitochondria per cell. This may range from one mitochondrion in a cell of *Micromonas,* a small flagellate, to 2,000 in a rat liver cell, but it would appear that each mitochondrion in a mouse cell has an average of five copies of mtDNA, whereas only one or two copies have been observed in sea urchin cells. Genetic analysis of mtDNA reveals little or no redundancy of nucleotide sequences, suggesting that the nucleotide sequence consists primarily of structural genes whose transcriptional or translational products are concerned primarily with the processes of aerobic respiration, ribosome formation, transcription of mitochondrial-specific tRNAs, and the structure of the mitochondrion. The mitochondrion, however, is not a totally independent organelle since several nuclear genes in a variety of organisms are known to have an effect on mitochondrial structure and function and to code for some of the mitochondrial enzymes.

A number of mitochondrial mutants in yeast, called *petites* because of the small size of plated colonies, are known, and they reveal a variety of metabolic deficiencies, accompanied by varying degrees of reduction in the amount of DNA in the remaining genome. A group of 13 petites showed a range in size of DNA circles from 0.13 μm to 5.5 μm, while the normal, or *grande,* yeasts revealed only linear molecules of greater length in their mitochondria. Other petites are known in which the entire genome has been replaced by DNA consisting almost entirely of AT nucleotide pairs or of tandemly repeated fragments of wild-type mtDNA, suggesting that in some instances a mechanism exists for keeping the amount of mtDNA constant even though it may be very different in nucleotide sequence from the original wild-type form. These observations reinforce the concept of mitochondrial dependence upon the nuclear genome, and that the crucial enzymes of mitochondrial metabolism, including the DNA polymerase for mtDNA replication, are coded in the nuclear genome.

Chloroplast DNA (chlDNA). The DNA that can be isolated from the chloroplasts of a variety of green plants, from algae to angiosperms, differs very little from mtDNA except in size. The usual form is circular, with a contour length of 37 μm to 44 μm, corresponding to a molecular weight of approximately 1×10^7 daltons (Fig. 2.19). Extraction of chlDNA from lettuce, peas, oats, *Euglena,* and *Chlamydomonas* reveal similarly sized circles but the marine alga, *Acetabularia mediterranea,* has yielded molecules up to 200 μm long and with a molecular weight of about 1×10^9 daltons. Whether such a long molecule represents a single chloroplastic genome or several tandemly arranged is not known. Linear molecules have been

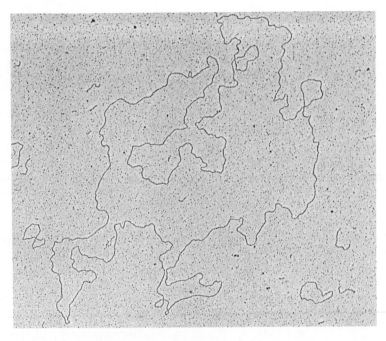

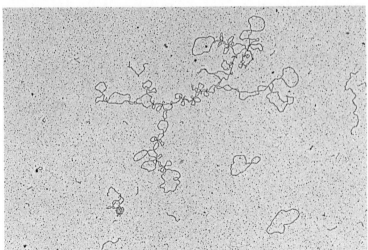

Fig. 2.19. Chloroplast genomes from oats. Top, open form; bottom, super-coiled form. The coiling is probably brought about by either RNA or protein connections or interactions, as indicated in Fig. 2.3 for *E. coli* chromosomes, and it may be similar, or at least analogous, to that which occurs in eukaryotic chromosomes (Fig. 3.56). The small circles in both figures are genomes of SV40 virus used for comparative purposes (courtesy of Dr. Richard Kolodner).

found in all species examined, but it seems likely that these result from fragmented circles.

The large size of these molecules, as compared to those of mtDNA, would suggest either that there are substantially more structural genes coding for proteins in the chloroplast genome, or, what is less likely, that there is a good deal of redundant, and non-transcribing, DNA present. A genome having a molecular weight of 10^8 daltons has a coding potential for approximately 250 proteins, each having an average molecular weight of 30,000 daltons, but the number of proteins known to be formed by normal chloroplasts remains to be ascertained. In barley, at least, more than 70 mutants of nuclear genes are known which affect chloroplast structure or function, indicating that chloroplasts, like mitochondria, are nuclear-controlled or nuclear-dependent organelles. The cooperation of chloroplast and nuclear genomes in the process of photosynthesis is dramatically demonstrated by the fact that of the two subunits of protein forming the photosynthetic enzyme ribulose -1,5-diphosphate carboxylase, the smaller one is nuclear and the larger one is plastid in origin. The enzyme ATPase has a similar dual origin.

Plastids, like mitochondria, have their own ribosomal-producing apparatus. In maize, the rRNA genes are duplicated and are part of a 22,000-nucleotide pair inverted repeat, with the 5S rDNA adjacent to the 23S rDNA, as it is in *E. coli*, and with both separated from the 16S rDNA by a spacer region (Fig. 2.20). In eukaryotes, by way of comparison, the 28S and 18S rDNA sequences are adjacent to each other, while the 5S rDNA sequences are located elsewhere in the genome. Transcription direction in plastid genomes appears to be initiated from the 16S end; it gives rise to a large precursor RNA which is later processed for inclusion into the plastid ribosomes.

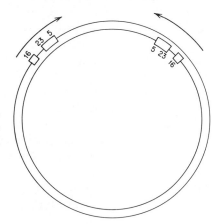

Fig. 2.20. The two sets of rRNA genes, located within a 22,000 nucleotide pair section of the plastid chromosome of maize. The reading frames for the two sets are on different polynucleotide strands, and hence are read in opposite directions, as indicated by the arrows. (Redrawn from Bedbrook, et al., 1977).

Mesokaryotes

It has been generally assumed that the prokaryotes were the progenitors of the more highly differentiated eukaryotes, an assumption originally based on the philosophy that complexity arises out of simplicity, but subsequently reinforced by the universality of the genetic code and its operation. Part of that evolution may have stemmed from the invasion of a host cell by other prokaryotes which, as a result of intracellular evolution, became symbiotically integrated into the life of the eukaryotic cell as the mitochondria and plastids of higher forms. This cannot be the whole answer, for the principal difference between the prokaryotes and eukaryotes relates not only to membrane enclosed cytoplasmic organelles but also more particularly to the organization, functioning, and perpetuation of the hereditary materials.

Whatever the relation of the two groups to each other (see Chapter 9), the evolution of the hereditary materials and the separation of the prokaryotes from the eukaryotes involves a number of distinct changes: (1) encapsulation of the genomes within a double-layered nuclear envelope; (2) a shift from a single, generally circular chromosome to a group of linearly organized, nonhomologous chromosomes constituting the so-called haploid number; (3) a shift from a naked DNA double helix to a set of chromosomes variously associated with several kinds of proteins and RNA which not only are involved in genetic regulation but are probably also instrumental in the maintenance of chromosome structure and the patterns of chromatin condensation and decondensation during the cell cycle; and (4) a shift from a cell division pattern where the DNA is associated with, and separated by, a mesosome to one where a spindle apparatus made up of microtubules governs chromosome separation and participates in segmentation of the cytoplasm. Coupled with the growing complexity of the eukaryotic cell is a comparably increased complexity of the genome, with various chromosome numbers, redundant as well as unique nucleotide sequences, differentiated sex-determining mechanisms, and the whole pattern of haploid–diploid alternation of generations that accompanies biparental inheritance.

No organism, or group of organisms, reveals a sequential pattern of changes that permit us to visualize the entire evolutionary picture (see Chapter 9). Some prokaryotes such as the thermophilic bacterium, *Thermoplasma acidophilum,* seem to have either anticipated or paralleled a eukaryotic state in having a histone-like protein associated with its DNA to serve, perhaps, as a protective device against the thermal denaturation of a portion of its DNA, but the amount of protein per cell is sufficient only to complex with no more than one-quarter of the genome at any one time. In addition, *T. acidophilum,* as well as the methanogens and halophilic bacteria, all reclassified by Woese and Fox as *archaebacteria* to distinguish them taxonomically from the *eubacteria* such as *E. coli,* have nucleotide sequences producing rRNAs which structurally are intermediate to those of the typical prokaryotes and eukaryotes. Further, the methanogen, *Methanobacterium thermoautotrophicum,* while having a genome similar to, but only half the size of, that of *E. coli*—a molecular weight of 1.1 × 10^9 daltons as compared to 2.7 × 10^9 for *E. coli*—also has about 6% of its DNA in a redundant form, a eukaryotic character.

Another group, however, that has been said to share the characteristics of both the prokaryotes and eukaryotes, and that might possibly warrant identification as a genuine intermediate form, or *mesokaryote,* is the family Dinophyceae or dinoflagellates (Pyrrophyta). The group consists of flagellated marine, sometimes freshwater, protists, some of which are free-living photosynthetic species, some are bioluminescent, e.g., *Noctiluca* and *Gonyaulax* (the latter responsible for the toxic "red tide"), while still others exist as symbionts in the cells of a number of marine invertebrates. Their seemingly intermediate state of differentiation is based on the fact that cytoplasmically they are eukaryotic in possessing mitochondria and chloroplasts enclosed by double membranes as well as ribosomes, endoplasmic reticulum and Golgi membranes, while their hereditary make-up is a combination of eukaryotic and prokaryotic features.

The dinophycean genome is contained within a defined nucleus whose membrane lacks annuli or pores, and which does not undergo dissolution at any time during cell division. The chromosomes, which may number as low as 24 and as many as 100 or more, have the appearance of ellipsoids within each of which a coiled arrangement of DNA-containing fibrils produces a banded periodicity (Fig. 2.21). In this form they are unlike any known eukaryotic chromosome, although they bear a superficial resemblance to the polytene chromosomes of the protozoan macronuclei just prior to the degradation of their DNA (p. 175). In addition, each chromosome seems to be attached to the nuclear membrane at one end in a manner reminiscent of the situation found both in bacteria and in the holomastigote protozoa (Fig. 2.22). At metaphase and anaphase a series of extranuclear microtubular structures form within invaginations of the nuclear envelope; these eventually penetrate across the nuclear membrane without disrupting it to function as a primitive form of a mitotic spindle. As anaphase proceeds, these membrane-covered

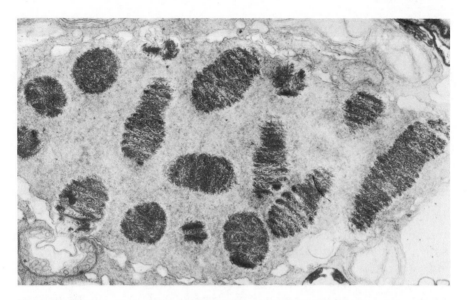

Fig. 2.21. Electron micrograph of the chromosomes of the dinoflagellate, *Prorocentrum mariaelebouriae* (strain 403), showing their compacted, banded and fibrillar appearance. The different fibrillar patterns are due to different orientations of the chromosomes in the nucleus (Loeblich, 1976, with the permission of the Society of Protozologists).

Fig. 2.22. Telophase stage in meiotic division in the holomastigote protozoan, *Barbulanympha*, showing the central spindle, the intact nucleus with its membrane, and the chromosomes attached to the nuclear membrane and in contact with the half-spindle fibers. The membrane will elongate as the spindle poles move apart, and eventually be pinched in two. One chromosome at the lower left is drawn to indicate the degree of coiling evident at this stage; the coiling is never so undone as to present the diffuseness characteristic of a typical interphase cell (Cleveland, 1954).

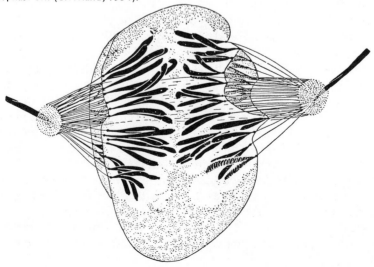

tubes elongate to convert the nuclear contents into a dumbbell shape; the membrane eventually constricts to cut the nucleus in two. The formation of V-shaped chromosomes suggests that the chromatids segregate as the nuclear membrane elongates. Although the spiral chromosomes of prophase in *Prorocentrum micans,* for example, are typically eukaryotic in appearance, the fact that the divisional apparatus does not respond to colchicine indicates that its organization may be distinctly different from the usual eukaryotic spindle. Presumably, the point of attachment of each chromosome to the nuclear membrane would be the position at the centromere, as it is in Fig. 2.22, but this may be true only in a functional rather than a structural sense. For example, chromosome fragments induced by X rays are not lost during cell division as they would be in a higher plant or animal cell; these fragments seem as capable of completing division as a normal chromosome and of having some way of engaging the divisional apparatus for purposes of segregation.

The chromosomes remain condensed at all times in some species (Fig. 2.21), while in others they may undergo a modified spiralization-despiralization cycle. Chemical analysis of the chromatin of a number of species of dinoflagellates reveals the presence of very little RNA in the chromosomes, and the chromosome-associated proteins constitute a much smaller proportion of the chromatin mass by weight than is typical of the eukaryotes. Thus, treatment of sectioned material with DNase effectively removes the DNA, but leaves virtually no protein residue behind. However, while low in amount, the proteins are basic and hence are like the eukaryotic histones in general. One of them has an electrophoretic mobility quite similar to histone 4 of maize. In addition, dinoflagellate DNA is quite unique in having a high, but variable, proportion (12% to 68%) of hydroxymethyluracil replacing thymine, a nucleotide which is known only in the DNA of certain bacteriophages which infect *Bacillus subtilis.* Also, some species contain substantial amounts of 5-methylcytosine—16% in *Exaviaella cassubica*—while another species, *Peridinium triquetrum,* contains 2% to 3% methyladenine. The last two nucleotides are not unknown in the eukaryotes, but they are nowhere found in the amounts indicated. A typical eukaryotic S period of DNA synthesis is evident in all species so far examined.

DNA fibrils can be detected by electron microscopy in the chromosomes of dinoflagellates, but their actual structure remains questionable. Oakley and Dodge have shown, by an electron micrographic analysis of serial sections, that the chromosomes of *Glenodinium hallii* at metaphase are coiled plectonemically, i.e., the two chromatids are wrapped around each other as in an electric cord while being thrown into the kinds of coils depicted in Figs. 2.23 and 2.24. The chromatids, however, seem to possess no detectable ends which strongly suggests that the chromosomes are circular, a more typical prokaryotic feature. How such a chromosome, condensed during most of the cell cycle, can replicate and separate into its respective chromatids remains an enigma, although the existence of Y and V configurations in prophase indicate no entanglement of circular structures.

There is no strong evidence that the chromosomes are polytenic. Mutation induction studies indicate that the dinoflagellates are haploid; sexual reproduction and meiosis have been observed, with the zygote being the only diploid stage in the life cycle. A further eukaryotic feature is the presence of repetitive DNA, amounting in some species to as much as 55% of the genome.

The term *mesokaryote* is, therefore, of dubious value if its purpose is to imply that these organisms are transitional between the

Fig. 2.23a-f. Electron micrographs of serial longitudinal sections through a chromosome of the dinoflagellate, *Glenodinium hallii*, which indicates that the compacted nature of these chromosomes results from the development of a regular coil. Sections *a* and *b* show the slant of the coil in one direction, while sections *e* and *f* show a slant of the coil in the opposite direction. Sections *c* and *d* appear to be through the middle of the coil since the slanting is not evident. The bar measures 0.5 μm (courtesy of Oakley and Dodge, 1979, copyright by Springer-Verlag, 1979).

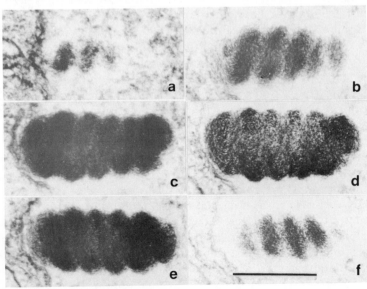

Fig. 2.24. Diagrams interpreting the coiled structures seen in Fig. 2.23. A: the chromosome in a fully contracted state; B: partially opened coils; C: more fully uncoiled. No ends have been detected in the electron micrographs, and this is indicated by the circular chromosomes depicted here. The occurrence of "Y" and "V" shaped chromosomes in mitotic prophases suggest that separation of chromatids is not a problem even if the chromosomes are circular (courtesy of Oakley and Dodge, 1979, copyright by Springer-Verlag, 1979).

prokaryotes and eukaryotes. Loeblich, in fact, puts the dinoflagellates in the eukaryotic camp, and suggests that they may have evolved prior to the full establishment of the main eukaryotic features of cell and chromosome structure. Possible support for this concept comes from the observation that in *Glenodinium foliaceum,* a binucleate form, the two nuclei are distinctly different from each other, one being typically mesokaryotic with condensed chromosomes, and the other eukaryotic, with a customary diffuse interphase stage, evident heterochromatin, and a nuclear envelope with pores. Whether one is derived from the other as the macronucleus is derived from the micronucleus in some protozoa is not indicated, but the balance of evidence would seem to place the dinoflagellates on the eukaryotic side of the taxonomic ledger, although it is equally obvious that they have followed an evolutionary track of their own.

BIBLIOGRAPHY

Achtman, M., "Genetics of the F. Sex Factor in Enterobacteriacae," *Curr. Topics Microbiol. Immunol., 60* (1973), 79–123.

Allen, J. R., "The Dinokaryon and the Mesokaryote Hypothesis," in *Chromosomes: Simple and Complex,* P. A. Roberts, ed. Corvallis: Oregon State University Press, 1974.

Bedbrook, J. R., et al., "*Zea mays* Chloroplast Ribosomal RNA Genes Are Part of a 22,000 Base Pair Inverted Repeat," *Cell, 11* (1977), 739–49.

Beers, F. R., Jr., and E. G. Bassett (eds.). *Recombinant Molecules: Impact on Science and Society.* New York: Raven Press, 1977.

Birkey, C. W., et al. (eds.), *Genetics and Biogenesis of Mitochondria and Chloroplasts.* Columbus: Ohio State University Press, 1976.

Cairns, J., et al. (eds.), *Phage and the Origins of Molecular Biology.* Coldspring Harbor, N.Y.: Cold Spring Harbor Laboratory, 1966.

Campbell, A. M., *Episomes.* New York: Harper & Row, Pub., 1966.

Cleveland, L. R., "Hormone-Induced Sexual Cycles of Flagellates. XII. Meiosis in Barbulanympha Following Fertilization, Autogamy, and Endomitosis," *J. Morph., 95* (1954), 557-619.

Delius, A., and A. Worcel, "Electron Microscope Studies on the Folded Chromosome of *Escherichia coli*," in *Chromosome Structure and Function.* Cold Spring Harbor, N.Y.: Cold Spring Harbor Symp. on Quant. Biol. *38*, 1973, 53-58.

Dodge, J. D., "Chromosome Structure in the Dinophyceae. I. The Spiral Chromosome," *Archiv. f. Mikrobiol., 45* (1963), 46-57.

——, "A Dinoflagellate with Both a Mesokaryotic and a Eukaryotic Nucleus. I. Fine Structure of the Nucleus," *Protoplasma, 4* (1971), 231-42.

Fiddes, J. C., "The Nucleotide Sequence of a Viral DNA," *Sci. Amer., 237* (1977), 54-67.

Gillespie, D., and R. C. Gallo, "RNA Processing and RNA Tumor Virus Origin and Evolution," *Science, 188* (1975), 802-11.

Gressel, J., et al., "Dinoflagellate Ribosomal RNA: An Evolutionary Relic?" *J. Mol. Biol., 5* (1975), 307-13.

Helinski, D. R., "Plasmids as Vectors for Gene Cloning," in *Genetic Engineering for Nitrogen Fixation,* A. Hollaender et al., eds. New York: Plenum, 1977.

Hershey, A. D., (ed.), *The Bacteriophage Lambda.* Cold Spring Harbor, N.Y.: Cold Spring Harbor Laboratory, 1971.

Landy, A., and W. Ross, "Viral Integration and Excision: Structure of the Lambda *att* Site," *Science, 197* (1977), 1147-60.

Loeblich, A. R., III, "Dinoflagellate Evolution: Speculation and Evidence," *J. Protozoology, 23* (1976), 13-28.

Meynell, G. G., *Bacterial Plasmids.* New York: Macmillan, 1972.

Mitchell, R. M., et al. "DNA Organization of *Methanobacterium thermoautotrophicium,*" *Science* 204:1082-84, 1979.

Oakley, B. R., and J. D. Dodge. "Evidence for a Double-Helically Coiled Toroidal Chromonema in the Dinoflagellate Chromosome," *Chromosoma* 70:277-291, 1979.

Pettijohn, D. E., and R. Hecht, "RNA Molecules Bound to the Folded Bacterial Genome Stabilizes DNA Folds and Segregate Domains of Supercoiling," *Cold Spring Harbor Symp. Quant. Biol., 38* (1973), 31-42.

Pigott, G. H., and N. G. Carr, "Homology Between Nucleic Acids of Blue-Green Algae and Chloroplasts of *Euglena gracilis*," *Science, 175* (1972), 1259-61.

Rae, R. M. M., "Hydroxymethyluracil in Eukaryotic DNA: A Natural Feature of the Pyrrophyta (Dinoflagellates)," *Science, 194* (1976), 1062-64.

Rizzo, P. J., and E. R. Cox, "Histone Occurrence in Chromatin from *Peridinium balticum*, A Binucleate Dinoflagellate," *Science, 198* (1977), 1258-60.

Schwartz, R. M., and M. O. Dayhoff, "Origins of Prokaryotes, Eukaryotes, Mitochondria and Chloroplasts," *Science, 199* (1978), 395-403.

Spector, D. H., and D. Baltimore, "The Molecular Biology of Poliovirus," *Sci. Amer., 232* (1975), 24-31.

Stein, D. B., and D. G. Searcy, "Physiologically Important Stabilization of DNA by a Prokaryotic Histone-like Protein," *Science, 202* (1978), 219-22.

Studier, F. W., "Bacteriophage T7," *Science, 176* (May 28, 1972), 367-76.

Temin, H. M., "RNA-Directed DNA Synthesis," *Sci. Amer., 226* (1972), 24-33.

——, "The DNA Provirus Hypothesis," *Science, 192* (1976), 1075-80.

——, "The Relationship of Tumor Virology to an Understanding of Nonviral Cancers," *BioScience, 27* (1977), 170-76.

Wolstenholme, D. R., et al., "Form and Structure of Mitochondrial DNA," *Oncology, 1* (1970), 627-48.

Zinder, N. D. (ed.), *RNA phages*. Cold Spring Harbor, N.Y.: Cold Spring Harbor Laboratory, 1975.

3

Architecture of the Chromosome: II. Eukaryotic Chromosomes

As we turn our attention to the chromosomes of eukaryotic species, we find that a number of changes have occurred which clearly distinguish the eukaryotic chromosomes from their prokaryotic counterparts. The most obvious difference is that of size, for virtually all eukaryotic chromosomes, after being appropriately fixed and stained, are sufficiently large to be discernible in the light microscope. The DNA of the haploid genome has been substantially increased so that even the fungi, which possess the lowest amount of DNA among the eukaryotes, have an amount that is at least an order of magnitude greater than that in *E. coli,* which has one of the largest of the prokaryotic genomes (Table 3.1). Other species have vastly greater amounts, a phenomenon to be discussed later in evolutionary terms (p. 494); it is clear, however, that although there is a relation of amount of DNA to taxonomic status, not all increases (or even decreases) in DNA parallel phylogenetic position, nor can these

Table 3.1. DNA per haploid complement in picograms (1 picogram equivalent to about 9.5×10^8 nucleotide pairs)

Mammals		*Amphibians*	
Human	3.65	*Amphiuma*	86.0
Beef	2.82	*Necturus*	24.2
Horse	3.18	Frog	7.5
Dog	2.93	Toad	4.5
Mouse	3.26	*Xenopus*	3.15
Marsupial		*Fish*	
Bandicoot	4.62	Carp	1.64
Kangaroo	3.13	Shad	0.91
		Lungfish	50.0
Birds		*Miscellaneous*	
Chicken	1.44	Maize	8.4
Duck	1.30	*Drosophila*	0.085
Goose	1.46	Jellyfish	0.33
Pigeon	1.72	Squid	4.50
Reptiles		Sea urchin	0.67
Snapping turtle	2.56	*Neurospora*	0.20
Alligator	2.94	*E. coli*	0.004
Water snake	2.51		
Black snake	1.48		

changes be accounted for solely in terms of morphological, physiological, or biochemical complexity or simplicity.

In addition, the eukaryotic genome is distributed among more than a single chromosome (the threadworm, *Parascaris equorum* var. *univalens,* is a possible exception if one considers its peculiar germ-line chromosome to be a single entity). The composite *Haplopappus gracilis* has a haploid number of two chromosomes, while the ancient adder's tongue fern, *Ophioglossum reticulatum,* has over 500. Equally variable numbers are to be found in the animal kingdom as well (Table 3.2), but while, as pointed out, DNA amount shows only a very general relation to phylogenetic position (Fig. 1.1), the number of chromosomes per haploid genome shows even less, unless comparisons are made within a genus or among closely related forms.

Increases in amounts of DNA and numbers of chromosomes in the eukaryotes are also accompanied by an increase in molecular complexity of the chromosomes and by an internal chromosomal differentiation that is evident structurally, behaviorally, and genetically. In addition, these changes are associated with a different kind of cell division process, one based on the formation of a spindle made up of microtubules interacting with the chromosomes at their centromeric regions, and with temporal patterns of chromosomal

Table 3.2. Range of haploid chromosome numbers found among the eukaryotes

Plants		Animals
Haplopappus gracilis	2	*Parascaris equorum* var. *bivalens* (horse threadworm)
Crepis capillaris	3	*Aedes aegypti* (yellow fever mosquito)
Tradescantia paludosa	6	*Drosophila willistoni*
Vicia faba (broad bean)		*Musca domestica* (housefly)
Brassica oleracea (cabbage)	9	*Chorthippus* sp. (grasshopper)
Citrullus vulgaris (watermelon)	11	*Cricetulus griseus* (Chinese hamster)
Camelia (Thea) *senensis* (Chinese tea)	15	*Mustela vison* (mink)
Arachis hypogaea (peanut)	20	*Mus musculus* (mouse)
Stipa spartia (porcupine grass)	23	*Homo sapiens* (human)
Magnolia cordata	38	*Ursus americana* (black bear)

contraction made necessary for maneuvering long strands of DNA within the limited confines of a cell, as well as for the selective storage and retrieval of coded information in cells of like genomic nature but of differential phenotypic expression. The eukaryotic chromosome is, therefore, a decidedly more complex organelle than is the relatively naked chromosome of a bacterium or a virus, but despite the complexity and the presence of RNA and proteins which perform necessary roles, DNA is the fundamental physical basis of the hereditary mechanism.

Chemical Constituents of Chromatin

The term *chromatin* is applied generally to the DNA-containing substance extractable from eukaryotic nuclei, and it should not be thought of as a macromolecular mass having a composition based on fixed ratios of constituent molecules. The sources of chromatin, even within a single species, as well as the methods of extraction and analysis, will have a determining effect on the kinds and relative amounts of the chemical species; even DNA, the basic molecule of inheritance, can vary in some organisms from one kind of nondividing cell to another within the same organism as the result of varying degrees of endomitosis (see p. 194); it can also vary between individuals of the same species. In dividing or physiologically active cells, however, the DNA as well as other chromatin-associated molecules usually vary in predictable fashion, particularly with the cell cycle.

Chromatin can be isolated from nuclei in various stages of cellular activity and either in an extended state or visibly compacted as chromosomes. In an extended state, the chromatin exists in the form of long strands having a diameter ranging from 10 nm to 100 nm, depending upon the state of the cells and the methods of extraction, and since the DNA in large chromosomes may be many centimeters long, fragmentation during the extraction process is customary and almost unavoidable. However, through gentle extraction procedures, it has been possible to recover, as intact structures in an extended state, the entire chromatin from single specified chromosomes in *D. melanogaster.* The molecular weight of the DNA of these long chromatin strands may be as high as 58×10^9 daltons (Table 3.3), and the correspondence of these weights with those calculated by other means shows quite conclusively that the DNA exists as an uninterrupted molecule from one end of the chromosome to the other. There are, in other words, no non-DNA linker molecules which serve to maintain the linear integrity of the chromosome.

An analysis of eukaryotic chromosomes indicates that, in addition to DNA, all contain three other kinds of macromolecules: RNA of varying molecular weights; low-molecular weight, acid-soluble, basic proteins, or *histones,* which in some sperm are replaced by a similar protein, *protamine;* and a group of *nonhistone,* or *acidic, proteins.* Calcium and magnesium have been thought by some to be necessary for maintaining the linear integrity of the chromosome,

Table 3.3. Comparison of the idealized chromosomes of *Drosophila melanogaster* with their measured and calculated molecular weights.

Strains	Idealized chromosomes at metaphase	Mol. wt. of[*] largest piece of DNA	Calculated mol. wt. of largest chromosome
wild-type	⟩⟩ ii ⟨⟨	$41 \pm 3 \times 10^9$	43×10^9
inversion	⟩⟩ ii ⟨	$42 \pm 4 \times 10^9$	43×10^9
translocation	⟩⟩ ii ⟨⟨	$58 \pm 6 \times 10^9$	59×10^9

[*]based on viscoelastic measurements (from Kavenoff, R., et al., 1973).

but from what is known of the nature of DNA, this seems most unlikely. Lipids have been detected in some chromosomes through the use of selective dyes such as Orange G, but it also seems unlikely that they perform any major structural role in chromatin. A number of enzymes are also known to be associated with chromatin, but since their presence is not constant, their role is physiological rather than structural.

The DNA-histone complex is the basic unit of structure in the chromosome. The two components are present in roughly equal amounts by weight, and they generally account for from 60% to 90% of the chromatin mass. Both are formed at the same stage of cell division—the S, or synthetic, period of interphase—and newly synthesized DNA is immediately complexed with either newly synthesized histone or that which has been recycled from the previous cell cycle. It had been thought that the histones have a turnover rate such that all, on average, are replaced within four cell divisions, but other evidence would suggest that the histones are distributed between old and new DNA rather than being replaced. DNA, on the other hand, as perhaps befits its role in inheritance, is a highly stable molecule, conserved from one cell generation to the next.

DNA is a polyanion by virtue of the continuous sequence of acidic phosphate (PO_4^-) groups, and these are thought to be neutralized and stabilized by the histones, which because of their basic nature act as polycations. The basicity of the histones derives from their high proportion of basic amino acids, namely lysine and arginine, which together with histidine, tend to be grouped at the amino-terminal end of the molecule, and which, because of the added $-NH_3^+$ groups (Fig. 3.1), can interact through ionic bonding with the

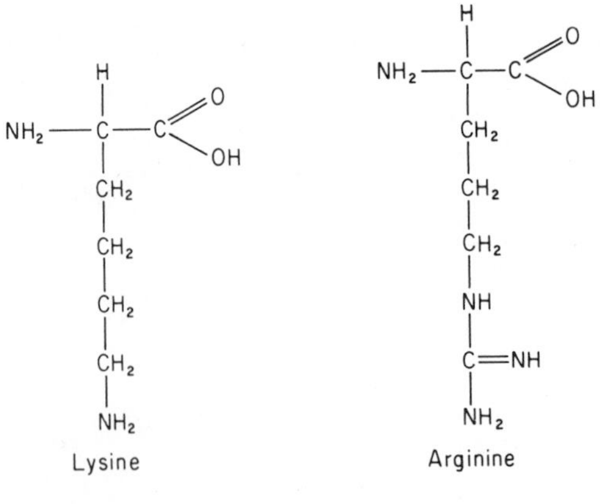

Fig. 3.1. Structure of lysine and arginine. Their basicity results from the number of NH_2 and NH-radicals in each molecule.

Table 3.4. The physical characteristics of the five major histones.

Class	Histone fraction	Structure	Total amino acid residues	mol. wt.
Lysine-rich	H1 (H5)	N⊢———◎————————⊣C	215	21,000
Slightly lysine-rich	H2A	N⊢—◎—⊣C	129	14,800
" " "	H2B	N⊢—◎—⊣C	125	13,800
Arginine-rich	H3	N⊢———◎—⊣C	135	15,300
" "	H4	N⊢———◎	102	11,300

Source: Various sources, but see particularly *Cold Spring Harbor Symp. Quant. Biology*, Vol. 42 (parts 1 and 2), 1977.

phosphate groups of DNA. The mid-region and carboxyl-terminal portions of each histone are less basic in nature, and under certain conditions of ionic strength they may adopt a globular form (Table 3.4). When DNA and the several histones, in appropriate concentrations, are added together in solution, they immediately and spontaneously interact with each other to form fibers of a very special character having a diameter of about 10 nm to 20 nm (100Å to 200Å).

Although the histones are considered to be a single class of proteins, 5 major types have been identified, with each type having several minor variants, i.e., minor in structure but not necessarily so in function. For example, in sea urchin embryos of blastula, gastrula, and pluteus stages, 11 histones have been identified, some of which are stage-specific, indicating a functional significance for each variant form. These fall into the five major types, and only type H4 (Table 3.4) is represented by but a single variant. Four of the five major types are extraordinarily similar throughout the eukaryotes; although they can exhibit cell-specificity, the variation even between species is strikingly low. Histone H4, for example, provides an instance of remarkable evolutionary conservatism; that extracted from such disparate organisms as cows and peas differs by only a couple of amino acid substitutions. Histone H3 is similarly conservative in structure, and it has been estimated that the rate of change is one amino acid substitution, per site, per 100 million years. Histone H1 is the least conservative of the five, and it exhibits more developmental, tissue, and species variation than all of the others combined, but the stability of the histones in general argues for a central role played by these molecules in the determination and regulation of eukaryotic chromosomal structure and function. The selection pressure to maintain their relatively unvarying structure must have been, and continues to be, severe indeed.

The physical characteristics of the histones are given in Table 3.4. H1 is the heaviest, being almost twice as large as any of the other four, and it is the most basic and variable as well; it is present in native chromatin only to about one-half the amount determined for the others, it may be replaced by another related histone, H5, and it is absent in some organisms, in particular micronuclei and in some forms of heterochromatin. The histones are nonspecific as to DNA nucleotide sequence or to species when mixed *in vitro*—it is possible to mix calf thymus histone with the DNA from SV40 virus and produce a nucleoprotein structure approximating very closely that of normal eukaryotic chromatin. In fact, SV40 viral particles in the nuclei of infected monkey cells take on histone from the host cell and have the appearance of miniature, circular eukaryotic chromosomes. This argues for an arrangement of DNA and histone which, while nonspecific in one sense, is not random, but rather one in which the histones are attached to DNA in such a manner as to give a regular repeating pattern of structure (Fig. 3.2). As the figure indicates, the prepared native chromatin, or nucleohistone, appears as a string of beads, or *nucleosomes,* closely appressed to each other, but which, possibly as an artifact of preparation, may appear to be connected by rather long interbead stretches of chromatin. The nucleosomes contain about 200 nucleotide pairs of DNA, linked into a globular structure by the molecules of histones, and having a diameter of 10 nm to 12.5 nm. The range in base pairs per nucleosome, as revealed so far, is from 154 in *Aspergillus* to 241 in sea urchin sperm. The number may even vary within the cells of an individual; in the rabbit, for example, cells of the cortical neurons have 162 base pairs per nucleosome while those of the cerebellar neurons have 200. Also, sea urchin gastrula cells have 218 as opposed to 241 in spermatozoa. Micrococcal nuclease, acting as an endonuclease, will segment extracted chromatin into particles of nucleosomal size, suggesting that the nuclease first digests the DNA between the nucleosomes but not within them. Trypsin can digest the tails of the histones in these isolated particles but not the histone complexed closely with the DNA. Further digestion by the nuclease reduces the nucleosomes to a basic "core" particle of about 140 nucleotide pairs regardless of the initial number present; the remaining nucleotide pairs (14 in *Aspergillus* to 101 in sea urchin sperm)

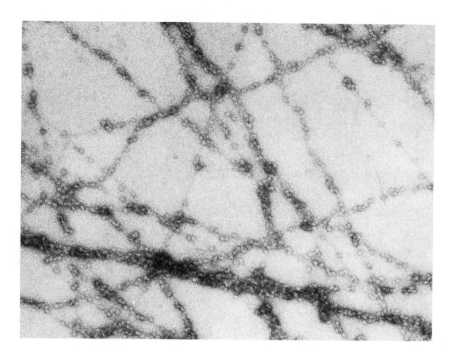

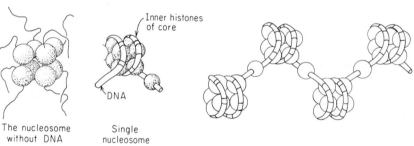

Inner histones
of core

DNA

The nucleosome
without DNA

Single
nucleosome

Fig. 3.2. Top, electron micrograph of isolated chromatin from chicken erythrocytes, show-ing the nucleosomes strung out in linear fashion. The nucleosomes consist of a core of eight histone molecules, made up of two molecules each of H2A, H2B, H3 and H4, around which the DNA is wrapped, and a stretch of linker DNA to which the histone H1 is attached. Bot-tom left, the core of the nucleosome formed by the eight histones joined at their globular portions with the ends of the proteins free; middle, a single nucleosome with its inner core of histones, the DNA wrapped around the core and bound by the free ends of the histones, and a stretch of linker DNA to which is bound a single molecule of H1; right, a chain of four nucleosomes (see Kornberg 1974, 1977 for details) (electron micrograph courtesy of Dr. C. L. F. Woodcock).

which, together with the core, make up the larger unit, are believed
to constitute the "linker" DNA that ties together adjacent nucleosomes. The length of the linker region is thought to be determined
by the size of the H1 histone.

Four of the five major histones are intimately involved in maintaining the structure of the nucleosomes. H3 and H4 exist as a
tetramer $(H3_2H4_2)$ and by themselves are capable of compacting
about 130 nucleotide pairs into an organized subnucleosomal particle. H2A and H2B are also present in each nucleosome in equimolar
amounts, so a nucleosome can be defined as a unit of about 200
nucleotide pairs of DNA organized into a globular structure by two
molecules each of the four major histones (Fig. 3.2). Experimental
and numerical verification of this as a basic unit is obtained by allowing the circular chromosome of SV40 to combine with the five
eukaryotic histones. This chromosome has 5,350 base pairs when in
double helical form and can form no more than 26 nucleosomes to
give rise to a "minichromosome." It would appear, then, that the
two arginine-rich H3 and H4 histones, in tetramer form, are critical
in establishing the central structure of the nucleosome, with two
molecules each of H2A and H2B acting to stabilize the unit and with
single molecules of H1 to bind the linker DNA to the cores to form
a linear series of beads. The histones are located internally in the
beads, with the DNA wrapped around the outside of the clustered
proteins. The remaining histone, H1, is thought to have a role in
developing a higher order of compaction or coiling (Fig. 3.2). It is
removed from chromatin by the action of nucleases when the 200-
base pair nucleosomes are reduced to their 140-base pair core. The
H1 histone, therefore, is not located internally within the beads; it is
external in the interbead stretches of DNA, and it may be added after
the beaded structure is formed. The nucleosome, therefore, is the
first stage of contraction above the level of the naked DNA double
helix, and its formation introduces a sixfold reduction in the length
of the DNA strand. An additional reduction in length by fivefold or
sixfold can be achieved by a grouping of the nucleosomes, aided very
likely by the binding action of histone H1. As stated earlier, H1 is
also the most variable of the histones in amino acid sequence; it can
possibly withstand mutational change more readily than the others
without suffering a loss in binding capacity.

It is possible to reconstitute chromatin *in vitro* so as to regain
the properties of native chromatin such as nuclease resistance as well
as the ultrastructure visible in the electron microscope. These studies,
most of them made within the last decade, leave little doubt that the

beaded structure of chromatin is a fundamental character of all eukaryotic genomes and that the histone-histone and histone-DNA interactions within and outside the nucleosomes are vital to the stability, maintenance, functioning, and maneuverability of the eukaryotic chromosome. There is no suggestive evidence that would point to the interaction of histone and DNA being nucleotide-specific, but the histone-histone binding seems to involve the globular portions of the internal histones, while their more linear basic ends engage the DNA. It would also appear that the newly formed histones are acetylated and phosphorylated as the DNA-histone complex is being formed during the S period; to what degree these processes alter the subsequent behavior and structure of the chromatin is not entirely clear.

Yeast chromatin is an exception to the above statement about DNA-histone relations. The yeast genome ($n = 7$) is quite small in size, and the chromosomes do not show a normal contraction as they go through the cell division cycle. There is very little repetitive DNA and histones H1 and H3 are absent, giving a ratio of histone to DNA that is less than one. (These same two histones are also absent from the micronuclei of *Tetrahymena*.) A repeated beaded structure is evident in isolated chromatin, however, but the nucleosomes are smaller than the usual eukaryotic ones and contain only 154 nucleotide pairs. How the histones are arranged within the nucleosomes is uncertain, and the absence of H1 and H3 may be an artifact.

It is known that both transcriptionally active and inactive DNA are organized into nucleosomes, even though it is also known that the nucleosomes must be modified before becoming transcriptionally competent. Any disturbance of the DNA-histone ratios, however, leads to a disturbed chromosomal behavior. For example, when the DNA-histone ratio is upset and histone synthesis appears to have been depressed relative to DNA, a situation found in the harlequin lobes of the testis of the house scorpion, *Scutigera forceps,* meiosis is abnormal in that synapsis is interfered with and asynapsis results. It is not known what appearance the nucleosomes exhibit in these cells, but with a wealth of asynaptic forms known among both plant and animal species, this problem seems worthy of further investigation.

The acidic proteins of chromatin are far more variable in character and in amount. They also tend to be tissue-specific and to show a variability in amount associated with the transcriptional capability of the cells from which they are derived. For example, when rat liver cells are stimulated by cortisol to engage in transcription, a 41,000-

molecular weight acidic protein of unique character makes its appearance and accompanies the chromatin when it is isolated. An acidic protein of molecular weight 42,000 is similarly associated with the chromatin of ecdysone-induced puffs of *Drosophila* and *Chironomus* polytene chromosomes. Although the role of the acidic proteins is still unclear, there is little doubt that they vary quantitatively and qualitatively with the physiological activity of the cells in which they are found. One suggestion is that they displace the histones, thereby rendering a particular segment of DNA transcriptionally competent, but this seems unlikely since nucleosomes are still present during transcription; a more likely suggestion is that they modify, possibly open up, the nucleosomes and thus allow the RNA polymerases better access to the DNA.

The RNA of isolated chromatin does not seem to play an integral part in the maintenance of structure, even though it is uniquely associated with chromatin and it is identifiably different in nucleotide content from the other RNA species within the nucleus, for example, the ribosomal RNA which is nucleolar related. This RNA, isolated from rat ascites tumor cells, has a high nucleotide sequence diversity internally, indicating that it is of mixed origin, but a high constancy of terminal nucleotides in that the 5' end is 90% cytosine and the 3' end 99% guanine. Furthermore, its half-life is about 17 hr, quite a bit shorter than the half-life of other rat RNA species. What its role might be, however, remains conjectural.

Linear Differentiation

As described earlier (Table 3.3), molecules of *D. melanogaster* DNA have been isolated which correspond in length to that calculated to be present in the largest single chromosome. Where variations in chromosome length were present, corresponding variations were found in the length of the longest isolated molecules. It is reasonable to assume, therefore, that one, and only one, double helix of DNA extends from one end of the chromosome to the other, and, furthermore, that any visible linear differentiation of the chromosome, or the manifestation of any equally evident genetic function, must depend upon particular sequences of nucleotide pairs even though other attached or associated molecular species of different character may relate to that differentiation or function. Just as genes, which are segments of DNA, give rise to different end products, so too does DNA have variable cytological expressions, some of which prove to be highly useful in the solution of cytogenetic problems.

Telomeres. Eukaryotic species exhibit characteristic numbers and sizes of chromosomes in their genomes. This observation suggests, but does not by itself prove, that each chromosome is a unique entity capable generally of retaining its identity through many cell generations and throughout each cell cycle; that is, chromosomes do not, as was once believed, spontaneously fuse together at their ends in interphase to form a so-called *spireme* and then break apart as they contract and enter prophase. The continued individuality of each chromosome is related to the fact that it is terminated at either end by a *telomere*, a term coined by H. J. Muller to indicate the uniqueness and stability of this portion of the chromosome.

There is nothing morphologically unique about the ends of isolated chromosomes to suggest that a telomere is anything more than a terminating sequence of nucleotides. The meiotic chromosomes of the tomato, for example, all terminate with a distinct and visible block of chromatin (Fig. 3.3), those of rye or sorghum trail

Fig. 3.3. Meiotic chromosomes from the microsporocyte of the tomato, *Lycopersicum esculentum,* showing the large chromomeres (masses of heterochromatin) concentrated at the centric regions of the paired homologues, and with smaller chromomeres located at the ends of the chromosomes (courtesy of C. M. Rick).

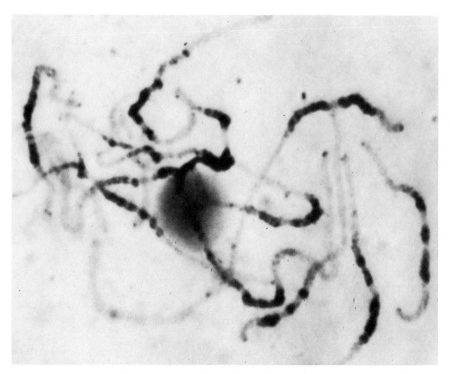

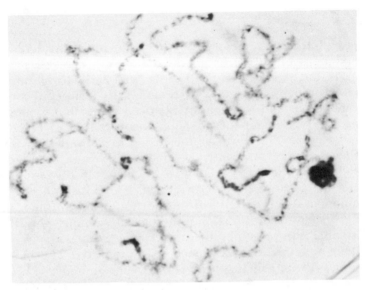

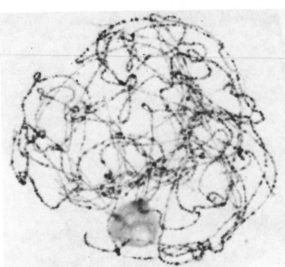

Fig. 3.4. Pachytene stages of meiosis in a grasshopper, *Chorthippus parallelus* (top), and in the lily, *Lilium longiflorum* (left), showing that the telomeres vary in degree of distinction from rather heavy terminal knobs to lightly staining ends. The darkly staining mass in the top figure is the highly contracted, heterochromatic X chromosome (courtesy of Drs. J. L. and Marta Walters).

off into ghost-like ends while those of grasshoppers and lilies seem to vary in degree of distinctiveness (Fig. 3.4). The uniqueness of the telomere, therefore, appears to lie in its characteristic behavior rather than in any constant structural feature, but a 3,000-base pair segment of DNA, labeled radioactively and possibly repeated several times, has recently been shown to hybridize with the five major ends of *D. melanogaster* polytene chromosomes, suggesting a chemical uniqueness yet to be determined.

A chromosome which, through the action of X-radiation or radiomimetic chemicals, has been fragmented into several pieces is an unstable structure. The only exceptions appear to involve chromosomes with diffuse rather than localized centromeres (see the next section). Broken ends can unite with other broken ends to create new chromosomal arrangements, but telomeres do not do this; that is, they do not unite with other telomeres or with newly broken ends of chromosomes. A telomere, therefore, cannot come to occupy any position in a chromosome except at an end, and an internal region of the chromosome, freshly broken, cannot, under ordinary circumstances, assume the role of a telomere. Presumably this behavior of a telomere, like the different behavior of a broken end, lies in the chemistry of DNA, or possibly of the histones, but if so we have no precise knowledge of what this might be.

Broken ends of chromosomes, whether arising spontaneously or induced by some kind of damaging treatment, can sometimes "heal" and function later as telomeres. In maize, for example, a broken end of a chromosome will "heal" in sporophytic tissue but not in the gametophyte or the endosperm. In *Parascaris equorum,* the horse threadworm, the large chromosomes of the zygote persist in the germline, but break up into heterochromatic ends and many smaller euchromatic entities in somatic tissues, with the heterochromatic ends being lost and each of the euchromatic elements retaining its individuality throughout an indefinite number of somatic cell divisions. The varied structural and behavioral aspects of telomeres preclude defining a telomere in very specific terms.

Other types of behavior also characterize the telomeres of particular species. In the salivary gland cells of *D. melanogaster,* the polytene chromosomes are often found exhibiting a telomere-to-telomere attraction of a nonrandom but also of a nonhomologous kind, while in the meiotic cells of some heteropteran insects the X and Y chromosomes do not synapse in prophase; instead, they undergo a brief "touch-and-go" pairing in their telomeric regions, thus providing a regular means of metaphase orientation and anaphase separation (Fig. 3.5). The telomeres of chromosomes in the spermatocytes of the mantid, *Stagomantis carolina,* exhibit a pronounced attraction to the nuclear membrane in the vicinity of the centrosome (Fig. 3.6), and it is at the juncture of telomere–nuclear membrane contact that the synaptinemal complex begins to bring pairing homologues into synaptic union. These observations, varied as they are in sporadic fashion, suggest that the DNA at the ends of chromosomes does possess properties different from the DNA located internally in the body of a chromosome, but at the moment the only information available is descriptive, and with no mechanism proposed.

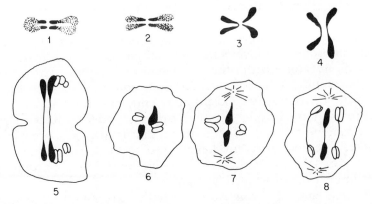

Fig. 3.5. Touch-and-go pairing in the hemiptern insect, *Rhytidolomia senilis*. In late meiotic prophase (1 and 2) the sex chromosomes, clearly split into their chromatids, come together at their euchromatic ends; in 6 and 7, the same chromosomes approach each other for second division orientation, but this time by their heterochromatic ends. The nature of the pairing force remains unknown (Schrader, 1940).

Fig. 3.6. Spermatocytes of the mantid, *Stagomantis carolina*, showing the polarization of the paired homologues, with their ends directed toward the centrosomic regions; some bivalents in the left figure have their ends directed toward different centers; others are looped so that both ends are directed toward the same center (Hughes-Schrader, 1943).

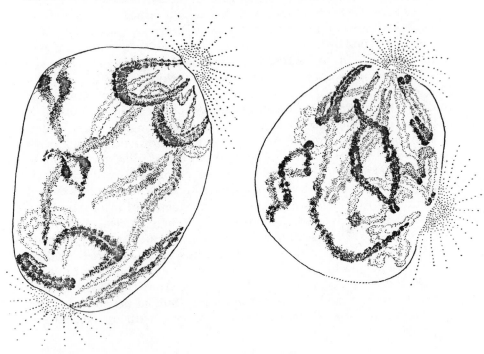

Centromeres. Most eukaryotic chromosomes possess a special structure whose presence is necessary if regular anaphase separation of chromatids is to take place in mitosis or if homologous chromosomes are to segregate from each other in meiosis. This is the *centromere,* a localized segment of the chromosome which exhibits the property of interacting with the microtubular proteins of the spindle, and thus of acting as the center of movement of the chromatids or chromosomes as anaphase separation takes place. When a centromere is missing from a chromosome and the chromosome is thus *acentric,* the chromosome lacks the capacity to segregate in a regular fashion, and eventually it is lost from the genome.

The centromere is known by a number of other names—*kinetochore* or *spindle fiber attachment point* being commonly used—but more recently the kinetochore has been redefined as a plate-like outer face of the centromere with which the microtubules of the spindle interact. The centromere is the functional, but apparently not the morphological, equivalent of the region of a bacterial chromosome that attaches to the mesosome and that region of a protozoan chromosome which is attached to the nuclear membrane during division. In the eukaryotes, the location of the centromere divides the chromosome into two arms. If centrally located, giving two arms of approximately equal length, the chromosome is said to be *metacentric* (Fig. 3.7). When one arm is considerably longer than the other, the chromosome is *submetacentric* or *acrocentric;* when the centromere is terminally located, the term *telocentric* is used. For example, the X chromosome of *D. melanogaster* is acrocentric, there being a very short arm on the right side of the centromere, while the second and third chromosomes are metacentric. There has been considerable debate in the literature as to whether a centromere, under

Fig. 3.7. The three major types of eukaryotic chromosomes, with the position of the centromere determining their shape—V, or J or rod shape—at anaphase. Mouse chromosomes are all telocentric while human chromosomes would exhibit both metacentric and acrocentric types, but no telocentric ones.

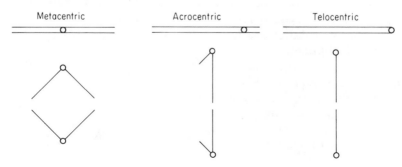

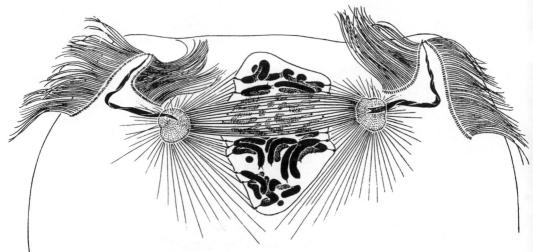

Fig. 3.8. Chromosomes of the protozoan, *Barbulanympha*, being arranged on the spindle forming between the two centrosomes. The chromosomes are telocentric, and are permanently attached to the nuclear membrane (Cleveland, 1938).

normal conditions, can be truly terminal, in which case it would presumably act as a telomere as well. This would seem to be the case in the protozoan Barbulanympha; its chromosomes seem to be permanently attached to the nuclear membrane, and hence unquestionably telocentric, but it may well be that such an attachment confers a degree of stability to an otherwise unstable element of the chromosome (Fig. 3.8). Artificially created telocentric chromosomes have been produced by fracture of the centromere, but such chromosomes seem to be unstable and to have difficulty in persisting as part of the genome. However, instances are known (see Chapter 7) where metacentrics have been converted into acrocentrics by a splitting of the centromere (a process known as *dissociation*), so telocentric chromosomes are not invariably unstable.

Chromosomes lacking a centromere are said to be *acentric;* by the same reasoning, those possessing two centromeres are said to be *dicentric*. Neither type is normally encountered in natural populations, but an aberrant dicentric wheat chromosome, with the two centromeres very close together, seems to be able to segregate properly; a similar instance has also been described for the Y chromosome in the human. Both centromeres of each chromatid of such dicentric chromosomes must, however, be oriented to the same pole in division if segregation is to be regular.

In addition to being the organelle of movement of chromatids or chromosomes during anaphase separation, centromeres can also exert an effect in meiotic prophase by influencing the rate of crossing over in their vicinity. This is described later in detail (p. 265), but it can be pointed out here that centromeres act as a sort of barrier in the sense that crossing over in one arm of a metacentric chromosome is without effect on crossing over in the other arm. A metacentric chromosome, therefore, at least in *D. melanogaster,* behaves as if it were composed of two independent units of recombination in meiosis. Whether the same can be said for chromosomes with arms of unequal length is not clear, but it is evident that crossing over between acrocentric homologues is sharply reduced in the neighborhood of the centromere as compared to other regions of the chromosome arm.

A comparable degree of arm independence is evident when contraction of the chromosomes takes place during prophase of cell division. The two arms of a chromosome contract at the same rate, and they are coordinated with the contraction of the other chromosomes in the same nucleus, indicating an overall control being exercised, but the arms show some independence of behavior since the direction of coiling in the arms—right-handed or left-handed—is at random. Contraction does not involve the centromeric region, which, as a result, can be identified in metaphase chromosomes as a constriction (Figs. 3.9 and 3.10). In the spermatocyte of the salamander, *Oedipina poelzi,* as well as in many other species, the centromere may stain differentially from the remainder of the homologous chromosomes (Fig. 4.22).

Fig. 3.9. Chromosomes from a roottip of the broad bean, *Vicia faba,* showing the position of the centromeres (Cm) in the large M chromosomes, together with the secondary constriction, or nucleolar organizer region. The M chromosomes are metacentric, the other ten chromosomes being acrocentric.

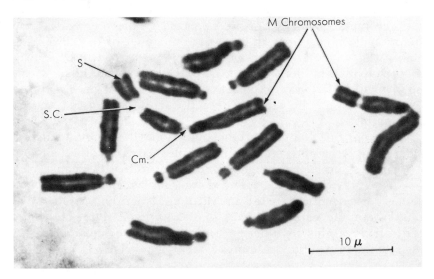

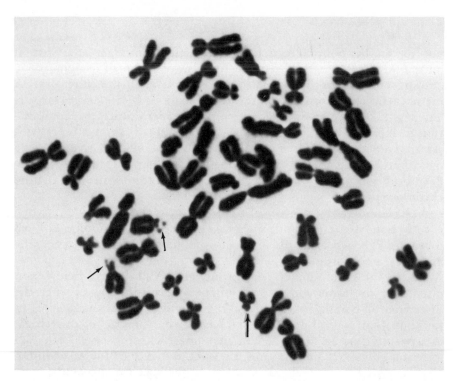

Fig. 3.10. Chromosomes from a cultured leucocyte of a human male; the division of the cell has been arrested by colchicine. The centric regions of every chromosome are readily visible as a definite construction. The lower arrow points to the Y chromosome; the X is more difficult to identify, being one of the middle sized chromosomes. The two arrows at the left point to the nucleolar organizers that are visible; of the ten possible nucleolar organizers, only these two show, and are in chromosomes 13, 14 or 15 (courtesy of Dr. Barbara Migeon).

Finally, the centromere serves as a locus on either side of which the chromatin in many species takes on a character and a genetic function different from that of the remainder of the chromosome. This problem of *heterochromatin* vs. *euchromatin* will be dealt with more fully in the next section of this chapter, but there seems to be a relation between the lack of, or reduction in, crossing over in the vicinity of the centromere and the presence of centric heterochromatin.

A centromere of a chromosome isolated in an extended state cannot easily be detected as a morphological entity; the sequence of nucleotide pairs runs through this region as it does through any other portion of the chromosome, and there is no evidence that the nucleosomes of the centromere are any different from those elsewhere. It is only when the chromosome is in a contracted state, or when the centromere begins to function as an organelle of movement, that a differentiated structure becomes evident. The relatively uncontracted region at meiotic prophase (Fig. 3.11) often reveals a beaded appearance which may be retained until metaphase, when the centromeric chromomeres may be part of a stretched-out region brought on by tensions imposed, presumably, by the contracting spindle fibers. It has recently been demonstrated that the centromeres of isolated chromosomes retain the ability, *in vitro,* of attracting and orienting microtubules into the beginnings of a spindle-like structure, much in the manner that they are presumed to do in a living cell. This organizing ability resides apparently in a kinetochore plate which forms on either side of the centromere; the half-spindle fibers extend, therefore, from the kinetochore plate to the poles. The fact that this occurs spontaneously would seem to rule out any enzymatic involvement at the face of the kinetochore plate, but this is only conjecture at the present time.

Fig. 3.11. Pachytene chromosomes of the normal genome of rye (*Secale cereale*) showing the chromomere patterns of the arms in the centromere region. Each centromere consists of three well-defined chromomeres, and since centromeres can be fractured, with each half functional, it would appear that each chromomere possesses centromeric properties. A similar pattern exists also in the B chromosomes (Fig. 3.74) (Lima-de-Faria, 1955).

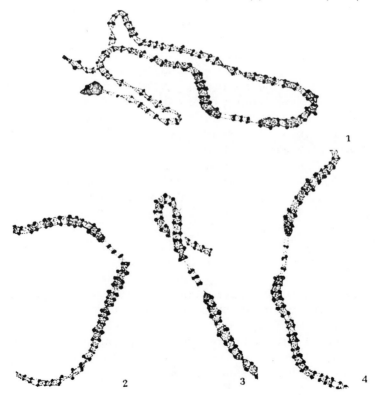

The centromeres just discussed are those located at particular regions of each chromosome and which divide the chromosome into two arms (telocentromeres would, of course, be exceptions). That localized centromeres embrace a stretch of nucleotide pairs is indicated by the fact that they can be broken in two, with both parts functional. Thus, in maize, chromosome 9 was broken by X rays in such a way that both products, in this instance a ring and a rod chromosome, were functional (Fig. 3.12). The chromosomes of some species, however, do not possess localized centromeres; rather, the whole face of a contracted metaphase chromosome engages the spindle, and movement of the chromatids to the poles is such that they remain at right angles to the spindle axis instead of assuming the V-, J-, or rod-shapes of the usual anaphase chromosome (Fig. 3.13). Such chromosomes characterize all of the species in the insect orders Homoptera and Heteroptera, and they are found very generally throughout the Lepidoptera, Dermaptera (earwigs), and Odonata (dragonflies), and in isolated genera such as Luzula of the wood rushes. The fact that the more primitive members of the rushes have localized centromeres would strongly suggest that the *holocentric* (holos = whole) condition is a more recently derived state.

The structural basis which causes a chromosome to act in a holocentric manner is unclear. The situation is *Parascaris equorum,* in which the larger germline chromosomes fragment into the numerous small ones of the soma, would suggest the existence of a polycentric condition, whereas the holocentric state, which could also exist in *P. equorum,* simply means that the chromatids as entities possess centric activity along their entire length. If holocentric chromosomes are fragmented, either spontaneously or by X-radiation, every fragment, regardless of size, can move in anaphase and persist from one cell generation to another, a circumstance not true of acentric fragments from chromosomes with localized centromeres. Whether a kinetochore plate comparable to that found associated with the usual chromosome exists is not known, but it does appear that broken ends of holocentric chromosomes can achieve telomeric stability without difficulty. The wide range of chromosome numbers in some genera of butterflies would suggest that such fragmentation is not uncommon.

Before closing this discussion, mention should be made of the fact that other regions of chromosomes, under special conditions, may function in a centromere-like fashion. In maize, for example, chromosome 10 exists in an abnormal form, being distinguished from its normal counterpart by the presence of a large heterochromatic knob readily visible during meiotic prophase. In a plant, heterozygous

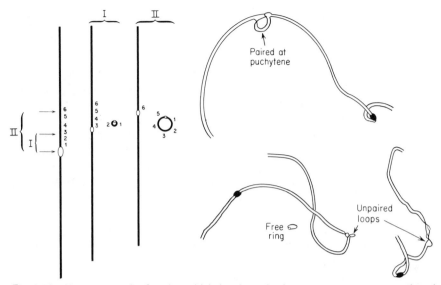

Fig. 3.12. Chromosome 9 of maize which has been broken on two separate occasions in such a manner as to produce a ring and a rod chromosome in each instance. One of the breaks had to be through the centromere, and the other in some other portion of one of the arms. The rings have a tendency to become lost during division, and if they carry recognizable genetic markers, their loss is readily recognized. The unpaired loop does not always indicate the region, adjacent to the centromere, from which the ring was derived. (McClintock, 1938).

Fig. 3.13. Somatic mitosis in the coccid Nautococcus, showing the metaphase and anaphase appearance of chromosomes possessing a diffuse centromere. The spindle fibers engage the chromosomes along their entire length, and the chromatids move away from each other in parallel fashion. As the spindle narrows at the poles, the separating chromatids are forced into a U-shape, with their ends directed toward the poles (Hughes-Schrader, 1948).

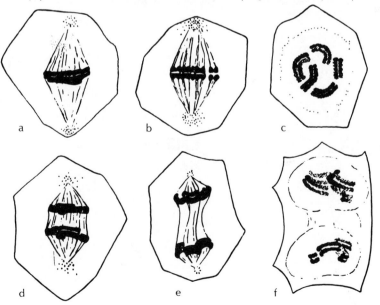

for chromosome 10, the knob on abnormal 10 acts as a *neocentromere,* precociously orienting the knobbed chromosome to the pole nearest to it in the late prophase of meiosis (Fig. 3.14); all other knobs on other chromosomes come under the influence of abnormal 10 and will behave similarly. Since the basal megaspore of the linear tetrad becomes the functional embryo sac, abnormal 10 and all other knobbed chromosomes, as well as all genes on these chromosomes, undergo preferential segregation to the inner and outer cells. The neocentric activity does not persist through anaphase, but it is sufficient to bring about distorted ratios of the affected genes. What is of further interest is that any knob, whether on abnormal 10 or any other chromosome, must be on a chromosome possessing a localized centromere in order to function as a neocentromere; knobbed acentric fragments do not respond in this manner. It would suggest that abnormal 10 induces localized centromeres to exert an influence which passes along the chromosome from the centromere to the knob and causes them to bring about a premature orientation of the bivalents. The mechanism remains unknown, but the phenomenon is not confined only to knobbed chromosomes, having been reported in rye and brome grass hybrids which lack maize-like knobs.

Fig. 3.14. Neocentric activity in maize as the result of the action of abnormal chromosome 10. A number of chromosomes have begun a precocious movement toward the poles, with the result that those regions of the chromosomes possessing knobs (see Fig. 3.15) lead the way in anaphase. Right, three metaphase II chromosomes, two of which show neocentric activity while the other does not, and below, two chromosomes in anaphase II, one of which is normal in behavior (on the left) while the other shows neocentric activity in one arm (Rhoades, 1942).

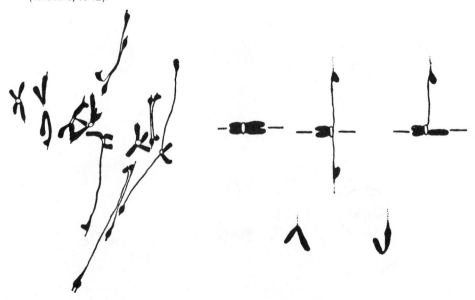

Nucleolar Organizers. Eukaryotic cells in interphase through mid- to late prophase contain one or more rounded bodies which are rich in ribonucleic acid (RNA). Failure to exhibit the Feulgen reaction indicates a general absence of DNA in these structures. These are the *nucleoli* (singular, *nucleolus*) which are formed by one or more specific regions in the haploid complement called *nucleolar organizers*. The term "organizer" was coined because it was believed at one time that all of the chromosomes were capable of forming nucleolar material and that the nucleolar organizer was a collecting or assembly point on one or more of the chromosomes. In maize, for example, loss of the nucleolar organizer regions causes nucleolar-like material to form into many small bodies unattached to any specific chromosomes, but rather scattered throughout the genome. From what is now known of nucleolus formation, it is most unlikely that all chromosomes of the genome contribute to the nucleolar mass, but it is also evident that as many as 40 or more micronucleoli can form during late prophase (diplotene stage) in human meiotic cells. These seem not to be associated with the normal nucleolar regions, but rather to be related to constrictions in the centric heterochromatin of chromosomes 1, 9, and 16. It has been assumed that this nucleolar material contributes to ribosome formation in a manner similar to that of the normal nucleoli, but this remains to be conclusively demonstrated. The nucleolar organizer, itself, is a differentiated region of chromatin DNA which, through transcription, produces the larger pieces of RNA (that is, the 18S and 28S rRNAs) which are incorporated into the ribosomes. Other regions of the genome, however, produce the 5S rRNAs and code for the ribosomal proteins; so the nucleolus, in addition to producing most of the rRNAs, is also a processing site where all of the rRNAs, together with the ribosomal proteins, are assembled into ribosomes which, subsequently, will be transported to the cytoplasm where protein synthesis will take place.

The more active a cell is metabolically the larger are its nucleoli in interphase. This is in keeping with the relation of the nucleolus to protein synthesis. For example, actively proliferating neoplastic cells tend to have larger nucleoli than the normal cells to which they are related. Also, such cells as vertebrate oocytes which, following fertilization, will undergo rapid developmental changes as embryogenesis proceeds, often evolve various modes of amplification of nucleolar regions to augment the need for enhanced protein synthesis (p. 337).

As prophase is initiated and the cell progresses toward metaphase, the nucleoli tend to diminish in size and eventually to disappear. When they do so, their former position on a chromosome is

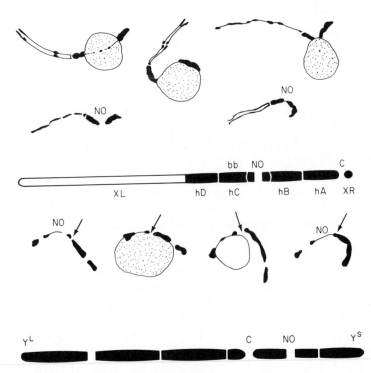

Fig. 3.15. Camera lucida drawings and diagrammatic representations of the X (above) and Y (below) chromosomes of *D. melanogaster*, to indicate the positions of the nucleolar organizers (NO). The position of the centromere (C) is identified by a small arrow in the camera lucida drawings of the Y. Heterochromatin is shaded, euchromatin is clear. The gene *bb* (bobbed) is located in hC of the X. The additional constrictions, particularly in the Y chromosome, are evident, but their cause is not certain (redrawn from Cooper, 1959).

generally revealed as a constriction in the metaphase chromosome (Fig. 3.9). This is called the *secondary constriction,* the primary constriction being the location of the centromere. However, just as centromeres do not produce a primary constriction if they are located at the ends of chromosomes, so nucleoli do not produce a secondary constriction if similarly placed. Thus, the vole, *Microtus agrestis,* forms nucleoli but produces no secondary constriction; the location of the nucleolar organizers at the ends of the particular chromosomes can be revealed, however, through the use of a specific silver staining method.

The nucleolar organizer is a necessary component of the eukaryotic cell; its absence in a homozygous state is a lethal condition. Such an anucleolate mutant is known in *Xenopus laevis,* the African clawed toad, with death occurring in early gastrular stages after the maternally derived ribosomes are exhausted; that is, the fertilized egg, even though homozygous recessive and unable to replenish its ribosomal population, can produce sufficient proteins on the ribosomes of the oocyte to reach the gastrula stage, but once the need for additional ribosomes becomes evident, death of the embyro ensues.

As we shall discuss later, the major genes which form rRNA are represented many times, in tandem array, in the nucleolar organizer region, and they can be increased or decreased in particular ways. Even among closely related species, the position, number, and distribution of rRNA forming regions may differ appreciably. In *D. melanogaster,* the nucleolar organizers are located in both the long arm of the X chromosome and the short arm of the Y chromosome, and they are closely associated with the *bobbed* locus (Fig. 3.15). The Y chromosome of *D. virilis* possesses two nucleolar organizer regions; the X chromosome possesses one. In maize and its near relative, teosinte, the nucleolar organizer is visible as a large knob on chromosome 6 (Fig. 3.16). This knob has been fragmented experimentally; both fragments are capable of forming nucleoli of somewhat smaller size, but indicating that both contain the appropriate

Fig. 3.16. The nucleolar organizers and nucleoli of maize and teosinte (*Zea mexicana*). Top row, two examples from maize showing chromosome 6 with the large knob attached to the nucleolus. The figure at the right lacks any other knobs on chromosome 6. Bottom row, Mexican (left) and Guatemalan (right) teosinte, each a large nucleolar knob, with other knobs differently placed (courtesy of Dr. T. A. Kato Y.).

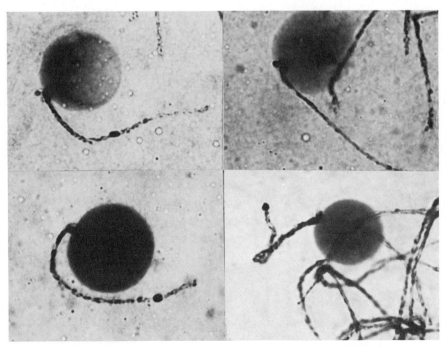

rRNA genes. In humans, there are nucleolar organizers on chromosomes 13, 14, 15, 21, and 22 (Fig. 3.10); a diploid cell, therefore, could have ten distinct nucleoli, but the tendency of nucleoli to fuse with each other results generally in far fewer being visible in any single cell. Among the primates related to the human, the chimpanzee, thought to be one of the closest of the human's relatives, has nucleolar organizers on chromosomes 13, 14, 18, 21, and 22, but the gorilla appears to have them only on chromosomes 21 and 22, while the gibbon and the catarrhine (Old World) monkeys have only a single nucleolar organizer per haploid genome.

The capacity of a nucleolar organizer to form a nucleolus, is, like all other metabolic events, under genetic control, and these control systems can vary in dominance. The genus *Crepis* is a member of the Compositae, and interspecific hybrids within this genus can be readily obtained through cross pollinations. Each species has its nucleolar organizer in a fixed position in its haploid complement of chromosomes, and the site is readily identifiable in somatic cells by the presence of a secondary constriction (Fig. 3.17). When interspecific hybrids are formed, however, their cells reveal that only one

Fig. 3.17. Diagrammatic representation of the haploid chromosomes of eight species of Crepis to indicate the use of a variety of landmark features for karyological analyses: number, position of centromere, size and position of the nucleolar organizer region (Babcock, 1947).

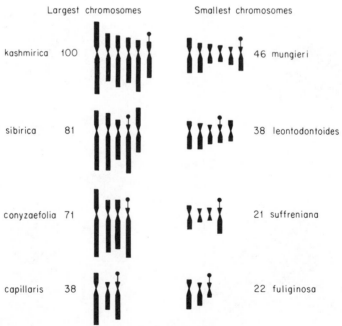

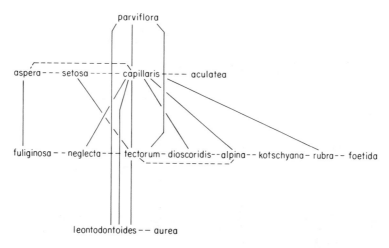

Fig. 3.18. Dominance relations of nucleolar formation in various species of Crepis. In species hybrids, the nucleolar organizer contributed by the parent occupying a higher position on the chart will suppress the action of any nucleolar organizer below it (solid lines); thus, the *parviflora* nucleolar organizer will suppress all others, that of *leontodotoides* can suppress none of those on the chart. Those occupying the same level exhibit a co-dominance (dotted lines) with both chromosomes producing nucleolar organizers (after Wallace and Langridge, 1971).

of the nucleolar organizers is capable of forming a nucleolus, the other being suppressed. When this happens, the secondary constriction also fails to appear, indicating that this constriction is simply an uncontracted region of the chromosome earlier occupied by the nucleolus. The species of *Crepis* can, therefore, be arranged in a sequence of decreasing dominance: *parviflora* $\longrightarrow$ *capillaris* $\longrightarrow$ *tectorum* $\longrightarrow$ *leontodontoides* (Fig. 3.18). That is, the presence of a *parviflora* nucleolar organizer can suppress the activity of any other organizer listed below it when both are in the same nucleus; *leontodontoides* would be incapable of suppressing any of those above it. How this control is exercised is not known, but a series of allelic genes of varying degrees of dominance has been suggested. A comparable situation, but better understood at the molecular level, also exists in hybrids of the toads, *Xenopus laevis* x *X. mulleri*. In the hybrid cells only the *laevis* rRNA genes can be transcribed (the two kinds of rRNAs are distinguishable), regardless of whether *laevis* is the male or female parent.

It has been stated above that the nucleolar organizers form rRNAs which are then processed within the nucleolus for inclusion into ribosomes. This has been demonstrated in *D. melanogaster* by a

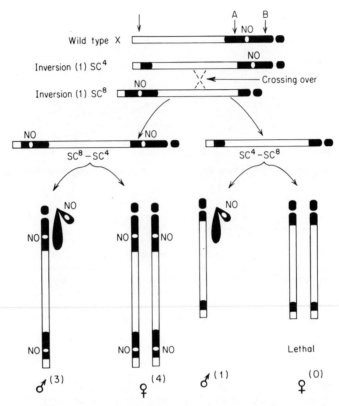

Fig. 3.19. Diagrammatic representation of the use of sc^4 and sc^8 inversions to obtain stocks of *D. melanogaster* containing one to four nucleolar organizer regions. In the wild type X chromosome (top), the left break point (arrow) was the same for both inversions; and was coupled with break point A to produce the sc^4 chromosome, and break point B to produce the sc^8 chromosome. The hybrid containing both the sc^4 and sc^8 chromosomes provided a stock with 2 nucleolar organizers; the others with 0, 1, 3, and 4 nucleolar organizers are shown at the bottom (redrawn from Ritossa and Spiegelman, 1965).

Fig. 3.20. Camera lucida drawings of the sc^4 and sc^8 chromosomes represented in Fig. 3.17. Left, sc^4 chromosome; middle, homozygous sc^8; right, the crossover product lacking a nucleolar organizer (redrawn from Cooper, 1959).

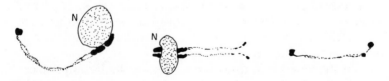

series of elegant experiments. As Figs. 3.19 and 3.20 illustrate, two inversions, *scute-4* (sc^4) and *scute-8* (sc^8), with their break points in euchromatin and heterochromatin, have led to a rearranged X chromosome such that when both inversions are present in the same individual, crossing over in the oocyte can produce two new chromosomes: one lacking a nucleolar organizer and the other having this region duplicated. By appropriate breeding procedures, and by using X chromosomes that are labeled with marker genes so that their presence or absence can be readily ascertained, individuals with one to four nucleolar organizers in the genome can be obtained; anucleolate individuals would, of course, fail to appear. Individuals with four nucleolar organizers should, therefore, have four times as much DNA producing rRNA as individuals have only a single organizer. Based on the technique of hybridizing RNA with the DNA from which it was transcribed (Fig. 3.21), the percentage of radioactively labeled rRNA isolated from cells should be in direct proportion to the number of nucleolar organizers in those cells. As Figs. 3.22 and 3.23 reveal, the proportionality holds. This means that virtually all of the rRNA is

Fig. 3.21. DNA/RNA hybridization. Left, The DNA extracted from the tissues of Drosophila is melted into single strands, and bound in place in a column. Middle, ribosomal RNA is added, and it hybridizes in complementary fashion with the DNA from which it is transcribed; the amount hybridized will be a function of the amount of DNA complementary to it, and this will depend upon the number of nucleolar organizer regions in the genome. Right, once the DNA is saturated with complementary RNA, no more can be hybridized. Single stranded DNA and RNA can be removed later, and the amount of DNA/RNA hybridization determined.

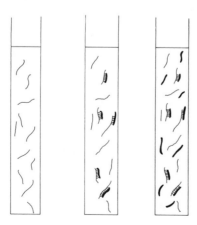

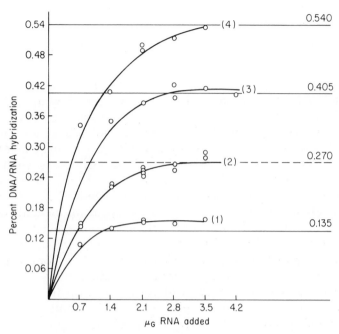

Fig. 3.22. The saturation of DNA/RNA hybridization as a function of the number of nucleolar organizers (indicated in parentheses) in *D. melanogaster*. The sex of the individual made no difference in saturation levels, only the number of organizers; it made no difference, therefore, as to whether the organizers were all on the X chromosome or distributed between the X and Y chromosomes (redrawn and simplified from Ritossa and Spiegelman, 1965).

Fig. 3.23. Relation between the number of nucleolar organizers in *D. melanogaster* and the percentage of DNA/RNA hybridization at saturation (redrawn from Ritossa and Spiegelman, 1965).

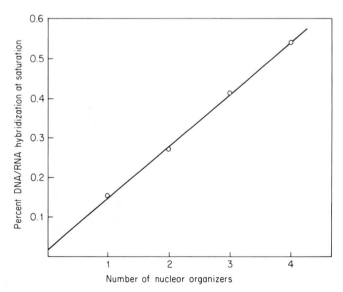

formed within the nucleolar organizer regions. This concept is further supported by the observation that DNA isolated from a number of different chicken cells—sperm, erythrocytes, embryo, liver, kidney —will all reach the same DNA/RNA hybridization plateau of saturation with rRNA even though these cells differ appreciably in their rates of rRNA synthesis. We shall see, however, that rDNA (that is, that DNA which is transcribed into rRNA) in some organisms and in some reproductive cells is capable of considerable amplification by means other than that of the linear repetition of genes (p. 337) and that rRNA production, and hence ribosome formation, can be abruptly altered in the higher plants as a shift takes place from a sporophyte to a gametophyte generation.

The structure of the genes which produce rRNA has been ascertained for two species of *Xenopus* and for humans, the latter through the use of Hela cells, a carcinoma strain maintained in tissue culture. Isolation of rRNA from ribosomes indicates that an 18S RNA (S = sedimentation value in ultracentrifugation) is associated with the smaller portion of the ribosome, a 28S RNA with the larger segment. Larger RNA segments, however, can be isolated from nucleoli: a 40S segment from *Xenopus* cells and a 45S one from Hela cells. These are rRNA precursor molecules from which the 18S and 28S rRNAs are derived during maturation and after the "spacer" segments have been removed (Fig. 3.24). Evidence for this has been obtained by degrading the precursor molecules by appropriate nucleases, a technique which has also revealed that the 28S rRNA is at the 5' end and that the 18S is at the 3' end of the precursor segment.

The segment of rDNA from which the precursor rRNA is derived has a molecular weight of 8.8×10^6 daltons in both *X. laevis* and *X. mulleri*. Those portions of the precursor segment which code for the 18S and 28S RNAs are found, by hybridization procedures, to be highly homogeneous in nucleotide content and sequence (Fig. 3.25); the species differ appreciably, however, in the spacer portion, long stretches of which fail to show close pairing on renaturation. The spacers in both species are similar in length, about 7,000 nucleotides long, and as Table 3.5 indicates, both are high in G/C content, but *laevis* more so than *mulleri*. Since, so far as is now known, the spacers serve only to keep the two coding sequences together, there would seem to be no obvious selection pressure to keep them free of nucleotide changes. This is indicated by the fact that the two species have diverged randomly in this respect, but the question of how the spacers are kept homogeneous within a species remains to be answered. Nucleotide changes in sequence within the 18S and 28S segments would probably be more rigorously selected against since such alterations could well affect the structure of the ribosomes, and hence be detrimental.

The structure of the rDNA sequences in *D. melanogaster* exhibits several differences when compared with those in *Xenopus*. Both the X and Y chromosomes possess nucleolar organizer regions, and the 18S and 28S RNAs produced by both chromosomes are similar, but the two chromosomes differ in the spacer regions, the Y chromosome having only a single, short spacer while the X has both long and short spacers. In addition, it appears that the rRNA segments with long spacers transcribe only, or primarily, during oogenesis (transcription occurs in the nurse cells, not in the oocyte), while those with short spacers do so during embryogenesis and in somatic cells. The reason for the differential transcription remains obscure, but as we shall see it is reminiscent of the three kinds of 5S rRNAs produced in *Xenopus* (Table 3.5).

Fig. 3.24. Sections of DNA in Xenopus coding for ribosomal RNA (see also Fig. 3.27). Top, a block of repetitious DNA coding for 18S and 28S rRNAs; the 18S and 28S regions are similar in both *laevis* and *mulleri*, while the spacer DNA is less homogeneous in the two species (see Fig. 3.25). The two bottom sequences are of the rDNA coding for 5S RNA; the genes (G) are similar in the two species, but the spacer DNA in *mulleri* (Sm) is much longer than it is in *laevis* (S1). As Table 3.5 indicates, the coding sequences differ in other respects as well (after Brown, 1973).

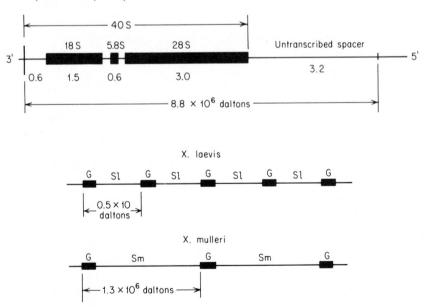

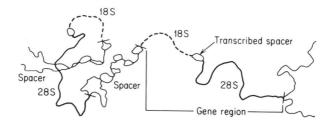

Fig. 3.25. Tracing of an electron micrograph of a strand of double stranded DNA, one strand of which consists of *Xenopus laevis* rDNA and the other of *X. mulleri* rDNA. The fact that the gene regions pair in well-matched fashion indicates their nucleotide homogeneity, but the spacer regions are but poorly matched, indicating their nucleotide heterogeneity (after Brown, 1973).

Table 3.5. Characteristics of rDNA and 5S DNA in *Xenopus laevis* and *X. mulleri*.

rDNA	laevis	mulleri
18S + 28S DNA		
Percent of total DNA	0.2	0.2
Number of repeated sequences	450	450
Location in genome	X	X
Repeat length in daltons	8.8×10^6	8.8×10^6
Gene: spacer ratio	1:1	1:1
Percent GC of spacer	73	69
*5S DNA**		
Percent of total DNA	0.7	0.7
Number of repeated sequences	24,000	9,000
Location in genome	Telomeres of most chromosomes	
Repeat length in daltons	$0.5\text{-}0.6 \times 10^6$	$1.2\text{-}1.5 \times 10^6$
Gene:spacer ratio	1:6	1:15
Percent GC of spacer	35	43

*Analysis of the 5S rRNAs derived from *X. laevis* indicates that there are three kinds of rRNA: (1) a major species formed in oocytes and derived from the 24,000 repeated sequences cited above, (2) another oocyte species differing in nucleotide content and represented by only 2,000 repeats, and (3) a species formed principally in somatic cells and represented by an unknown number of repeats. How the three kinds of repeats are arranged in respect to each other is not known (after Brown, 1973, 1977). Further evidence indicates that the structure of the major oocyte rDNA gene is: 5'—longspacer—gene—linker—pseudogene—3'. This evidence is from the plus or transcribing strand of the rDNA gene; the pseudogene has a nucleotide structure similar to the transcribing gene, and while its function is not known for certain, it is presumed to have once been a normal gene but is now only a relic of evolution (Jacq et al., 1977).

Fig. 3.26. Three electron micrographs of *Drosophila melanogaster* rRNA genes in the process of transcription. Top, both sequences are transcribing in the same direction, from left to right, and a spacer separates them; middle, a greater magnification of a similar pair of tandem sequences, with the lengths indicated by a marker bar; bottom, an adjacent pair of sequences which are transcribing in opposite directions (courtesy of Dr. Y. W. Chooi).

As Fig. 3.24 indicates, the *Xenopus* 28S rDNA is at the 5′ end of the ribosomal gene and the 18S portion is at the 3′ end. If all of the ribosomal genes are in tandem array in the nucleolar organizer region, they would probably be oriented in the same direction, and hence transcribe in the same direction (Fig. 3.26). Generally this is so, if one is to judge from the many electron micrographs of transcribing genes, but occasionally adjacent ribosomal genes transcribe in the opposite direction. This must mean that in some regions of the nucleolar organizer both strands of the double helix are transcriptionally competent and that such adjacent segments are reverse repeats of each other.

Hybridization and renaturation kinetics and calculations similar to those performed for the analysis of the ribosomal genes in *D. melanogaster* show that *Xenopus* has about 450 of these identical sequences arranged in tandem in the nucleolar organizer region. These are all located in a single chromosome of the haploid set. The 500 ribosomal genes in *D. melanogaster* are about equally distributed

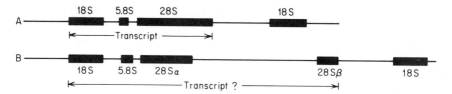

Fig. 3.27. Structure of the two kinds of rRNA genes in *D. melanogaster*. A. the sequence found only in the Y chromosome, constituting about 1/3 of the sequences in the X; the spacer between the 28S and 18S sequences is not transcribed. B. the sequence in which an insertion of variable size is found toward the right hand half of the 28S region; this region is believed to be transcribed and then excised as a post-transcriptional event, after which the two halves of the 28S rRNA are spliced together (after Glover and Hogness, 1977; White and Hogness, 1977; Wellauer and Dawid, 1977; and Pellegrini, et. al., 1977).

between the X and Y chromosomes, but some of them differ structurally from those in *Xenopus* (Fig. 3.27). The coding direction in the two genera is the same, as is the spacer between the 18S and 28S sequences which codes partially for a 5.8S rRNA which seems to be associated with the larger half of the ribosome, but more than half of the 28S sequences have a variably sized noncoding spacer inserted at a particular location in its right-hand half. The insertion ranges in size from 500 to 6,000 base pairs, with each insertion being some multiple of a 500-base pair unit. When 28S rRNA is reassociated back on to denatured rDNA, a loop forms corresponding to the location of the inserted spacer (Fig. 3.28). It would appear that both sections of 28S rDNA, plus the insertion, are transcribed but that, as a post-transcription event, the insertion is excised and the two halves of 28S rRNA are spliced together to form a functional unit. This splicing phenomenon is typically and uniquely eukaryotic in nature and involves other kinds of RNA as well (see Chapter 5).

Fig. 3.28. Above, a stretch of rDNA which has been denatured and reassociated with sequences of both 28S and 18S RNAs, both of which are drawn in with heavy lines. Below, a diagrammatic representation of the difference between 28S rDNA and a finished and functional piece of 28S rRNA (redrawn from White and Hogness, 1977).

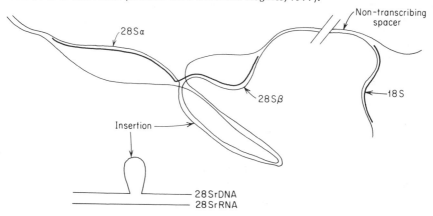

The 28S sequences with insertions seem to be confined to the X chromosome of *D. melanogaster* and to be dispersed rather than clustered within the nucleolar organizer region. Since the Y chromosome lacks the inserted sequences, and since the rRNA genes are about equally shared by the two chromosomes (the Y actually has somewhat fewer), those with insertions are about twice as numerous in the nucleolar organizer region of the X chromosome as those without.

The 220 rDNA loci thought to be present in the human haploid complement must be distributed among the five nucleolar organizer regions, with an average of 44 in each one. The genes coding for rRNAs in the fungus, *Aspergillus nidulans,* are about 11,000 base pairs long, are repeated about 60 times to make up approximately 2% to 3% of the genome, and are the only significantly reiterated sequences in the haploid complement. Mitochondria and chloroplasts possess ribosomes peculiar to themselves, and with their rRNAs coded for by their own ribosomal genes. As Fig. 2.20 indicates, the genome of maize chloroplasts possesses only two copies of rDNA while the normal nuclear genome has about 3,100 copies (Table 3.6). The number of such copies in other angiospermous species may differ by over an order of magnitude.

The X chromosome mutant *bobbed* of *D. melanogaster* is of interest here. Its phenotypic expression, when homozygous, includes

Table 3.6. Number of rRNA genes and the amount of rDNA in higher plants based on DNA/RNA hybridization studies

Species	Percent of rDNA in Haploid Genome	Number of rRNA Genes in Haploid Genome
Beta vulgaris v. *cicla*	0.20	1,150
Cucumis sativum	0.96	4,400
Allium cepa	0.09	6,650
Pisum sativum	0.17	3,900
*Zea mays**		
Growing plant tissue	0.18	3,100
Dry embryo	0.17	3,100
48-hr embryo	0.15	3,100
48-hr roottip	0.19	3,100
*Helianthus tuberosum**		
Growing plant tissue	0.022	260
Dormant tuber	0.032	260
Young leaves	0.022	260

*Data from several types of tissue to indicate that very little, if any, amplification of rRNA genes takes place during developmental stages.

Source: J. Ingle and J. Sinclair (1972).

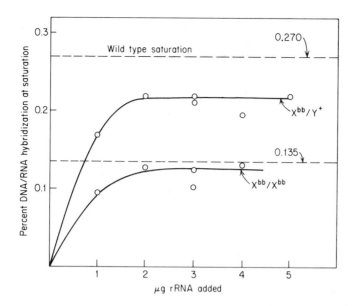

Fig. 3.29. Saturation kinetics of DNA/RNA hybridization for *bobbed (bb)* males and homozygous *bb* females of *D. melanogaster*. The normal Y^+ chromosome is insufficient to bring the saturation level up to that of the wild type individual (0.270), while the homozygous *bb* female saturates at less than that of a male with only one nucleolar organizer (see Fig. 3.22) (after Ritossa and Spiegelman, 1965).

short bristles, slow development, low viability, and low fecundity; in other words, it is a *hypomorph* which exhibits subnormal characteristics. It is also located in the nucleolar organizer regions of both the X and Y chromosomes, and it now seems clear that when the number of ribosomal genes falls below 130, for whatever reason, the *bobbed* phenotype, with varying degrees of expressivity, makes its appearance. This is borne out by the data presented in Fig. 3.29 and obtained from *bobbed* males or homozygous *bobbed* females, as well as by the observation that a culture of *bobbed* flies will, over the course of several generations, regain a normal, or wild-type, phenotypic appearance. This would suggest that there is a strong selection pressure to maintain a specific, or at least an adequate, number of rRNA-producing genes, and this shift toward normality is probably accomplished by the process of unequal crossing over (Fig. 3.30). It is also possible that increases or decreases can occur during the process of replication, as indicated in Fig. 3.31. If this is so, then one could predict that flies with extra sets of nucleolar organizer regions would have reduced numbers of rRNA-producing genes per set, and further, that by rigorous selection, it should be possible to obtain *bobbed* with different degrees of phenotypic expression. Both of these predictions have been realized experimentally.

99

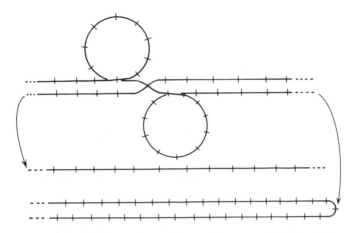

Fig. 3.30. Unequal crossing over within repeated sequences. Beginning with two hypothetical, but homologous, chromosomes having an equal number of repeats, it is possible to increase or decrease the number by a process of unequal crossing over, but a process that is always between homologous regions. This could readily account for the tendency of a population of *bb* individuals of *D. melanogaster* to approach gradually a wild type phenotype.

In *D. melanogaster,* an increase in rRNA-producing sites is a property of the X but not of the Y chromosome, while in *D. hydei,* which has one nucleolar organizer in the X and two in the Y chromosome, both chromosomes show compensatory ability. A comparable ability to regulate the number of rRNA-producing sites has also been found in the ciliate *Tetrahymena* and in rye and wheat. In the latter this aspect is governed by group 1 chromosomes. It would appear, then, that eukaryotic genetic systems can regulate the number of their ribosomal genes independently of the remainder of the genome, but it is not an exclusive property of this kind of gene. Studies of increased drug resistance in both prokaryotes and eukaryotes have revealed that the phenomenon is accompanied by increases, through duplication, in the genes responsible for the resistance. The larger question of gene amplification is dealt with at greater length on p. 337.

Before leaving consideration of the nucleolar organizer and ribosome formation, we should mention the fact that the larger subunit of ribosome contains a piece of 5S rRNA in addition to the 28S segment. The anucleolate mutant of *Xenopus,* for example, shows no deficiency of 5S rRNA despite its inability to form the 18S and 28S rRNAs. The genes coding for the several kinds of 5S for rRNA are, therefore, located elsewhere in the genome, and *in situ* hybridization studies show that the 5S rDNA of *Xenopus* is located at the ends of at least 15 of the 18 chromosomes of the haploid complement. This

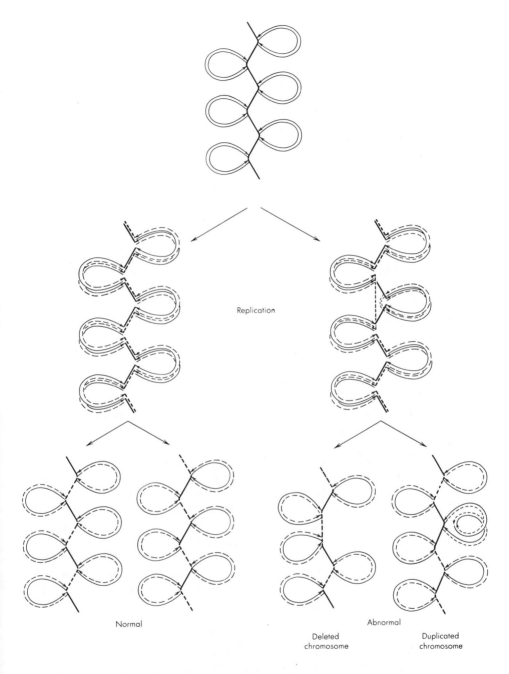

Fig. 3.31. Normal and abnormal DNA replication during the S period of cell division. When abnormal, particular regions in the chromosome can be deleted or duplicated; if this were to occur among repeated tandem sequences of DNA their number could be increased or decreased (after Keyl, 1965).

Replication

Normal

Abnormal

Deleted
chromosome

Duplicated
chromosome

is in contrast to the situation in *E. coli* and in maize chloroplasts in which the 5S rDNA is adjacent to the segment coding for the 23S rRNA (Fig. 2.20). Additional studies show that *X. laevis* and *X. mulleri,* while producing 5S rRNAs of approximately the same length and nucleotide content, have spacers of very different length and nucleotide sequence as well as numbers of 5S rRNA-producing sites. As Table 3.5 indicates, the percentage of total DNA related to 5S rRNA production is the same for both species, 0.7%, but the longer spacer regions of *mulleri* mean that there are fewer copies of the genes coding for the major oocyte 5S rRNA: 9,000 in *mulleri* to 24,000 in *laevis.*

In contrast to the 18S and 28S ribosomal genes which, within each species, are highly similar to each other in nucleotide sequence, the 5S DNAs within a single species show a good deal of nucleotide heterogeneity. This is revealed by the fact that previously dissociated *X. laevis* 5S DNA does not renature to produce perfect duplexes; only about 70% to 80% of the expected fit is obtained, and since most of the 5S DNA is spacer material, it is likely that it is here that the nucleotide heterogeneity exists. Based on the fact that 1.5% nucleotide mismatching leads to a drop of 1°C in melting temperature, it would appear that there is approximately 5% intraspecific mismatching. This indicates that there is a reduced selection pressure for 5S DNA homogeneity as compared to that for the 18S and 28S rDNAs, and it may well be that the mechanism that preserves genetic homogeneity in a nucleolar organizer region is inadequate to cope with the random mutational changes when the numerous 5S genes are distributed among so many different chromosomal sites.

The two species of *Xenopus* each produce several distinct kinds of 5S rRNA (Table 3.5). Whether, as species, they are unique in this respect is not known. In *X. laevis,* somatic cells produce a 5S rRNA that is 120 nucleotides long and that differs from comparable mammalian 5S rRNAs by 8 nucleotides. Two 5S rRNAs are also produced in the early oocytes of *Xenopus,* the major one of which differs from the somatic type by 7 nucleotides. The two genes coding for oocyte rRNA are distinguishable from each other by 5 or 6 nucleotide changes in the spacer region, thus emphasizing the heterogeneity that exists among the 5S DNA sites. Questions arise as to how many of each kind of 5S RNA genes exist and how these are distributed among the *Xenopus* chromosomes. Since there are 450 ribosomal genes producing 18S and 28S rRNA in the nucleolar organizer region, there would presumably be a need for only 450 5S genes of the somatic type to meet the demands of somatic cells for ribosome formation; so it may well be that most of the *Xenopus* 5S genes are of the oocyte variety, being active only when the great needs of early embryonic development must be met. We shall return again to this question when the problem of gene amplification is discussed in a later chapter (p. 337).

Chromomeres. The hereditary material of both prokaryotes and eukaryotes is too long to be conveniently contained within the cell without some form of packaging to reduce its length, to increase its maneuverability during cell division, and presumably also to control its genetic activity. Among the prokaryotes, the unitary chromosome is folded into loops numbering, in *E. coli*, between 12 and 80, and with each loop supercoiled as in Fig. 2.2. There is some protein, mostly RNA polymerase, associated with the folded chromosome, but the folded nature is maintained by and is dependent upon the presence of RNA as evidenced by the fact that unfolding occurs when the chromosome is exposed to RNase.

As previously explained, the basic structural organization of the eukaryotic chromosome involves the nucleosomes which are made up of DNA and histones to give an approximate 1:1 weight ratio of DNA:histone. This leads to a packing ratio of about 6:1, but each kind of chromosome, as it prepares to go through its particular cell cycle maneuvering, undergoes further packing and/or coiling to reduce the overall length, as in a somatic or a compacted meiotic chromosome, or is extended to different degrees, as in the diplotene lampbrush chromosome of an amphibian oocyte, or the polytene chromosome of a dipteran salivary gland cell. Several stages of compaction of a human metaphase chromosome have been recognized, until at full metaphase a compaction ratio of approximately 5,000:1 is achieved. As discussed earlier, the four major histones are responsible for the first order of compaction into nucleosomes, and histone H1 carries the process forward into a second order of compaction. The mechanisms for the latter stages of compaction are unknown.

The packing, however, is not random along the length of the chromosome, for depending upon the species, the type of cell, and the stage of cell division, the chromatin can undergo localized compaction to produce characteristic patterns of differentiation visible at the level of light microscopy. These are the *chromomeres*, a term which includes such varied structures as those found in early prophase of lily microsporocytes and grasshopper spermatocytes (Fig. 3.4), the knobs of maize and teosinte chromosomes at meiotic prophase (Fig. 3.32), the aggregate masses of chromatin from which the loops of lampbrush chromosomes arise (Fig. 3.33), the heavy telomeres of tomato chromosomes (Fig. 3.3), or the beaded structures,

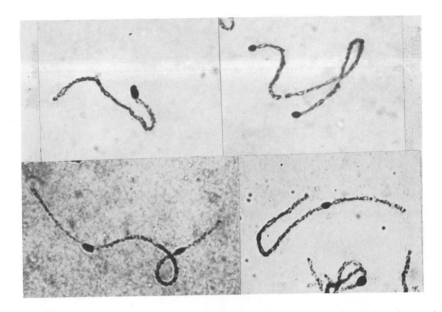

Fig. 3.32. Knobs of maize and teosinte meiotic chromosomes. Top left, chromosome 7, Guatemalan teosinte; top right, chromosome 4, Guatemalan teosinte; bottom left, chromosome 2, Mexican teosinte; bottom right, chromosome 3, maize (courtesy of Dr. T. A Kato Y.)

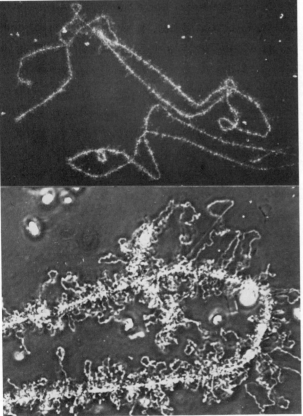

Fig. 3.33. Lampbrush chromosomes from the oocyte of an amphibian. Top, two of the very long chromosome pairs photographed at a relatively low magnification; the fuzziness of the strands of chromatin is due to the formation of lateral loops which arise from the individual chromomeres. Bottom, the lateral loops at a higher magnification (courtesy of J. Gall).

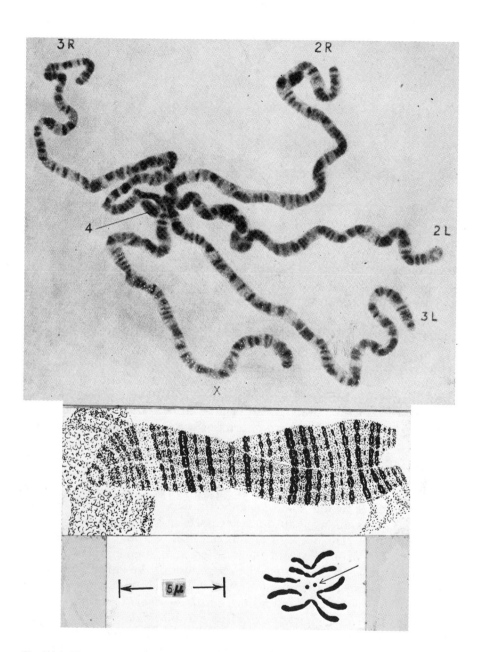

Fig. 3.34. The polytene chromosomes from a salivary gland cell of *D. melanogaster* (top), an enlarged drawing of the small chromosome 4 (middle), and, for comparative purposes, a drawing of the metaphase chromosomes from a ganglionic cell (bottom, with arrow pointing to chromosome 4). The top figure shows the X chromosome, the arms of the two autosomes (2L, 2R, 3L, and 3R), and chromosome 4, linked together at the chromocenter, the latter formed by an aggregation of the centric heterochromatic regions of all of the chromosomes. The diploid number of chromosomes is present, but the homologues are in intimate synapsis (top, courtesy, B. P. Kaufmann; middle and bottom, from Bridges, 1935).

which, coalescing laterally, form the bands of polytene chromosomes (Fig. 3.34). The progress of continuous compaction of the chromosomes through both mitotic and meiotic cell divisions eventually obliterates the chromomere pattern.

The size, number, and position of chromomeres are so consistent for a given chromosome at a given stage of the cell cycle that they can be regarded as a genetic expression of some underlying chromosomal organization, the nature of which is not yet understood but which very probably depends upon some specific kinds of DNA-protein or protein-protein interaction. The cytogenetic usefulness of the chromomeres, quite apart from their intrinsic nature, lies in the fact that in the few organisms that have been carefully studied they provide a series of convenient and recognizable landmarks within the chromosomal map of a genome.

Euchromatin and Heterochromatin. The expression of a linear differentiation of chromatin just described—centromeres, telomeres, chromomeres, and nucleolar organizers—involves small, discrete, and generally limited areas of the chromosome, readily identifiable, at the level of light microscopy and with proper staining, in terms of position, structure and/or behavior (nucleosomes are not included here since they, at a much lower level of organization, are basic to all types of chromatin). The contracted state of the chromosomes during cell division brings these structures into more obvious prominence, but throughout the cell cycle, in most cells and in virtually all eukaryotic species, another and more evident pattern of linear differentiation is discernible in that large blocks of chromatin, and in some instances, whole chromosomes, show a greater degree of compactness and staining than do other parts of the genome (Figs. 3.3, 3.4, and 3.35). The deeply stained regions of chromosomes are made up of *heterochromatin,* and their differential stainability is referred to as *positive heteropycnosis.* The remainder of the chromatin, which stains more lightly with some stains and which undergoes a regular cycle of contraction and extension, consists of *euchromatin.*

A further cytological distinction can be made in that there are two kinds of heterochromatin: *constitutive* and *facultative.* Constitutive heterochromatin is a permanent element of the genome, may involve part or all of a chromosome, and is heteropycnotic in most types of cells. So far as is known, constitutive heterochromatin is not convertible or revertible to a euchromatic status in a genetic sense, although in such rapidly dividing nuclei as are found in the early phases of embryogenesis of *D. melanogaster,* the deeply staining and compacted nature of heterochromatin is not readily

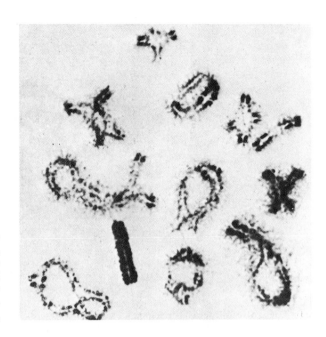

Fig. 3.35. A prophase stage of a spermatocyte of the grasshopper, *Schistocerca gregaria,* showing the highly compacted state of the X chromosome, and the variously compacted states of the autosomes (Tjio and Levan, 1954).

identifiable cytologically. To speak of reversion is to imply that heterochromatin is derived from euchromatin; this is obviously the case for facultative heterochromatin, and, in evolutionary terms, it is undoubtedly so for the constitutive kind as well. Facultative heterochromatin, on the other hand, is not a permanent part of the genome, and it consists of euchromatin which becomes heteropycnotic, that is, it takes on the staining and compactness characteristics of heterochromatin. This generally occurs on a regular basis during some phase of development, and it generally involves whole chromosomes rather than selected parts of them. Since compacted chromatin is depressed in its transcribing ability, the formation of facultative heterochromatin appears to be a way of removing portions of the genome from participating in the metabolism of the cell.

Constitutive heterochromatin can be characterized in a variety of ways. Cytologically, and in addition to its heteropycnosis, it reacts preferentially to certain fluorescent dyes and to some of the treatments preceding Giemsa staining. The basis of this selectivity is not known, but it must reside in a biochemical difference present in either the DNA or the protein of the chromatin, or in some aspect of their interaction. The Y chromosome of humans, for example, reveals an intense fluorescence in the distal half of its long arm when stained with quinacrine mustard; a short region near the centromere of chromosome 3 and the short arms of chromosomes 13, 14, and 15 behave similarly. Constitutive heterochromatin can also show a

107

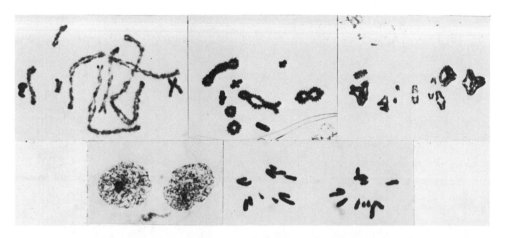

Fig. 3.36. The changing pycnosis (allocycly) of the X chromosome of the grasshopper, *Chorthippus parallelus.* Top row left, positive pycnosis in pachytene (X at left of figure); middle, isopycnosis in late prophase (X at bottom of figure); right, fuzziness, or somewhat negative pycnosis at metaphase (X at left of figure). Bottom row left, positive pycnosis in meiotic interphase; right, isopycnosis in second meiotic anaphase (courtesy of D. J. L. Walters).

negative as well as a positive heteropycnosis, and this behavior can take place in specific cells of a given individual. Thus, the spermatogonial, pre-meiotic metaphases of a number of grasshoppers and locusts show the X chromosome, and in specific instances an autosome as well, to be negatively heteropycnotic while the same chromosomes in the following meiotic divisions will show a variable behavior, or *allocycly.* The X chromosome, positively heteropycnotic in early prophase, becomes *isopycnotic* in late prophase, fuzzy, or even pale, in appearance in metaphase and anaphase, returns to a positive heteropycnosis in interphase, and again to an isopycnosis in the second meiotic division (Fig. 3.36). M. J. D. White has also pointed out that in spermatogonia that, on occasion, are tetraploid, only one of the X chromosomes will be negatively pycnotic while the other behaves as do the autosomes. The cells, obviously, have some process of selection which causes the two identical X chromosomes to behave differently. Another characteristic of constitutive heterochromatin, as well as of facultative heterochromatin, is that it invariably replicates late in the S period.

Much of the detailed information on the nature of constitutive heterochromatin is derived from studies of *D. melanogaster.* In this species, which is representative of most of the other species of *Drosophila,* constitutive heterochromatin is present in all of the four haploid chromsomes, comprising one-third of the mitotic length of the X chromosome (Fig. 3.15), most of the fourth chromosome, all of the Y, and substantial blocks of chromatin on either side of the centromeres of the second and third chromosomes. These blocks of chromatin behave similarly in a number of ways: they are condensed in all stages of the somatic cell cycle, appearing as masses of chromatin lying against the nuclear membrane in interphase (Fig. 3.37); they

Fig. 3.37. A chief cell from the stomach lining of a bat, showing how the heterochromatin comes to lie against the inner membrane of the nucleus, suggesting the "bodyguard" hypothesis put forth by Hsu to explain, in part, the function of heterochromatin (courtesy of Keith Porter).

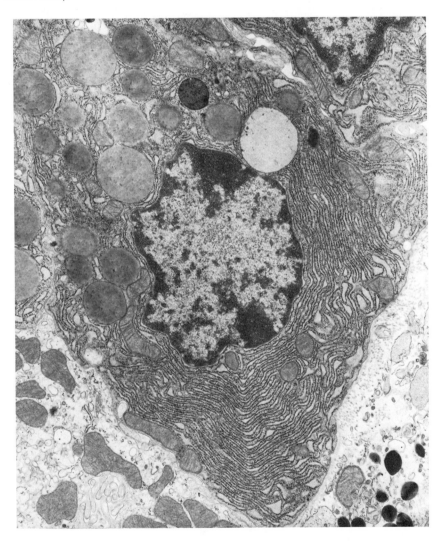

replicate late in the S period, a phenomenon due, in all probability, to their condensed state and to the difficulties of uncoiling as a prelude to replication, a suggestion supported by the observation that the rapidly dividing nuclei in early embryogeny mentioned earlier show no such preferentially condensed masses of chromatin and no comparable pattern of late replication; they show negative pycnosis in the cells of the salivary glands and are underreplicated as compared to the highly polytene nature of the euchromatin; they mass together into a more-or-less structureless *chromocenter* in these same salivary gland cells; they stain similarly with Giemsa stains; and as shall be discussed later, they are the sites of highly repetitive DNA.

Evidence from other species as well as *Drosophila* indicates, however, that there are several kinds of constitutive heterochromatin and that distribution can vary widely as to site and amount. In a great many unrelated species, constitutive heterochromatin tends to be found in blocks adjacent to the centromere, as it is in *Drosophila* (Figs. 3.15 and 3.20), mice, *Oenothera* and Seba's fruit bat (Fig. 3.38), but it can also be at the ends of chromosomes, as it is in tomato (Fig. 3.3), variously positioned throughout the genome, as are the knobs of maize and teosinte (Fig. 3.32), the entire arm of some chromsomes (Fig. 3.39), the Y chromosomes of many heterogametic species such as *Drosophila* (Fig. 3.15) and humans, the B chromosomes of maize, and the microchromosomes of avian species such as the quail, but interestingly enough, not the chicken. It seems highly unlikely that these are all alike as to function and structure.

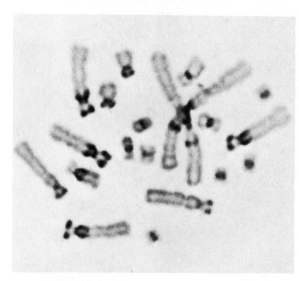

Fig. 3.38. Centric heterochromatin in the somatic metaphase chromosomes of Seba's fruit bat, *Carollia perspicillata*. Some of the short arms of the acrocentric chromosomes are also entirely heterochromatic (courtesy of Dr. T. C. Hsu).

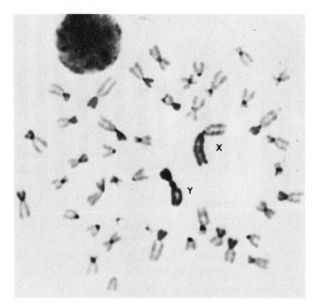

Fig. 3.39. Somatic metaphase chromosomes of the Syrian hamster showing the many arms which are entirely heterochromatic. The Y chromosome is completely heterochromatic, as is one of the arms of the X chromosome (courtesy of Dr. T. C. Hsu).

The seemingly random position of the knobs in different strains of maize would suggest that small blocks of constitutive heterochromatin can be interspersed throughout the euchromatic regions of the genome. The answer would seem to depend upon the techniques used for the identification of heterochromatin and hence how heterochromatin is defined. As evidenced by the formation of chromocenters in *Drosophila,* heterochromatin has a tendency to pair with other regions of similar character and in a nonspecific and nonhomologous manner; this has been termed *ectopic pairing* when it has involved internal regions within euchromatin, and it has been used as a criterion for the existence of interstitial blocks of heterochromatin. Similarly, the interchromosomal variations found in such plants as *Trillium* and *Scilla sibirica,* brought on by prolonged cold treatment, have also been cited as further evidence for the dispersion of constitutive heterochromatin throughout the euchromatic arms (Fig. 3.40). Such cytological evidence, however, has not been supported by more critical data of a genetical nature; the cytological similarities, in fact, tend to obscure the diversity of function that is now known to characterize constitutive heterochromatin from a variety of sources, and the initial belief that such chromatin was genetically inert is now no longer valid. As shall be pointed out in the next section, the heterochromatin from one chromsome to another in *D. melanogaster* can similarly be shown to be chemically distinct as well.

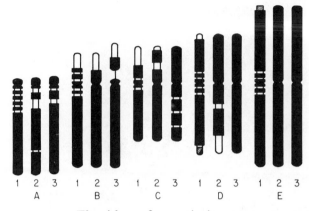

Fig. 3.40. Schematic representation of the size and position of intra-arm heterochromatin in the five haploid chromosomes of *Trillium* when cell division takes place at low temperatures. 1, *T. erectum*; 2, *T. grandiflorum*; 3, *T. undulatum*. Letters indicate the different chromosomes of the haploid set; cross-hatched regions appear only rarely (Wilson and Boothroyd, 1941).

The idea of genetic inertness stemmed from the early observation that individual flies of *D. melanogaster* lacking an entire Y chromosome (XO males), or substantial portions of the centric heterochromatin of the X chromosome, possessed a fully normal phenotype, were as viable as their XY siblings, and revealed no abnormal behavior; they were, however, sterile. A search for genes in the Y was, at that time, unsuccessful. The diversity of function now revealed in constitutive heterochromatin rivals that of euchromatin, but with a good many of the activities being concerned with the events of reproduction. As stated earlier, nucleolar organizer regions of the X and Y chromosomes, located in heterochromatin, behave differently, those in the X chromosome of *D. melanogaster* having amplification or deamplification capabilities, as related to rRNA-producing sites, which those in the Y lack; in *D. hydei*, however, both the X and the Y possess such adjusting mechanisms. In addition, and again in *D. melanogaster*, two closely linked euchromatic genes in the second chromosome, *abnormal oocyte (abo)* and *daughterless (da)*, have been shown to control either the amount, or the nature, of a product produced by the heterochromatin of the X but not of the Y, while the cr^+ gene, located just distal to the cluster of rRNA genes in the X chromosome, can step up the rate of rRNA production in the genes contiguous to it when the nucleolar organizer regions are underrepresented in number or size. The undetermined maternally produced substance is necessary for normal zygotic development which, when absent or in deficient amounts, leads to unisexual male progenies from *da/da* mothers and an excess of females from *abo/abo* mothers. Both deleterious effects are mitigated by extra X or Y heterochromatin. The *abo* and *da* genes produce no morphologically discernible phenotypes, so they behave as two regulatory genes governing two different structural genes in X heterochromatin; whether these loci are related to rRNA-producing sites and to the *bobbed* phenotype remains conjectural.

The relation of heterochromatin to reproductive events involves meiotic and post-meiotic phenomena. In *D. melanogaster,* some regions of the centric heterochromatin are very much concerned with meiotic pairing and disjunction between the X and Y in males, while other regions are uninvolved; the proper maturation of spermatids in this species is also impaired when X heterochromatin is lost through deletion, while recombination between the X chromosomes in females is governed in some way by X heterochromatin. In maize, the levels of recombination, chromosome segregation, and chromosome elimination are affected by the presence of the kind of heterochromatin found in knobs. Thus, abnormal 10 in maize (p. 84) owes its peculiar behavior apparently to the extra knob that it has acquired, while knobs in general cut down on crossing over in their vicinity, although the latter influence may be due simply to the mechanical interference brought on by the presence of the knob rather than to any intrinsic influence of knob heterochromatin.

Constitutive heterochromatin, in addition to the fact that it can control a variety of functions, also contains genes having identifiable phenotypic expressions. The Y chromosome of *D. melanogaster,* for example, has seven separate loci concerned with male fertility, and it has been shown in *D. hydei,* and presumably in *D. melanogaster* as well, that these are actively transcribing just before and after meiosis (Fig. 3.41); much of the Y, however, is genetically silent, being free of mutations induced by chemical agents effective as mutagens in euchromatin. In the second chromosome of *D. melanogaster,* the genes *light* and *rolled* lie in centric heterochromatin, while three additional loci have been identified between them through the use of overlapping deficiencies, and 13 recessive lethals—6 in the right arm and 7 in the left—have been induced in the same region by EMS (ethyl methylsulphonate). It was once thought that the nondisjunctive action of the B chromosome in maize was due to its heterochromatin, but more recent evidence indicates that it is the result of loci in the adjacent euchromatin.

Finally, constitutive heterochromatin is notably low in crossover frequencies as compared to the frequencies in comparable lengths of euchromatin. The left arm of the second chromosome in *D. melanogaster* is approximately 55 map units long. The gene *purple (pr),* 0.4 genetic units from the centromere, is located one-fourth of the distance from the centromere to the end of the chromosome as judged by somatic chromosome lengths. *Black (b),* 6.4 units from the centromere, lies in the middle of the arm. The distal half of the

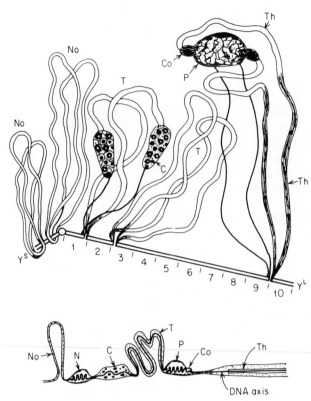

Fig. 3.41. Seven actively synthesizing lampbrush loops that appear from the Y chromosome of *D. hydei* during spermatogenesis (top). Bottom, a linear representation of the sequence as it would be in a single chromatid. The centromere in the bottom figure would lie between C and N. Th = threads; Co = cones; p = pseudonucleolus; T = tubular ribbons; C = clubs; N = nucleolus; No = nooses (redrawn from Hess, 1973; and Henning, et. al., 1973).

chromosome arm, therefore, has a map length of 48.2 units, the proximal half only 6.8 units. Similar discrepancies between crossover frequencies and chromosome length can be cited for the other chromosome arms. It is always possible, of course, that the depression of crossing over in heterochromatin is really a centromeric effect, but it is far more likely to be due to the condensed state of heterochromatin in pachytene which prevents the enzymes involved in crossing over from coming into close contact with the condensed chromatin.

Euchromatin does not ordinarily exhibit the staining characteristics of heterochromatin, nor its general paucity of structural genes, but it is apparent that, under certain circumstances, euchromatin can become heterochromatinized and take on a genetic inertness (Fig. 3.42). Such chromatin is referred to as *facultative heterochromatin*. Thus, during the early development of the human female embryo (and other mammals as well), one of the X chromosomes in each cell of the soma, but not in meiotic cells, undergoes condensation and comes to lie against the nuclear membrane as a deeply staining heterochromatic body, a process called *lyonization* after Mary Lyon who correctly interpreted the phenomenon (Fig. 3.43). Also called sex chromatin, or the *Barr body* after its discoverer, this chromosome

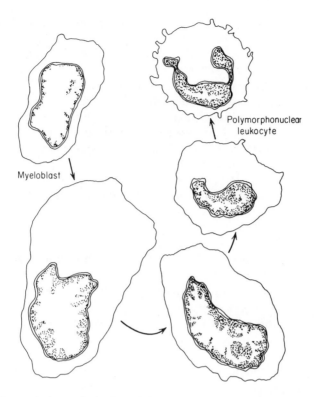

Fig. 3.42. The gradual transformation of the nucleus of a myeloblast in the bone marrow of a rabbit into the lobed structure characteristic of a mature polymorphonuclear leukocyte. The other organelles of the cells have been omitted. As differentiation proceeds, more and more of the euchromatin is condensed against the nuclear envelope and, presumably, is placed in a non-transcribing (non-functional) state. If such cells are to fulfill their differentiated function as well, such condensation must be carried out on a highly selective basis, and hence under genetic control (modified from Bainton and Farquhar, 1966).

Fig. 3.43. Nuclei from cells of the human male (left) and female (right). The Barr body, or sex-chromatin body (arrow), appears only in cells having more than one X chromosome, and it has been identified as the inactivated X chromosome.

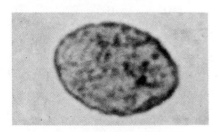
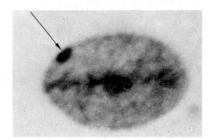

will retain its heterochromatic character for the remainder of its existence, and the structural genes in it will no longer be expressed phenotypically. In the polymorphonuclear neutrophils, one class of white blood cells, and in other white blood cells, the inactivated X is sometimes discernible as the so-called *drumstick*. The particular X chromosome to become heterochromatinized seems to be random if both X's are normal; if one of the X's is seriously abnormal, however, such as a ring chromosome, an isochromosome or one which is chromatin deficient, it is the abnormal one which will form the *Barr body*. This is understandable since cells carrying a deficient chromosome and an inactivated X would be inviable because of the loss of crucial genes. On the other hand, if one of the X chromosomes has been involved in a reciprocal translocation (see p. 476) with an autosome, and is chromosomally heterozygous but phenotypically normal, it is the normal X which appears to be inactivated. The logical argument for this is that lyonization of the X chromatin, now split between two chromosomes, would tend to spread the heterochromatic effect to the adjacent autosomal regions, thereby shutting these genes off and inducing a detrimental or even a lethal effect. The basis, however, may not be preferential inactivation of a particular X chromosome, but rather preferential cell death, that is, certain types of inactivation may be incompatible with cell viability, leading to cell death, and thus leaving only those cells which have an appropriate and active genome. Selection operating at the level of the gene may, therefore, have a bearing on which X chromosomes can be detected in an active or an inactive state when cell samples are being screened. This is made clear by a consideration of two X-linked enzyme mutants: G6PD (glucose-6-phosphate dehydrogenase) and HGPRT (hypoxanthine guanine phosphoribosyl transferase). G6PD exists in two allelic forms distinguishable from each other by their electrophoretic mobility ($G6PD_A$ and $G6PD_B$), and skin fibroblasts taken from females heterozygous for the variants reveal that inactivation of the two X chromosomes is quite random: the two forms of the enzyme, while different, are equally viable, and in cells in which both X's are functional, a heteropolymeric enzyme can be found. The mutant HGPRT produces an enzyme deficiency rather than an enzyme modification, but in heterozygous females no enzyme-deficient cells have been detected in red blood samples. This could result from either a selective inactivation of the X's or a selective cell survival. That it is the latter can be determined from an examination of females heterozygous for both genes. When the normal allele of HGPRT is linked on the same X with $G6PD_B$, both G6PD variants were found in skin fibroblasts but only $G6PD_B$ in red blood cells. A selection of cells in specific tissues, therefore, favored the retention of the red blood cells containing the normal allele of HGPRT, and the elimination at some stage of development of those cells having a mutant HGPRT allele and $G6PD_A$ on the same X chromosome. The $G6PD_B$ allele preferentially survived in red blood cells simply by "hitching a ride" with the normal allele of HGPRT.

In a female mule, leukocyctes show either a horse or a donkey X chromosome to be active, but not both (they are readily distinguishable from each other). However, a strong selective pressure of some sort leads to the horse X being active in nine out of every ten cells. Lyonization of the X chromosomes of cats is also responsible for the tortoise shell pattern or hair color that shows up very rarely in males. Males, which are XY, are generally only black (B) or orange (b), while a heterozygous female shows a mixture of cells producing hair of either black or orange color. The rare tortoise male turns out to be XXY, with the sex-linked genes being heterozygous, one or the other of the X's being inactivated, and with the Y chromosome exerting no influence other than that of determining maleness.

The time of X chromosome inactivation in human females is not known, but a variety of tissues derived from a heterozygous 5-week-old embryo showed that inactivation had already occurred. A 16-week-old fetus had both X's active in its developing oocytes, suggesting that the X's had not been inactivated in the germline, but an equally plausible explanation would be that the onset of meiosis has activated a previously inactivated X.

A more refined system of X chromosome inactivation is operative in the marsupials. In the kangaroo as well as some other species in the group, it is the paternally-derived X that is selectively heterochromatinized in females. A related marsupial, the bandicoot, has evolved a mechanism which has carried the process still further: one X chromosome, presumably the paternal one, is eliminated from female somatic cells, and the Y is eliminated from male somatic cells, thereby ridding the soma of genetically inert structures which, in placental mammals, must be replicated at each cell division. The unwanted chromosomes are extruded through the nuclear membrane during interphase, after which they disintegrate in the cytoplasm.

The selective inactivation of one of the human X chromosomes would predict that there should be females heterozygous for colorblindness who exhibit a bilateral mosaicism, that is, one eye normal and the other eye colorblind. The most likely source of such mosaicism should be among the daughters of colorblind fathers and normal mothers, but none seems to have been identified. The most probable explanation for this lack of mosaicism is that heterochromatinization of one of the X's occurs before the formation of the definitive tissue (anlage) giving rise to the eyes, and that this tissue is uniform in having one or the other of the X's inactivated. If this is so, the daughters of colorblind fathers should reveal a 1:1 ratio of normal to colorblind vision. However, if genetic silencing of one of the X's takes place at a later date, the mosaicism may be so fine-grained, and the cells so spaced, that colorblindness would not be evident.

A somewhat more complicated expression of chromatin inactivation occurs during embryonic development of the male mealy bug, a member of the coccid (lecanoid) group of homopteran insects. Males and females have the same number of chromosomes, but in the males the entire set of chromosomes contributed by the paternal parent becomes heterochromatic at the time when the fifth and sixth

Fig. 3.44. The differential behavior of male and female chromosomes in the mealy bug. The chromosomes of the female behave similarly in all somatic divisions (middle row), while those of the male have half of the chomosomes (those from its paternal parent) become heterochromatinized at an early embryonic stage, and remain so throughout development (top row). In meiosis (bottom), the chromosomes of the spermatocytes separate into euchromatic and heterochromatic sets, and only the euchromatic ones (those from the maternal parent) enter into functional sperm (Brown and Nur, 1964).

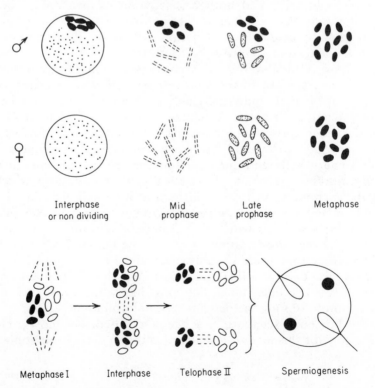

| Interphase or non dividing | Mid prophase | Late prophase | Metaphase |

| Metaphase I | Interphase | Telophase II | Spermiogenesis |

cleavage nuclei migrate to the periphery of the egg; there they clump together as a conspicuous chromocenter, in which form they remain throughout the life cycle of the insect (Fig. 3.44). During spermatogenesis, these chromosomes are selectively segregated into heavily condensed nuclei which eventually disintegrate. The male mealy bug, therefore, can only contribute maternal genes to the next generation. No such differential behavior of maternal and paternal chromosomes is apparent in the female, for condensation of all chromosomes follows a normal course, and their segregation in meiosis seems to conform to Mendelian expectations. In an advanced group of the coccid insects, the armored scales, or Diaspididae, the males, like the bandicoots among the marsupials, simply eliminate the entire paternal set, and hence are haploid for the remainder of the life cycle.

Search for an explanation of facultative heterochromatinization or chromosome elimination must take into account the fact that the event occurs in the presence of a homologue, or a set of homologues, which continues on a genetically normal course. When paternal chromosomes are selectively silenced or eliminated in the early cleavage divisions, it might seem reasonable to assume that these chromosomes entered the egg already preconditioned in some manner and thus scheduled for genetic limbo. However, the fact that these same chromosomes behave differently when in the egg, coupled with the random inactivation of the human X chromosome in females, renders such a suggestion untenable. A plausible but untested explanation has been offered by Sager and Kitchin (Fig. 3.45). This hypothesis suggests that silencing or elimination occurs through the process of enzymatic modification and restriction of DNA. There are specific sequences of DNA, 4 to 8 nucleotides long, which can be recognized and acted upon by both modification and restriction enzymes. In prokaryotes, for example, the modification process, perhaps through enzymatic methylation of DNA, protects these specific sequences from the nuclease action of restriction enzymes. It is presumed that the modifying action occurs for each cell cycle, or that methylation (if this is what protects the DNA) immediately follows upon replication since the restriction enzymes will attack unmodified DNA within these specific sequences in a way similar to the action of the bacterial restriction enzymes (Table 5.1), cut the polynucleotide strands by acting as endonucleases, and open the way for less specific nucleases to further the degradation of DNA. Thus, in prokaryotes, the restriction enzymes are especially effective in destroying foreign DNA while leaving the host DNA intact. The Sager-Kitchin hypothesis proposes that in eukaryotes the modification enzymes act, not to protect the DNA against destruction, but rather to act as precondi-

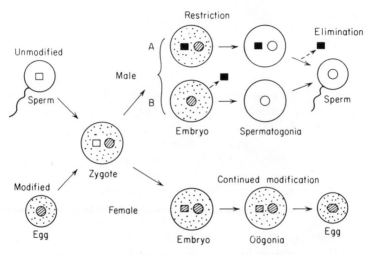

Fig. 3.45. The Sager-Kitchin scheme to account for the selective restriction, elimination and/or modification of some chromosomes in a diploid set. The fertilized egg is assumed to receive a modified set of chromosomes from the maternal parent, and an unmodified set from the paternal parent. Modified chromosomes in the female behave normally throughout development and gamete formation, with the mechanisms of modification active throughout the life cycle. Unmodified chromosomes become heterochromatinized, or restricted, and in the male remain in this condition until spermatogenesis when they are eliminated, as happens in the mealy bug system (Fig. 3.44). Various modifications of this scheme are possible, and can theoretically account for the several types of selective behavior of chromosomes (Sager and Kitchin, 1975, copyright 1975 by the American Association for the Advancement of Science).

tioning mechanisms which will eventually lead to the silencing or elimination of particular DNAs. The random silencing of one X chromosome or a set of chromosomes in the presence of another similar chromosome or set possibly occurs through the sequence of events depicted in Fig. 3.45.

It is apparent, then, that the process of facultative heterochromatinization, however it is determined and achieved, is a means of selective inactivation, or silencing, of certain dispensable genes: condensed chromatin is nontranscribing chromatin at the same time that it retains its replicative ability. In some respects, facultative heterochromatinization is a dosage compensation mechanism; that is, in mammals it is a means of equalizing the somatic cells of both male and female by the removal of one X chromosome while maintaining the sex-determining and reproductive mechanisms intact in the respective germlines. In another sense, the process may well be a prelude to the establishment of a haploid-diploid sex-determination mechanism; that is, the functional silencing of the paternal genome in the male mealy bug may be but a step toward male haploidy, a condition found in some genera of coccids.

That facultative heterochromatinization is a highly regulated and specific process is indicated by the fact that if a piece of an inactivated X chromosome in the mouse becomes attached by translocation to an autosome, that fragment will become genetically inert and condensed along with the remainder of the X chromosome, even while the autosome itself remains genetically functional. However, a dominant autosomal gene translocated to a heterochromatinized X chromosome of a heterozygous mouse may have its action suppressed in some cells, thereby permitting expression of the recessive gene. This means that the process of heterochromatinization creates a condition of transcriptional instability which can alter the expression of neighboring euchromatic genes. There is evidence that the degree of instability can be increased or decreased by selection, pointing to the presence of other genes which can modify the influence which heterochromatin exerts on nearby loci. We shall return to this topic in the final chapter when we discuss the observations that the expression of a gene is based not only on its intrinsic nature but also on the position it occupies in relation to other genes and that some facets of the phenomenon of *position effect* are clearly related to the presence and influence of heterochromatin.

Unique and Repetitive DNA

The previous sections of this chapter have dealt with differentiated units of the chromosome that are visually discernible; it is apparent, then, that a long duplex of helical DNA, composed of a continuous sequence of nucleotide pairs, is capable of varied, but very regular, expression and behavior at the levels of both light and electron microscopes. It is now recognized that visual uniformity masks another kind of differentiation of the DNA, that, namely, which distinguishes between stretches of DNA which code for proteins and which are present generally in single-copy, or unique, editions—these include the so-called structural genes which can be readily mapped in both prokaryotic and eukaryotic systems—and those stretches of DNA which tend to be regularly repetitive in nucleotide character and whose function may possibly be regulatory or even neutral. Repetitive DNA, with the exception of a limited repetition of those genes that code for ribosomal RNA, is relatively unknown in the prokaryotes, but in eukaryotes it varies in amount from about 1% to over 75% of the total genome. It is now clear that each eukaryotic species is characterized by repetitious DNA which varies in frequency of repetitions, length and nucleotide sequence of the repetitive units, location, and, possibly, function. DNA can now be divided into single-copy sequences, middle-repetitive sequences which may have up to 1,000 copies in the genome, and highly repetitive sequences which may be represented in the genome by millions of copies. The latter two categories probably have no sharp limits which separate them absolutely.

The discovery of repetitive sequences of DNA arose out of studies dealing with the reassociation of melted DNA. If DNA is isolated, freed of any bound protein, and raised to a temperature between 70°C and 100°C, the hydrogen bonds between nucleotide pairs are broken and the double helix dissociates into single-stranded stretches of DNA. The higher the G-C content of the DNA, the greater the bonding strength, and the higher the temperature required for melting. If the temperature is now lowered to approximately 60°C, reassociation (also referred to as reannealing or renaturation) of the DNA takes place to reestablish the double-stranded condition by complementary base pairing. It is a process dependent upon a random collision of molecules in solution, and hence it is a function both of the concentration of molecules capable of base pairing and of the time available for reassociation. The reassociation can be either DNA with DNA or DNA with RNA, that is, with that RNA which would be complementary to the DNA from which it was initially transcribed. The degree of reassociation is generally expressed as a C_0t value, which is equal to moles of nucleotides $\times$ seconds/liter. (A C_0t of 1 mole-second/liter results when DNA is reassociated for 1 hr at a concentration of 83 μg/ml, a solution which gives an optical density value of 2.0 at 260 mμ.) Another factor of importance is the length of the DNA fragments, and methods are now available, either through regulated shearing forces or by means of restriction enzymes, for obtaining DNA in fragments of desired length, thus enabling one to deal with molecules of varying lengths, but all of relatively uniform size.

More recently it has been found that some of the repetitive sequences, and particularly those of rather short length, have a buoyant density different from that of the main bulk of genomic DNA and can, therefore, be separated out through density gradient centrifugation. Because they do differ from the bulk of the DNA, these are referred to as *satellite DNAs*. In addition, it is also possible to prepare slides of mitotic, meiotic, or polytene chromosomes, subject them to temperatures which will denature the DNA *in situ,* and then, at a renaturation temperature, expose them to a solution of radioactively labeled single-stranded DNA or RNA of appropriate kind. Since the RNA or DNA in solution, if complementary to the DNA of the prepared slide, will reassociate, the radioactive label will indicate the site at which such complementarity exists. Thus, radioactive 18S rRNA of *D. melanogaster,* if handled in this way, would reassociate with the DNA in the nucleolar organizer regions of both X and Y chromosomes, and the radioactive label would be sharply localized in the centric heterochromatin of these chromosomes.

Mention has already been made that some genes are multiply present in tandem array in eukaryotic systems. Those coding for 18S and 28S rRNAs seem invariably to be associated with the nucleolar organizer regions, while the 5S coding sites are widely distributed in the telomeric regions of most *Xenopus* chromosomes, but they are more narrowly localized in chromosome 1 of humans and in the form of about 160 copies in two bands in the 56EF region of the right arm of chromosome 2 of *D. melanogaster.* One might expect that there would be a numerical ratio between the repeats coding for 5S RNA and those coding for 18S and 28S RNAs since the three rRNAs are present in ribosomes in a fixed ratio, but as Table 3.5 indicates, this is not so, at least in *Xenopus.* The discrepancy in numbers of the rRNAs relates to the fact that in the oocyte the 18S and 28S genes exhibit the phenomenon of gene amplification (p. 337) while the 5S genes do not, making their greater numbers of copies comprehensible. However, the situation in *Xenopus* is even more complicated since it produces three kinds of 5S RNAs. That described in Table 3.5 is the most common in *X. laevis,* but these are transcribed in the oocytes; another 5S RNA, represented by about 2,000 copies, transcribes only in the ovaries, while a third one, with an unknown number of copies, is found in somatic cells and contributes about 10% of the 5S RNAs in the oocytes. How their separate appearances are governed is not known, as are their possible distinctive roles.

Other genes of the middle-repetitive kind such as those coding for rRNA include those which code for transfer RNAs (tRNAs) and those whose mRNA will eventually be translated into histones. The tRNAs, which, theoretically, were thought to be of as many kinds as there are amino acid-determining triplet codes, but which, according to Jukes (see below), now are known to number no more than 54 instead of 61, can be isolated in bulk as a 4S RNA fraction through ultracentrifugation, while the mRNA to be translated into histones is isolated as a 9S RNA fraction.

In *Xenopus,* the DNA complementary to unfractionated tRNA is equal to 0.0015% of the total haploid genome. Since each tRNA molecule consists of about 80 nucleotides, this DNA represents an amount equal to a maximum of about 8,000 cistrons coding for tRNA, and since some 43 families of tRNA sequences are recognizable in *Xenopus* bulk DNA, each detectable family should be represented on the average by about 180 cistrons. In *D. melanogaster*, similar studies have shown that the tRNA cistrons are equal to about 0.015% of the genome, and since evidence indicates that some 92 different genes (56 major ones) are coding for tRNA in this species, each one is represented on the average by 8 copies. When radioactively labeled tRNA is reannealed *in situ* to polytene chromosome preparations, bulk tRNA shows an affinity to all of the chromosomes. Some are more heavily labeled than others, but it appears that the tRNA genes are distributed throughout the genome. However, when specific tRNAs of relatively high purity were used, for example, a lysine-accepting tRNA ($tRNA^{Lys}$), it showed a high affinity for the 56EF locus which is also associated with the 5S RNAs. Also, $tRNA_5^{Lys}$ and a phenylalanine-accepting one ($tRNA_2^{Phe}$), produced a high radioactive count in the 48F-49A region of the second chromosome. The affinity for the 56EF region may possibly be due to 5S RNA contamination, but the relation of the two tRNAs to the 48F-49A regions suggests that at least two of the gene families are contiguous, with each calculated to be respresented by at least a 10-fold repetition.

The evidence from *Xenopus* indicates that the various tRNAs are not uniformly redundant. Each of four leucine-accepting tRNAs are, on average, represented by about 90 repeated sequences, but the two methionine-accepting tRNAs are not identical in numbers of repetitive units, one having 330 copies and the other having only 170.

The triplet code concept suggests that there should be 61 different DNA sequences in every genome to code for the 61 possible tRNAs, the remaining three being terminating codons, but, as noted above, the number of tRNAs is less than expected. A variation exists from one species to the next. *E. coli*, for example, contains only 30 to 40 different tRNAs in its genome, each represented singly, while eukaryotic organisms have somewhat more, each reiterated a number of times. The absence of some tRNA sequences from the genome has a possible explanation in Crick's "wobble" hypothesis which states that the first two nucleotides of an mRNA triplet code pair precisely, in an antiparallel fashion, with their complementary nucleotides in a tRNA triplet (anticodon), while the first nucleotide of the anticodon may be "wobbly" in its pairing with the third nucleotide of the mRNA codon (Figs. 3.46 and 3.47), thus making one tRNA serve several purposes. A different explanation which would also bring about a reduction in the need for all of the possible tRNAs is the post-transcriptional modification and maturation of the tRNAs. According to Jukes, the total number of tRNA anticodons would be 54, not 61 (Table 3.7). Thus, unusual bases are often inserted into the tRNAs as a replacement for a normal nucleotide, and there are no known anticodons which have adenine as an initial nucleotide. By the enzymatic action of anticodon deaminase, adenine is converted to hypoxanthine, which in turn serves as the base in the nucleotide inosinic acid (I). Inosine is very similar, chemically, to guanine, and it can pair in complementary fashion quite well with cytosine and

Fig. 3.46. Generalized structure of a tRNA molecule in its cloverleaf form (left), and two specific tRNAs from yeast: tRNA^{ala}, top right, and tRNA^{lys}, bottom right. There is in all tRNAs a U flanking the anticodon on the 5' side, and an A or a modified A (e.g., I) on the 3' side; other similarities are also found among the many tRNAs, regardless of source. The figures in the loops of the generalized model indicate the number of nucleotides contained in each. Modified nucleotides, usually methylated, are indicated by a superscript (+). T = ribothymidine; I = inosine; D - dihydroxyuridine; ψ = pseudouridine; X = unknown nucleotide.

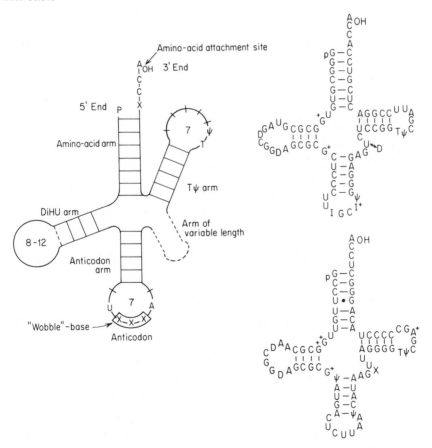

somewhat less so with adenine and uracil (Fig. 3.45). Where there is degeneracy in the code, i.e., proline is coded for by any codon starting with two cytosines (*CCC, CCU, CCA, CCG*), an *IGG* anticodon could pair with the first three and fulfill a normal function. Degeneracy, however, seems to have a role since different tRNAs handling the same amino acid may behave differently in different circumstances. Thus, two lysine-accepting tRNAs insert lysine into different sites in hemoglobin as it is being translated in rabbit reticulocytes, and with only about a 4% overlap. A reduction in the kinds of tRNA genes, while conservative and possibly advantageous from an evolutionary point of view, may also introduce a higher degree of error into the translation process, suggesting that there may be a fair degree of selection against too much "wobble" pairing.

Fig. 3.47. Using the anticodon of tRNA^{ser} in Fig. 3.39, the pairing of inosine with cytosine is normal, but a "wobble" occurs when pairing takes place with either U or A. (Redrawn from Goodenough and Levine, 1974, after Crick, 1966).

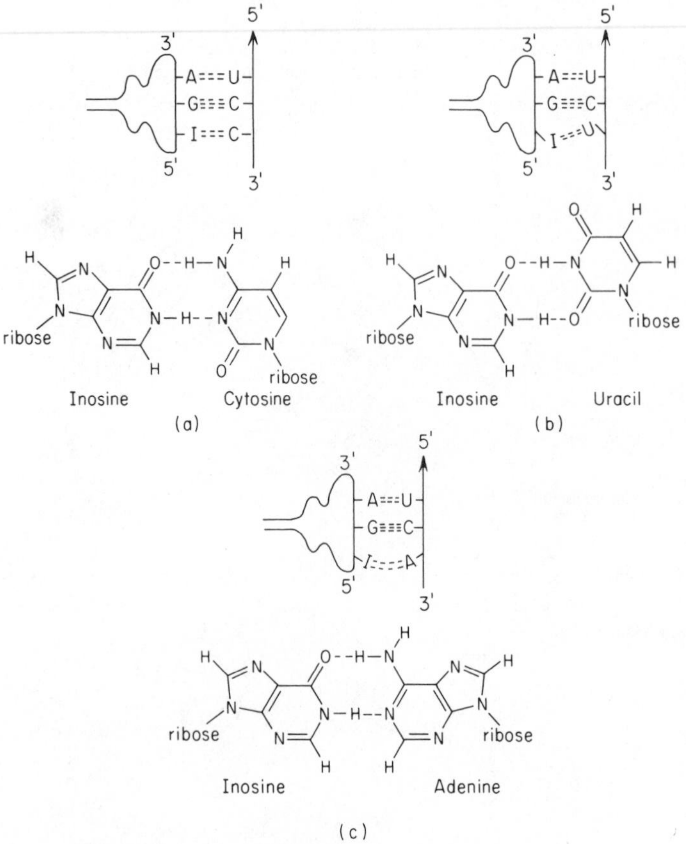

Table 3.7. The 54 anticodons in the tRNAs needed to pair with the 61 possible codons in mRNAs, a reduction resulting from the fact that the base A (adenine) is not found in the first position of the anticodons.

GAA	Phe	*IGC*	Ala
UAA	Leu	GGC	Ala
CAA	Leu	*UGC*	Ala
IAG	Leu	CGC	Ala
GAG	Leu	*GUA*	Tyr
UAG	Leu	*GUG*	His
CAG	Leu	*UUG*	Gln
IAU	Ile	*CUG*	Cln
GAU	Ile	*GUU*	Asn
UAU	Ile	*UUU*	Lys
CAU	Met	*CUU*	Lys
IAC	Val	*GUC*	Asp
GAC	Val	*UUC*	Glu
UAC	Val	CUC	Glu
CAC	Val	*GCA*	Cys
IGA	Ser	*CCA*	Trp
GGA	Ser	*ICG*	Arg
UGA	Ser	GCG	Arg
CGA	Ser	UCG	Arg
IGG	Pro	CCG	Arg
GGG	Pro	*GCU*	Ser
UGG	Pro	*UCU*	Arg
CGG	Pro	GCU	Arg
ICU	Thr	ICC	Gly
GGU	Thr	*GCC*	Gly
UGU	Thr	*UCC*	Gly
CGU	Thr	*CCC*	Gly

Source: Jukes, 1977, © 1977, by American Association for the Advancement of Science.

The structure of a *Xenopus* tRNA gene seems to be similar to that found in rDNA in that each copy of a gene alternates with spacer DNA (Fig. 3.48). The spacer DNA has been judged to be about 11 times longer than the coding sequence. Reassociation techniques also suggest that the spacer sequences of the different tRNAs are of similar size but of different nucleotide composition. Being at different locations in the genome, such tRNA spacer heterogeneity might well be expected to arise through random nucleotide change, as it has in 5S spacer sequences in both *Xenopus* and *Drosophila,* since the spacer has probably a less crucial role to play than does the tRNA gene itself, but until it is known fully what the spacers do, any suggestion can only be provisional.

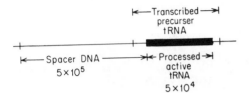

Fig. 3.48. Structure of a tRNA gene with its spacer DNA. The transcribed molecule is larger than the active tRNA, which is processed into an active state post-transcriptionally (redrawn from Clarkson and Birnstiel, 1973).

A suggestion has been made that a class of mutant genes in *D. melanogaster* known as *Minutes* may be due to some kind of deficiency in either the number of sites of a particular tRNA gene, or to its absence. Some 40 or more *Minute* loci are known, randomly distributed throughout the genome, and all show a similar phenotype of a kind which might well be due to a deficient or an inefficient protein synthesis—small size, shortened bristles, reduced wings, and slow development. Many of the *Minutes* are also associated with recognizable deficiencies of chromatin; others, however, are not. The number of *Minutes* is somewhat less than the calculated number of tRNA genes, but the hypothesis is still a reasonable one, although it needs to be pointed out that the number of *Minutes* approximates the number of ribosomal proteins, and it may well be that impaired ribosomal activity resulting from an absent or malformed protein could lead to impaired synthesis, and hence to the *Minute* phenotype.

A 9S RNA, translatable into histones, can be resolved into 5 major components, corresponding to the coding sequences for the 5 major histones. In *D. melanogaster* all reassociate with about 12 chromomeres in the 39D-E region in the left arm of the second

Fig. 3.49. The sequence of histone genes in *D. melanogaster* (a) and in the sea urchin (b). The direction of transcription is indicated by the pointed end of each gene, and the number of nucleotides in both genes and spacers is given for the *D. melanogaster* unit. It is clear that while the nucleotide sequence of each of the histones is rigidly conserved, their serial order in each repeated unit in the chromosome is not. It has been estimated that it would take five breaks and rearrangements of chromatin to derive *a* from *b*, or vice versa; the insertion of 270 base pairs into the spacer between H1 and H3 of the sea urchin could have occurred some time after the two forms separated from their common ancestor, i.e., some 600×10^6 years ago (after Birnstiel, et. al., 1978; and others).

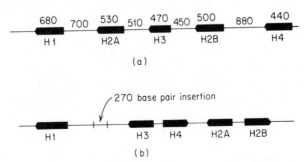

chromosome. The number of coding sequences in *melanogaster* is about 100 as compared to approximately 1,000 cistrons in the sea urchin genome, with each coding sequence containing the loci for all 5 histones, but the 2 organisms differ as to the serial relation of the loci, one to the other, as to the direction of transcription, and as to the variant forms that make their appearance during development (Fig. 3.49). *D. melanogaster* has two sequences which differ from each other in that one has a block of 270 base pairs inserted into the spacer between H1 and H3. Of those in the sea urchin, sperm H2B differs from embryonic H2B both in numbers and in kinds of amino acids, and H2B of cleaving eggs is different from that of the gastrula. Gonadal H2A differs similarly from the H2A of somatic cells. In fact, about one-fourth of the reiterated sea urchin sequences code for variant types of histones, indicating considerable nucleotide divergence over time, but a divergence that seemingly has a fixed relation to tissue specificity.

The large amount of histone needed to maintain the nucleosomic organization of the chromosome, and the demand for histone by newly synthesized DNA during the S period, makes the redundancy of the histone genes logical, as does the fact that they transcribe together even though as five separate loci in each sequence; equimolar ratios of all histones are thereby maintained.

Simple-Sequence, or Highly Repetitive, DNA. As techniques for the handling of DNA improved, it became apparent, as a result of density gradient centrifugation, that isolated DNA is not of uniform buoyant density and that various fractions would come to equilibrium at various levels in a density gradient according to their nucleotide content; G-C pairs are heavier than A-T pairs and, therefore, stretches of DNA rich in A-T or G-C pairs can be readily separated from each other. These peaks (Fig. 3.50) consist of a major one representing the bulk of the genomic DNA and various satellite peaks

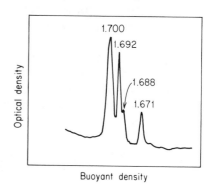

Fig. 3.50. The buoyant density patterns obtained when DNA from brains and imaginal discs of *D. virilis* was centrifuged to equilibrium in neutral CsCl. The main peak at 1.700 is largely made up of unique, or single copy, DNA, while the satellite peaks of 1.692, 1.688, and 1.671 consist of highly repetitive, simple-sequence DNA which is concentrated in constitutive heterochromatin. The different satellite peaks result from repetitive sequences of similar length, but which differ in G + T content and hence in molecular weight (redrawn from Gall, et. al., 1973).

which analysis has shown to consist of what has been termed *simple-sequence DNA.* Forty-one percent of the genome of *D. virilis* and a lesser percentage of that of *D. melanogaster* consist of simple-sequence DNA. That of *virilis* is made up of three kinds of hepta-nucleotides, the light strands of which have the following sequences and sedimentation peaks (given in grams of DNA/CC) when centri-fuged in neutral CsCl:

1.692 peak	5′ · · · · –A–C–A–A–A–C–T · · · · –3′
1.688 peak	5′ · · · · –A–T–A–A–A–C–T · · · · –3′
1.671 peak	5′ · · · · –A–C–A–A–A–T–T · · · · –3′

These sequences are obviously related to each other, and their physical characteristics are due to weight changes within very short stretches of DNA. Being as short as they are, yet constituting 41% of the *virilis* genome, each of them must be present as several million copies in the haploid set of chromosomes. Reassociation techniques show that all three are in the form of long homogeneous blocks of repeats in the centric heterochromatin in all of the chromosomes, although binding *in situ* to the heterochromatin in the Y chromosome is low. No satellite is confined to any single chromosome.

Simple-sequence DNA in *D. melanogaster* is smaller in amount than in *D. virilis,* comprising about 20% to 25% of the genome. At least six different satellite DNAs have been identified, and, as in *virilis,* they are A-T rich but variable in length. Of these, the 1.672 peak satellite is a mixture of five and seven base pair repeats (5′ · · · –AATAT · · · · –3′ and 5′ · · · · –AATATAT · · · –3′), consists of between 1.0 and 1.5 × 10⁶ copies, and is found in the centric hetero-chromatin of all chromosomes, but it is most evident in the Y and fourth chromosomes. The 1.686 satellite is made up of nearly one million repeated units of 10 base pairs, two of which are G-C pairs; it is found in all chromosomes, but most prominently in the Y, second, and third chromosomes. The other satellites exhibit their own char-acteristic patterns of structure, size, and distribution. A comparison of the *melanogaster* satellites with those of its sibling species, *D. simulans,* shows that they share some satellites while possessing some that are species-specific, but the most significant change in satellite DNA between these two very similar species is in the repeat length, not in the nucleotide content; both within and between species the unit repeat is as conservative in nucleotide content as is euchromatin.

The *Drosophila* group of species is not unique, of course, in possessing simple-sequence DNA; virtually all species examined have given evidence of its existence, and the concepts derived from the

Drosophila studies seem, in general, to apply to other species, that is, that the satellite DNAs are A-T rich, are represented by many copies, and are found in heterochromatin. That in the mouse is 7 or 8 base pairs long; that in the guinea pig 6 pairs long. The red-backed salamander, *Plethodon cinereus cinereus,* has repetitive sequences which are about 100 pairs long and make up about 30% of the total genome. In the cereals, rye possesses 2 satellites, 80% of which reassociates with the large blocks of heterochromatin at the ends of the chromosomes; the remainder is scattered over the genome, with some evidence that there are 3 major interstitial sites where concentrations occur. A single satellite has been isolated in hexaploid Chinese Spring wheat, *Triticum aestivum,* with a 5 to 10 base pair repeat, which reassociates with all 7 chromosomes of the B genome and 2 of the A genome (see p. 437 for a discussion of the composite nature of the wheat genome).

The repetitive nature of the DNA of the African green monkey, *Cercopithecus aethiops,* reveals additional facets of simple-sequence chromatin. The repeat unit is larger than usual, being a 172-base pair segment that is repeated several million times in the genome and that constitutes 7% to 10% of the haploid genome. The monkey DNA differs from the shorter, simpler repetitive sequences of *Drosophila* in exhibiting extensive nucleotide diversity within the segments, with certain nucleotide sites showing a greater degree of variance than others in a nonrandom manner. Further, some of the monkey sequences show a striking homology with the genome of the simian virus 40 (SV40), suggesting that SV40 is, possibly, a derived element from the monkey genome, and/or that this centric DNA is the site within which, through recombination, the virus can be integrated into the monkey genome and from which it can be removed.

The function of satellite DNA, if any, is not known. The length of most repeat units in the species which have been thoroughly investigated seems too short to code for any protein of significance or to regulate genetic expression in any controlled manner. The considerable variability in the amount of centric heterochromatin—chromosome 1 in the human is an example—would tend to support this point of view. The effects of the loss of centric heterochromatin from the X chromosome in *D. melanogaster,* however,–upsets in meiosis, fertility, and spermiogenesis in males, and reduced recombination in females–advise caution in dismissing simple-sequence DNA as of no phenotypic importance.

It is of some interest to inquire into the distribution of "centric" heterochromatin when an organism lacks a localized centromere and has instead a diffuse or holokinetic centromere. This has been looked into in the milkweed bug, *Oncopeltus fasciatus.* The total DNA is divided into 10% to 15% of a fast-renaturing, or simple-sequence DNA, 45% of a middle-repetitive, intermediately renaturing type, and 45% of unique sequence DNA. The simple-sequence DNA consists of units about 400 nucleotide pairs long, not necessarily contiguous with each other, and apparently scattered throughout the genome. Are these the blocks of chromatin which contact the spindle and act in a centromere-like fashion during cell division? The neo-centric action of the knob of abnormal chromosome 10 in maize would indicate that this a possibility (p. 84).

The Interspersed Nature of Euchromatin. The short-length, tandemly repeated, rapidly reannealing sequences of DNA concentrated in constitutive heterochromatin, and usually identifiable through centrifugation as satellite DNAs in density gradients, do not represent the majority of the repetitive sequences found in eukaryotic genomes even though these sequences may be represented by a million or more copies. Euchromatin also contains a variety of repetitive sequences, some of which have previously been discussed, i.e., those coding for the several kinds of rRNAs, the tRNAs, and the histones. Others include the *412* and *copia* genes recently discovered in *D. melanogaster,* which code for copious amounts of Poly (A)-containing RNA found in the cytoplasm, and for which no function is yet known. Both of these genes are found at about 30 scattered locations in the euchromatin, and their lengths are between 4,000 and 7,000 nucleotide pairs long. Other euchromatic genes, not generally grouped with repetitive units, are doubly or triply represented as a result of one or more duplications (Figs. 6.10, 6.11, 6.12), but in addition to these varied examples of repeated sequences it is now evident that the euchromatin of eukaryotic species is organized in such a manner that nonrepetitive DNA, i.e., unique, or single-copy, sequences which may or may not code for particular proteins, alternates in a more-or-less regular way with noncoding DNA, repetitive to varying degrees. The latter, making up the bulk of repetitive DNA in most organisms, consists generally of sequences ranging between 200 and 500 nucleotides in length, and includes related families of repeated sequences, each family consisting of a number of similar copies ranging from a few in number to many thousands.

The repeated sequences in euchromatin, unlike those in constitutive heterochromatin, cannot be identified by patterns of late replication, differential contraction or C-banding; their presence is revealed only through reannealing techniques (Fig. 3.51). When isolated DNA is fragmented into units 2,000 to 3,000 nucleotides

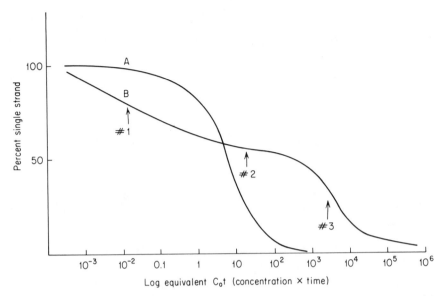

Fig. 3.51. Plot of the time course of single-stranded DNA reannealing to form double-stranded duplexes. Curve A; an idealized curve where every single-stranded fragment has an equal opportunity of re-establishing the duplex nature of DNA through complementary pairing. In the initial state, all strands are single; in the final state, all duplexes possible have formed. Such a curve would be realized with *E. coli* DNA where no repetitive sequences are found, and every fragment is unique. Curve B; a typical curve of a species in which several families of repetitive DNA establish their own patterns of interaction. Arrow #1 points to that portion of the curve where short-length, highly repetitive sequences reanneal rapidly; arrow #2 indicates that portion of the time sequence when moderately repetitive DNA is reannealing; arrow #3 indicates that portion of the curve which is related to single-copy DNA and which follows a time sequence comparable to that in Curve A.

long, denatured into single polynucleotide strands, and then allowed to reanneal under conditions which permit only repetitive sequences to react with each other in a complementary way, the simple-sequence, highly repetitive DNA undergoes rapid reannealing, and can be readily separated from the remainder of the genome by passage through hydroxyapatite columns. The latter DNA, which reanneals more slowly and over a long period of time, is known as *single-copy* DNA; it represents a substantial portion of the euchromatin, making up, as Table 3.8 indicates, approximately 70% of the genome of most animal species, and a variable portion of the genomes of higher plants.

If the 2,000 to 3,000 nucleotide fragments are given sufficient time to interact, most will show some degree of reannealing. Typically, they will form double-stranded structures, terminated at both

Table 3.8. Haploid genome size, percentage of single-copy DNA in the genome, and single-copy complexity as measured in nucleotide pairs $\times 10^8$ (compiled and modified from Bennett and Smith, 1976; Davidson, et. al. 1975; Thompson, 1977–78; Belford, 1979).

Species	Common Name	Genome Size in pg.	% of Single-copy DNA	Single-copy Complexity (n.p. $\times 10^8$)
Animals				
D. melanogaster	fruit fly	0.12	75	0.82
Crassostrea virginica	oyster	0.69	60	3.8
Aurelia aurita	jellyfish	0.73	70	4.7
Strongylocentrotus purpuratus	sea urchin	0.89	75	6.1
Spisula solidissima	surf clam	1.2	75	8.2
Cerebratulus lacteus	worm	1.4	60	7.7
Aplysia california	sea cucumber	1.8	40	10.7
Xenopus laevis	clawed toad	2.7	75	18.5
Limulus polyphemus	horseshoe crab	2.8	70	17.9
Rattus novegicus	Norway rat	3.2	75	22.3
Plants				
Vigna radiata	mung bean	0.48	70	4.0
Spinacia oleracea	spinach	0.76	26	2.07
Gossypium hirsutum	cotton	0.77	68	5.0
Petroselinum sativum	parsley	1.83	30	6.0
Nicotiana tabacum	tobacco	1.93	45	9.0
Avena sativa	oats	4.15	25	11.0
Pisum sativum	garden pea	4.63	30	14.0
Triticum aestivum	wheat	5.4	25	14.0
Hordeum vulgare	barley	5.5	30	16.0
Secale cereale	rye	8.3	25	22.0
Vicia faba	broad bean	12.5	20	26.0

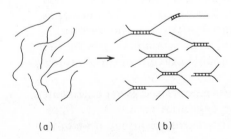

(a)   (b)

Fig. 3.52. A. Single polynucleotide fragments of DNA in solution, with each fragment being 1,000 to 2,000 nucleotides long. B. Duplexes formed when reannealing is permitted to take place. The paired regions are repetitive sequences about 300 nucleotides long, while the unpaired tails and internal sections are the unique sections which have difficulty finding a complementary pairing partner because of their low concentration.

ends by single or double stretches of DNA which are unpaired (Fig. 3.52). The central double-stranded section will be the repetitive element paired with a complementary member of its own family of repeats, while the tails at either or both ends will be the unpaired, nonrepetitive, unique genes, or parts thereof, which would have difficulty in finding complementary sequences. If allowed to reassociate with fragments of DNA that are 5,000 to 7,000 nucleotides long, two or more double-stranded sections may be formed within this length. From information obtained in this manner and from a wide spectrum of species, the organization of euchromatin that emerges is of unique, nonrepetitive coding sequences flanked by noncoding, moderately repetitive units averaging around 300 nucleotides in length. Very few of the repeats are as long as 1,000 nucleotides; a randomly selected piece of euchromatic DNA 2,000 nucleotides long has, therefore, a high probability of containing a repeat of some sort.

This pattern of alternating repetitive and nonrepetitive stretches of DNA is known as the *Xenopus* pattern, named after the genus in which it was first discovered; it is characteristic of most animal and plant euchromatin, even though in higher plants the percentage of single-copy DNA may be much lower than it is in animals (Table 3.8). In the sea urchin, for example, it has been estimated that interspersed among the single-copy sequences there are several thousand families of repeated sequences, some represented by less than a thousand copies each, others by many more copies. In the euchromatin of the rat, a single repeat is followed, in about 60% of the cases, by another repetitive sequence, but only rarely are long stretches of repetitive sequences encountered (Fig. 3.53). As in the sea urchin, a good deal of diversity exists among the repeats in rat euchromatin, but clustered repeats need not necessarily be of the same related family.

In wheat, rye, barley and the broad bean (*Vicia faba*), the percentage of single-copy DNA, with interspersed noncoding repeats, is much lower than it is in the animal group, but the pattern of interspersion is similar: the repeated units, averaging around 300 nucleotides, are grouped into many diverse families, and alternate with the nonrepetitive DNA which extends for 1,000 to 2,000 nucleotides before another repeat occurs. Occasionally, repetitive sequences greater in length than 2,000 nucleotides are found. They constitute only a very small proportion of the interspersed repeats, and in terms of nucleotide content, they seem unrelated to the shorter classes.

In the garden pea, as Murray and Thompson have shown, the degree of interspersion is even more pronounced, with the majority of unique sequences being only 300 to 400 nucleotides long before

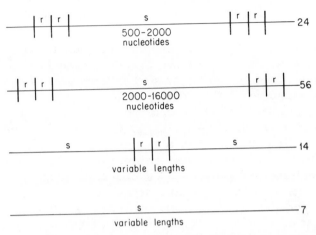

Fig. 3.53. The sequence organization of rat DNA with middle repetitive DNA (r) alternating with single copy DNA (s). The percentage of DNA of various single copy lengths is given by the figures at the right end of each line, but the 2,000 to 16,000 nucleotide single copy sequences are open to question since other investigators have rarely found such stretches uninterrupted by repetitive sequences (after Bonner, 1977).

being interrupted by a repeat unit. If the average gene coding for a protein has a length of 1,000 to 1,200 nucleotides, this means that the repetitive units break the genes into several separate segments as well as separating them one from the other. As shall be discussed in Chapter 5 (p. 349), this seems to be an organization characteristic of many, perhaps most, eukaryotic genes. It has also been estimated that in the total pea genome there are approximately 10 different repetitive elements, grouped into many families, each of which is made up of copies spread over a continuous range from a few hundred to about 10^5.

The *Drosophila* pattern of interspersed sequences differs from that in *Xenopus,* and appears to characterize species with very low amounts of DNA per haploid genome. The interspersed repeats occur at less frequent intervals such that about 90% or more of the fragments over 2,500 nucleotides long show no repeated sequences under conditions favorable for reannealing. Similar patterns are found in the honeybee and the slime mold, *Dictyostelium,* raising the question as to the relation of the two patterns to each other, and if they had a common origin but diverged during their course of evolution.

A further word about the nature of single-copy sequences. As implied earlier, the term is not to be equated exclusively with unique coding sequences. As Tables 3.8 and 3.9 and Fig. 3.54 reveal, the amount of single-copy DNA, calculated in terms of nucleotide pairs,

Table 3.9. Haploid genome size, percentage of single-copy DNA in the genome, and single-copy complexity as measured in nucleotide pairs × 10⁰ in several species of the plant genus Atriplex (family Chenipodiaceae) (modified from Belford, 1979).

Species	Genome Size in pg.	% of Single-copy DNA	Single-copy Complexity (n.p. × 10^8)
A. fruticulosa	3.41	40.1	1.42
A. serenana	3.93	30.4	1.24
A. rosea	3.89	36.7	1.48
A. sabulosa	3.95	38.9	1.59
A. phyllostegia	4.48	41.0	1.91
A. truncata	5.44	38.0	2.14
A. triangularis	6.61	30.8	2.14
A. hortensis	7.03	28.1	2.05

Fig. 3.54. Relation of haploid genome size to the complexity (i.e., diversity) of the single copy component in a variety of plant (x) and animal (o) species (see Table 3.8). Picograms and nucleotide pairs are interconvertible on the basis of 1 pg. = 0.965 × 10^9 nucleotide pairs (data from Bennett and Smith, 1976; Davidson, et. al., 1975; Thompson, 1977-1978).

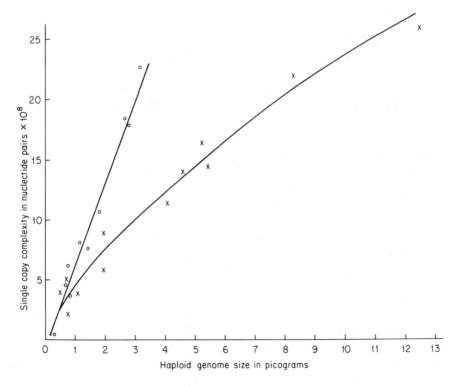

shows a linear relation with the haploid amount of DNA in both plants and animals, although the slopes of the two curves diverge once the amount of single-copy DNA exceeds 10^9 nucleotide pairs. The correlations observed are probably to be expected. Over a period of time random nucleotide changes, as well as chromatin deletions, duplications, and rearrangements, will gradually transform sets of similar repeated sequences into a number of dissimilar, unique units, in other words, shunting blocks of nucleotide pairs from a repetitive into a single-copy category. This topic will be dealt with further when the evolution of the genome is considered (see p. 500).

Different Forms of Chromosomes

As we have seen, the eukaryotic chromosome is a molecule of extraordinary length, and it is equally extraordinary in its varied morphology and genetic potential. Despite its seemingly uniform molecular nature, repeated in the form of nucleotide pairs from one end of the chromosome to the other, it is differentiated in both form and function: each of its parts plays a role in the life of the cell, it can respond genetically in a highly selective and discriminating manner in a variety of cells, and its response to varied cellular states is often reflected, in a distinctive way, through an altered morphology of the chromosome or some of its parts.Thus, the same chromosome will behave and appear differently when in a rapidly dividing somatic tissue such as a roottip from when it is in a sporogenous tissue whose cells are undergoing meiosis; a given dipteran chromosome will have a distinctively different appearance when it is in a salivary gland cell from when it is in an early cleavage stage. Similarly, the same chromosome in a micronucleus as opposed to a macronucleus of a protozoan can take on quite different morphological and functional characteristics. Here we wish to examine several of the forms of chromosomes which the cytogeneticist has found useful.

Somatic Metaphase Chromosomes. The chromosomes of the roottip cells of the broad bean, *Vicia faba,* those from the haploid microspores of *Trillium sp.* or *Tradescantia paludosa,* and those from the cultured cells of a variety of mammals, including the human, have been used extensively for experimental purposes because of their ready availability, large size, small number per nucleus, distinctive characteristics and, in the case of the human, obvious intrinsic interest (Figs. 3.9, 3.10, and 3.55). The chromosomes of these species, prepared by a variety of techniques, and stained with Feulgen,

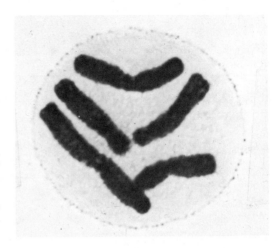

Fig. 3.55. The five haploid chromosomes from a microspore of *Trillium erectum*. The longest one is nearly 30 μm in length; the location of each centromere is indicated by a constriction (courtesy of Dr. A. H. Sparrow).

orcein, or acetocarmine, have provided a great deal of quantitative as well as qualitative data on the effects of various radiation and chemical agents, and hence have proven to be most useful in screening procedures to test for the cytological effects of noxious treatments or drugs suspected of inflicting genetic damage. The somatic metaphase chromosome is also that one customarily used in cytotaxonomic studies, and, more recently, for pre- and post-natal genetic evaluations in humans, with the karyological analyses including length and number of chromosomes, position of centromeres and secondary constrictions, and any other topographical peculiarities which could serve as distinctive landmarks for identification (Fig. 3.10). The constancy of somatic chromosomes from one cell to the next makes such comparative studies feasible.

As observed through the light microscope, the metaphase chromosome is a highly compacted structure, the compaction taking place through the progressive development during prophase of a series of coils (Tables 3.10 and 3.11). From earliest prophase to

Table 3.10. Progressive decrease in the number of coils (gyres) in the haploid set of chromosomes in the microspores of *Trillium* as a function of the stage of division (total chromatid length in micrometers is also given).

Stage of Division	Number of Coils	Total Length, μm
Earliest prophase	ca. 600	ca. 600
Early prophase	554	346
Mid-prophase	259	203
Late prophase	169	156
Metaphase	120	90
Anaphase	130	95

Source: A. H. Sparrow, *Can. J. Res., D20* (1942), 257–66.

Table 3.11. Compaction of DNA into a metaphase chromosome based on calculations made on chromosome 1 of humans.

Stages of Compaction	Diameter of Linear Unit	Physical Appearance	Degree of Compaction
I DNA double helix	30 Å		—
II Nucleosomes	100 Å–125 Å (see Fig. 3.2)		7-fold
III Solenoid	300 Å		6.28-fold
IV Tube (minor coil?)	2,000 Å		17.8-fold
V Final (major) coil	6,000 Å (see Fig. 4.32)		6.28-fold

Total compaction: 4914-fold.

DNA in chromosome 1 of humans: 17,199 μm in length of extended DNA, consisting of 8.6×10^7 base pairs, and measuring about 3.5 μm at mitotic metaphase.

Sedat and Manuelidis also consider two other models of contraction, based on somewhat different assumptions, and these give final compaction figures of 5,176-fold and 8,400-fold.

Source: Based in part on data and redrawn figures from Sedat and Manuelidis (1977).

metaphase more than a 6-fold reduction in length has occurred (600 μm to 90 μm), but as shall be pointed out below, this reduction is but the last in a series of contraction stages.

The 6-fold contraction factor is insufficient to convert the extended DNA molecule in each chromosome into its metaphase dimensions. The haploid genome of *Trillium luteum* ($n = 5$), for example, contains 1.3×10^{11} nucleotides, or about 13×10^9 nucleotides per chromosome. Each extended double helix of DNA is, therefore, several meters in length, which from late interphase to metaphase is compacted into a unit about 20 μm long. Other levels of compaction below the limits of the visible coils must be involved in order to achieve this degree of reduction. Sedat and Manuelidis have considered this problem as it relates to the three-dimensional nature of the somatic human chromosome, making use of a variety of light and electron microscopic methods in arriving at their conclusions (Table 3.11). As discussed earlier, the formation of the nucleosomes is the first order of compaction, transforming a 30-Å diameter duplex DNA molecule into a sequence of nucleosomes, each having a diameter of 100 Å, and bringing about a 7-fold reduction in length. The four major histones—H2A, H2B, H3, and H4—are instrumental in achieving this step (Fig. 3.2). Histone H1, occupying a site in the linker region between nucleosomes, and tied in possibly with other linkage elements such as Ca^{++}, has been implicated in bringing on the next order of length reduction from the 100-Å diameter nucleosomes to a coiled helix 300 Å in outside diameter. The coil, or *solenoid,* includes about six nucleosomes per turn and induces a contraction factor of 6.0 to 12.5, depending upon the assumptions made and the model used. The solenoid coils would be well below the limits of light microscopy, but the transformation of the solenoid to the tube stage increases the diameter of the chromosome to about 2,000 Å, and brings evidence of a series of coils close to the limits of resolution of the light microscope. A major reduction in length would occur, with a reduction factor ranging from 9.42-fold to 40-fold being possible. The next and final stage of compaction would bring an additional 5.0-fold to 6.28-fold reduction. Whether the final pattern of coiling, identified in Table 3.11, apparently corresponds to the *major coils* identified by the early chromosome cytologists (see Fig. 4.30) or whether the previous stage corresponds to what were once called *minor coils* is not certain. If these figures are put in terms of chromosome 1 of humans, its approximately 10^8 nucleotide pairs have been reduced to a length of about 10 μm at metaphase, and a total reduction factor of between 5,000 and 8,000 would be achieved.

The transformation of a chromosome, therefore, from interphase to metaphase is the imposition of a sequential series of helical coils upon a previously extended strand of duplex DNA, with the diameter of the coils increasing significantly at each step. The first step in the process, that of forming the nucleosomes, is most clearly understood; beyond that very little is known for certain as to how the regularity of order is initiated, developed, and maintained. Acidic proteins, and probably histones as well, are undoubtedly involved at all levels above that of the nucleosomes, since removal of the histones from a metaphase chromosome transforms it into a central structure (or core, or scaffold) held together by the acidic proteins and surrounded by a halo of DNA loops, each about 10 μm to 30 μm in length (Fig. 3.56). However, it seems clear that two kinds of controls are operative: that functioning at the level of the nucleus as a whole, since all of the chromosomes are compacted at the same rate, and that functioning locally to define the more obvious structural features of the chromosomes such as knobs, chromomeres, and constrictions, as well as the banded structures revealed through the use of special staining techniques (see later). In the region of the centromere, for example, and possibly also where the nucleolar organizers are located, the final stage of coiling does not take place, leaving a constricted aspect to the chromosome, and thus defining

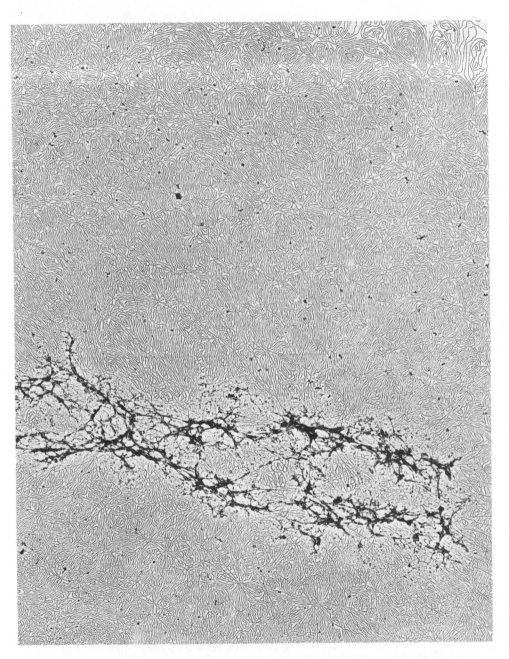

Fig. 3.56a. Electron micrographs of metaphase chromosomes from human Hela cells which have had their histones removed by exposure to dextran sulfate and heparin, or to 2M NaCl, and then spread on a grid for observation. A: the chromosome consists of a central core of nonhistone (scaffolding) proteins running the length of each chromatid, and from which loops of DNA, free of nucleosomes, extend outward to form a dense halo of strands. The intricate patterns of strands presumably reflect previous patterns of packing as the chromosome underwent prophase condensation, with some kind of molecular crosslinking taking place to form swirled configurations.

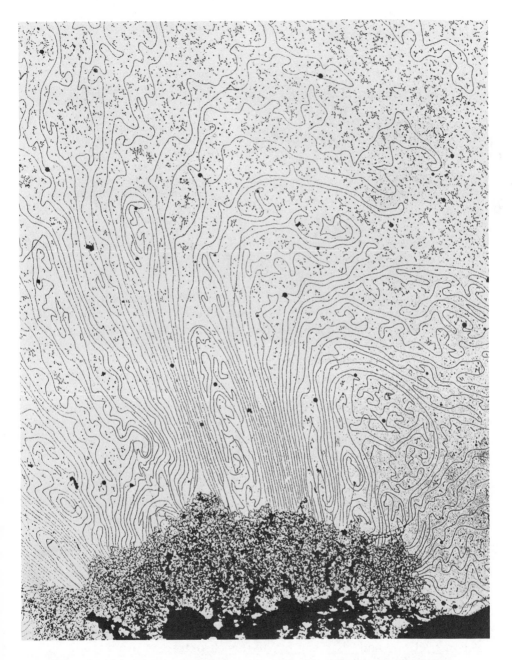

Fig. 3.56b. A higher magnification of a portion of a more relaxed chromosome. Most of the loops fall into a 10 μm to 30 μm range, but some are shorter while others may be nearly 50 μm in length. Both ends of each loop can be traced back to adjacent points within the scaffolding of the chromatid, and adjacent loops are presumed to be continuous with each other within the scaffolding. The scaffold can be isolated independently of the histone-depleted DNA, so it has a structural identity of its own (Paulson and Laemmli, 1977; copyright, 1977, by MIT Press).

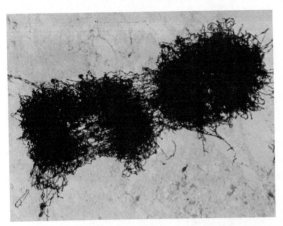

Fig. 3.57. Two metaphase chromosomes isolated from disrupted cells of a bovine tissue culture, spread on a Langmuir trough, and dehydrated prior to being photographed in the electron microscope. The chromosome at the right is divided into its two chromatids. The basic strand is about 250 to 280 Å wide and appears to be quite irregularly folded rather than coiled, but its nature is not yet certain particularly since the same kinds of strands seem to connect the two chromosomes to each other. S. L. Wolfe, *Exp. Cell. Res.*, 37 (1965), 45–53.

the two arms of the chromosome. In any event, the unit of coiling is the chromatid and the duplex strand of DNA found within it, not the chromosome as a whole. The independence of sister chromatids in coiling is indicated by the fact that while they contract at the same rate, they may, in fact, have their coils randomly oriented either in a left- or right-handed direction.

Figure 3.57 would suggest that not all chromosomes contract by means of the coiling process just described. Although the method of preparation undoubtedly determines the appearance of the extracted chromosomes and chromatids, those of some species resemble more a ball of yarn than a regularly coiled structure such as a door spring, with the double helix of each chromatid extended out from, and drawn back into, a central core, thus creating a fuzzy appearance instead of a regularly ordered coil. The mechanics of cell division, however, demand that the chromosome, whether coiled or folded, must be greatly compacted if it is to participate normally during the cell cycle.

Fig. 3.9, a roottip cell of the broad bean at metaphase, reveals the salient features of the somatic metaphase chromosome. Any chromomeric pattern or chromocentric blocks of heterochromatin have been largely obliterated by the contracting process, leaving as landmarks only the position of the centromeres as primary constrictions, the position of the nucleolar organizer region as a secondary constriction, and the lengths of the chromosome arms relative to each other. The latter point is made because, from cell to cell and due to variables of fixation, the chromosomes may not always have the same total length. They will, however, retain a constant length ratio relative to each other, and this holds both for whole chromosomes and for chromosome arms.

The secondary constrictions show no particular structure at somatic metaphase; they are simply regions of the chromsome where the earlier presence of the nucleolus prevented the normal coiling process from taking place. Whether the primary constriction is caused by the presence of the kinetochore plate interfering with the coiling process is not known, but it is the region to which the half-spindle fibers of microtubules are attached. Chromosomes possessing diffuse centromeres also lack primary constrictions, and the spindle fibers engage the entire face of each chromatid, causing them to separate from each other in parallel fashion (Fig. 3.13).

The region of the chromosome distal to the secondary constriction is referred to as a *satellite.* Whether it is small or large depends upon the position of the nucleolar organizer regions relative to the end of the chromosome, but when very small the satellite may often be drawn into the main body of the chromosome and thus obscured from view. This is generally the case for the tiny satellite of *Tradescantia paludosa* which is rarely visible except in pollen tube mitoses, but such nucleolar organizer positions, as well as those terminally located, as is the case for the vole, *Microtus agrestis,* can be identified by selective staining with silver-containing reagents. The human karyotype has five such satellited chromosome pairs, not all of them visible in every cell. Fig. 3.10 indicates their general appearance.

The rather uniform appearance of metaphase chromosomes resulting from conventional Feulgen or acetocarmine staining masks a considerable array of heterogeneous segments or bands which can now be revealed by other techniques. It has long been known that when roottip cells of a plant such as *Trillium* undergo cell division at approximately $0°C$, the chromosomes show a pattern of staining quite different from that obtained at higher temperatures (Fig. 3.40). Furthermore, within a population of a single species, the same chromosome may exhibit a variety of banding patterns, indicating that the population is heterozygous for these patterns. Once it was interpreted as being due to "nucleic acid starvation," but now it seems more likely to be due to differential expressions of heterochromatin and euchromatin at low temperatures, expressions which now can more adequately be revealed by fluorescent or Giemsa stains. Four major banding patterns are now recognized, and so precise are these patterns that every human chromosome, as well as those of other mammals, can be unmistakenly identified (Fig. 3.58). A discussion of these patterns will also enable us to consider, in some detail, the remarkable progress that has been made in the area of human cytogenetics, progress that is all the more surprising in that it was only in 1956 that Tjio and Levan ascertained the correct diploid number of the human karyotype ($2n = 46$).

It is not surprising that we have an intrinsic interest in the cytogenetics of our own species, but there is an added dimension in that specific genes can now be assigned to particular chromosomes and that a wide variety of clinically important human variations, both normal and pathological, can be correlated with variations in chromosome structure and number. The explosion of knowledge in this area permits us now to rank the human karyotype along with that of *D. melanogaster* and maize as being among the most thoroughly investigated of eukaryotic systems. However, human, even mammalian, cytogenetics was hardly possible until recently, when the culturing of leukocytes, fibrocytes, and other kinds of dividing cells was developed, the examination of chromosomes was made feasible by exposing dividing cells to solutions of low osmotic content, thereby causing the cells to burst and the chromosomes to spread widely, and the staining techniques which revealed hitherto unsuspected variations which allowed each chromosome to be individually identified in an unequivocal manner. Much of this refinement has occurred during the last decade.

The human karyotype was first separated into seven distinguishable groups, identified by letters, and then subsequently into a numbered series, with chromosome 1 being the largest and 22 the smallest (but see below): A(1-3); B(4-5); C(6-12,X); D(13-15); E(16-18); F(19-20); G(21-22); and Y. Of these, and with conventional techniques, chromosomes 1, 2, 3, 16, and Y were earlier identifiable because of their distinctive morphologies, while the D and G groups could be recognized if their satellites were visible, which was not always the case. The C group presented the greatest difficulty, and even the important X chromosome could not be identified with any degree of certainty until, with the use of radioactive labels, it was recognized as a late-replicating unit. Eventually other chromosomes of the genome could be characterized by their sequential pattern of replication during the S period, but the method was laborious and not always consistent since homologous chromosomes might differ as to their labeling pattern.

The first of the successful staining methods was developed by Caspersson and his colleagues. By exposing chromosomes treated with the acridine dye, quinacrine mustard, to ultraviolet light, characteristic sequences of light and dark banding were revealed in each of the members of the genome (Fig. 3.58). In particular, the distal part of the long arm of the Y chromosome, a region near the centromere in the long arm of chromosome 3, and a spot near the

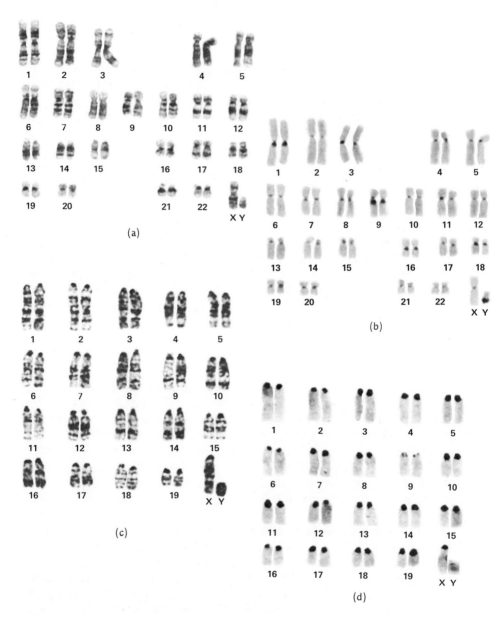

Fig. 3.58. Human and mouse chromosomes at mitotic metaphase showing the distribution of G-bands and C-bands, the latter revealing the location of constitutive heterochromatin. A, human chromosomes with G-bands; B, human chromosomes with C-bands; C, mouse chromosomes with G-bands; D, mouse chromosomes with C-bands. The human chromosomes are numbered from largest (1) to smallest (22), plus the X and Y, according to the provisions of the 1971 Paris Conference (A and B, courtesy of Dr. S. Pathak; C and D, courtesy of Dr. A. Markwong).

centromere of chromosome 4 fluoresced brilliantly, but the other chromosomes were also identifiable. The Q-bands, as these were called, became excellent cytogenetic markers, and even the elusive X chromosome proved to have a unique banding pattern. The technique also revealed an error in earlier thinking. Down's syndrome (or mongolism as it is often, but unfortunately, called) was thought to be due to the presence of an extra chromosome 21, that is, the next to smallest member of the genome. The Q-banding technique, however, showed the offending member to be the smallest chromosome, or chromosome 22, but the association of Down's syndrome with "trisomy-21" was so fixed in the literature that it was allowed to remain.

From one individual to another, the Q-bands may be absent and may vary in size and fluorescent intensity. A striking variability has been observed in the long arm of the Y chromosome which fluoresces so brilliantly. It may be reduced in size to the point where it is almost indistinguishable, or it may in some individuals show a 3-fold to 4-fold enlargement. Japanese males, for example, generally tend to have a longer long arm of the Y chromosome than do males of other populations, but so far as can be determined, this variation is without effect on either fertility or virility.

Several quinacrine derivatives and a dye, Hoechst 33 258, have also been successfully used in fluorescent studies, the latter revealing other markers, notably in chromosomes 1, 9, and 16, which were unstained by quinacrine mustard. The fluorescent dyes provide, therefore, sensitive probes for the identification of particular chromosomes or some of their parts, but these are not suitable for routine studies; the quartz lenses required for the passage of ultraviolet radiation are expensive, and the fluorescence fades rapidly, necessitating prompt photographic recording of the chromosomes under investigation.

The C-bands are so labeled because the procedure employed— pretreatment in NaOH, incubation in trisodium citrate, followed by staining with Giemsa—preferentially darkens the centric heterochromatin, although the banding pattern of the remaining chromosome arms corresponds well with that induced by quinacrine mustard, plus that revealed by Hoechst 33 258. Thus, the distal arm of the Y chromosome stains intensely, as do the secondary constrictions of chromosomes 1, 9, and 16 (these are not associated with nucleolar formation as are the secondary constrictions in chromsomes 13, 14, 15, 21, and 22). The staining of centric heterochromatin has proved to be especially useful since it has revealed an astonishing level of morphological polymorphism in human populations. That found in chromosome 1 has already been mentioned in another connection (p. 85), but the same holds true in chromosomes 9 and 16; those in chromosome 9, for instance, may vary greatly in size, and may indeed be inverted, as can be ascertained by a comparison of the homologues in the same individual. The frequency of polymorphisms associated with C-bands is as high as 4% in certain populations.

G-bands are produced by a pre-treatment of the chromosomes in an alkaline-saline solution, coupled sometimes with enzymatic exposure to pronase or trypsin, and followed by Giemsa staining. The banding sequence that results is comparable to that induced by fluorescent dyes, but the staining is relatively permanent, and the technique is the one most commonly used for clinical screening and evaluations. Some 80 major and 250 minor bands are distinguishable in metaphase chromsomes, but if late prophase or early metaphase chromosomes are examined before they are fully condensed, up to 1,250 bands can be located with some degree of precision. The significance of this for chromosome mapping and for the study of chromosome polymorphisms is obvious.

R-bands are obtained by increasing the temperature (up to 87°C) during incubation in buffer and then followed by Giemsa staining. These are the reverse of the G-bands, hence their name.

Just as the banded polytene chromosomes of dipteran species permit close identification and comparisons to be made from one individual to another, and from one species to another, so do the bands described above. In 1971 the Paris Conference of human cytogeneticists developed a nomenclatural system for identifying each human chromosome as to region and band. As stated above, the chromosomes are numbered from 1 to 22 on the basis of length, with the X and Y chromosomes separately identified. Short arms are indicated by the letter p, long arms by the letter q. As Fig. 3.59 indicates, each arm is divided into regions and then into bands. Thus, $12p14$ identifies the short arm of chromosome 12, region 1, band 4; $9q34$ identifies the long arm of chromosome 9, region 3, band 4, the latter being that band associated with the ABO blood type genes. The system, therefore, accommodates for any variations or additions that are encountered, enhancing the degree of sophistication and precision with which chromosomal problems, particularly those of a clinical nature, can be managed. For example, in ascertaining the course of bone marrow transplants, the use of a band variant between host and donor chromosomes permits an accurate assessment of the proportion of the host vs. donor cells at any given time. Also a number of malignant conditions have been correlated with specific chromosomal changes. Chronic myelogenous leukemia (CML) was

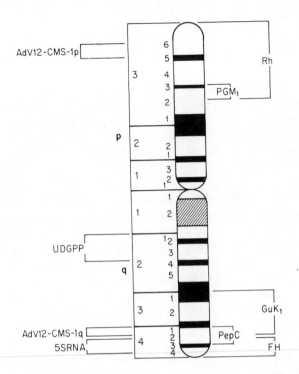

Fig. 3.59. Chromosome 1 of humans, showing the major bands as revealed by Giemsa banding of metaphase chromosomes, and as lettered and numbered according to standards set by the Paris Conference of 1971. The short arm is designated p, the long arm q, with each arm further subdivided twice into identifiable numbered regions. Compare this figure with Fig. 3.60 which reveals even more distinctive bands, and with the high-resolution banding that Yunis, et. al., (1978) have demonstrated in prophase chromosomes. Some gene locations obtained by linkage analyses are indicated, with the scarcity of linkage data meaning that locations can be only approximately designated. The location of the Rh gene indicates another facet of uncertainty; this gene is located approximately 66 map units from the centromere in the short arm of the male, whereas the same gene appears to be over 130 map units away in the female. Presumably, the crossover frequency is higher in the female than it is in the male (for details see McKusick and Ruddle, 1977; Schnedl, 1974; Yunis, et. al., 1978).

the first malignancy shown to be related to a characteristic chromosomal defect. Prior to the banding techniques, CML was assumed to be due to the absence of the short arm of chromosome 22, the so-called Philadelphia (Ph') chromosome. G-banding has revealed, however, that the arm is not missing but has been translocated to the end of 9q as the result of a translocation. Interestingly enough, however, the 22/9 translocation is apparently a consequence, not a cause, of the malignancy, since, in remission, the altered chromosomes cannot be recovered. In cases of retinoblastoma, a loss of chromatin in 13q14 has been found, while an 8/14 translocation is regularly associated with Burkett's lymphoma.

The Q-, C-, G-, and R-bands are clearly the result of qualitative and, possibly, quantitative differences along the length of the chromatids. The association of C-bands with centric heterochromatin suggests that they identify sequences of DNA consisting of highly repetitive units, more densely packed perhaps, and hence more resistant to the drastic pre-treatments that are used. It has also been suggested that these repetitive units are more tightly bound to acidic proteins than are the paler staining regions elsewhere. The Q-bands, based on several pieces of evidence, seem to be rich in A-T pairs of nucleotides; if this is so, then the R-bands are associated with regions high in G-C pairs, since G-C rich DNAs tend to quench fluorescence. However, this suggestion should be viewed with caution since the two X chromosomes in the human female should be identical to each other in nucleotide content, yet the inactivated X shows a greater degree of fluorescence in interphase than does the presumed active X. Thus, the state of compaction, or the chemical changes associated with compaction, may be the critical factor in determining some banding patterns.

Extension of the banding techniques to other species has revealed much the same kinds of patterns as those uncovered in humans. No species has been discovered which lacks differentially stained regions, but plants are not as revealing of subtle differences in the manner of animal species, particularly the mammals (Figs. 3.60 and 3.61). As one might expect, the primates, and in particular the anthropoid apes, show both similarities and differences (Fig. 3.62). The gorilla, however, was the only primate possessing an intensely fluorescent Y chromosome, but comparisons of overall patterns are in agreement with the generally accepted phylogenetic relationships.

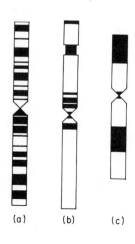

Fig. 3.60. A comparison of the details revealed by the newer staining techniques in human chromosomes (a) and those of higher plants (b and c). The chromosomes are not drawn to scale, those of the plant species being much larger than those of humans. a, chromosome 1 of humans; b, the large M chromosome of *Vicia faba*; c, one of the members of the genome of *Scilla sibirica*. Compare these with Figs. 3.59 and 3.61 (a, redrawn from Schnedl, 1974; b and c from Vosa, 1976).

(a)   (b)   (c)

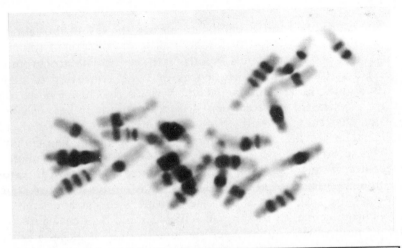

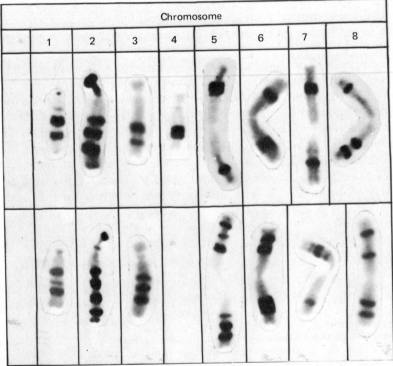

Fig. 3.61. The somatic chromosomes of *Anemone blanda* stained with Giemsa to show the banding patterns. Top, one of the karyotypes from a colchicined metaphase. Bottom, two of the karyotypes to show the variations found among seedlings; only chromosome 4 showed no variation from plant to plant (Marks, 1976, with the permission of Keter Publishing House, Jerusalem, Ltd.). Similar patterns of banding have also been identified in several cultivars of the tulip (Filion, 1974).

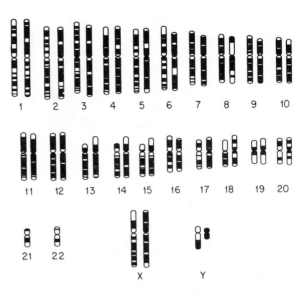

Fig. 3.62. Comparison of the banding patterns found in humans and in the Old World monkey, *Macaca mulatta*. For each pair of chromosomes, the left member belongs to humans, the right to Macaca. Similarities are obvious, but so too are the differences between the genomes of these two species of primates (Seth, et. al., 1976).

The banding patterns in the chromosomes of the laboratory mouse, as well as those of the rat, the hamster, the vole, *Microtus agrestis*, and Seba's fruit bat, are very similar to those in humans in that the Q- and C-bands largely coincide and the C-bands correspond to the positions previously identified as having constitutive heterochromatin (Figs. 3.38 and 3.39). This similarity might be expected to be true for the X chromosome in particular, since throughout the mammalian group many of the same genes are identifiably sex-linked. However, genes and banding patterns need not necessarily coincide from one species to the next, as Fig. 3.62 indicates that they do not. Bovine chromosomes, on the other hand, show a typical C-banding pattern when handled in one way and a lack of centric staining under other conditions.

Final mention may be made of a different technique for revealing chromosomal variations, a technique that also makes use of the Hoechst 33 258 dye. When cells in culture are grown in a thymidineless medium, but in the presence of 5-bromodeoxyuridine (BrdU), this nucleotide enters into the replication process as a replacement for thymidine. When cells are grown for two rounds of replication in the presence of BrdU, stained with Hoechst 33 258, and then exposed to blue light, the two chromatids are clearly differentiated from each other (Fig. 1.6). This is in accordance with the semiconservative expectations of the Watson-Crick model of the double helix of DNA; that is, one chromatid has both polynucleotide strands containing BrdU while the other has only one. Since BrdU quenches the fluorescence of Hoechst 33 258 to about one-fifth the level it would have in the presence of thydimine, the doubly substituted

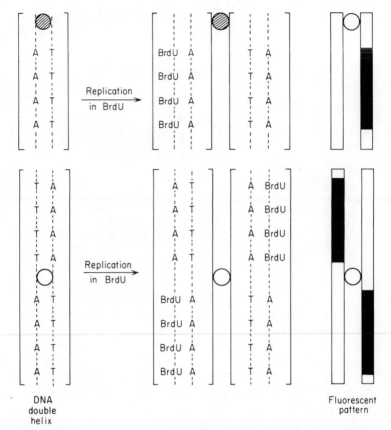

Fig. 3.63. Diagrammatic representation of the aymmetrical fluorescence that results from an asymmetrical distribution of thymine in the polynucleotide strands of a single double helix of DNA, and after replication in the presence of BrdU and the absence of thymidine (above). The centromere is indicated by a circle, the fluorochrome used was Hoechst 33 258, and the dark areas indicate the absence of BrdU and, consequently, the quenching of the fluoresence. Below, the asymmetrical fluorescence on either side of the centromere in an isochromosome in either the mouse or human (Yq•Yq in the latter species) (after Latt, 1977, and Lin and Davidson, 1974).

chromatid will be pale staining, the singly substituted one dark. This kind of "harlequin" staining can be made permanent by Giemsa staining following the fluorochrome.

The "harlequin" chromosomes have proven to be useful probes in a number of different ways. The first is in terms of a lateral distinction that can be made between the polynucleotide strands of a single chromatid. If one strand of a double helix contains more thymine than the other, for example,

$$-\ldots T-T-T-T-T-\ldots-$$
$$-\ldots A-A-A-A-A-\ldots-$$

and the thymine is replaced during replication by BrdU, one chromatid should fluoresce more brightly after replication than the other (Fig. 3.63). This has been shown to hold for certain regions of mouse satellite DNA which have been known, from chemical analyses, to have twice as much thymine in one strand as in the other, and it can be demonstrated as well for the human Y chromosome. An additional piece of information, related to the fluorescent picture and derived from such observations, that emerges is when two mouse acrocentrics have joined centrically to produce a metacentric chromosome or when an abnormal human Y chromosome consists of the two long arms joined at the centromere (i.e., an isochromosome consisting only of the long arms of the Y). Since the polynucleotide strands of each double helix on either side of the centromere must join in such a manner as to preserve their antiparallel nature with respect to each other, the BrdU-rich areas will be asymmetrical in position relative to the centromere. This indicates that there is no change in DNA polarity across the centromere as well as the fact that there is DNA continuity through the centromeric regions, thus corroborating the viscoelastic evidence derived from studies of whole chromosomes of *Drosophila* (Table 3.3).

A continuation of such studies also reveals, quite unexpectedly, that there is a lateral and directed polarization of the newly replicated products; thus, if two successive replications take place in the presence of BrdU, but no cell division is allowed to intervene, the four chromatids at the second metaphase will lie close to each other as *diplo-chromosomes*. The lightly fluorescing chromatids will always occupy the outside positions (Fig. 3.64). Why this is so remains unclear, but the observations make it apparent that the template and

Fig. 3.64. Diagrammatic representation of the composition of DNA after two replications in BrdU, but without an intervening cell division (above) and the fluorescent patterns below. The BrdU-containing chromatids are light and always on the outside of the endoreplicated chromosomes (after Latt, 1976).

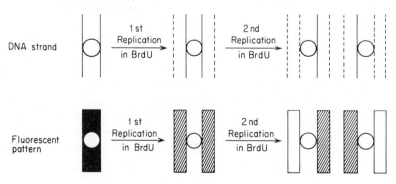

newly formed daughter polynucleotide strands of DNA must be so disposed that they come to occupy fixed rather than random positions relative to each other and to the centromeres.

Final mention can be made to the effect that BrdU has been particularly useful in confirming and extending observations made earlier with radioactively labeled chromosomes, especially those relating to the phenomenon of sister chromatid exchange (Fig. 1.6). The ratio of these exchanges to broken or rearranged chromosomes visible at metaphase is about 200 to 1. This indicates that the DNA strands exhibit a high degree of instability in that they break and rejoin fairly frequently, but the existence of numerous sister chromatid exchanges also indicates that the repair processes in the nucleus are highly efficient in putting the broken ends back together again, even though in illegitimate ways to produce the exchanges. That there are actual breaks capable of rejoining with other broken ends of chromatids is demonstrated through the use of agents such as mitomycin C, which react with the open breaks, prevent them from rejoining, and thus increase significantly the number of altered chromosomes at metaphase. A high incidence of altered chromosomes, whether of exchanges or rearrangements, is coupled in humans with an increased incidence of malignancy, and it is clear that this can result either from increased fragility of the chromosomes or from flawed repair mechanisms.

Several human diseases, including Bloom's syndrome, ataxia telangiectasia, xeroderma pigmentosum, and Franconi's anemia, have clinical manifestations which include a varying incidence of aberrations as well as a varying number of sister chromatid exchanges, and all are under genetic control since they are inherited in Mendelian fashion. Bloom's syndrome is characterized by sun-sensitivity and dwarfism, while the cultured cells, particularly the leukocytes, show a high level of aberrations as well as of sister chromatid exchanges; patients with Franconi's anemia, on the other hand, show the highest numbers of broken chromatids, but the frequency of sister chromatid exchanges is not significantly increased. Xeroderma pigmentosum exhibits a variety of expressions, but all variants are light-sensitive, particularly as regards ultraviolet light, and all seem to share characteristic defects in certain of the enzymes involved in the normal repair of disrupted chromatids. The break or exchange points along the chromatids seem not to be randomly distributed but to occur within the interbands or at boundaries of bands in quinacrine-stained chromosomes. The significance of this observation is not clear, and there is no certainty that such a distribution is found in all organisms.

Meiotic Prophase Chromosomes. It is the events of homologous pairing and crossing over, coupled with the formation of the synaptinemal complex and a much slower tempo of cell division, which distinguish meiosis from mitosis, and since these events are largely chromosomal in nature (but involving cellular phenomena as well), we might expect that meiotic chromosomes would exhibit a behavior and an appearance that set them apart from their somatic counterparts. This expectation is fully realized, particularly in the prophase stages of division.

Meiosis as a process will be dealt with more fully in the next chapter, but in contrast to somatic chromosomes during comparable prophase stages, the meiotic prophase chromosomes in a few favorable species possess patterns of linear morphology which allow one to distinguish one chromosome pair from another. Figures 3.3, 3.4, 3.65, and 3.66 show meiotic chromosome pairs at about the middle of prophase, or the pachytene stage (also Figs. 4.13, 4.17, 4.19, and 4.21). The paired chromosomes are closely appressed to each other and show similar morphologies. Those of rye reveal a more or less uniform series of chromomeres which gradually become larger in size as they near the centromere and with the centromeres represented by several distinct chromomeres (Fig. 3.11). There is every reason to believe that the meiotic chromomeres are the result of localized patterns of contraction brought on by circumstances dictated by the meiotic cell and reflecting chemical differences which, in turn, lead to differences in the packing of the DNA. The chromomeres of maize fall into two size categories: the typical small chromomeres found along the length of all chromosomes and the occasional large knobs which are known to vary in size, position, and number in the different strains of maize. A very large knob is on chromosome 6 and is associated with the nucleolar organizer region (Fig. 3.65). The knobs, in fact, can vary from 0 to 16 in single plants and occupy at least 22 different positions in the genome (Figs. 3.16 and 3.32). An equal variability is found in teosinte, a relative of maize. The chromomeres of *Luzula* are distinct, with some terminal ones of large size (Fig. 3.66), while those of tomato show large, irregularly shaped chromomeres adjacent to the centromere (and representing, in all probability, centric heterochromatin), as well as the smaller, less deeply staining chromomeres along the arms of the chromosomes and in the telomeric regions (Fig. 3.3). The centromere itself, at least in maize and in tomato, is an understained region in meiotic prophase, although somewhat later it can assume a deeply stained appearance and at metaphase be distinctly drawn out from

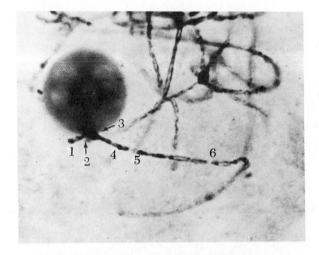

Fig. 3.65. Chromosome 6 of maize in the pachytene stage, and the details that can be seen at this time. Number 1 is a terminal chromomere, 2, a chromomere adjacent to the large knob (3) which is the nucleolar organizer as well; 4 and 5 are recognizable chromomeres, and 6 is the centromere region which lacks any visible structure (courtesy of B. McClintock).

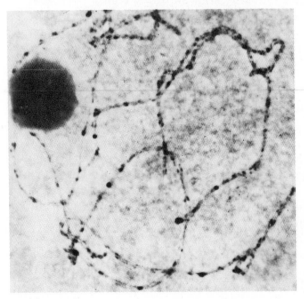

Fig. 3.66. Meiotic pachytene chromosomes of the woodrush, Luzula. At least one of the paired homologues is attached to the nucleolus. The larger chromomeres are scattered throughout the chromosomes, with some of the largest being located at the ends of the chromosomes (courtesy of Dr. S. Brown).

the main body of the chromosome, as though it were under strong tension from the spindle poles.

The *lampbrush* chromosome is a special kind of mid-prophase, or diplotene, chromosome found in the oocytes of fish, birds, amphibians, reptiles, and some invertebrates, but those of the urodeles among the amphibia, and particularly those of the newt, *Triturus,* have been most extensively examined (Fig. 3.33). Being at the diplotene stage of meiosis, they have previously undergone homologous synapsis and crossing over before being spread throughout an enlarging nucleus, that in *T. viridescens,* may reach a diameter of about 500 μm. These chromosomes may remain in this state for six months or more before maturation of the egg is achieved and fertilization stimulates the cell to proceed with the meiotic process. In the human, this stage lasts for many years, from about six months in the gestation period to the time when the last egg is shed and menopause brings an end to the fertile period.

In the frog, *Rana temporaria,* the 13 pairs of chromosomes are individually distinguishable, each pair having a constant pattern of chromomeres, from some of which emerge loops of individual uniqueness. At the time of their greatest extension, each chromosome pair is about 800 to 1,000 μm in length. In *T. viridescens,* the 12 pairs of chromosomes range in length from 350 μm to 800 μm, with the total length being 5,900 μm or 5.9 mm. By the usual meiotic or mitotic standards, these are enormous chromosomes, but when the actual length of the DNA contained in these chromosomes is calculated, these figures have to be revised upward to a substantial degree.

Figure 3.67 illustrates the general appearance of a single pair of homologous chromosomes held together by several chiasmata at the points of crossing over and shows an enlargement of one of the loop areas. Figure 3.68 provides an interpretation of the chromomere and loop structure. Each pair of homologues, therefore, consists of a row of variously sized chromomeres separated by interchromomeric stretches of DNA. Most of the chromomeres have a single pair (occasionally several pairs) of loops extending out from the longitudinal axis, giving the chromosomes their fuzzy, or "lampbrush," appearance. The crested newt, *T. cristatus,* has about 5,000 chromomeres among its 11 pairs of chromosomes, and a minimum of 3,300 show projecting loops during the diplotene stage. The size of the loops varies considerably, some being only a few micrometers in length and others being as long as 100 μm. Some are straight and others are twisted, but all seem to be thinner at one end that at the other. The reason for this is that the loops are actively transcribing, as evidenced by the accumulation of radioactive uridylic acid, which as uracil becomes incorporated only into RNA. From a variety of pieces of both observational and experimental evidence, it appears that the loop axis is progressively being paid out from the chromomere at the thin end and just as continuously being wound back into the chromomere at the other end. That region of chromatin exposed in the loop begins to transcribe as soon as it is released from the chromomere, continues to transcribe for a short period while it is a loop, and then stops as it is wound up at the other end.

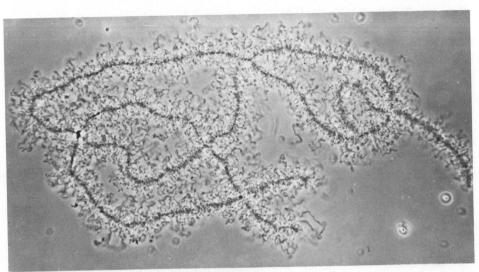

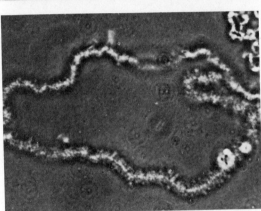

Fig. 3.67. Above, a meiotic bivalent of the newt, Triturus, consisting of two homologues held together at four points by chiasmata, and in the diffuse diplotene, or lampbrush, stage. The fuzziness results from the projection of loops of chromatin from the chromomeres; the loops are in active states of RNA synthesis. Left, an enlargement of a single loop (courtesy of Dr. J. Gall).

An interpretation of the structure of the lampbrush chromosome is that each chromomere is a region of localized coiling or compaction and the interchromomeric stretches of chromatin are two closely appressed chromatids, each with a diameter of about 40 nm. This diameter is consistent with the idea that each chromatid is a continuous double helix of DNA. This is borne out by stretching experiments. If microneedles are placed on either side of a double loop and tension is brought to bear on the loop region, the loops finally yield and become stretched, thereby showing themselves to be a portion of the linear axis of the chromosome.

The above information can now be translated into actual length of the DNA strands. The number of loops, their average length, plus the interchromomeric regions, gives a figure of about 50 cm as the minimal length of the haploid set of *Triturus* chromosomes. It is estimated, however, that only one-twentieth of the DNA is exposed as loop material at any one time, so the total computed length of haploid DNA is about 10 m. Once meiosis proceeds beyond the diplotene stage, the loops are drawn back into the body of the chromosome and disappear as entities.

The amphibian lampbrush chromosomes exhibit an additional peculiarity. The nuclear sap of developing and mature oocytes of amphibians shows the presence of many unattached nucleoli, in fact, as many as 1,000 in some nuclei. These are formed by a nucleolar organizer region that is the equivalent in structure and function to the nucleolar organizer regions of maize or *Drosophila,* but instead of forming a single nucleolus per organizer region per cell cycle, the amphibian nucleolar organizer region, during maturation of the oocyte, produces a succession of DNA fragments by some method of replication which free themselves from the parental chromosome and move into the nuclear sap where they proceed to form nucleoli. These can be readily isolated, and the DNA attached to them—

Fig. 3.68. An interpretation of the loops of a lampbrush chromosome. Top right, the chromomeres are of varying sizes and shapes, as are the loops that spin out from them. Below, an interpretation, with RNA synthesis proceeding as the loop spins out from the chromomere at the top and is then compacted back into the chromomere at the bottom; each loop is a portion of a single chromatid (courtesy of Dr. J. Gall).

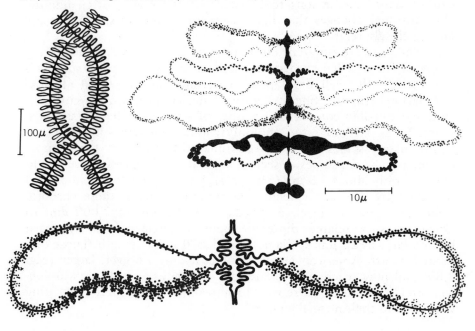

constituting about 70% of the total DNA of the nucleus of a mature oocyte—can be freed as rings of varying circumference and as linear units ranging from a few micrometers to as much as 40 μm. As shall be discussed in a later chapter (p. 337), this is a process of genetic amplification at a time when massive amounts of ribosomal RNA are being assembled for the subsequent embryonic development. Once the oocyte is mature, fertilization occurs, and development is initiated, nucleolar formation ceases and does not begin again until differentiation of cells takes place in the early gastrular stage.

Polytene Chromosomes. Balbiani is generally credited with the discovery, in the 1880s, of the giant chromosomes in the cells of the salivary glands of dipteran species (Fig. 3.34). For many years they constituted only an interesting chromosomal variation, but their cytogenetic significance was realized when the similarities between their band structure and the linearly ordered sequence of genes on the chromosome were recognized and when it was established that each of the large visible chromosomes actually consisted of a pair of homologous chromosomes intimately synapsed. Since the 1930s the use of these giant chromosomes has enabled *Drosophila* cytogeneticists to approach a degree of precision in gene location scarcely considered possible when only the usual mitotic and meiotic chromosomes were available for study; in addition, it has permitted detailed evolutionary comparisons to be made of the many, often minute, chromosomal changes that have taken place during speciation and among closely related species.

The dipteran polytene chromosomes are the largest ones readily available for detailed cytogenetic studies (the genetics of amphibia is not sufficiently well developed to permit the same kind of close correlation with the fine structure of the lampbrush chromosomes, although detailed comparative studies of structural changes are possible). In *D. melanogaster,* the polytene chromosomes in late larval (third instar) stages are over 100 times the length of the somatic metaphase chromosomes as seen in a larval ganglion cell; these would measure only about 7.5 μm (Fig. 3.34). When prepared for study and stretched by the smear technique commonly used, these chromosomes in aggregate reach a length of 1,180 μm to 2,000 μm. In *Chironomus,* a related dipteran genus, a single pair of synapsed homologues in 20 μm in diameter and 270 μm in length. Those of a related South American genus, *Rhyncosciara,* are even larger (Fig. 3.69) and may reach even greater dimensions as a result of pathological infections which cause the chromosomes to continue replicating well beyond their normal amount.

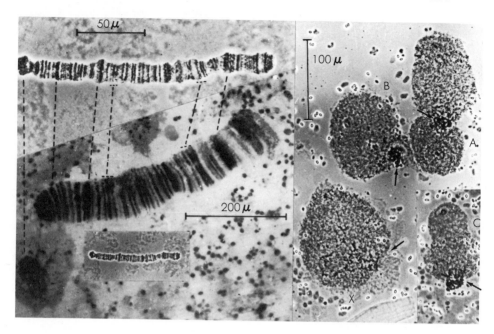

Fig. 3.69. Polytene chromosomes of the South American dipteran, *Rhynocosciara angelae.* The chromosomes are not linked together at a chromocenter, as they are in Drosophila, but lie separate in the nucleus. Left, chromosome C; the top and bottom figures are of normal polytene chromosomes, but at different magnifications, while the large middle and small bottom figures are at the same magnification, the very greatly enlarged chromosome resulting from a protozoan infection which induces many more rounds of replication, a many-fold increase in size, and a nucleolus-like development of the end of the chromosome. The vertical dotted lines connect homologous regions in the two different sized chromosomes; the infective protozoa are seen as small dark spheres in the cytoplasm. Right, continued infection by the protozoa leads to further replication, causing the chromosomes to lose their banded structure and to assume a pompom-like appearance; arrows identify the centromeric regions of the pairs of chromosomes which seem not to replicate to the same extent (Pavan and Basile, 1966).

The polytene chromosomes, even in the same cell, need not be of identical diameter throughout their length. Those of *D. melanogaster* show a variety of pinched-in regions as well as large puffs, the latter an indication of RNA transcriptional activity. In the cecidomyid fly, *Camptomyia,* the ends of the chromosomes often flare out like worn-out toothbrushes; the heterochromatin of the nucleolar organizer regions is loosely arrayed, and the D chromosome has a much greater diameter than any of the other chromosomes and presumably has undergone additional rounds of replication on a selective basis (Fig. 3.70). Those of the collembolan species, *Bilobella*

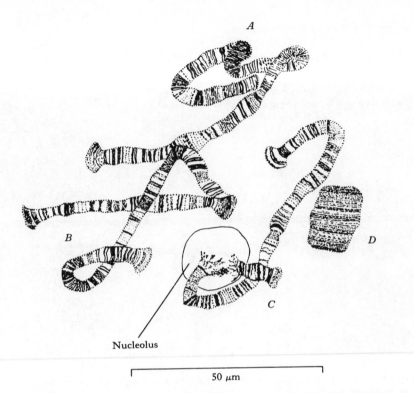

A

B

C

D

Nucleolus

50 μm

Fig. 3.70. Salivary gland chromosomes of a cecidomyid fly *Camptomyia sp.,* showing that the chromosomes differ appreciably in diameter, and presumably in degree of polyteny (White, 1948).

massoudi, are almost caricatures of a typical polytene chromosome, so strangely varied are they along their length (Fig. 3.71). Such variations, however, wherever found, provide convenient landmarks for the cytogeneticist.

More important than size are two other characteristics which the polytene chromosomes almost invariably demonstrate. First, the homologous chromosomes show a type of synapsis (somatic pairing) at least as intimate as that characteristic of meiotic chromosomes in mid-prophase. In *B. massoudi,* however, synapsis does not always occur so that polyteny and somatic synapsis are independent of each other, although associated in most polytene cells. Second, the chromosomes reveal a distinctive traverse pattern which consists of alternate chromatic and achromatic bands. The bands differ in thickness and other structural features in a manner so specific, and so repetitive from cell to cell and individual to individual, that an accurate and detailed mapping of each chromosome from end to end is possible. The dual significance of the synaptic behavior and band

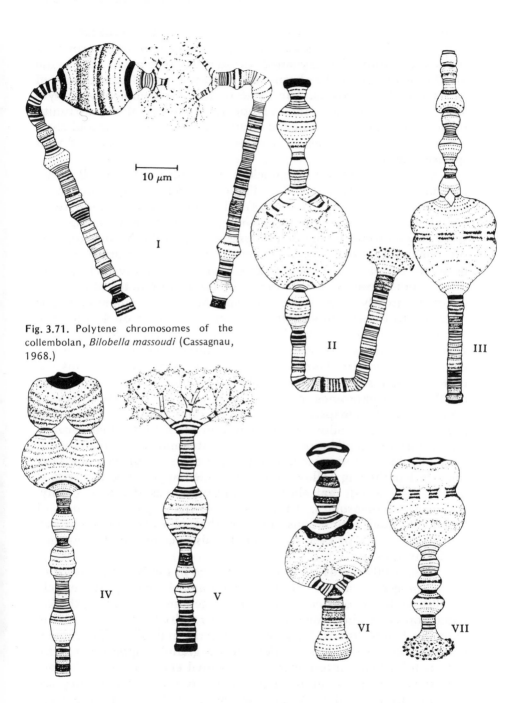

Fig. 3.71. Polytene chromosomes of the collembolan, *Bilobella massoudi* (Cassagnau, 1968.)

10 μm

I

II

III

IV

V

VI

VII

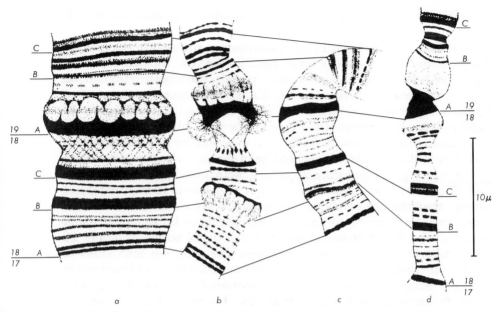

Fig. 3.72. A portion of one of the polytene chromosomes of *Chironomus tentans* showing how both size and appearance differ in different tissues: a) salivary gland; b) Malpighian tubules; c) rectal tissue; and d) midgut (Beerman, 1952).

region, this means that it is near the juncture with heterochromatin in the left arm of chromosome 2 but clearly within the euchromatic area. Centric heterochromatin is not distinguishable in these cells, and hence is not numbered or lettered.

The polytene chromosomes are most commonly studied in the cells of dipteran salivary glands, but they are not peculiar to this tissue. They have also been found in the cells of Malpighian tubules, fat bodies, ovarian nurse cells, and gut epithelia, but these tissues are not so convenient for study and the chromosomes do not commonly reach the large size found in the salivary glands. However, while the size of the chromosomes and the appearance of the individual bands may differ from tissue to tissue, there is an essential similarity to the banding pattern (Fig. 3.72), indicating that the pattern is a reflection of a basic linear differentiation of chromatin that becomes visible when polyteny intervenes, while the tissue specificity of appearance is more related to tissue physiology and function. Attention has been particularly directed at the puffs that form in these structures (Figs. 3.72 and 3.73). The incorporation of radioactive uridylic acid into these areas points to the puffed regions as sites of active transcription, and hence of gene activity, while the nonpuffed bands are genetically inactive. There is recent evidence to show that transcription can take place in some interband regions and in the absence of puffing, but the degree to which puffing occurs is closely correlated with the amount of RNA produced. Since the third instar larvae are not only the best source of these chromosomes but are also in the stage during which the insect is preparing for pupation, the puffing provides evidence of developmental patterns of change. Thus, while the banding pattern is constant, the puffing pattern is not, changing as different genes come into play to accomplish their developmental roles. The hormone ecdysone, for example, induces molting in insects, a phenomenon requiring a sequence of physiological activities occurring on a time schedule. The hormone, therefore, triggers some genes into action while turning off other genes in sequential fashion; the timing and appearance of puffing reveal the activation or deactivation sequence of different genes involved in the physiology of pupal development. Since the puffs are both band- and tissue-specific, their presence provides visible manifestation of the fact that different genes come into play in different tissues and at different times. Even within a given tissue, some cells may show activation by puffing and others may not, and since the RNA from such puffs can be isolated, the sequential appearance of immediate gene products is a step toward an understanding of the enigma of development and differentiation.

Fig. 3.73. A portion of a polytene chromosome of Chironomus in an unpuffed and a puffed state, together with Beermann's (1952) interpretation of such a puffed region.

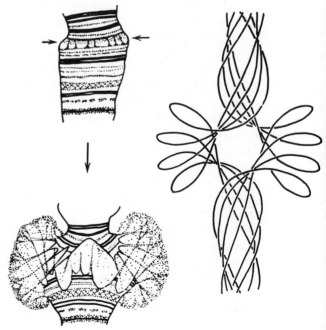

B Chromosomes. Every eukaryotic species has a basic and essential set of chromosomes which constitutes its haploid genome. These are the A chromosomes, and gains or losses of them even in the diploid state generally induce a deleterious condition as a result of genetic balance. Many species, however, possess additional chromosomes—the so-called B chromosomes—which are dispensable, unstable in mitosis and meiosis, variable in number and structure from one individual to another within the species, and as a rule smaller in size than those of the A set (Fig. 3.74). While not necessarily heterochromatic in their entirety, they are typically devoid of mendelizing genes, and hence exhibit little or no influence on the segregation of characters of a qualitative sort. They do, however, exert a quantitative genetic effect when their numbers are large, an effect which differs from one species to another. In the plantain, *Plantago coronapus,* for example, a serious disturbance results from the presence of but one of these chromosomes; all plants carrying a B chromosome are male sterile. In other species, in which the numbers of B chromosomes may be as high as 16, such quantitative characters as fertility, seed set, germination of seed, vigor, number of tillers, and straw weight are often reduced in proportion to the number of Bs present. The duration of the cell cycle is also increased in proportion to the number of B chromosomes, but this may be due simply to the increased amount of DNA.

In some grasshoppers and maize, however, the presence of such chromosomes signals not only an increase in chiasma frequency but also a redistribution of the chiasmata, thus bringing about an increase in genetic recombination and, consequently, an increase in variability among the offspring, a distinct advantage to short-lived, outbreeding species. In the perennial rye grass, *Lolium perenne,* the effect is the opposite, with increased numbers of B chromosomes depressing the chiasma frequency. Additional evidence on their quantitative influence is also obtained from studies on *Lolium*; under stress conditions such as sowing density, plants with B chromosomes occurred more frequently the higher the population density.

The B chromosomes, with rare exceptions, do not appear to be homologous with any portion of the A set as judged by pairing relationships at meiosis, although they must obviously have been derived from the A set since there seems no other source from which they could have come. If this is so, they have diverged structurally to a considerable extent, although the nucleotide ratios of their DNA are generally indistinguishable from that of the A set. The suggestion that they may have been derived from some heterochromatic element(s) of the A set is meagerly supported in one instance by the

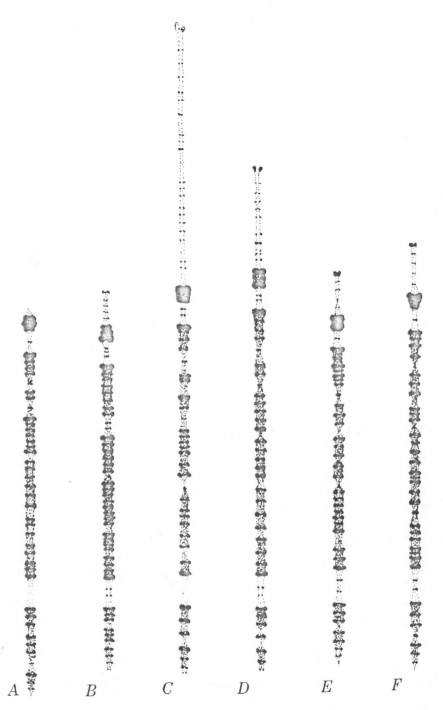

Fig. 3.74. Different forms of the B chromosome of rye found in strains from Sweden (A and B), Turkey (C), Afghanistan (D), Transbaikalia (E) and Korea (F). The "gene" governing non-disjunction is located in the large terminal heterochromatic knob; the greatest degree of variation among these Bs is distal to the large knob (Lima-de-Faria, 1963).

observation that in one grasshopper species the B chromosomes have a higher A-T content than the A set, while some pairing does occur at meiosis in the flowering plant *Clarkia.* The nonrandom pairing of the B chromosomes of *Tradescantia paludosa* with one or more members of the A set at their centromeres and their ends can be explained by inversion crossing over in both arms which deletes all of the chromatin except a small portion at the centromere and at the ends of both arms, but the evidence is insufficient to be considered anything more than merely suggestive.

The B chromosomes are found in a wide variety of angiosperm species and families, and in six different gymnosperms, but they are apparently absent from other plant groups. Of the 591 examples known in the angiosperms, 354 are found in diploid species, with almost all of these in outbreeding groups. Among animals, B chromosomes have been identified in flatworms, mollusks, most insect orders, and especially in the Coleoptera and Orthoptera, and all vertebrate groups except the fishes and primates. Most of these B chromosomes are structurally similar to those in the A set in that they possess normal centromeres, and hence have a regular means of segregating during cell division, but the presence of B chromosomes in organisms with diffuse centromeres suggests that these could have been derived from some fragment of the A set having an initial survival value. The hemipteran insects *Acanthocephala (Metapodius) terminalis* and *Cimex lectularius,* the bedbug, have such B chromosomes, and it has been suggested that they were derived from the heterochromatic Y in the former species and from the heterochromatic X in the latter.

The B chromosomes of maize and rye are representative of B chromosomes in general. Figure 3.75 illustrates that found in maize at the pachytene stage of meiosis. Except for nontranscribing euchromatin proximal to the centromere, the remainder of the chromosome is heterochromatic. It is widespread in strains of Indian maize, but it has largely been bred out of commercial strains. Since it is not possible to distinguish individual plants which possess or lack one or more B chromosomes, they are without significant genetic effect of a detectable phenotypic sort, but their presence causes any knobbed chromosome of the A set to behave irregularly, and when

Fig. 3.75. Diagram of the B-chromosome of maize in the pachytene stage of meiosis. The centromere is represented as terminal although it may be slightly subterminal; the segment to the left with the six small chromomeres is euchromatic, the remainder largely heterochromatic.

present in large numbers an increase in chiasma frequency and a decrease in vigor and fertility are detectable. In meiosis, the B chromosomes will pair with each other, but apparently without the formation of chiasmata; they often fall apart before metaphase and undergo erratic distribution.

The lack of genetic effect of low numbers of B chromosomes of maize suggests that they are of a "parasitic" nature and that they ought to be eliminated rapidly from a population. Their continued presence, as well as their widespread occurrence in the native populations of other species, would seem to discount this hypothesis. The fact is that these chromosomes have a system not only of self-preservation but also for insuring an increase in numbers. For example, self-fertilization of a plant of maize with 1 B chromosome ($n = 11 \times n = 11$) will yield offspring possessing 0, 1, 2, 3, and 4 B chromosomes. The B chromosomes were obviously eliminated in the 20-chromosome offspring, but those with 3 and 4 B chromosomes possess more chromosomes than the two haploid genomes added together. The answer lies in the phenomenon of selective distribution coupled with that of selective fertilization.

The B chromosomes of maize, like those of the A set, behave more or less normally when paired with each other at meiosis; if present singly, they may become lost, or, by chance, be distributed to one or the other of the two telophase I nuclei. The second division of meiosis in the microsporocytes, and the first microspore mitosis, are normal, but at the second microspore division, when the generative nucleus divides to produce the two sperm cells, the two chromatids of the B chromosomes fail to separate cleanly, and both pass into one of the two sperm cells (Fig. 3.76). This is a form of nondisjunction and of meiotic drive as well (p. 273), but the process of selective retention of the B chromosomes is enhanced by the fact that the sperm cell carrying the two B chromosomes will preferentially fertilize the egg 60% to 70% of the time. Since, in the above cross, some offspring with four B chromosomes appear, nondisjunction must also take place in the megasporocyte as well.

The ability of the B chromosomes of maize to behave in this fashion resides in a "gene" located in the euchromatin proximal to the large blocks of heterochromatin. It exerts its effect by delaying replication of the DNA of its own heterochromatin (and in that of the knobs of chromosomes of the A set), and thereby interfering with the division process. Cytogenetic advantage can be taken of this circumstance because any gene from the euchromatic portion of an A chromosome, if translocated to the end of the B chromosome (as is the *Su* gene in Fig. 3.76), can readily be manipulated in

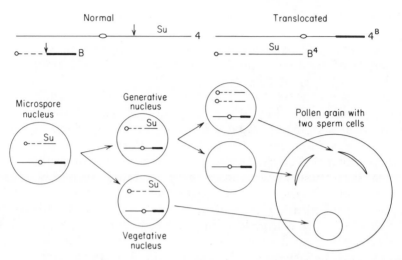

Fig. 3.76. Schematic representation of the nondisjunction of the B chromosomes of maize in the second microspore division. The B chromosome can be best followed when it possesses a marker gene, in this instance the *Su,* which causes the formation of starchy endosperm. One of the sperm cells will lack a B chromosome, the other will possess two of them; the latter sperm is the one most likely to fertilize the egg, thus perpetuating the B chromosomes through nondisjunction and selective fertilization.

dosage study experiments. The B chromosomes of rye, *Secale cereale,* behave somewhat differently from those in maize. Their variable structure is indicated in Fig. 3.74, and as their relatively greater euchromatic content might suggest, they are not tolerated as well by an individual as are their maize counterparts. Three or four of these in a rye genome will have a definite depressing effect on vigor and fertility, although, oddly enough and for reasons unknown, plants with even numbers of B chromosomes do better than do those with uneven numbers.

The B chromosomes from different wild populations of rye differ in the magnitude of their genetic effect. They persist in populations, however, because they undergo nondisjunction at the first microspore division, both chromatids passing to the generative nucleus, and from there to both sperm cells. As in maize, nondisjunction is governed by a "gene" which, in this instance, is located in the large terminal heterochromatic knob in the long arm. When the knob is missing, the chromosome behaves as do the members of the A set.

The B chromosomes of other species do not necessarily behave in the manner of those of maize or rye. In *Sorghum* and *Xanthisma* species, the B chromosomes tend to be eliminated from cells of the roots and leaves but to persist in those of the stem and flowers; in *Poa alpina,* they remain in primary roots and stems but are eliminated from leaves and adventitious roots; in *Crepis capillaris,* nondisjunction at the time of floral initiation directs them preferentially into the germline. How the selective process works in all instances is not clear, although delayed replication is obviously implicated.

The persistence of B chromosomes in natural populations, therefore, provides some species with an added genetic dimension for increased variability, enabling them to cope better with a diversity of environments. Their deleterious effects must be kept to a minimum and must be offset by advantageous features if they are to persist. Among populations of the flowering plant, *Centauria scabiosa,* in Scandinavia and Finland, one-quarter of the plants had B chromosomes, the number per individual ranged from 0 to 16, and the majority of B-containing individuals had one or two Bs. But those with and those without Bs seemed to prefer different environmental conditions. A comparable situation has been observed in the grasshopper, *Myrmeleotettix maculatus,* with the B-containing individuals preferentially found under conditions which were optimum for the species. It seems unlikely that the term "parasitic" is appropriate for these particular situations.

Chromosomes of Ciliated Protozoans. The ciliated protozoans contain a micronucleus and a macronucleus, the latter derived from the former. The chromosomes of the micronucleus are, in most ways, comparable to those of any eukaryotic species: they are diploid in number, visible by the usual microscopic techniques, and can divide by mitosis or meiosis as the circumstances of the life cycle dictate, but they produce only traces of RNA, indicating that their transcriptional abilities are very much repressed. By way of contrast, the macronucleus has its chromatin dispersed in such a manner that its chromosomes are not discernible, but it produces virtually all of the RNA needed for the day-to-day existence of the organism. The micronucleus, therefore, is the genetic repository of the species, the macronucleus the physiological unit. It is the structure of the latter with which we are here concerned.

Ciliated protozoans are of two types as regards the macronucleus. In the holotrichous genus *Paramecium,* the macronucleus is a highly polyploid version of the micronucleus. In the hypotrichous genera *Stylonychia* or *Oxytricha,* a single cell has four micronuclei and two macronuclei, the latter having been derived from the former by a series of mitoses and are very much larger. The transformation to a macronuclear status involves an intricate series of events. A

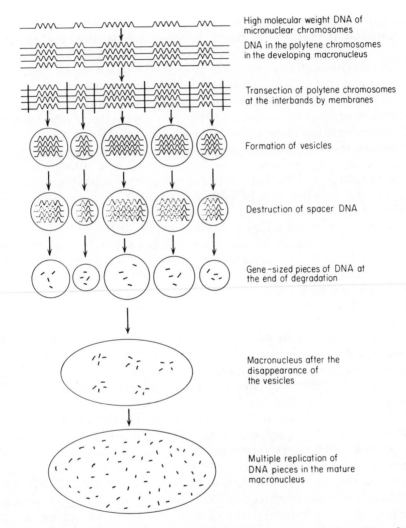

High molecular weight DNA of micronuclear chromosomes

DNA in the polytene chromosomes in the developing macronucleus

Transection of polytene chromosomes at the interbands by membranes

Formation of vesicles

Destruction of spacer DNA

Gene-sized pieces of DNA at the end of degradation

Macronucleus after the disappearance of the vesicles

Multiple replication of DNA pieces in the mature macronucleus

Fig. 3.77. The gradual transformation of the micronuclear chromosomes into those of the macronucleus. Several rounds of DNA replication produce a polytene condition in each chromosome; these are then transected by membranes to form a series of vesicles; within the vesicles the spacer DNA is degraded, leaving only pieces of euchromatin (genes?); the vesicles then merge into one single large vesicle by coalescence; other rounds of replication increase the number of copies of each gene in the mature macronucleus (Fig. 14 from Prescott, 1974, and with the permission of the Oregon State University Press.)

period of DNA replication in the macronucleus lasting about 40 hr produces a group of polytene chromosomes somewhat reminiscent of those in dipteran flies in being banded; in other respects, their appearance is reminiscent of the kind of chromosome found in the dinoflagellates (Figs. 2.21 and 2.23). Following the replicative period, each band becomes isolated by membranes which cut through the chromosome in the interband regions. Each band, therefore, together with a segment of interband chromatin on either side of it, becomes enclosed as a vesicle (Fig. 3.77). The chromosomes lose their identity as such, and the macronucleus eventually consists of a series of vesicles numbering around 10,000. There then follows a period of DNA degradation, with over 90% of the DNA in each vesicle being degraded. The remaining DNA is presumed to be band material. When micronuclear DNA is tested for buoyant density peaks, four make their appearance, with the peaks being at 1.699, 1.701, 1.704, and 1.709 g/cm^{-3}. The macronuclear DNA corresponds only to the 1.701 peak, and it is believed that the degradative process eliminated the other three DNA components, which may well have consisted only of repetitive or spacer DNA.

Three rounds of DNA replication in the macronucleus occur about 30 hr after the completion of the degradative process, bringing the cell to a mature vegetative state. The macronucleus, therefore, contains not whole chromosomes but rather DNA fragments, each having an average length of 0.75 μm (the range is from 0.2 μm to 1.6 μm). Each remaining DNA fragment is presumably a functioning gene to which RNA polymerase can attach for transcribing purposes. All unnecessary DNA has been removed, and only that needed for the continuing function of the cell is retained. Since the micronucleus takes care of the genetic future, the organization of the genetic material into maneuverable chromosomes is no longer required. The macronucleus can divide, however, at each cell generation, and by an amitotic process it may even regenerate itself from fragments after it has undergone apparent dissolution following meiosis.

BIBLIOGRAPHY

Babcock, E. B., *The Genus Crepis*, Vols. I and II. Berkeley: Univ. Calif. Publ. Bot. 21 and 22, 1947.

Beermann, W., "Chromosomenkonstanz und spezifiche Modifikation der Chromosomenstruktur in der Entwicklung and Organdifferenzierung von *Chironomus tentans*," *Chromosoma*, 5 (1952), 139-98.

————, (ed.), *Developmental Studies on Giant Chromosomes*. New York: Springer-Verlag, N.Y., 1972.

Belford, H. S., "Single-Copy DNA Comparisons Within the Genus *Atriplex*." Ph.D. thesis, Univ. Mass., Amherst, 1979.

Berlowitz, L., "Chromosomal Inactivation and Reactivation in Mealy Bugs," *Genetics, 78* (1974), 311–22.

Bennett, M. D., and J. B. Smith, "Nuclear DNA Amounts in Angiosperms," *Phil. Trans., Roy. Soc. London* B 274 (1976), 227–74.

Bloom, S. E., and C. Goodpasture, "An Improved Technique for Selective Silver Staining of Nucleolar Organizer Regions in Human Chromosomes," *Hum. Genetics, 34* (1976), 199–206.

Bonner, J., "Complex Chromosomes: Structure and Function," in *Chromosomes: From Simple to Complex*, P. A. Roberts, ed. Corvallis: Oregon State University Press, 1974.

Bridges, C. B., "Salivary Chromosome Maps," *J. Heredity, 26* (1935), 60–64.

Britten, R. J., and D. E. Kohne, "Repeated Sequences in DNA," *Science, 161* (1968), 529–40.

Brown, D. B., "The Isolation of Genes," *Sci. Amer., 229* (1973), 20–29.

————, et al., "The Isolation and Characterization of a Second Oocyte 5S DNA from *Xenopus laevis*," *Cell, 12* (1977), 1045–56.

Brown, S. W., and U. Nur, "Heterochromatic Chromosomes in the Coccids," *Science, 145* (1964), 130–36.

Caspersson, T., et al., "Differential Binding of Alkylating Fluorochromes in Human Chromosomes," *Exptl. Cell Res., 60* (1970), 315–19.

————, "The 24 Fluorescence Patterns of the Human Metaphase Chromosomes —Distinguishing Characters and Variability," *Hereditas, 67* (1971), 88–102.

Cassagnau, P., "Sur la Structure des Chromosomes Salivairies de *Bilobella Massoudi* Cassagnau (Collembola: Neanuridae)," *Chromosoma, 24* (1968), 42–58.

Catcheside, D. G., "The Genetics of B Chromosomes in Maize," *Heredity, 10* (1956), 345–51.

Chooi, W. Y., "RNA Transcription and Ribosomal Protein Assembly in *Drosophila melanogaster*," in *Handbook of Genetics*, Vol. 5, R. C. King, ed. New York: Plenum, 1976.

Clarkson, S. G., and M. L. Birnstiel, "Clustered Arrangement of tRNA Genes of *Xenopus laevis*," *Cold Spring Harbor Symp. Quant. Biology*," *38* (1973), 451–59.

Cleveland, L. R., "Origin and Development of the Acromatic Figure," *Biol. Bull., 74* (1938), 41–55.

————, "Hormone-Induced Sexual Cycles of Flagellates. XII. Meiosis in *Barbulanympha* Following Fertilization, Autogamy, and Endomitosis," *J. Morph.*, *95* (1954), 557-619.

Cooper, J. E. K., and T. C. Hsu, "The C-band and G-band Patterns of *Microtus agrestis* Chromosomes," *Cytogenetics, 11* (1972), 295-304.

Cooper, K. W., "Cytogenetic Analysis of Major Heterochromatic Elements (Especially Xh and Y) in *Drosophila melanogaster,* and the Theory of Heterochromatin," *Chromosoma, 10* (1954), 535-88.

Crick, F. H. C., "Codon-Anticodon Pairing: The Wobble Hypothesis," *J. Mol. Biol.*, *19* (1966), 548-55.

————, "Split Genes and RNA Splicing," *Science 204* (1979), 264-71.

Davidson, E. H., et al., "Comparative Aspects of DNA Organization in Metazoa," *Chromosoma 51* (1975), 253-259.

Dickinson, H. G., and J. Heslop-Harrison, "Ribosomes, Membranes, and Organelles During Meiosis in Angiosperms," *Phil. Trans. Roy. Soc. London, B.*, *277* (1977), 327-42.

Dubochet, J., and M. Nöll, "Nucleosome Arcs and Helices," *Science 202* (1978), 280-86.

Filion, G. W., "Differential Giemsa Staining in Plants. I. Banding Patterns in Three Cultivers of *Tulipa Chromosoma 49* (1974), 51-60.

Fitch, W. M., "Is There Selection Against Wobble in Codon-Anticodon Pairing?" *Science, 194* (1976), 1173-74.

Fox, G. E., and C. R. Woese, "The Architecture of 5S rRNA and its Relation to Function," *J. Mol. Evol., 6* (1975), 61-76.

Gall, J. G., et al., "Repetitive DNA Sequences in *Drosophila,*" *Chromosoma, 33* (1971), 319-44.

————, "The Satellite DNAs of *Drosophila virilis,*" *Cold Spring Harbor Symp. Quant. Biol., 38* (1973), 417-21.

Georgiev, G. P., et al., "Isolation of Eukaryotic DNA Fragments Containing Structural Genes and the Adjacent Sequences," *Science, 195* (1977), 394-97.

Glover, D. M., and D. S. Hogness, "A Novel Arrangement of the 18S and 28S Sequences in a Repeating Unit of *Drosophila melanogaster* rDNA," *Cell, 10* (1977), 167-76.

Goodenough, U., and R. P. Levine, *Genetics.* New York: Holt, Rinehart & Winston, 1974.

Goodpasture, C., and S. E. Bloom, "Visualization of Nucleolar Organizer Regions in Mammalian Chromosomes Using Silver Staining," *Chromosoma, 53* (1975), 37-50.

Hahn, W. E., and C. D. Laird, "Transcription of Nonrepeated DNA in Mouse Brain," *Science, 173* (1971), 158-61.

Harford, A. G., "Ribosomal Gene Management in *Drosophila:* A Chromosomal Change," *Genetics, 78* (1974), 887-96.

Hennig, W., et al., "Structure and Function of the Y Chromosome of *Drosophila hydei,*" *Cold Spring Harbor Symp. Quant. Biology, 38* (1973), 673-84.

Hess, O. "Local Structural Variation of the Y Chromosome of *Drosophila hydei* and Their Correlation of Genetic Activity," *Cold Spring Harbor Symp. Quant. Biology, 38* (1973), 663-72.

Hsu, T. C. "A Possible Function of Constitutive Heterochromatin. The Bodyguard Hypothesis," *Genetics, 79* (1975), 137-50.

————, and F. E. Arrighi, "The Distribution of Constitutive Heterochromatin in Mammalian Chromosomes," *Chromosoma, 34* (1971), 243-53.

Hughes-Schrader, S., "Polarization, Kinetochore Movements, and Bivalent Structure in the Meiosis of Male Mantids," *Biol. Bull., 85* (1943), 265-300.

————, "Cytology of the Coccids (Coccoidae, Hemiptera)," *Adv. Genetics, 2* (1948), 127-203.

Ingle, J., and J. Sinclair, "Ribosomal RNA Genes and Plant Development," *Nature, 235* (1972), 30-32.

Jacq, C., et al., "A Pseudogene Structure in 5S DNA of *Xenopus laevis,*" *Cell, 12* (1977), 109-20.

Jones, R. N., "B-chromosome Systems in Flowering Plants and Animal Species," *Int. Rev. Cytology, 40* (1975), 1-100.

Jukes, T. H., "How Many Anticodons?" *Science, 198* (1977), 319-20.

Kato, H., "Spontaneous and Induced Sister Chromatid Exchanges as Revealed by the BUdR-Labelling Method," *Int. Rev. Cytology, 49* (1977), 55-98.

Kavenoff, R., et al., "On the Nature of Chromosome-Sized DNA Molecules," *Cold Spring Harbor Symp. Quant. Biology, 38* (1973), 1-8.

Keyl, H.-G., "Duplikation von Unterheiten der chromosomalen DNS während der Evolution von *Chironomus thummi,*" *Chromosoma, 17* (1965), 139-80.

Kornberg, R. D., "Chromatin Structure: A Repeating Unit of Histones and DNA," *Science 184* (1974), 868-71.

————. "Structure of Chromatin," *Ann. Rev. Biochem. 46* (1977), 931-54.

Kunz, W., and U. Schäfer, "Variations in the Number of the Y Chromosomal rRNA Genes in *Drosophila hydei,*" *Genetics, 82* (1976), 25-34.

Laird, C. D., "DNA of *Drosophila* Chromosomes," *Ann. Rev. Genetics, 7* (1973), 177-204.

Latt, A. A., "Longitudinal and Lateral Differentiation of Metaphase Chromosomes Based on the Detection of DNA Synthesis by Fluorescence Microscopy," in *Chromosomes Today,* P. L. Pearson and K. R. Lewis, eds. New York: John Wiley, 1976.

Lima-de-Faria, A., "Structural Differentiation of the Kinetochore of Rye and *Agapanthus*," *Chromosoma, 7* (1955), 78–89.

———, "The Evolution of the Structural Pattern in a Rye B Chromosome," *Evolution, 17* (1963), 289-95.

Lin, M. S., and R. L. Davidson, "Centric Fusion, Satellite DNA, and DNA Polarity in Mouse Chromosomes," *Science, 185* (1974), 1179-81.

Marks, G. E., "Variation of Giemsa Banding Patterns in the Chromosomes of *Anemone blanda*," in *Chromosomes Today*, P. L. Pearson and K. R. Lewis, eds. New York: John Wiley, 1976.

McClintock, B., "The Production of Homozygous Deficient Tissues with Mutant Characteristics by Means of the Aberrant Mitotic Behavior of Ring-shaped Chromosomes," *Genetics, 23* (1938), 315-76.

Migeon, B. R., "Clonal Analysis of Development: X-inactivation and Cell Communication as Determinants of Female Phenotypes," in *The Clonal Basis of Development*, S. Subtelny and I. M. Sussex, eds. New York, Academic Press, 1978.

Pardue, M. L., "Repeated DNA Sequences in the Chromosomes of Higher Organisms," *Genetics, 79* (1975), 159-70.

Paulson, J. R., and U. K. Laemmli, "The Structure of Histone-Depleted Metaphase Chromosomes," *Cell, 12* (1977), 817-28.

Pavan, C., and R. Basile. "Chromosome Changes Induced by Infections in Tissues of *Rhynchosciara angelae*," *Science, 151* (1966), 1556-58.

Pellegrini, M., et al., "Sequence Arrangement of the rDNA in *Drosophila melanogaster*," *Cell, 10* (1977), 193-212.

Prescott, D. M., Genetic Organization of Eukaryotic Chromosomes, in *Chromosomes: From Simple to Complex*, (P. A. Roberts, ed.) Corvallis: Oregon State University Press, 1974.

Quigley, G. J., and A. Rich, "Structural Domains of Transfer RNA Molecules," *Science, 194* (1976), 796-806.

Rhoades, M. M., "Preferential Segregation in Maize," *Genetics, 27* (1942), 395-407.

———, "The Cytogenetics of Maize," in *Corn and Corn Improvement*, 2nd ed., G. Sprague, ed. New York: Academic Press, 1977.

———, and E. Dempsey, "The Effect of Abnormal Chromosome 10 on Preferential Segregation and Crossing Over in Maize," *Genetics, 53* (1966), 989-1020.

———, et al., "Chromosome Elimination in Maize Induced by Supernumerary B Chromosome," *Proc. Nat. Acad. Sci., 57* (1967), 1626-32.

Ritossa, F. M., and S. Spiegelman, "Localization of DNA Complementary to Ribosomal RNA in the Nucleolus Organizer Region of *Drosophila melanogaster*," *Proc. Nat. Acad. Sci., 53* (1965), 737-45.

Rosenberg, H., et al., "Highly Reiterated Sequences of SIMIANSIMIANSIMIAN-SIMIANSIMIAN," *Science, 200* (1978), 394–402.

Sabour, M., "RNA Synthesis and Heterochromatinization in Early Development of a Mealy Bug," *Genetics, 70* (1972), 291–98.

Sager, R., and R. Kitchin, "Selective Silencing of Eukaryotic DNA," *Science, 189* (August 8, 1975), 426–33.

Sandler, L., "Evidence for a Set of Closely Linked Autosomal Genes That Interact with Sex-Chromosome Heterochromatin in *Drosophila melanogaster,*" *Genetics, 86* (1977), 567–82.

Schnedl, W., "Banding Patterns in Chromosomes," *Int. Rev. Cytology, Suppl. 4,* (1974), 237–72.

Schrader, F., "Touch-and-Go Pairing in Chromosomes," *Proc. Nat. Acad. Sci., 26* (1940), 634–36.

Sedat, J., and L. Manuelidis, "A Direct Approach to the Structure of Eukaryotic Chromosomes," *Cold Spring Harbor Symp. Quant. Biol., 42* (1977), 331–50.

Seth, P. K., et al., "Comparison of Human and Non-Human Primate Chromosomes Using the Fluorescent Benzimidozol and Other Banding Techniques," in *Chromosomes Today, 5* (1976), 315–22.

Smith, J. D., "Genetics of Transfer DNA," *Ann. Rev. Genetics, 6* (1972), 235–56.

Söll, D., "Enzymatic Modification of Transfer RNA," *Science, 173* (1971), 293–99.

Stein, G., and J. Stein, "Chromosomal Proteins: Their Role in the Regulation of Gene Expression," *BioScience, 26* (1976), 488–97.

Tjio, J. H., and A. Levan, "Some Experiences with Acetorcein in Animal Chromosomes," *Ana. Est. Exp. Aula Dei, 3* (1954), 225–28.

Vosa, C. G., "Heterochromatin Classification in *Vicia faba* and *Scilla sibirica,*" in *Chromosomes Today, 5* (1976), 185–92.

Wallace, H., and W. H. R. Langdridge, "Differential Amphiplasty and the Control of Ribosomal RNA Synthesis," *Heredity, 27* (1971), 1–13.

Weintraub, H., and M. Groudine, "Chromosomal Subunits in Active Genes Have an Altered Conformation," *Science, 193* (1976), 848–56.

Wellauer, P. K., and I. B. Dawid, "The Structural Organization of Ribosomal DNA in *Drosophila melanogaster,*" *Cell, 10* (1977), 193–202.

White, M. J. D., "The Cytology of the Cecidomyidae (Diptera). IV. The Salivary Gland Chromosomes of Several Species," *J. Morph. 82* (1948), 53–80.

White R. L., and D. S. Hogness, "R Loop Mapping of the 18S and 28S Sequences in the Long and Short Repeating Units of *Drosophila melanogaster* rDNA," *Cell, 10* (1977), 177–92.

Wilson, G. B., and E. R. Boothroyd, "Studies in Differential Reactivity. I. The Rate and Degree of Differentiation in the Somatic Chromosomes of *Trillium erectum* L.," *Canad. J. Res. (C), 19* (1941), 400–12.

Woodcock, C. L. F., "Reconstitution of Chromatin Subunits," *Science 195* (1977), 1350–52.

Woodward, W. R., and E. Herbert, "Coding Properties of Reticulocyte Lysine Transfer RNAs in Hemoglobin Synthesis," *Science, 177* (1972), 1197–99.

Yunis, J. J., and M. E. Chandler, "The Chromosomes of Man—Clinical and Biologic Significance," *Amer. J. Pathology, 88* (1977), 466–95.

Yunis, J. J., and W. G. Yasmineh, "Heterochromatin, Satellite DNA, and Cell Function," *Science, 174* (1971), 1200–09.

Yunis, J. J., et al., "The Characterization of High-resolution G-banded Chromosomes in Man," *Chromosoma 67* (1978), 293–307.

4

Cell Division: The Basis of Genetic Continuity and Transmission

The dictum of Rudolph Virchow that cells arise only from pre-existing cells holds true for all organisms, regardless of the simplicity or complexity of their cellular contents. Cell division in the prokaryotes follows a seemingly less complicated course than that of the eukaryotes, but in both the result is the same: the production of a population of cells of equivalent genetic endowment. Among unicellular forms, such cell division is a means of increasing the number of individuals in a population. Among multicellular organisms, however, cell division serves two purposes. First, during the growth phase of an individual, and coupled with the phenomenon of cell differentiation, it is a process of building the organism to its mature proportions. A whale and a mouse differ in cell numbers, but not so much, if at all, in the size of individual kinds of cells (nerve and muscle cells may be a significant exception to this statement). Second, cell division is also a process of maintaining mature proportions once these are achieved. In mammals, for example, the cells of the skin, blood, and digestive tract have rather limited life spans, and as these cells die they must be replaced at a comparable rate if the body is not to be depleted of particular cell types. If somatic cell division is to serve these two purposes of growth and maintenance, it must, therefore, be a conservative process, preserving and transmitting the genetic endowment from one cell to another without serious deviation or error. As we have seen, the structure and behavior of the chromosomes insure that this will occur.

Somatic cell division is, in a sense, uniparental: One cell gives rise to another cell. The introduction of sexuality among organisms meant that biparentalism had become inserted into the life cycle, and a number of cytogenetic innovations were introduced as a result. Without trying to ascertain the sequential occurrence of all of the steps during evolutionary time, the fusion of gametes and their nuclei during fertilization required the existence of a compensatory type of cell division if the number of chromosomes were not to be doubled at each generation. Such specialized cell division, or meiosis, leads to the formation of haploid eggs and sperm in animals and to haploid spores in most plant species while at the same time insuring that each gamete or spore has a complete set of chromosomes even though the maternal and paternal chromosomes are randomly distributed. Since the same chromosomes in different individuals are likely to differ among themselves because of structural and genetic changes that are constantly occurring, meiosis is a means both of introducing genetic diversity into a population and of transmitting to the next generation essential genetic information necessary for species preservation. Meiosis, as a consequence, is a far more complicated type of cell division than is somatic cell division.

Mitosis

The basic elements of cell division include replication of the hereditary material of the cell and the subsequent segregation of this DNA into two genetically equivalent cells. There may be added features which make the process more complicated, but they are secondary in that their insertion into the division cycle is to insure that the basic processes are accomplished. The term mitosis is now generally used to cover the entire process of cell division (other than meiosis), although in a strict etymological sense, it refers only to the nuclear division of eukaryotic cells.

Cell division in a prokaryotic cell such as a bacterium is a relatively simple process, with replication of its DNA and division of the cell usually coordinated in time. If one is inhibited so is the other, with cell division following replication in about 20 min at 37°C. However, cells with two or more nuclei are found often enough to suggest that the coordination is not exact. The single, naked chromosome is associated with an infolded portion of the bacterial membrane called a *mesosome* (Fig. 2.1), with the point of attachment also being the site at which replication will be initiated. Although the details of the process remain obscure, the two halves of the replicating chromosome come to lie on either side of the mesosome, and as the latter forms a septum to divide the cell, each half will receive a single chromosome.

There is some indication that the sex (*F*) factor is similarly associated with the bacterial membrane and that its replication and separation are coupled to membrane elongation, but this is not so for the numerous plasmids found in many bacterial cells. Their replication, and possibly that of the *F* factor as well, proceeds independently of that of the bacterial chromosome, so that replication and cell division, as coordinated events, apply only to the host chromosome and not to other DNA inclusions.

The process of mitosis in the eukaryotic cell is a far more complex affair, with the conventional stages in a plant tissue depicted in Fig. 4.1. If the dividing cell contains centrioles (Fig. 4.2), these had previously been divided in the preceding cell generation. If there is a comparable structure in cells lacking a centriole, as is true for all of the flowering plants, it has gone unrecognized. The next conspicuous event is the replication of the chromosomes, but prior to this the cell has increased in size, and presumably, has brought its metabolic machinery up to the proper level for division. Division is a tumultuous affair in the life of a cell, and time for recovery between successive divisions is necessary. It is only in the early cleavage divisions of a sea urchin egg, for example, that the initial cellular mass is divided, successively, into halves, quarters, eighths, etc., with no intervening growth periods between divisions. The necessity for a recovery period is indicated by the fact that a human cell in tissue culture takes about 18 hr to complete an entire division cycle, yet only 45 min is spent in going from earliest prophase through telophase. The remaining 17 hr or more are in interphase, with the cell preparing for the replication of its hereditary material and the partitioning of the cytoplasm. A roottip cell of the onion, or the broad bean, *Vicia faba*, would have a cell cycle of comparable length.

Fig. 4.1. The stages of eukaryotic cell division. As the cell enters division, having previously spent a long time preparing for it, the chromosomes become more and more visible as they shorten by compaction, and a longitudinal split appears along their length except at the position of the centromere. The spindle appears at metaphase, and the two chromatids are separated by the action of the spindle at anaphase (e, f, g). The cell plate (in plant cells) or a furrow (in animal cells) cuts the cell into two daughter elements. Karyokinesis, or mitosis, refers to the nuclear events; cytokinesis to the partitioning of the cytoplasm (from Swanson & Webster, 1977).

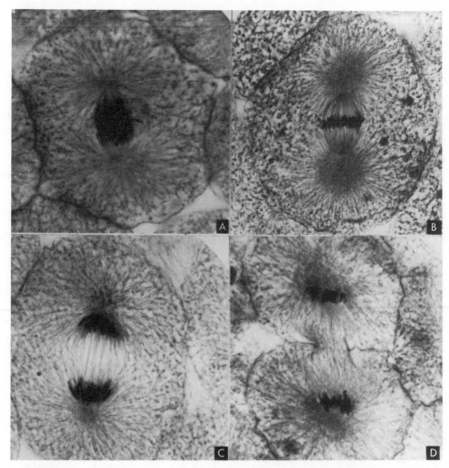

Fig. 4.2. Mitotic division in the cells of the whitefish embryo. The centriole, while not visible at this magnification, lies at the center of the centrosome, from which the spindle fibers and astral rays radiate. A furrow instead of a cell plate divides the cell in two.

The ususal cell cycle can be conveniently divided into four recognizable periods: G_1, S, G_2, and M (Fig. 4.3). The first three make up interphase and have pertinence with respect to DNA synthesis. S is defined as the period of DNA synthesis as judged by the incorporation of DNA precursor molecules into the chromosome. It is also the period of histone formation and its association with DNA. The S period may occur during early, mid, or late interphase, or, as in some cells such as those in the rapidly growing grasshopper embryo, it may occupy nearly the entire interphase period. G_1 and G_2 (G = gap) refer, respectively, to the portions of the interphase period preceding or following DNA synthesis, and M (also labeled D for division by some authors) is the division phase of the cell cycle beginning with the earliest recognizable stage of prophase and terminating with the end of telophase and the formation of two daughter cells.

In a population of dividing cells, whether in tissue culture or in a rapidly growing tissue, G_1 can be thought of as a preparatory period for S. A sufficient pool of nucleotides for DNA synthesis, and a similar pool of amino acids for protein, and particularly histone formation, must be synthesized and assembled; the appropriate enzymes and energy sources must be present and available; and the DNA of the chromosome must be primed for replication. In bacterial cells, a *replicator,* which is at the point of origin of replication and is at or near the point of attachment to the bacterial membrane, must be activated in order to initiate DNA synthesis. The activator appears to be a gene product, but what prompts this gene to release its product is not known. Even less is known about what triggers mitosis in eukaryotic cells, but it is becoming evident that the cell membrane and microtubule proliferation and disposition within the cell are somehow involved. With DNA synthesis providing an initial and necessary commitment for cell division, the stimulus seems to be cytoplasmic rather than nuclear. Thus, nuclei such as those from cells in the frog brain, and which would not normally synthesize DNA during their mature lifetime, can be induced to do so by being inserted into mature oocytes which are ready to begin development (immature or unfertilized oocytes do not possess such an inductive influence).

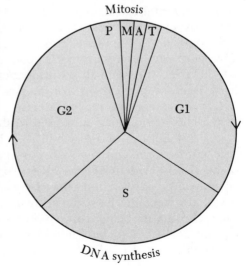

Fig. 4.3. A division of the cell cycle into recognizable stages. The period of synthesis is determined by the time taken to incorporate radioactive thymidine into DNA; G1 is that portion of interphase before DNA synthesis, G2 that after DNA synthesis, with G1, S and G2 constituting the whole of the interphase stage. The stages of mitosis are relatively short compared with the long duration of interphase, but all of the stages can be of varying lengths, depending on the organism, type of cell, temperature, and other external variables (from Swanson & Webster, 1977).

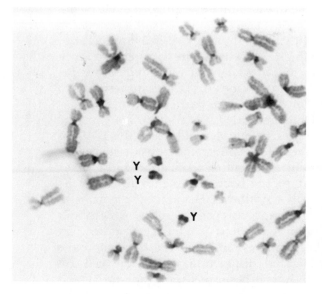

Fig. 4.4. Human cell at meta-phase. The cell was exposed during the S period to bromode-oxyuridine, which when suitably stained, is darker than the chromatin lacking this substance. The late-replicating heterochromatin is found adjacent to the centromeres of many of the chromosomes, and particularly in the Y chromosomes. This cell was derived from an individual having an aberrant sex chromosome complement of XYYY (courtesy of T. C. Hsu).

The S period can be readily followed through the use of radio-actively labeled DNA precursors such as thymidine. Its duration in eukaryotes is about 30% to 40% of the cell cycle, but its absolute length can vary enormously. In human and hamster cells in culture, 6 hr out of 18 hr are spent in S; in broad bean roottip cells, 7.5 hr out of 13.0 hr. Lily microspores spend 15 hr in S. In the salivary gland cells of *Drosophila melanogaster,* the S period is over a 12-hr span even though only euchromatin is being replicated, while in *Drosophila* embryos, the nuclei divide every 9 min. Furthermore, the chromosomes, or parts of them, do not replicate synchronously, or from one end to the other. Some chromosomes replicate early, others late, with the heterochromatic regions being generally delayed (Fig. 4.4). During this time the chromosomes are not visible as such in the light microscope, except for masses of chromatin, probably of a heterochromatic nature, but electron micrographs often show much of the chromatin being appressed to the nuclear membrane.

Prophase is a period when the chromosomes undergo consider-able contraction and the two chromatids of each chromosome be-come visibly evident (Fig. 4.5). In some forms it would appear that chromosome contraction (the chromatid is the contracting unit and it will also become the segregating unit) occurs by the double helix in each chromatid being thrown into a series of coils; in others the chromatids seem to be folded back and forth, or even haphazardly arranged as a tight ball of strands. Whatever the manner of contrac-

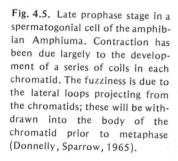

Fig. 4.5. Late prophase stage in a spermatogonial cell of the amphibian Amphiuma. Contraction has been due largely to the development of a series of coils in each chromatid. The fuzziness is due to the lateral loops projecting from the chromatids; these will be withdrawn into the body of the chromatid prior to metaphase (Donnelly, Sparrow, 1965).

tion, the mechanism remains unknown, although some kind of DNA-protein interaction is involved. As contraction proceeds, the nucleolus or nucleoli diminish in size, and by late prophase they have disappeared; in a few forms the nucleolus may persist into anaphase and telophase stages.

By late prophase, preparations for the segregation of the chromatids and the partitioning of the cytoplasm are underway. If the cells possess centrioles, the two units have migrated from each other until they lie at opposite sides of the nucleus, with a spindle being formed between them and embracing the nucleus (Fig. 4.2). This orientation of centrioles determines the orientation of the spindle, and the latter in turn determines the plane of cytokinesis. When centrioles are not a normal part of the divisional apparatus, two aggregations of microtubules develop next to the cell membrane and from previously synthesized units of *tubulin,* the protein from which the spindle forms. These aggregations determine the planes of cytokinesis since the axis of the spindle will be perpendicular to the plane connecting the two aggregations.

Metaphase follows prophase and is a state of momentary equilibrium (Figs. 4.6 and 4.7). The nuclear membrane has dissolved, and the microtubules have established an orientation running from pole to pole (spindle fibers) and from pole to the kinetochore portion of the centromere (half-spindle fibers). A functional division of the centromeric region of each chromosome takes place (presumably this

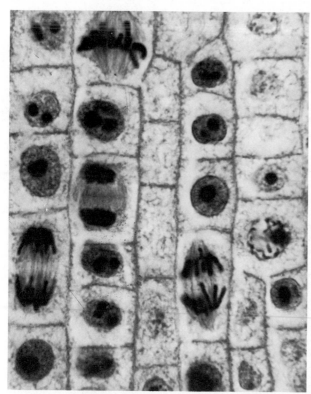

Fig. 4.6. (Left) Longitudinal section through the roottip of onion, stained with haematoxylin to reveal the spindle elements, which are clearly visible in the metaphase, anaphase and telophase cells.

Fig. 4.7. (See below) Metaphase and anaphase cells from the whitefish embryo. Contrast the character of the spindle in these cells with those of the onion roottip cells in Fig. 4.6.

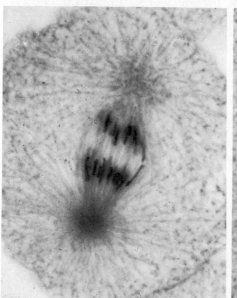

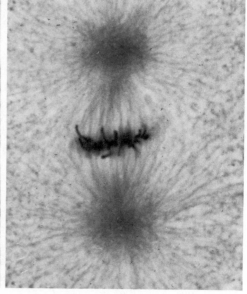

region had been replicated along with the remainder of the chromosome during the previous S period), and the separated sister (identical) chromatids move to opposite poles. This period of movement is the anaphase portion of the mitotic cycle, and in animal cells in particular, the separation of the chromatids is further increased by an elongation of the spindle in its central region.

In the next stage, telophase, the nuclear membrane reforms, the chromatids (now more correctly called unreplicated chromosomes) lose their compacted structure and elongate to attain an interphase state, the nucleoli make their appearance, and the centrioles, if present, undergo replication in "anticipation" of the next cell division. The latter accomplish this by growing a new centriole perpendicular to the old one. Cytokinesis then divides the cell into two daughter cells. In animal cells this is accomplished by a process of *furrowing* and in plant cells by the formation of a *cell plate*. Interphase then ensues.

The nuclear membrane is not totally formed as a new structure. Elements of the endoplasmic reticulum and undissolved fragments of the previous nuclear membrane aggregate around the polar group of chromosomes and contribute to the nuclear membranes of the daughter cells. The remainder of these membranes, however, must be reformed from the fatty acids and proteins present in the cytoplasm.

In a genetic sense, cell division, as just described, can be viewed as a process evolved to provide a group of cells each with an identical genetic endowment. All of the events of division contribute to that goal, and the exactness of replication of the chromosomal DNA and the separation of chromatids insures that the daughter cells will possess, quantitatively and qualitatively, the same genetic constitution as the cell from which they arose. Cell division is, therefore, a conservative process, both in a morphogenetic and an evolutionary sense.

Variations in Cell Division. The type of cell division just described is characteristic of large groups of eukaryotic organisms: all of the photosynthetic land plants and virtually all of the multicellular animals that have been examined cytologically. This suggests, of course, that the significance of cell division is so fundamental to genetic continuity that it evolves but very slowly. This conservatism extends to the molecular level as well; in such diverse and distantly related organisms as the pig and the unicellular alga, *Chlamydomonas,* the tubulin of their spindle microtubules shows extreme conservatism and similarity in amino acid sequences and in biochemical reactivity. Structural diversity, however, exists, and the fungi, some algae, and numerous unicellular species display their own processes of cell division even though the genetic consequences are still the same.

The most primitive form of cell division is that of furrowing, which is characteristic of animal cells. The rigid wall of higher plants places mechanical constraints on a dividing cell, and a form of cell division evolved which makes use of a *phragmoplast* to assist in the formation of the cell plate, as well as the new cell membranes which arise from the fusion of Golgi-derived vesicles. The phragmoplast forms from microtubules at the outer edges of the spindle, and then it moves outward toward the cell wall, forming the cell plate as it goes. Some algal groups have a similar kind of division, but with the phragmoplast less well developed; others, however, form a *phyco-plast,* an assemblage of microtubules with the individual tubules oriented in the plane of cytokinesis. The formation of the phycoplast brings the nuclei, widely separated by anaphase movement, very close to each other. The phycoplast presumably is involved in membrane and wall formation since, like the phragmoplast, it moves outward until it reaches the edges of the cell.

The dinoflagellates and some protozoans differ not only in the structure of their chromosomes but also in that the nuclear membrane does not, at any time, break down (Fig. 2.22). The spindle forms externally to the nucleus, with the microtubules running between two polar, strap-like centrioles. The nucleus invaginates and forms a channel in which the spindle lies. The chromosomes are attached to the nuclear membrane, and from the point of attachment to the polar centrioles. As the centrioles move apart to their respective poles, the segregation of chromatids occurs. Elongation of the spindle takes place to move the chromatids farther apart, followed by a pinching of the nucleus into two parts and then furrowing of the cell to effect cytokinesis.

Endomitosis and Polyteny. The belief that all cells of an organism, barring the rare accident, must have the same genic and chromosomal constitution was a logical outgrowth of the chromosome theory of inheritance, reinforced as it were by the regularity and accuracy of mitotic processes. The occurrence of varying chromosome numbers and of irregular mitoses in neoplastic tissues were viewed as abnormalities brought on by diseases—as indeed they are—but it gradually became evident that differentiation and histological change were often accompanied by an orderly change in the chromosomal situation of constituent cells, and that each tissue had its own characteristic pattern of nuclear behavior. This was most readily observable in some insects where the cells of some tissues became very much enlarged during larval life, with the nuclei enlarging in parallel fashion. Growth in these instances was by cell enlargement rather than by an increase in cell number.

In proliferating tissues such as root tips or the vertebrate blood-forming centers, DNA synthesis, mitosis, and cytokinesis follow each other in orderly sequence. Cytokinesis, however, may often be delayed until after a number of nuclear divisions has taken place, or it may be entirely omitted from the sequence. For example, endosperm formation in maize or in coconut, or early embryogeny in a Drosophila egg, consists of many nuclear divisions in rapid sequences, after which the nuclei migrate toward the nuclear membrane, where cell membrane formation cuts the common cytoplasmic mass into uninucleate cells. When cytokinesis is entirely omitted, a coenocytic mass of cytoplasm is achieved with few or no cross walls, a characteristic condition in some algae and in the water molds, *Phycomycetes.* When nuclear division is absent, division of the cell cannot take place. But replication can occur independently of mitosis and cytokinesis, and it obviously means that the DNA content of the cell is increased, usually doubled. Thus, in 1925, Jacobj showed that when nuclear volume in some tissues was plotted against frequency, the curve was not a continuous one, but rather was made of a series of maxima which corresponded to exact doubling of the nuclear volume. Since nuclear volume was known to be related to the amount of DNA, the doublings of volume must, therefore, be related to replication events, but without accompanying cytokinesis. The phenomenon took place in cells which were no longer destined to divide, and it resulted in two different nuclear results: 1) if DNA synthesis took place, and mitosis was initiated but was aborted before anaphase separation of the chromatids could take place, cells with double the number of chromosomes would result (Fig. 4.8); a process which the German cytologist, Geitler, termed *endomitosis*; and 2) if DNA synthesis took place without contraction of the chromosomes or any evidence of mitosis, the chromosomes became multi-stranded, or *polytene,* in a lateral direction.

The male water strider, *Gerris lateralis,* with an XO genome, has a chromosome number of 21, with the odd chromosome being the X which stains deeply in interphase cells. It is possible, consequently, to determine the number of chromosomes in the cell by simply counting the number of deeply stained X's in the nucleus. Giant cells in the salivary glands were 512-ploid, and an occasional cell was 1024- or 2048-ploid. This would mean that 8, 9, or 10 rounds of replication has taken place without intervening mitoses, but the separateness of the X chromosomes from each other seemed to indicate that at least some kind of partial mitosis was involved. Geitler showed that the formation of polyploid cells parallels the mitotic behavior of diploid cells, to a degree. Replication obviously takes place during the S period, and the cells enter prophase. The chromo-

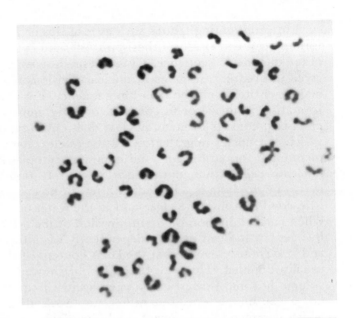

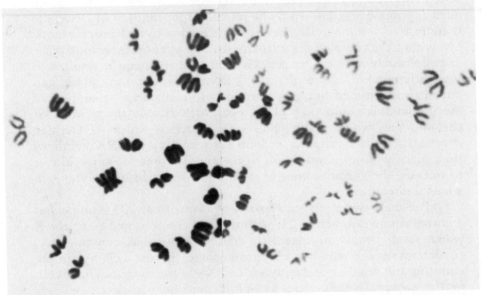

Fig. 4.8. A normal (top) and an endomitotic (bottom) metaphase obtained from the leucocytes of a freemartin calf. The chromosomes in the endomitotic cell have come to metaphase but the failure of spindle formation would eventually lead to a return to an interphase state and the formation of a restitution nucleus with double the number of chromosomes (courtesy of Marie Tompkins).

somes contract normally and become more stainable, but the process is interrupted in late prophase before the nuclear membrane breaks down and the spindle forms. The chromatids of each chromosome separate from each other, uncoil, and then return to a diffuse interphase state. The chromosome number has been effectively doubled, and the nuclear volume is enlarged to accommodate the increased number of chromosomes. To acquire a higher level of endopolyploidy, a cell must be capable of undergoing a series of endomitoses. The most extreme example known has been described in the mollusk, *Aphysia*. The nucleus in the giant neuron has a volume of about 70×10^6 μm^3, and contains 75,000 times the haploid amount of DNA.

A similar series of events also takes place in the tapetum of higher plants, a tissue surrounding the sporogenous cells in the anther, and in the antipodal nuclei of the embryo sac. Fig. 4.9 illustrates the appearance of several antipodal nuclei in the embryo sacs of the ranunculaceous species, *Papaver rhoeas* and *Acontium ranunculifolium*. As many as six doublings of the DNA content can occur, thus raising the nuclei to 128-ploid. The degree to which the chromosomes are distinctly separate from each other is related to the degree to which the chromosomes were contracted in the early stages of the endomitoses. The bundles of chromonemata are held together at their centromeric regions, or by attachment to the nucleolus at the nucleolar organizer region.

The polytene chromosomes of dipteran species have been described (Figs. 3.55–3.58), and their cytogenetic usefulness has been discussed. No mitoses of even a partial sort have occurred to separate the replicated strands, with the result that the chromosomal diameters increase as the number of strands increase. The replications, however, are only partial since the constitutive heterochromatin adjacent to the centromeres, and the Y chromosome, are underreplicated, appearing as a diffuse mass called the chromocenter.

Cells which undergo endomitoses to become highly polyploid, or whose chromosomes become polytenic, are generally those in a terminal state of differentiation; no further mitoses would normally ensue. Endopolyploidy, however, is not a bar to subsequent mitoses (although polyteny of the sort encountered in *Drosophila* probably is). The diploid number of chromosomes in the mosquito, *Culex pipiens*, is 6. During late larval, and early pupal, life, the iliac epithelium is significantly increased in size, primarily by cell enlargement. Nuclear volumes increase by about 4-fold. Between the 8th and 14th hours of pupal life, nuclear volumes return to a diploid volume by a series of reduction divisions. It appears that a succession of

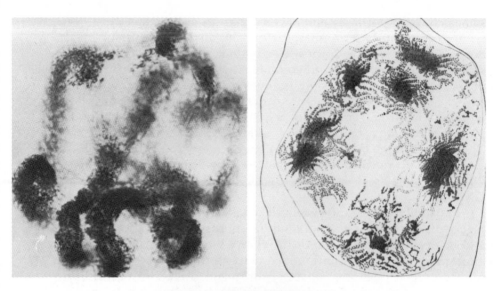

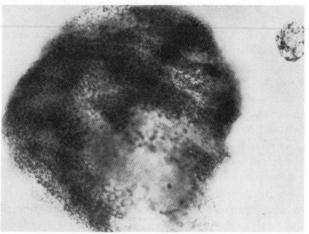

Fig. 4.9. Endomitotic nuclei from the antipodals of angiosperm embryo sacs. Upper left, *Aconitum ranunculifolium,* indicating that in some cells the highly replicated chromosomes take on the appearance of dipteran polytene chromosomes, whereas in other cells (bottom) the same degree of polyploidy may be expressed as a greatly enlarged interphase nucleus. Upper right, *Papaver rhoeas,* in which the individual haploid chromosomes are separate from each other, but with each subdivision of a chromosome joined with others at their centromeres (Aconitum figures from Tschermak-Woess, 1956, that of Papaver from Hasitschka, 1956).

endomitoses increases the amount of DNA per cell, with each chromosome ending up as a bundle of strands similar to those in Fig. 4.9, and with each cell being 4-, 8-, 16-, or 32-ploid. The chromosomal situation falls somewhere between the state of dipteran polyteny and the free chromosome state in the water striders. During metamorphosis, the chromosomes enter prophase, condense, and separate into individual units, with each one reaching the metaphase plate as longitudinally dual structures (chromatids) that have synapsed. These then segregate in anaphase to bring about a reduction in the number of chromosomes per nucleus. Later divisions follow a similar pattern, that is, the chromosomes condense as separate strands without previous replication, synapse to form homologous structures, and segregate in anaphase. All cells of the adult iliac epithelium are eventually 4- or 8-ploid instead of the earlier much higher polyploid state. Similar reduction divisions in somatic cells are known to transform the chromosome number from $2n$ to n in the cottony-cushion scale, but in both species the phenomenon is one of unknown significance.

Endomitosis and polyteny are not haphazard occurrences, and the distinction between them is a superficial one, resting on the degree to which mitoses are, or are not, involved. In *Lestodiplosis,* a gall midge, White has shown that all cells except one of the salivary glands are polytene; this is a single giant cell in each gland that is both polytene and polyploid, suggesting that several Gerris-like endomitoses preceded the institution of polyteny. In attempting to assess the significance of both endopolyploidy and polyteny, it seems clear that these two phenomena, together with gene amplification (p. 337), gene dosage compensation (p. 120), and chromosome elimination or inactivation (p. 456), are examples of the ability of the genome to regulate its genic output at the cellular or tissue level. Although the mosquito example seems to be an exception to the rule, the fact that the two phenomena are associated generally with terminally differentiated, metabolically active, and often secretory, cells or nuclei supports this contention: the macronuclei of ciliates (p. 175); the hypopharyngeal gland cells of the honey bee which secrete royal jelly; the cells of the spinning gland of web-forming spiders; the nurse and salivary gland cells of dipteran species; and the cells of the fat bodies and Malpighian tubules of many insects. Endopolyploidy and polyteny would obviously pose problems for any actively dividing cells, and one can only wonder why the epithelial cells of the mosquito need to go up and then down the polyploid scale to achieve their final differentiated ends.

Meiosis

Cells dividing mitotically are generally thought of as undifferentiated units producing cells of similar nature which subsequently may either retain their meristematic or embryonic character or become differentiated into a variety of cell types performing a variety of specialized activities. Meiotic cells, by way of contrast, are already differentiated and are genetically programmed to carry out a series of activities in a particular sequence and in such a manner as to produce a group of daughter cells—gametes or spores—quite different, genetically and structurally, from the meiotic mother cells. These activities—pairing of homologous chromosomes, crossing over between homologues, the reduction in chromosome number, the slow pace of meiotic prophase, the requirement of two cell divisions instead of one to complete the process, and the lack of an S period between the two divisions—clearly set meiosis apart from mitosis. One or more of these processes can occur sporadically in some mitotic cells, and at times on a regulated basis, e.g., the reductions in chromosome number which take place in the gut epithelium of the mosquito, so it is most reasonable to assume that meiosis evolved from mitosis, but it is the entire complex of phenomena, occurring in tissues or organs set aside during development for this purpose, that characterizes meiosis. Meiosis, therefore, made necessary by biparentalism, is not only the counterpart of fertilization, in that meiosis reduces chromosome number, it is also a significant means for increasing diversity among related individuals in a species through the processes of chromosome segregation and gene recombination.

The German cytologist Van Beneden, in 1883-1884, demonstrated that equal numbers of chromosomes were contributed to the offspring at the time of fertilization. Prior to this time it had been recognized that the dividing nuclei of a particular individual, or individuals in a species, possessed a characteristic number of chromosomes and that the essential feature of plant and animal fertilization was the fusion of paternal and maternal nuclei after the sperm had penetrated into the egg. It followed, therefore, that the contribution of each parent to its offspring was a single, or haploid, set of chromosomes to provide the zygote with a double, or diploid, set. All cells of the offspring that were subsequently derived from the zygote by mitosis possessed a diploid set of chromosomes, with each chromosome represented twice. August Weismann, another German biologist, recognized in 1887 that the doubling of the chromosome number that occurs in fertilization would lead to unmanageable numbers of chromosomes unless a means was at hand to counteract this buildup. Meiosis, of course, is that process.

There is, as might be expected, a parallel between the degree of ploidy (haploidy, diploidy, etc.) of a nucleus and the amount of DNA present; the two vary directly. If C represents the amount of DNA in a haploid sperm, egg, or spore, a diploid cell such as a zygote, or a somatic cell at the end of mitosis, would contain a $2C$ amount of DNA. (Note that the C value refers to the amount of DNA per nucleus and bears no necessary relation to chromosome number, the latter being expressed by the value of n.) As the cell prepares to divide, replication of the DNA during the S period would double the amount to $4C$, with the amount again being reduced to $2C$ with the anaphase separation of chromatids. Cells in a dividing somatic tissue would show a regular cycle of DNA changes (Fig. 4.10), with meiosis reducing the amount of DNA to C during gamete formation.

In broad terms, the first division of meiosis separates the homologous chromosomes, which had paired with each other in prophase, into two $2C$ cells. The second division, which is not preceded by an S period, separates the sister chromatids, with the result that four $1C$ (and also $1n$) nuclei are produced. In animals, meiosis immediately precedes fertilization and produces sexual cells, the eggs and sperm. Their union in fertilization gives a diploid zygote, which, through numerous mitoses coupled with the processes of growth and cell differentiation, develops into an individual characteristic of each species.

Most plant groups differ from the animals in exhibiting an alternation of diploid and haploid generations. The time relationships between meiosis and fertilization vary widely in plants, as do the size

Fig. 4.10. Sequences of stages in the life cycle of a sexually reproducing animal correlated with the changes in the amount of DNA per cell. S, period of synthesis in interphase; A, mitotic anaphase; A_1, first meiotic anaphase; A_2, second meiotic anaphase.

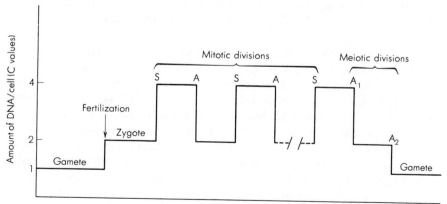

and duration of the sporophytic (diploid) and gametophytic (haploid) generations. In some algae and fungi, meiosis immediately follows fertilization, and the resultant cells, which are haploid asexual spores, germinate to produce a haploid thallus. This structure, the gametophyte, produces gametes by mitosis, and the zygote derived from their union undergoes meiosis without any intervening mitotic divisions. There develops, as a consequence, no diploid body form. Since direct phenotypic expression of all genes occurs, the question of recessiveness or dominance is not normally a factor in genetic analysis.

The higher plants, including here the mosses, ferns, seed plants, and some of their ancient allies, as well as some of the more advanced algae, possess a similar life cycle, except that the diploid zygote produces a diploid sporophyte through mitotic division and development. Meiosis occurs in specialized structures of the sporophyte: sporangia in the ferns and conifers, ovules and anthers in the flowering plants. Thus, a diploid plant body intervenes between fertilization and meiosis, as it does in animals, giving the forms of plant life with which we are most familiar. With each gene and each chromosome represented twice in each cell, the question of the dominance and recessiveness of alleles emerges when genetic analyses are made.

For convenience, meiotic prophase, which is unique for its duration and sequence of events, is separated into various stages, each of which possesses certain characteristics of structure and behavior permitting ready identification. DNA and histone synthesis takes place in pre-meiotic interphase, but a difference from the typical somatic interphase is evident in that a unique meiotic histone is also synthesized, at least in lily and tulip microsporocytes. This histone, which differs in amino acid composition from histones of somatic tissues, persists throughout meiosis, microsporogenesis, and pollen maturation, but its relation to meiotic chromosomes or its significance to the meiotic process itself remains unknown.

Preleptonema. This stage, corresponding generally to G_2 of a mitotic division, is a time when preparations are being made to switch from a mitotic to a meiotic sequence of events. That preleptonema is not just another G_2 period, however, is indicated by the fact that in some species a distinct timing adjustment is made that requires a number of the preceding cell divisions for completion. Thus, in wheat anthers, it has been shown that the three somatic cell cycles immediately preceding meiosis in the sporogenous tissue are gradually lengthened from 25 hr to 35 hr to, finally, 55 hr in duration, bringing the cells into line with the slower pace of meiosis. How this

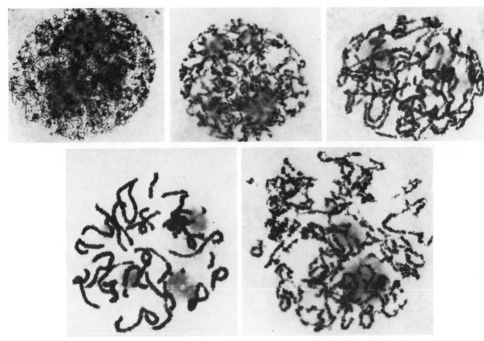

Fig. 4.11. The spiralization-despiralization events taking place in the microsporocytes of *Lilium longiflorum*. A, a premeiotic cell. B - D, stages of spiralization which progress to what would be the equivalent of a late mitotic prophase with the nucleoli still attached to their respective chromosomes (D). E, beginning of the despiralization process which returns the nucleus to a state similar to that in A, but with the nuclear volume much greater (courtesy of Marta S. Walters, 1970, and with permission of Springer-Verlag).

regulation is achieved remains unknown, but a prior commitment to meiosis seems evident some time before meiosis actually is initiated.

On the other hand, some premeiotic cells, particularly those in the lily anther, but also observed in humans and a number of other mammals, appear to initiate a mitotic-like contraction of the chromosomes, advance to a point where individual chromosomes are recognizable, and then undergo a despiralization to lose their individual identity and return to a diffuse, preleptotene state (Fig. 4.11). Only then do they shift their pathway of development and enter the leptotene stage of meiosis (Fig. 4.12). There is no indication that the spiralization–despiralization events are in preparation for meiosis: the chromosomes are randomly dispersed throughout the nuclear area, there is no semblance of an alignment of homologues, no crossing over has occurred, and the events are not uniformly present in all

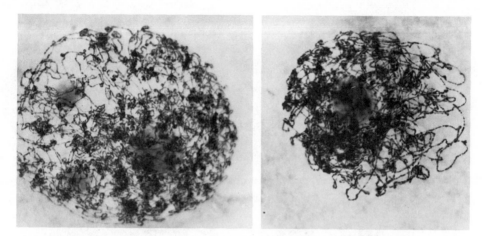

Fig. 4.12. Early (left) and late (right) leptotene stages in the microsporocytes of *Lilium longiflorum*. These cells have presumably gone through the spiralization-despiralization events depected in Fig. 4, and have now entered the normal pathway of meiosis. The chromosomes show no evidence yet of homologous pairing which will take place in the zygotene stage (courtesy of Marta S. Walters, 1970, and with permission of Springer-Verlag).

microsporocytes in a single anther. Such mitotic-like steps are often seen in cultured lily anthers, suggesting that slight environmental changes can push the cells, which are ready to divide, in a mitotic or a meiotic direction; in the anther, however, some as yet undetermined influence exerts its effect such as to make the meiotic direction a dominant one. In those species where the transition between mitosis and meiosis is sharp and unequivocal, the pattern of spiralization and despiralization does not occur.

Leptonema (Thin-Thread Stage). This stage differs from the earliest prophase stages of mitosis in that the chromosomes are longer and more slender; it is impossible to follow a single chromosome throughout its entire length (Figs. 4.12 and 4.13). Although DNA synthesis has occurred, no longitudinal doubleness of the chromosomes is visibly evident in the light microscope as it is in early mitotic prophase. However, the chromosomes already show a substantial degree of contraction, exhibiting a series of chromomeres which, at a later stage in prophase, become more evident in size, character, and position. The chromosomes in the spermatocytes of animals behave similarly (Fig. 4.14). As pointed out in the previous chapter, the chromomeres are a reflection of localized coiling, or contraction patterns, and as such can serve as landmarks for the identification of particular chromosomes. It would appear that the pattern of

Fig. 4.13. Microsporocytes of *Lilium longiflorum*, in late leptotene to very early zygotene stage (left), and with the possible beginnings of homologous pairing showing at the upper left where the chromomeres seem matched, and zygotene (right) with paired and unpaired regions clearly distinguishable (courtesy of Marta S. Walters).

Fig. 4.14. Spermatocyte nucleus of *Amphiuma means tridactylus* in early zygonema. Some homologous regions are synapsed, others are not. The chromomeres are quite regular in size and distribution at this stage; they tend to become irregularly so, but in characteristic fashion, in pachynema (see Figs. 4.13 and 4.17 (Donnelly and Sparrow, 1965).

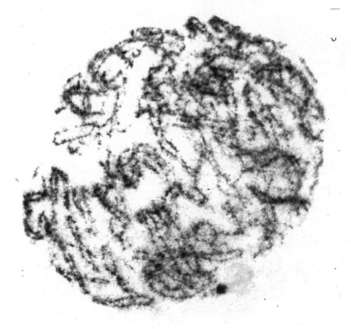

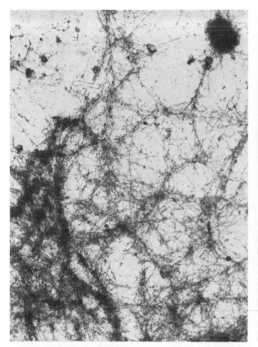

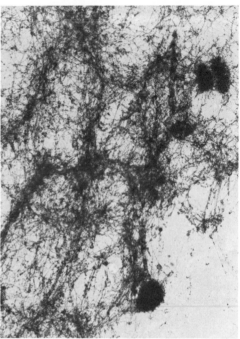

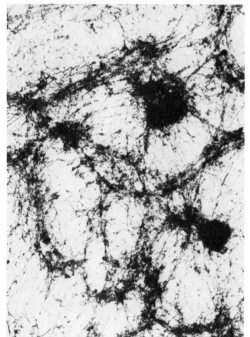

Fig. 4.15. Electron micrographs of spermato-cyte nuclei in the milkweed bug, *Oncopeltus fasciatus*, prepared by being first floated onto a water surface, lifted off, and then dried by the critical point method, a technique which de-hydrates the specimen without serious distor-tion. Top left, leptotene stage; top right, zygotene stage; and left, pachytene stage. The denser clumps of chromatin are sex chromo-somes which condense earlier than the auto-somes and the typical chromomere pattern of meiotic prophase seems to be absent. (courtesy of Dr. S. Wolfe).

contraction in the small chromosomes of some insects may be quite different; distinctively localized chromomeres are absent (Fig. 4.15), but the appearance of the chromosomes may be due to the method of preparation rather than to some intrinsic difference in contraction.

The strikingly larger size of both cell and nucleus of the typical meiotic cell is also evident in leptonema, and the chromosomal changes that will take place will proceed at a leisurely pace; in the lily somatic cell, for example, mitotic prophase is completed in 30 min to 60 min, while a meiotic prophase in the microsporocyte is from 4 days to 6 days in duration. Each species, however, has its own meiotic rate, a topic to which we will return later.

A cell destined to enter leptonema is not irrevocably committed to do so; in the lily it can be diverted experimentally so that it undergoes a more or less typical mitotic division. The timing, however, is critical. If the diversion by means of a temperature manipulation, or by explanting the microsporocytes from anther to a cell culture, is carried out immediately prior to the leptotene state, the cell will enter a mitotic prophase, progress to metaphase, but fail to go through anaphase because the centromeric regions, for some reason, do not separate properly. That region of the chromosome, but not the cell itself, reflects a prior commitment to a meiotic behavior. A leisurely progression of the preleptotene cell up to and into leptonema is necessary if the subsequent events are to follow in proper order.

Leptonema is not ordinarily a point of stoppage in the meiotic sequence of events, but such a stoppage for a period of time occurs in the embryo sac mother cell of the orchid, *Dactylorhiza.* The inhibition is overcome only by a successful pollination, after which meiosis proceeds normally.

Zygonema (yoked-thread stage). The active pairing, or synapsis, of homologous chromosomes into intimate association takes place in zygonema (Figs. 4.16 and 4.17). Because each parent has contributed a haploid set of chromosomes to a diploid offspring, and because the chromosomes contributed by the sperm—the sex chromosomes may be an exception—are usually identical to those contributed by the egg, all diploid cells possess pairs of morphologically and genetically similar, or homologous, chromosomes. These chromosomes pair lengthwise with each other, with pairing being initiated at one or more sites along the length of the chromosomes. When once initiated, synapsis proceeds to bring the homologous pairs into intimate chromomere-by-chromomere alignment along their entire length, the progress of pairing proceeding in zipper-like fashion. There is some

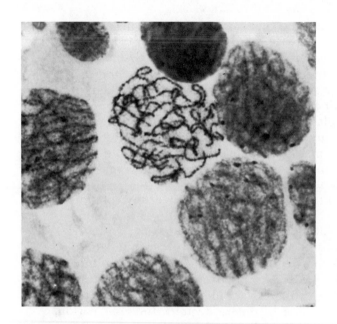

Fig. 4.16. A group of spermatocytes from the salamander, *Plethodon vehiculum*. The central one is in pachytene, all of the others in zygotene stages (courtesy Dr. J. Kezer).

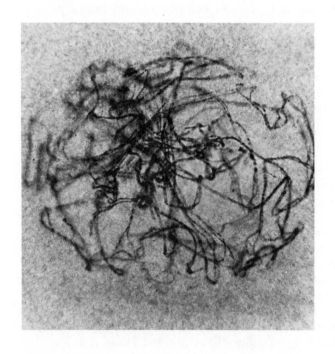

Fig. 4.17. Zytogene stage of meiosis in the lily, *Lilium regale*. In the bottom and upper right of the figure, some portions of the homologues still remain unpaired (courtesy of Dr. J. McLeish).

suggestion in certain organisms that a specific arrangement of the chromosomes resulting from the preceding mitotic division facilitates pairing, but other than the chromosomes being attached to the nuclear membrane by their telomeres or heterochromatic masses there is no recognizable pattern of homologues that is invariable and peculiar to all meiotic cells at this stage.

By the end of leptonema each homologue has developed a lateral component of nucleoprotein between its two chromatids, and as synapsis takes place the lateral components combine with a central core to form the *synaptinemal complex* (Fig. 4.18). The "force" or process that promotes synapsis and keeps it going remains unknown, but the central core, possibly binding the homologues in tight association, is believed to be a ribonucleoprotein derived from the nucleolus. The enzymatic action of both ribonuclease and trypsin can prevent the formation of the synaptinemal complex, or bring about its dissolution, with both the central core and the lateral components being open to attack.

Meiosis can proceed in the absence of synaptinemal complexes, and the complex can apparently form between nonhomologues in haploid cells, but all evidence points to the necessity of the complex for subsequent crossing over and chiasma formation. An interpretation of the structure of the complex, as set forth in Fig. 4.15, indicates that the complex is very much shorter in length than the calculated length of the total DNA of the haploid nucleus. This is borne out by several sets of measurements. For example, the total complex in *Drosophila* is 110 μm long as compared to its 61 mm of DNA; in *Neurospora* the comparable figures are 50 μm vs. 16 mm. Thus, only a very small proportion of the DNA (0.2% in *Drosophila*, 0.3% in *Neurospora*, and between 0.017% and 0.014% in maize) is in intimate synapsis within the complex and with its opposite homologue; the remainder of the DNA is looped or coiled to form the heavier chromatin boundaries of the complex. What determines the particular segments of DNA to be encased in the complex remains unknown. The fact that recombination can occur between any two genes would suggest that the encased DNA is randomly determined, but it may well be that within that randomnesss selected nucleotide sequences perform a critical role.

The initiation of pairing and the formation of the synaptinemal complex are also correlated with an additional series of cellular changes: the appearance of a new lipoprotein whose function is not yet ascertained, the parallel appearance of a DNA-binding protein which catalyzes the reassociation of single-stranded DNA (and hence

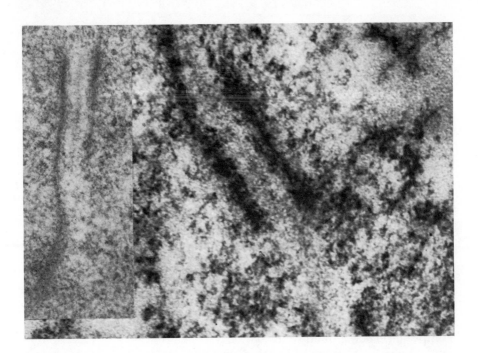

Fig. 4.18. Electron micrographs and interpretations of the synaptinemal complex. Right, the longitudinal structure running from top to bottom is a synaptinemal complex, a bivalent in early meiotic prophase, probably pachytene, of the rooster. The complex consists of two dense lateral elements flanking a less dense central core through which another longitudinally oriented central element is faintly visible. Both the lateral elements and the central core seem to be formed of ribonucleoprotein, the former derived from the paired homologues and the latter from material transported from the nucleolus. The continuity of the synaptinemal complex is interrupted by what has been interpreted as a chromosome puff or a lateral loop; it may also be due to the complex dropping out of the plane of sectioning. Upper left, a comparable axial section of a bivalent in a rat spermotocyte. Below, interpretations of the complex; left, cross-section of two homologous chromosomes, each of which has formed a lateral element between the sister chromatids; middle, the paired homologues with the lateral elements indicated as projected structures of ribonucleoprotein, with the central core formed between them; right, the formation of a chiasma within the central core (electron micrographs, courtesy of Dr. M. Moses; interpretations redrawn from D. von Wettstein, see Riley, Bennett and Flavell, 1977).

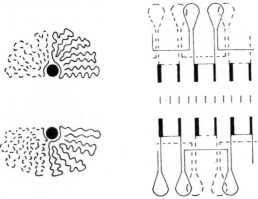
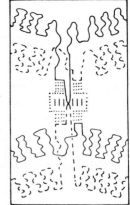

known as the reassociation protein or r-protein), and a pattern of DNA synthesis. The amount of DNA synthesized at this time is only 0.3% to 0.4% of the total DNA of the cell, but it is unrelated to rDNA and its composition differs from that of bulk DNA in having a higher G-C content. It now appears that this is a delayed replication phenomenon and that small blocks of DNA, scattered widely throughout the chromosome set, pass through the S period in an unreplicated state and are only replicated later in zygonema. These blocks of DNA are about 10^4 base pairs long and are found in unique rather than repetitive DNA. If inhibitors of DNA synthesis, such as hydroxyurea or 5-aminouracil, are given to leptotene cells, they will not pass over into the zygotene stage; if the same inhibitors are given to the cells in zygonema, meiosis is blocked as is the formation of the synaptinemal complex. In addition, the inhibition of DNA synthesis in zygonema also leads to the subsequent appearance of broken chromosomes, with the breaks being formed, presumably, at the sites of zygotene replication. It would appear, therefore, that the integrity of the chromosomes during meiosis is dependent upon the successful completion of DNA synthesis in zygonema. However, no free single-stranded ends of DNA can be detected in the zygotene nucleus, and the newly synthesized DNA is readily released from the bulk DNA and sedimented as a 15S to 20S component. Consequently, it would seem that this DNA, although synthesized in zygonema, is not immediately incorporated into the chromosome by the action of a ligase. Incorporation, in fact, does not occur until 5 or 6 days later, as the meiotic prophase is terminating.

It is the belief of Stern and his colleagues, who have carried out much of the work on zygotene DNA synthesis, that the unreplicated blocks of DNA serve to hold sister chromatids close together from the preceding S period to zygonema, that these regions are also the first to pair in zygonema, and that the completion of DNA synthesis at these sites is necessary for the formation, but not the structure, of the synaptinemal complex, and, hence, for the retention and progression of pairing. A carry-over of the phenomenon into the next stage of meiosis suggests that these sites are possibly the locations at which crossing over will take place. There is, on the other hand, no critical supporting evidence for this suggestion.

The situation as described for the lily does not hold for all species, even for those in the monocots. There is evidence that DNA synthesis continues throughout meiosis in wheat and rye, that the DNA synthesized does not exhibit distinctive base ratios, and that an inhibition of this synthesis does not lead to an impairment of the synaptic process or to the production of chromosome breaks.

Pachynema (thick-thread stage). The persistence of pairing, maintained by the presence of the synaptinemal complex and by a continuing protein synthesis, is carried over into pachynema. This is a stable and easily identified stage of meiotic prophase; the chromosomes, visibly thicker not only because they are paired but also because considerable contraction has occurred, appear to be present in a haploid number, but each can be recognized as two closely appressed homologues (Figs. 4.19 and 4.20; see also Fig. 3.4). These pairs are now referred to as *bivalents*. Any unpaired homologues, or regions of homologues, will remain so since active pairing does not

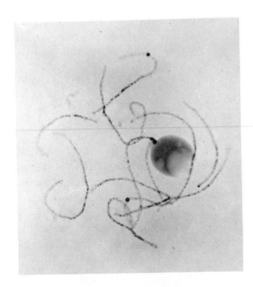

Fig. 4.19. Pachytene chromosomes of maize ($2n = 20$). The nucleolus is attached to chromosome 6 (see Fig. 2.15), and dark-staining knobs are present on chromosome 7 (top) and chromosome 5 (lower center). (Courtesy of Dr. M. M. Rhoades.)

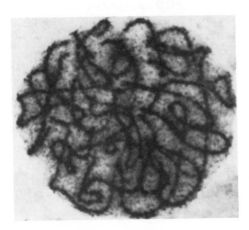

Fig. 4.20. Pachytene stage in a spermatocyte of a salamander. The chromomeres are very regularly spaced along the length of the paired homologues, no prominent knobs are present as in maize or teosinte, and pairing appears to be complete for all of the chromosomes (courtesy of Dr. J. Kezer).

take place during pachynema. As Figs. 3.3 and 3.66 reveal, each species has a distinctive chromomere pattern, while strains of maize or its close relative, teosinte, also show distinctive patterns of knob number, size, and distribution among the members of the chromosome set (Fig. 3.32). Nucleoli are particularly evident at this stage, being large and attached to specific chromosomes at their nucleolar organizer regions (Figs. 3.4 and 3.16).

Additional DNA synthesis also occurs during pachynema, but several features distinguish this from that taking place in zygonema. It is much less in amount; it is associated with the appearance or the activation of a pachytene-specific endonuclease; the composition of the synthesized DNA is comparable to that of the total cellular DNA; its formation is not essential for the continuation of the meiotic process; and it occurs preferentially in regions of the genome containing moderately repeated nucleotide sequences about 200 base pairs in length rather than in larger blocks of unique DNA, as happens in zygonema. Inhibition of pachytene DNA synthesis does cause some chromosome damage in the forms of breaks, but the aberrations are only one-tenth as frequent as those resulting from zygotene inhibition, and they do not make their appearance until the second meiotic division. As nearly as can be ascertained at the moment, pachytene DNA synthesis is not the continuation of a delayed semiconservative replication of zygonema but rather is related to a repair-replication system which closes up the broken ends of chromosomes formed as a result of crossing over, an hypothesis supported by the observation that chiasmate and achiasmate cells differ as to the amount of DNA nicking and repair than occurs during pachynema.

The differences in DNA biochemistry which distinguish zygonema from pachynema point to the fact that there are specific regions of the chromosomes, of a specific nucleotide character and length, which carry out specific meiotic functions at discrete periods during the meiotic process. The changing morphological character of the chromosomes is paralleled by a changing biochemistry, and this argues for an internal kind of chromosomal differentiation comparable to that known to exist at the genic level.

Diplonema (double-thread stage). The termination of pachynema coincides with the dissolution of the synaptinemal complex, the lapsing of the synaptic force, and the beginning of diplonema, so named because the longitudinal duality of each chromosome into two chromatids becomes clearly evident. DNA synthesis in interphase and zygonema was responsible for this duality, but it did not reveal itself visually until the attraction of the homologues to each

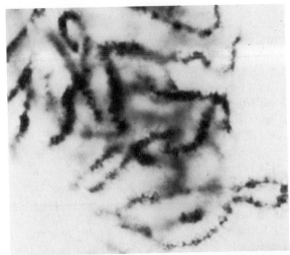

Fig. 4.21. Early diplonema in *Amphiuma* as the chromosomes are beginning to contract. The lateral loops projecting from the chromomeres suggest that they are in an actively transcribing state (Donnelly and Sparrow, 1965).

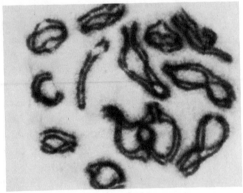

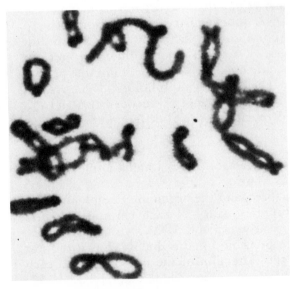

Fig. 4.22. Diplotene stages later than that in 4.17: above, salamander; left, *Amphiuma*. The fuzziness due to the projection of the lateral loops is still evident (top, courtesy of Dr. J. Kezer; bottom, Donnelly and Sparrow, 1965).

other lapsed and they fell apart. The separation is generally not complete, however, for the paired chromosomes are held together at one or more points along their length. The changes taking place in diplonema are striking, depending upon the particular organism and on the progress of contraction (Figs. 4.21, 4.22, and 4.23). The paired homologues, or bivalents, take on the appearance of a cross if there is one point of contact, a loop if there are two points of contact, or a series of loops if there are three or more points of contact (Fig. 4.22). The longer the paired chromosomes, the greater, in general, is the frequency of contact points (Fig. 4.23).

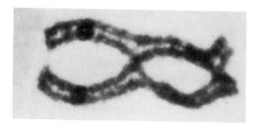

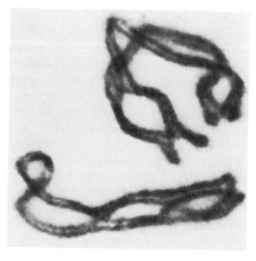

Fig. 4.23. Individual bivalents from the diplotene stage of a salamander. Above, two chiasmata are present in the right arm (the centromeres are visible as the dark areas to the left), but whether there was a chiasma in the left arm is debatable; if present earlier, it is now terminalized. Right, both bivalents show four and possibly five chiasmata (courtesy of Dr. J. Kezer).

Each point of contact is a *chiasma* (pl., chiasmata), and in clear preparations of diplotene stages it can be seen that only two of the four chromatids at any given point, one from each homologue, are involved and have exchanged pairing partners (Figs. 4.23). Because the two sister chromatids of any chromosome do not separate laterally to any great extent, and since they are closely united at their centromeres, the chiasma is a point of chromatin exchange that helps to preserve the bivalent structure.

The equivalence of a cytologically detected chiasma and a genetically determined crossover is taken for granted by many, and it is a reasonable assumption to make on the basis of bivalents such as those in Fig. 4.23, but no unequivocal proof of this is available. However, there are numerous pieces of circumstantial evidence pointing to such an equivalence and there is no single piece of contradictory information; therefore, the stance here will be that they are the same phenomenon viewed through two different techniques.

The number and approximate position of the chiasmata are variable, being under genetic control and dependent upon the species in question and the length of the chromosomes. In those species whose chiasmata are somewhat randomly distributed along the chromosome arms, longer chromosomes will naturally have more chiasmata than the shorter ones, but even short bivalents appear to form at least one if only as an aid to normal disjunction. If two or more chiasmata are present in a single chromosome arm, they are generally evenly spaced rather than being clustered in given regions. However, a random distribution of chiasmata, as determined cytologically, may obscure patterns of crossing over that are distinctly nonrandom. The synthesis of DNA during meiotic prophase would suggest such a nonrandomness, but it is obvious that support for this can be obtained only when good cytological and genetical maps of particular chromosomes can be interpreted on a biochemical basis.

Extreme localization of chiasmata occurs in a number of species. Localization may be either proximal (procentric) or distal, and in some instances it may be distinctly different in the two sexes. For example, in the spermatocytes of the grasshopper, *Stethophyma grossum*, where all of the chromosomes are acrocentric, the chiasmata at meiotic metaphase in spermatocytes are completely procentric, while those in the oocytes are either interstitial or distal. A similar situation is found in the liliaceous form, *Fritillaria meleagris*; here two of the pairs of chromosomes are metacentric while the remainder are acrocentric or subterminal. The opposite obtains in the palmate newt, *Triturus helveticus;* a distal localization is characteristic of the spermatocytes while being interstitial in the oocytes (Fig. 4.24). The pattern of localization, or even lack thereof, seems not to be a function of the degree of synapsis or of the position of the centromere; on the other hand, if the localization is an accurate reflection of the position of earlier crossing over, then significantly large blocks of chromatin (and genes?) are tightly linked and relatively free from recombination.

When procentric localization is encountered, the question arises as to whether the chiasmata form only in or near centric heterochro-

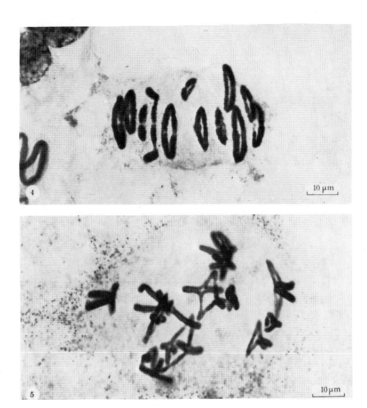

Fig. 4.24. Meiotic chromosomes at metaphase I in *Triturus helveticus.*
Top, a spermotocyte showing all of the chiasmata located at the distal
ends of the chromosomes; bottom, an oocyte with the chiasmata at
more proximal positions (Watson and Callan, 1963).

matin, a suggestion made more provocative by the discovery in the
plant, *Scilla campanulata,* that more chiasmata form in late than in
early replicating regions. Late replication, of course, is generally
characteristic of heterochromatin. The same argument relating late
replication to chiasmata localization could not be made for distal
distributions since heterochromatin is rarely distally located, and it
could not apply to species in which localization is different in the
two sexes. What these examples do reveal, however, is that chiasma
formation, like any other cellular process, is under genetic control
and that the controlling system in one sex can differ from the controlling system in the opposite sex.

The distribution and localization of chiasmata at the first meiotic metaphase do not necessarily indicate that the chiasmata were
first formed in these regions. As the chromosomes continue to contract during meiotic prophase, whether by coiling or by folding, the
chiasmata have a tendency to move toward the ends of the paired

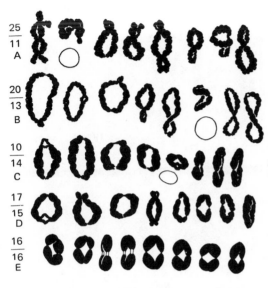

25/11 A

20/13 B

10/14 C

17/15 D

16/16 E

Fig. 4.25. Bivalents of *Campanula persicifolia*, showing the changes in chiasma distribution and number as terminalization takes place during meiotic prophase from diplonema (A) to metaphase (E). The total number of chiasmata (above) and terminal below is given for each nucleus; the nucleolus is indicated as an open circle.

homologues, a process called *terminalization* (Fig. 4.25). Movement of chiasmata is more pronounced the shorter the chromosomes, probably because of the torsional strains developed in the chromosomes by the contraction process; terminalization may, in fact, be sufficiently great to cause the chiasmata to be lost at the ends of the bivalents, thereby reducing the chiasma frequency in metaphase vs. diplotene stages.

Although spermatogenesis in animals and microsporogenesis in the higher plants are continuous processes when once initiated, diplonema in the oocytes of vertebrates can be a stage of temporary stoppage. In frogs, for example, diplonema may last a matter of a year or more, while in the human female the stage may persist for 12 to 50 years. The human oocytes are formed during fetal life and have entered meiosis and stopped in diplonema prior to birth. Many more oocytes are formed than are ovulated during the reproductive periods, but they do not proceed through meiosis unless fertilized. During this prolonged diplonema the chromosomes, which have previously synapsed and undergone crossing over, assume a diffuse state, that is, the lampbrush stage described earlier (Fig. 3.23), or as that which occurs in some insects (Fig. 4.26). Although meiotic progression ceases, the chromosomes are in an active state of synthesis, and in the frog oocyte at least, the extended loops are transcribing RNA and the nucleolar organizer region is replicating itself on a selective basis. This point will be discussed at length in the next chapter.

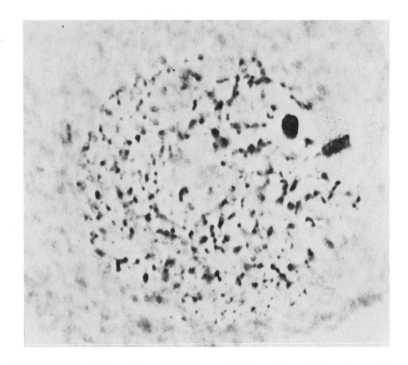

Fig. 4.26. The diffuse stage of diplonema in the milkweed bug, *Oncopeltus fasciatus,* taken through both light (top) and electron (bottom) microscopes. The highly condensed bodies are the sex chromosomes, which undergo condensation earlier than the remainder of the chromatin (courtesy of Dr. S. Wolfe).

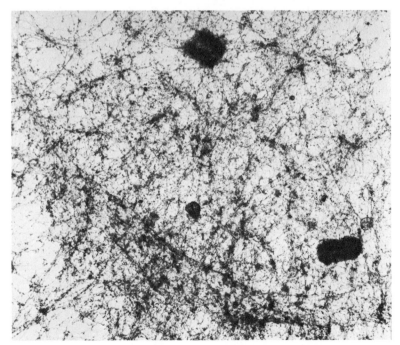

The prolonged diplonema, often called the *dictyotene* stage in order to distinguish it from the typical diplonema, has been suggested as the possible cause of the increased number of chromosomal abnormalities among the offspring of older mothers. Five of the chromosomes of the haploid human set, numbers 13, 14, 15, 21, and 22, have nucleolar organizer regions, and it may well be that segregation of the homologues in meiosis becomes impeded by the sharing of a common nucleolus for a period of many years. Such a suggestion should be viewed with caution, however, because while the nucleolar organizer-bearing chromosomes are involved in abnormalities, and with increasing frequency in older mothers, even more abnormalities are related to the sex chromosomes, and these do not possess nucleolar organizer regions. In addition, recently obtained evidence indicates that the meiotic error leading to the occurrence of Down's syndrome may be spermatogenic as well as oogenic. The use of chromosome banding techniques can pinpoint the origin of the extra members of the genome, and it appears now that a substantial proportion of such trisomics are due to paternal meiotic errors, and hence may be unrelated to the phenomenon of dictyotene delay.

Diakinesis. The distinction between diplonema and diakinesis is not a clear one, at least as judged from the study of spermatocytes and pollen mother cells, although diakinesis is generally characterized by a more contracted state of the bivalents, their wide dispersal throughout the cell as the nuclear envelope breaks down, and by the vacuolization, dissolution, and eventual disappearance of the nucleolus (Fig. 4.27). No obvious cytogenetic events of interest occur in this stage, but cytoplasmic preparations for the formation of the spindle are obviously going on, and there is evidence of a renewed activity of the nucleolar organizer regions. Thus, the meiocytes, and particularly the pollen mother cells of higher plants, have had their cytoplasm "scrubbed" of ribosomes and RNA (resulting in a loss of basophilia) during the period between leptonema and late diplonema. The loss of basophilia is paralleled to an increased presence of ribonucleases, acid phosphatases, and other hydrolases, these presumably arising from pre-packaged lysosomes which release their contents into the cytoplasm. Such RNA as can be detected in the cells has an A/G ratio characteristic of chromosomal but not ribosomal RNA. The plastids and mitochondria are also affected during this period, becoming reduced in size and losing their internal membranes and ribosomes.

A subsequent return of basophilia in diakinesis is brought about by the renewed activity of the nucleolar organizers which shed a succession of small nucleoli (nucleoloids) into the cytoplasm as the old nucleolus disappears. These newly formed nucleoli will be the source of the ribosomes necessary for the functioning of the gametophyte. The plastids and mitochondria also return to their former state. This system differs, therefore, from that in the amphibian oocyte where the enrichment of the ribosomal population of the cytoplasm is achieved by gene amplification of the rRNA loci, followed subsequently by nucleolar formation (p. 337). In the simplest terms, the changes taking place in higher plant cells would seem to be a means whereby the protein-synthesizing machinery of the sporophyte gives way to the protein-synthesizing machinery of the gametophyte. Since the sporophytic ribosomes could presumably have

Fig. 4.27. Diakinesis in rye *(Secale cereale)* (top) and the Lily *(Lilium longiflorum)*, showing the highly contracted bivalents, the nucleoli attached to the nucleolar organizer bearing chromosomes, and the bivalents dispersed throughout the cell as a result of nuclear envelope dissolution (top, courtesy of Dr. R. Nilan; bottom, courtesy of Dr. Marta S. Walters).

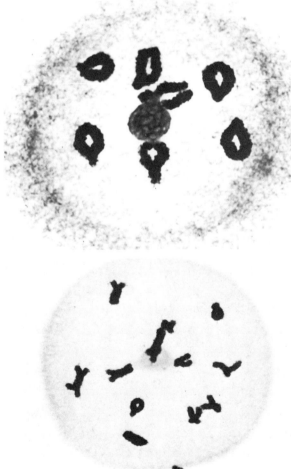

served equally well as a site for the translation of gametophytic mRNAs, it must be assumed further that the genetic systems of higher plants cannot direct the selective removal of sporophytic mRNAs without affecting the ribosomes, and hence must scrub the cytoplasm of all RNA and then repopulate it with new ribosomes to serve as sites for the gametophytic messages.

The continued contraction of the chromosomes, which is greater than it is in mitosis, can be viewed not only as a mechanical necessity for transforming the long and tenuous chromosomes into more maneuverable units within the confines of a cell but also for disentangling homologues which have been twisted about each other as a result of crossing over and chiasma formation. The disentanglement had been begun earlier by the lapsing of homologous attraction in the diplotene stage; complete separation will not occur until anaphase disjoins the homologues.

Metaphase I. Once the bivalents have been moved to the center of the cell, metaphase is a period of relatively little movement of chromosomes. The nuclear envelope has disappeared, the spindle has been fully formed, and the bivalents have ceased their jockeying for position at the equatorial plate (Figs. 4.28, 4.29, and 4.30). The homologues of each bivalent are co-oriented in a position midway between the poles; that is, the paired homologues, still joined by the previously formed chiasmata, have their functionally undivided centromeres equidistant from the metaphase plate and oriented in the long axis of the spindle. The distance between the homologous centromeres is determined by the distance from the centromere to the nearest chiasma, and hence is not the same for all bivalents.

Fig. 4.28. Metaphase I of meiosis in *Amphiuma* with the bivalents arranged on the metaphase plate, and with their centromeres co-oriented and directed towards opposite poles. The chiasmata are not strongly terminalized, but this is more or less typical of large chromosome pairs (Donnelly and Sparrow, 1965).

Fig. 4.29. Metaphase I (left) and anaphase I (right) in maize. The stretching of the chromosomes to the poles can be seen in the left figure while the right one indicates that at times the homologues may have difficulty separating from each other (courtesy of Dr. M. M. Rhoades).

Fig. 4.30. Metaphase I and anaphase I in the milkweed bug, *Oncopeltus*. Top left, a side view of metaphase I, with the sex chromosomes only loosely paired with each other; bottom left, a polar view of metaphase I, with the sex chromosomes occupying the center of the metaphase plate, and with the X chromosome to the right and the lighter Y chromosome to its left; right, anaphase I (courtesy Dr. S. Wolfe).

The differences between the metaphase chromosomes of mitosis and meiosis is emphasized by the manner by which each engages the spindle. The functionally single mitotic centromere, located at the metaphase plate, is dual-faced, each chromatid having a kinetochore which interacts with the nearest pole and establishes microtubular connections. Its meiotic counterpart, however, presents but one face to the spindle even though both kinds of chromosomes are presumably fully replicated through the centromeric region as well as through the arms. This difference can be demonstrated experimentally. It is possible, by micromanipulation, to insert a meiotic bivalent from a metaphase stage into a mitotic cell of comparable divisional status, or the reverse, a mitotic metaphase chromosome into a meiotic cell at metaphase. In each instance, the inserted chromosomes retain a capacity for division and behave as they would have in their own cellular environment, indicating that the behavior of the centromeres is conditioned by their previous exposure and is not immediately influenced by their new environment. The unique character of the meiotic chromosome is also revealed by the behavior of univalents. Should a pair of homologues fail to be united by one or more chiasmata, they arrive at the metaphase state as unpaired univalents. In this state they have difficulty in achieving a proper orientation on the metaphase plate, and consequently, difficulty in segregating at anaphase. Nondisjunction frequently ensues, although distributive pairing may aid in some instances (see p. 239). On the other hand, single X chromosomes, as in XO males, exhibit an evolved behavior which circumvents this difficulty; without loss of their meiotic character, they undergo precocious contraction and precocious movement, passing to one or the other of the poles prior to the regular segregation of the other bivalents.

During metaphase, or even before the spindle is fully formed, there is evidence, derived particularly from an examination of the spermatocytes of mantids, that the homologous centromeres are being "pulled" toward opposite poles where the centrosomes are located. The phenomenon, not evident in all organisms, has been referred to as a *pre-metaphase stretch*; it illustrates the manner by which the microtubules of the spindle bring the bivalents into a state of co-orientation. In some meiotic cells, as in the spermatocytes of the lobster (Fig. 4.31), the bivalents become connected and drawn to the poles by the half-spindle fibers first, after which a connection to the other pole is established, and the bivalents are moved to a position of equilibrium on the metaphase plate. How the microtubules, formed of the protein tubulin, bring the bivalents into position remains unknown, but it is possible, by micromanipulation,

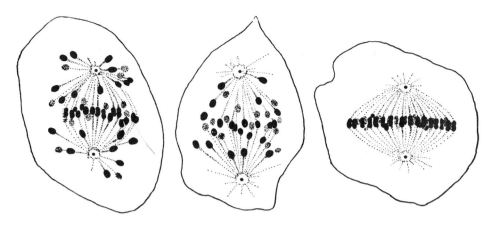

Fig. 4.31. Spermatogenesis in the lobster, illustrating how the bivalents become attached first to one pole of the spindle, and eventually to both poles, by the microtubules, or fibers. This process brings the bivalents on to the metaphase plate prior to their anaphase separation.

to displace the bivalents repeatedly from the spindle; on release they quickly assume a normal position on the spindle by forming new connections with the poles.

Anaphase. The movement of chromosomes from the metaphase plate to the poles constitutes anaphase (Figs. 4.29, 4.30, 4.32 and 4.33). The centromeres of the homologues in a bivalent remain functionally undivided as they are moved poleward, with the result that whole chromosomes instead of chromatids segregate. Each

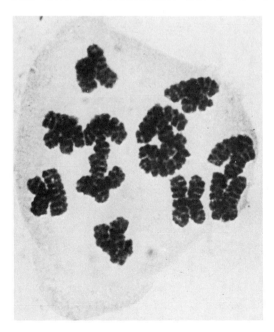

Fig. 4.32. The meiotic chromosomes of *Tradescantia sp.* at anaphase I, showing the major coils in each of the chromatids. The sister chromatids are held together only at their centromeres.

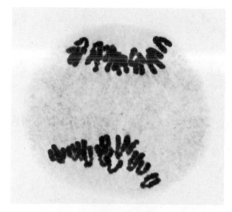

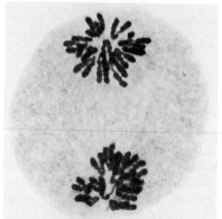

Fig. 4.33. Mid and late anaphase I in the microsporocytes of the lily (courtesy of Dr. Marta S. Walters).

Fig. 4.34. Anaphase I and prophase II in a salamander to show that there is relatively little change in the character of the chromosomes during the interval between the two meiotic stages (courtesy of Dr. J. Kezer).

anaphase group, therefore, is made up of a complete haploid set of chromosomes instead of a diploid number of chromatids, as would be the case in mitosis. In this way, a reduction in chromosome number is achieved at the first meiotic anaphase. Each anaphase group of chromosomes, however, has a 2*C* DNA content, the same amount characteristic of a similar stage in diploid mitosis.

As the chromosomes begin their poleward movement, the chiasmata lose their retentive influence and free the separating homologues. The chromatids of each homologue, now greatly contracted, are also free of each other and tend to flare apart widely, being held together only at their centromeres (Figs. 4.32 and 4.34).

Telophase. Telophase and the succeeding interphase may not necessarily be integral stages in the meiotic cycle. In some organisms a nuclear membrane is formed in telophase to produce a dyad, the spindle disappears, and the chromosomes go into a diffuse interphase stage as the chromosomes relax their highly compacted structure. The monocots in general follow this pattern, with a delay of varying duration existing between the first and second divisions, and a cell wall, or cell membrane, forms to divide the meiotic cell into two distinct daughter cells, or a dyad. In *Trillium,* also a monocot, and in most animals, telophase and interphase are absent, and the anaphase chromosomes pass directly to late prophase of the second meiotic division. The contraction of the chromosomes is retained; in fact, it persists until the whole meiotic cycle is completed. Whatever procedure is followed, however, no S period of DNA synthesis occurs between the two meiotic divisions. Meiotic interphase is consequently quite different from that of mitosis.

Second Meiotic Division. The meiotic cycle is completed when the two nuclei, each of which has a 2*C* amount of DNA, divide by a process which is essentially mitotic in character. Four haploid nuclei, each now having a 1*C* amount of DNA, result when the chromatids of each chromosome are distributed to their respective poles (Figs. 4.35, 4.36, and 4.37).

Four differences, then, serve to distinguish the second meiotic division from a typical mitotic division. First, the chromosomes in each nucleus are present in a haploid instead of a diploid number; second, and most importantly, no period of DNA synthesis precedes the division; third, the chromatids of each chromosome are widely separated from each other instead of being appressed as they are in mitotic prophase; and fourth, again, most significantly, each chromatid at the end of meiosis may be genetically different from its state

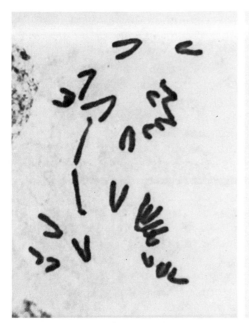
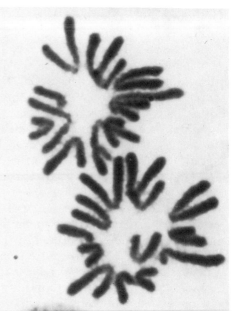

Fig. 4.35. Side and polar views of anaphase II in a salamander. The chromatids are now separating, and each nucleus will round up and become, in this instance, the nucleus of a sperm. Notice that the distinctiveness of the chromosomes can be seen in that they are not of the same size, and their centromeres are not similarly located (courtesy of Dr. J. Kezer).

Fig. 4.36. Metaphase II and late telophase II of meiosis in the milkweed bug, *Oncopeltus fasciatus*. In the metaphase I cell in Fig. 4.30, each bivalent shows four chromatids; in this metaphase, only two chromatids are evident in each chromosome (courtesy of S. Wolfe).

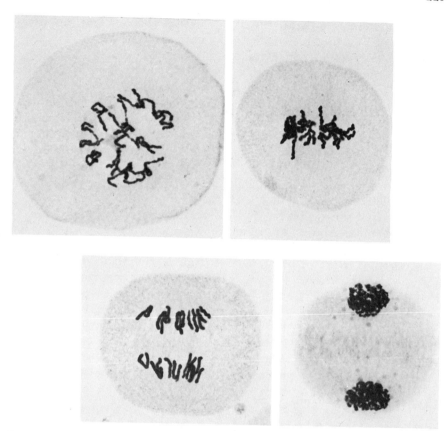

Fig. 4.37. Second meiotic division in the lily. Upper left, prophase II; upper right, metaphase II; lower left, anaphase II; lower right, telophase II. These cells represent only a half of an original microsporocyte which was cleaved by a cell division following telophase I, after which the two halves separated. The beginnings of a cell plate can be seen forming in the lower right figure (courtesy of Dr. Marta S. Walters).

at the onset of the process (see Fig. 4.4). This will depend, of course, upon the number of times each chromatid was involved in chiasma formation and genetic recombination and to the extent that the two homologues were allelically or structurally different from each other.

Oogenesis. Meiosis to produce the haploid eggs of animals differs from the description above in several ways. Mention has already been made of the prolonged nature of diplonema in the human female and in amphibia. A difference also exists in the end result. In spermatogenesis and microsporogenesis, meiosis gives rise to four haploid cells, all of an equal size and all potentially functional.

Meiosis in female animals usually gives rise to only one functional haploid cell and to three smaller ones which are discarded.

The oocytes in which meiosis takes place are relatively few in number compared with the spermatocytes, and, except in placental mammals, are large cells filled with reserve nutritive materials. Meiosis takes place close to the cell membrane, and the first division buds off a small nucleated cell, the first *polar body* (Fig. 4.38). The second division produces a similar cell, the second polar body. The remaining haploid nucleus sinks back into the egg where it will eventually be fertilized by an entering sperm. Meanwhile, the first polar body may have gone through a second meiotic division to produce two haploid cells; these, together with the second polar body, remain on the surface of the egg where they eventually disintegrate.

The polar bodies, or *polocytes,* usually behave as described above, but certain modifications are known. In eggs with a mass of yolk, such as those of insects and crustacea, the polar bodies do not form as such but rather remain as unextruded polar nuclei embedded in the periphery of the egg cytoplasm. Sometimes, as in parthenogenetic forms of coccids, one of the polar nuclei may act as a fertilizing agent, fusing with the innermost nucleus to restore the diploid chromosome number by forming a zygote. In other coccid insects, the polar bodies become infected with a fungus to form an organ called a *mycetosome,* whose function is not known but which might

Fig. 4.38. Polar body formation in the eggs of the whitefish, *Coregonus sp.* Left, anaphase of the first meiotic division, with the first polar body being pinched off. Right, metaphase II within the egg, and the first polar body still in position; the latter may or may not divide again.

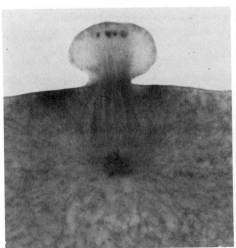
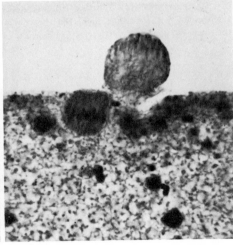

serve in a symbiotic role. Other polar bodies may be fertilized by incoming sperm and undergo abnormal and generally short-lived development, but it seems clear that they are vestigial sexual cells resulting from a process (meiosis) that must occur to bring about the chromosomal changes needed, but the products of which, in females, are only partially of use.

The process of meiosis in megasporogenesis in plants is similar to that described above, but the products vary so much from one species to another, and are complicated further by double fertilization and endosperm development, that the student is referred to elementary botany texts for details.

Duration of Meiosis. As pointed out earlier, the mitotic cell cycle in mammalian tissues at 37°C takes about 18 hr; a similar division in a roottip cell of *Vicia* at 25°C would take 3 hr or 4 hr less (Table 4.1). The duration of meiosis, on the other hand, is usually a much lengthier process when comparisons are made within a single species; meiosis in the rat at body temperature takes 18 days for completion while an epithelial cell in a seminiferous tubule completes its mitotic cycle in about 13 hours. A substantial portion of the drawn-out meiotic process is spent in the prophase of the first division. For example, in the snapdragon, *Antirrhinum majus,* the total time spent in male meiosis adds up to 24 hr; 15 of these hours are in early prophase (6 hr in the leptotene stage, 3 hr in the zygotene stage, and 3 hr in the pachytene stage). The remaining 9 hr take the process from the diplotene to the tetrad stage. Comparable figures for the much lengthier process in *Trillium erectum* would be 70 hr in leptonema, 70 hr in zygonema, 50 hr in pachynema, and 64 hr from diplonema to the tetrad stage, for a total of 254 hours.

Meiosis is also a much more variable phenomenon, ranging from a brief 6 hr in yeast to as long as 40 years in the human female. When unusual periods of time are involved, it generally means blockage of the process at some particular stage—the diplotene stage (dictyonema) in the human female and in *Larix decidua,* pre-meiotic interphase in *Trillium,* or the leptotene stage in the orchid *Dactylorhiza* —with blockage being undone by the entry of a sperm into the egg or by effective pollination. Even when no developmental block occurs, the time span can vary significantly, indicating a genotypic control by each species and, in some species, a differential timing in the different kinds of meiotic cells. In the cereals, *Triticum aestivum* and *Hordeum vulgare,* for example, the pollen mother cells and the embryo sac mother cells go through meiosis in roughly the same span of time; in *Tradescantia paludosa,* male meiosis is considerably longer than that in the embryo sac mother cell (120 hr vs. 80 hr), while in a lily hybrid the reverse is true (180 hr vs. 252 hr). In animals, female meiosis is generally of greater duration than in males, but in the mouse this is due largely to blockage at some stage since the earlier prophase stages—leptonema to pachynema—in the male are much more drawn out in time than in the female (2 days to 3 days as compared to 3 hr to 6 hours).

Table 4.1. Duration of male meiosis (in hours for plant species, days for animal species) and the amount of DNA/cell in picograms. Mitotic data from Van't Hof and Sparrow, 1963, at 23°C; meiotic data from various sources and at several temperatures, but see Bennett, 1977.

	Duration of Meiosis or Mitosis	DNA/Cell in pg
Mitosis		
Helianthus annus	9	9.85
Pisum sativum	10	11.67
Vicia faba	13	38.4
Tradescantia paludosa	18	59.4
Tulipa kaufmanniana	23	93.7
Trillium erectum	29	120.0
Meiosis		
Plant species		
Vicia sativa	24	8.2
Pisum sativum	30	14.8
Triticum monococcum	42	21.0
Vicia faba	72	44.0
Tradescantia paludosa	126	54.0
Lilium henryi	170	100.0
Trillium erectum	274	120.0
Fritillaria meleagris	400	233.0
Capsella bursa-pastoris (4X)	18	2.6
Veronica chamaedrys (4X)	20	2.8
Triticum dicoccum (4X)	30	38.5
Hordeum vulgare (4X)	31	40.6
Secale cereale (4X)	38	56.8
Triticum aestivum (6X)	24	54.3
Triticale "Rosner" (6X)	35	66.3
Triticale Genotype A (8X)	21	82.7
Animal Species		
Drosophila melanogaster	1–2	0.085
Locusta migratoria	7–8	12.8
Chorthippus brunneus	8.5	20.0
Triturus viridescens	12–13	72.0
Mus musculus	12	5.0
Rattus sp.	17	5.7
Homo sapiens	24	6.0

Within a species, temperature clearly affects the rate of meiosis; for example, in *Secale cereale,* at temperatures of 15°C, 20°C, and 25°C, the duration of meiosis in the pollen mother cells was 3.65 days, 2.12 days, and 1.63 days, respectively. In *Trillium erectum,* at temperatures of 1°C, 2°C, 5°C, and 15°C, the times were 90 days, 70 days, 40 days, and 16 days, while in the grasshopper, *Schistocerca gregaria,* male meiosis took 14.0 days, 11.0 days, 7.5 days, and 6.0 days at temperatures, respectively, of 30°C, 32°C, 38°C, and 40°C.

Table 4.1 indicates that within broad limits, as well as within specific ploidy levels, the duration of meiosis is directly correlated with the amount of DNA per cell. Comparable figures for plant mitosis are included to indicate the slower tempo of the meiotic process. Among diploid species the greater the amount of DNA per cell, the greater is the duration of both meiosis and mitosis. This also holds true when comparisons are made within a random group of tetraploids and hexaploids, but when polyploidy within a genus is involved, the situation reverses itself, the higher polyploids completing their cycles of meiosis in shorter time periods than their relatives with lower chromosome numbers (compare the three species of *Triticum* and the two cultivars of *Triticale,* the latter a hybrid between wheat and rye). The reason for the reversal of the trend within a genus is not clear, and there is no suggestion that the number of chromosomes in either the diploid or polyploid forms is an influencing factor. For those species in which meiosis is a continuous process without blockage at any stage, all stages of meiosis, and particularly those in prophase, are proportionally affected; the duration of the process is, therefore, not selectively determined by the slowing down or speeding up of any single facet of the meiotic cycle.

Because many of the determinations of meiotic timing were made at different temperatures and under different circumstances, caution must be exercised in drawing comparisons that are too close, but there seems little doubt that animals, as a rule, have a slower paced meiotic process than plant species having comparable amounts of DNA. Thus, *Pisum sativum* and *Locusta migratoria* have similar amounts of DNA per cell, as do *Triticum monococcum* and *Chorthippus brunneus,* but the time durations for the two animal species are substantially longer. The trends among the diploid animal species are, nevertheless, similar to those among the plants in that the more DNA, the slower the meiotic cycle, but it is also evident that the trend within the insect species is not the same as that within the mammalian group. The amount of DNA per cell also has evolutionary ramifications, a topic pursued further in Chapter 9.

The Mendelian laws of genetics form the backbone of classical genetic analysis; the basic tool is the inheritance test, in which phenotypes, produced by pairs of alleles, are followed through successive generations. In haploid organisms such as *Neurospora*, 1:1 ratios among the offspring are regularly observed for single pairs of alleles, and the phenotypes of all genes are directly observed. In diploid organisms such as the garden pea used by Mendel, the phenomena of phenotypic dominance and recessiveness are encountered, and the familiar 3:1 ratio for single pairs of alleles, and the 9:3:3:1 ratio for two independent alleles, are customarily observed.

Mendel was aware of the role of eggs and sperm in fertilization but quite unaware of meiosis as it is known today, a fact that only serves to emphasize the remarkable ingenuity he displayed in analyzing the problem of character inheritance. It remained to relate the abstract factor, or gene, to some cellular structure. As pointed out in Chapter 1, this was done by Sutton and Boveri in the period 1902–1904, more than 35 years after Mendel had set forth his genetic principles. The following facts, as pointed out earlier (p. 13), indicate that the behavior of the Mendelian genes in inheritance is mirrored in the behavior of the chromosomes in fertilization and meiosis.

1. Fertilization in both plants and animals involves the union of maternal and paternal nuclei, providing a means for the union of parental characteristics in the offspring. The contribution of the sperm consists primarily of nuclear materials, so the sperm nucleus is the source of all paternal genetic contributions, just as the egg is the source of all maternal genetic contributions. The sperm and egg are, therefore, genetically equivalent in terms of chromosomal contribution to the zygote (the sex chromosome would be an exception to this statement), despite the fact that egg and sperm differ radically in size and morphology.

2. Meiosis, by means of its two consecutive divisions, provides for a reduction in the number of chromosomes in the egg and sperm, with fertilization restoring the somatic number in the zygote. The somatic, or diploid, chromosome number is therefore made up of two equivalent haploid sets of chromosomes, one of maternal and the other of paternal derivation. In the usual instance, every chromosome has a mate with which it is linearly and genetically homologous,

and with which it synapses at zygonema and pachynema; the only exceptions are the X and Y chromosomes, which generally are only partially homologous with each other.

3. A mechanism for the segregation of the maternal and paternal derivatives of every chromosome pair is provided through the process of synapsis. The two members of every pair synapse in meiotic prophase, separate from each other, and pass to opposite poles at anaphase, and thus are incorporated into the nuclei of different gametes.

If we substitute "genes" or "alleles" for "chromosomes" in the above statements, we are describing the inheritance and transmission of Mendelian factors. The first critical demonstration of the relationship of a particular character to a particular chromosome was that involving sex determination. Although chromosomes now known to be sex or X chromosomes were first found, in insects, in 1891, it was not until 1901–1902 that a particular chromosome was actually shown to possess a sex-determining role. This stemmed from the observation that two types of spermatids are produced in equal numbers by an XO male of the insect *Protentor,* and that the two sexes are produced in equal numbers. The two types of spermatids differ only in that one type has an X chromosome, seen as an heteropycnotic body in the interphase nucleus, while the other lacks one; so the presence or absence of the X chromosome must be influential in determining the sex of the offspring. Each egg receives a single X chromosome as the result of chromosome segregation; whether the resulting zygote will be male or female is determined by the type of fertilizing sperm.

It remained for T. H. Morgan and C. B. Bridges to demonstrate, in a classical series of studies, that a particular gene was to be found in a particular chromosome. Morgan had shown that the transmission of *white,* a recessive eye color in *Drosophila melanogaster,* depends upon which sex carries the allele initially. For example, if a white-eyed male is crossed with a homozygous red-eyed female, the F_1 flies of both sexes are red-eyed, the F_2 females are all red-eyed, but F_2 males are red- and white-eyed in equal numbers (Fig. 4.39). With the sexes being produced in equal numbers, this gives a normal 3:1 ratio of the eye colors but with the added qualification that all of the white-eyed flies are male. When the reciprocal cross is made, using a white-eyed female and a red-eyed male, the F_1 males are white-eyed and the F_1 females are red-eyed. In the F_2 generation, half of the males and the females are white-eyed and the other half red-eyed.

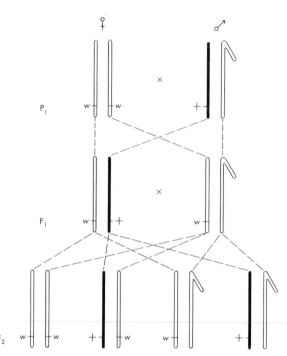

Fig. 4.39. Transmission of red and white eye color when the female is homozygous white (ww) and the male hemizygous red ($+Y$) in the P_1 generation. The transmission of eye color follows the transmission of the X chromosomes. Chromosomes bearing the white (w) allele, unshaded; those bearing the red ($+$) allele, solid black.

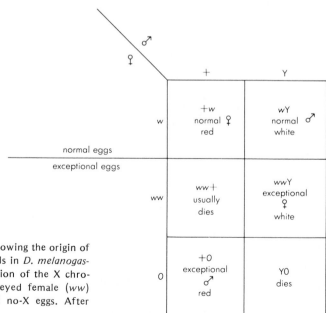

Fig. 4.40. Diagram showing the origin of exceptional individuals in *D. melanogaster* when nondisjunction of the X chromosome in a white-eyed female (ww) gives rise to 2X and no-X eggs. After C. B. Bridges.

This type of inheritance paralleled the transmission of the X chromosome; Bridges showed conclusively that the gene white was located on the X chromosome. When making a cross between a white-eyed female and a red-eyed male, he noted the appearance of *exceptional individuals* whose genotype was such as to suggest a failure of the usual type of inheritance of the X chromosomes. Thus, in an F_1 population that should have contained only red-eyed females and white-eyed males, there appeared an occasional white-eyed female or red-eyed male, with the exceptional females occurring with a frequency of one in 2,500. The frequency of exceptional males was about one in 1,200.

Bridges' explanation of the phenomenon, known as *primary nondisjunction,* is diagrammed in Fig. 4.40. The exceptional females result from the fact that in the meiotic divisions of the egg, both X chromosomes are included in the egg nucleus instead of being segregated from each other. That all white-eyed exceptional daughters were XXY instead of XX was verified cytologically. The exceptional red-eyed sons resulted from the fertilization of a no-X egg by X-bearing sperm, and they are sterile, owing to the absence of a Y chromosome. The Y chromosome in *Drosophila* controls male fertility but not the phenotypic expression of maleness.

Bridges also found the process of primary nondisjunction to be more general, because it could be demonstrated for a variety of other genes known to be on the X chromosome as well as for those found on the tiny IV chromosome, which can occur in a haploid, diploid, or triploid condition. In each instance, the correlation between genetical and cytological inheritance was exact, and these observations constituted the first critical experimental evidence that genes are on chromosomes.

L. V. Morgan discovered, in 1922, a strain of *Drosophila melanogaster* that apparently gave 100% nondisjunction. As in the case of Bridges' exceptional individuals, the daughters appeared to receive both X chromosomes from their mother, and the sons their single X chromosome from their father. Examination of the chromosomes of the female showed two X chromosomes attached to each other in the neighborhood of their centromeres, with the result that they always passed to the same pole. A Y chromosome is present in these females, and it segregates from the joined X chromosomes. The inheritance of attached-X flies, as these have become known, is outlined in Fig. 4.41. It is the exact opposite of the pattern of inheritance with normal X and Y chromosomes (Fig. 4.39); that is, in the attached-X stock the males receive their Y chromosome from their mother and their X chromosome from their father.

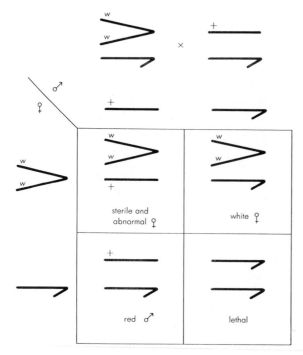

Fig. 4.41. Inheritance of the attached-X chromosomes in *D. melanogaster* and the transmission of red (+) and white (*w*) eye color. The mother transmits her X chromosomes to her daughter, the father his X chromosome to his son, the reverse of the usual pattern of X-chromosome inheritance.

Evidence that the random assortment of two pairs of alleles (Mendel's second law) is paralleled by the random assortment of paired chromosomes was provided by Eleanor Carothers in 1913. Sutton had previously postulated that a random distribution of chromosomes occurred in meiosis such that the resultant haploid gametes received all possible combinations of paternal and maternal chromosomes. To have this condition hold true would mean that the attachment of any given centromere to a given pole by means of half-spindle fibers is a random process, and, similarly, that the orientation of a bivalent relative to the poles is a matter of chance. Conclusive proof was supplied by Carothers from a study of the distribution of the members of a heteromorphic pair of autosomes in the male orthopteran *Brachystola*.

The members of this pair of chromosomes, which regularly synapsed and segregated from each other, differed visibly in size. By using the X chromosome of the XO male as a basis for comparison, the distribution of the heteromorphic homologues (capable of synapsis and chiasma formation but distinguishable in form) could be shown to be random, so that four classes of sperm were produced in equal numbers. In two other orthopterans, *Trimerotropis* and *Circotettix*, the number of heteromorphic pairs in various individual males ranged from one to eight. By comparing the distribution of two or more heteromorphic pairs with each other, or with the X chromosome, Carothers again demonstrated that the Mendelian expectancy of random assortment was always achieved. The behavior of the genes is therefore correlated with the behavior of the chromosomes insofar as segregation and independent assortment are concerned. There are, however, situations in which random assortment in meiosis is not achieved. These have been found in the mouse and in the mole-cricket. The passage of one member of a heteromorphic pair to the same pole as the single X in an XO male may possibly be explained as an adjustment of a sex-determining mechanism, but the pattern of linkage of a group of genes, apparently on different chromosomes in the mouse, appears to be an example of directed rather than random segregation.

An additional bit of experimental data reveals that segregation may be more complex than previously realized. From the above discussion it is clear that crossing over and chiasma formation accomplish two things: (1) they provide a means for the exchange of chromatin between homologues and (2) the act of exchange binds the pair of homologues together until the end of metaphase and thereby facilitates their regular segregation to opposite poles in anaphase. Recent studies by R. Grell, however, strongly suggest that these two events may be separated in time, that is, that the pairing which results in exchange need not be the same as, or occur simultaneously with, the pairing that leads to distributive, or segregational, events. The chromosomal and genetic nature of her experimental flies, indicated in Fig. 4.42, permits the detection of the presence or absence of crossing over at the same time that the distribution of each chromosome can be followed. The data accompanying the figure show that when crossing over takes place, the Y chromosome is randomly distributed (48:65; 53:53), but that when no crossing over occurs, the Y chromosome tends to segregate from chromosome 2 (387:109 for one class of offspring, 153:734 for the other). Also, when no Y chromosome is present (control experiment), the rate of crossing over is the same as when it is present; the Y chromosome, therefore, exerts no influence on the process of crossing over. Grell has consequently interpreted these data in the following manner: Exchange pairing between homologues, presumably arising from a synaptinemal complex, occurs first in time. If the pairing is effective and crossing over takes place, the paired homologues remain together until disjoined at anaphase. If crossing over does not take place, the paired homologues fall apart and are free to synapse, homologously or nonhomologously but presumably without a synaptinemal complex formed, with any other free chromosomes for purposes of anaphase disjunction. Because the Y chromosome has no effect on

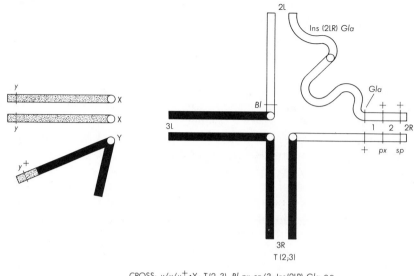

CROSS: $y/y/y^+ \cdot Y$; $T(2;3)$, $Bl\ px\ sp/3$; $\text{Ins}(2\text{LR})\ Gla$ ♀♀
×
y/Y; $px\ sp/px\ sp$ ♂♂

	$T(2;3)Bl$		$3; \text{Ins}(2\text{LR})Gla$		Percent crossing over	
	$y^+ \cdot Y$	No Y	$y^+ \cdot Y$	No Y	Region 1	Region 2
No crossing over	387	109	153	734		
Crossing over	48	65	53	53	9.2	4.5
Control experiment (no Y)		292		566	9.2	4.6

Fig. 4.42. Experimental design used by Grell in females of *D. melanogaster* to follow the distribution of the Y chromosome relative to the distribution of chromosome 2 (Ins(2LR)*Gla*). Two classes of viable offspring are produced: those possessing a normal chromosome 3 plus the inverted chromosome 2 and identified by the dominant gene *Gla*; and those possessing the two translocated chromosomes 2L:3L plus 2R:3R and identified by the dominant gene *Bl*. The presence or absence of the Y chromosome is determined by the fact that it carries the dominant allele of *y* at its tip. Crossing over is detected by the recombinational events taking place at the right end of chromosome 2, which is heterozygous for the linked genes *Gla*, *px*, and *sp*. (Baker, 1965), after Grell.

crossing over, it apparently does not attempt to pair with chromosome 2 at the time when crossing over is occurring and thereby reduce the crossover frequency through interference. When no crossing over takes place, both the Y and chromosome 2 are free, and the fact that they tend to segregate from each other rather than behaving in random fashion indicates that they have paired and disjoined from each other. Grell has called this *distributive pairing* and considers it distinct from, and occurring after, *exchange pairing*.

Novitski has offered an alternate explanation to account for Grell's observations. He suggests that anaphase distribution in the mitotic division immediately preceding should bring all centromeres in close proximity to each other and that nonhomologous pairing of heterochromatic elements, which differs only in intensity and duration from homologous pairing, should keep the chromosomes together until synapsis frees pairs of homologues from the mass. Those which are not paired homologously would then tend to remain paired nonhomologously until the end of metaphase, at which time they would disjoin and accomplish the same end result as that observed by Grell. At the present time it can only be stated that both hypotheses offer solutions to a group of perplexing data. Our ignorance of meiotic events does not permit us to make an unequivocal choice as yet.

Meiosis, Linkage, and Crossing Over

If the segregation and transmission of chromosomes in meiosis provides a physical explanation for the segregation and transmission of mendelizing genes, the events of meiosis ought also to account for the phenomenon of genetic crossing over and its consequences.

The number of pairs of chromosomes of an organism or species is constant. In *Drosophila melanogaster,* for example, the number is 4, in human beings 23, in the mouse 20. With the discovery of an increasing number of mutant genes came the realization that some of these genes must occupy positions on the same chromosome. Each of these, tested individually for transmission, segregated from its allele in conformity with Mendelian expectations, but when studied in nonallelic pairs or groups of three or more, they yielded ratios that frequently departed from the distributions expected on the basis of random assortment. A general acceptance of the chromosome theory of inheritance was, in fact, delayed by such findings until the elucidation of linkage (also called synteny) and crossing over by the *Drosophila* geneticists under T. H. Morgan (1910–1915) made it quite clear that the Mendelian law of random assortment, although not invalid, was not as universally applicable as was once believed; it held for genes on different nonhomologous chromosomes but not for all combinations of genes. Where the number of known genes exceeds the haploid number of chromosomes, some of the genes must inevitably show linkage; that is, they tend to be inherited as a group rather than individually.

Complete linkage is relatively rare among organisms, but the fact that it is characteristic of male *Drosophila,* the genus from which so much information has been obtained, makes the phenomenon of some importance. The following data are typical of complete linkage:

Purple (pr) (an eye mutant) and *vestigial (vg)* (a wing mutant) are recessive characters whose genes are on chromosome 2 of *D. melanogaster.* Crossing a wild-type (red-eyed, normal wing) female (+ +/+ +) to a homozygous recessive male (*pr vg/pr vg*) yields heterozygous progeny (+ +/*pr vg*), all of them wild-type. When a heterozygous + +/*pr vg* male is then test-crossed to a homozygous recessive female (*pr vg/pr vg*), the possible progeny and the values observed by Bridges are as follows (because the genes are not sex-linked, data on the sexes are not included; males and females, as expected, would be equally represented):

		Observed	Expected on Basis of Random Assortment
Parental types	{ Wild-type	519	268
	{ Purple, vestigial	552	268
Recombinant types	{ Red-eyed, vestigial	0	268
	{ Purple, normal wing	0	268
		1,071	1,072

No recombinant progeny are found among the offspring; the original combinations are inherited together (preserved) whenever the male is the heterozygous individual in such a test cross. This is also true of other linked genes, whether on this or other chromosomes.

A slight diversion is needed here to consider a number of problems relating to complete linkage. It is now recognized, if one equates crossing over with chiasma formation, that the lack of crossing over (achiasmate meiosis) is widespread in the animal kingdom, being reported in a number of dipteran species in addition to *Drosophila,* and in a variety of species of protozoa, mollusks, scorpions, copepods, grasshoppers, mantids, and worms. It has been reported only once among species of higher plants, in the pollen mother cells of the genus *Fritillaria.* Obviously, the absence of crossing over and the absence of chiasmata can be correlated with absolute certainty only in a genetically and cytologically well-investigated species, and the *Drosophila* species provide us with the best information, but where the sex chromosome situation is known, it is always the heterogametic sex which is achiasmatic. In most groups this would be the males, but in the copepods it is the female that is heterogametic and achiasmatic.

The existence of achiasmatic meiosis poses a number of cyto-genetic problems: What is the relation of the synaptinemal complex to crossing over? And what keeps achiasmate bivalents intact so that proper segregation can ensue at anaphase? *D. melanogaster* males show complete linkage, lack synaptinemal complexes, but exhibit homologous pairing and regular segregation. In several species of achiasmate mantids, synaptinemal complexes are clearly evident. *D. melanogaster* females that are homozygous for the C_3G gene which suppresses crossing over, lack synaptinemal complexes. From these data one may conclude that the formation of a synaptinemal com-plex is highly correlated with but is not a sufficient condition for synapsis and crossing over and that crossing over occurs only if subsequent processes come into play. However, even the relation of the complex to synapsis is suspect since some species without com-plexes can pair regularly, and have their bivalent structure preserved sufficiently well, in the absence of chiasma, to allow for regular segregation. It may well be that there are degrees of synapsis, with that leading to crossing over being more intimate than that which only guarantees bivalent formation, but if so, the force keeping homologues together without chiasmata remains unknown.

To return to the *pr vg* cross mentioned earlier, if the identical parental cross is made again, but in this instance the F_1 heterozygous females are test-crossed to homozygous recessive males, the following data are obtained:

		Observed	Expected on Basis of Complete Linkage	Expected on Basis of Random Assortment
Parental types	{ Wild-type	1,339	1,420	710
	{ Purple, vestigial	1,195	1,420	710
Recombinant types	{ Red-eyed, vestigial	152	0	710
	{ Purple	154	0	710

All four possible phenotypes, which because of the nature of the test cross accurately reflect gametic genotypes, are recovered among the progeny. The departure from values expected from either complete linkage or from random assortment is obvious. It is equally clear that the parental types, + + and *pr vg,* are much more numer-ous than the recombinant types, + *vg* and *pr* +. This, of course, means that the linked genes did not undergo recombination in all oocytes of the female and that the majority of recovered chromatids preserved intact their original gene combinations. The percentage of recombination, or crossover frequency, is found by adding the new types and dividing by the total number of offspring; that is, (152 + 154)/2,840. This figure, 9.3% or 9.3 map units, is a measure of the genetic distance separating the two genes on the chromosome, the assumption being that the probability that two genes will recombine increases as the distance between them increases. This is true only to 50% recombination, because random recombination, as in independent assortment, also produces 50% recombinants. The same percentage, within the range of experimental error, would be obtained if the original parental stocks had been + *vg*/+ *vg* and *pr* +/*pr* +, indicating that the frequency of recombination is not conditioned by the particular parental combination but by the spatial relationships of the genes on the chromosome. Thus, a similar cross, using the genes brown (*bw*) and speck (*sp*) on the same chromosome, would yield a recombination percentage of 2.5. In a genetic sense, these two genes are closer to each other than *pr* is to *vg*, since recombination frequencies represent the probabilities that they will separate from each other.

Linear Order of the Genes and Map Distances. Morgan postulated from such studies of genes in the X chromosome of *Drosophila* that the genes are arranged in a linear order, each gene having a fixed position along the length of the chromosome. Its allele occupied a corresponding position in an homologous chromosome. Such a postulate arose out of the observations that (1) the many genes in *D. melanogaster* could be divided into four, and only four, linkage groups corresponding to the four haploid chromosomes and (2) the evidence that linkage with respect to two pairs of genes, when incomplete, was incomplete in a characteristic frequency that could be interpreted as a function of the constant spatial relationship between linked genes. A. H. Sturtevant devised a test whereby the position and linear order of a third gene could be determined with reference to two other genes on the same chromosome. Because three genes are used simultaneously, the test has become known as a *three-point cross.*

If it is assumed that the correct serial order of three hypothetical genes is *abc,* and the distances between the genes are reflected in the frequencies with which crossing over takes place between them, then a test cross of the heterozygote + + +/*a b c* to the triple recessive *a b c/a b c* can yield the following possible phenotypes, the complementary types being grouped:

	Region I	Region II				
			Parentals	Noncrossovers	+ + +	
+	+	+			a b c	
				Single crossovers, region I	+ b c	
					a + +	
			Recombinants	Single crossovers, region II	+ + c	
					a b +	
				Double crossovers	+ b +	
a	b	c			a + c	

If it is assumed that crossing over occurs as indicated in the diagram, the noncrossover group would be the largest in number. A region I single crossover would arise from a chromatid exchange in the region between genes *a* and *b*, region II single crossovers from an exchange in the region between genes *b* and *c*.

The frequencies of the two types of single crossovers will, in general, depend upon the linear distance between *a* and *b* and between *b* and *c*. The double-crossover group will contain the smallest number of individuals. This results from the fact that the occurrence of a crossover between any two genes is a statistical function of distance. The simultaneous occurrence of crossovers in two adjacent regions would thus be the product of the individual single-crossover probabilities and would therefore be much less than either alone.

An experiment carried out in maize will illustrate the actual use of the three-point cross. The genes *brown midrib (bm)*, *red aleurone (pr)*, and *virescent seedling (v)* are located in chromosome 5, and the data obtained from linkage studies are as follows:

$$\text{Parents} \; \frac{+ \quad + \quad +}{+ \quad + \quad +} \times \frac{bm \quad pr \quad v}{bm \quad pr \quad v}$$

$$F_1 \; \frac{+ \quad + \quad +}{bm \quad pr \quad v}$$

Test-cross progeny:

+	+	+	232	Noncrossovers = 42.1% Parentals
bm	pr	v	235	
+	pr	v	84	Single crossovers between
bm	+	+	77	*bm* and *pr* = 14.5%
+	+	v	201	Single crossovers between
bm	pr	+	194	*pr* and *v* = 35.6% } Recombinants
+	pr	+	40	Double crossovers between
bm	+	v	46	*bm* and *pr* and between *pr*
			1,109	and *v* = 7.8%

From the above groupings, each made up of equally occurring complementary types, certain facts are readily deduced. The non-crossover types, of course, are those that preserve the original parental gene grouping; the least frequent groups represent the double crossovers, representing those having the lowest probability of occurrence. This group also provides information on the linear order of the genes. With respect to *bm* and *v,* the gene *pr* has shifted positions with its allele and is therefore between *bm* and *v.* The order of the three genes is *bm pr v.* The two remaining classes constitute the single-crossover types, one class representing the crossovers taking place between *bm* and *pr* and the other class representing those between *pr* and *v.*

In calculating the genetic distance between two genes, *all* crossovers (single and double) that have occurred in a particular region must be taken into consideration, because we are attempting to construct an accurate representation of the genetic map from crossover data. The map distance between *bm* and *pr,* then, is not 14.5, but 14.5 + 7.8, or 22.3. Similarly, the total crossover frequency for the *pr-v* region is 35.6 + 7.8, or 43.4 The total distance from *bm* to *v* would be 65.7, because units of the map can be successively added to indicate genetic distance. So, except for very short regions in which double crossing over is rare, the map distance is greater than appears from the single-crossover frequency.

Through the accumulation of genetic data on the frequencies of crossing over, it has been possible to construct genetic, or chromosome, maps in which the serial order of the genes and their genetic spacing have been accurately determined. A portion of X chromosome of *D. melanogaster* is given in Fig. 4.43. In addition to a mutant name and an abbreviated symbol, each gene, or locus, is designated by a number, which is obtained by summing up the crossover values for all known intervals to its "left." Thus, the gene *facet* (*fa*) on the X chromosome of *D. melanogaster* is located at position 3.0. As *yellow* (*y*) is known from other evidence to be at the extreme left end of the X chromosome, it is given a zero (0.0) designation. *Yellow* and *white* (*w*) show a crossover frequency of 1.5; *white* and *facet* show a similar frequency. *Facet* therefore lies 3 map units from the left end of the X chromosome.

Only those regions of the chromosomes that possess detectable mutant genes can be mapped. It has also been found that, as a general rule, the shorter the map distance between two genes, the more accurately the two genes can be located with respect to each other. The reason for this lies in the fact that when two genes are more than 10 to 20 map units apart, double crossovers can occur, and these will

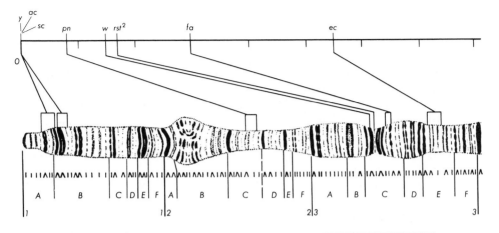

Interval	Approx. no. of bands between genes	Map distance	Percent crossover per band
y-pn	57	0.8	0.014
pn-w	18	0.7	0.038
w-rst²	2	0.2	0.1
rst²-fa	2	1.3	0.65

Fig. 4.43. Portion of the left end of the X chromosome of *D. melanogaster* showing the location of certain genes, their relation to specific bands, and their position on the linkage map (y = 0.0; pn = 0.8; w = 1.5; rst^2 = 1.7; fa = 3.0). Below, data showing the relation between the number of bands between genes and the map distances. After C. B. Bridges.

result in a spuriously low frequency if they are not taken into account. A more accurate genetic map can be made, therefore, by adding successive short intervals than by utilizing intervals between genes that are widely separated on the chromosome.

Interference. If genes *a, b,* and *c* are linked in the order given, it should be theoretically possible to predict the frequency of double crossing over when the map length of the *a-b* and *b-c* intervals are known. If it is assumed that the *a-b* interval is 10 map units long and that the *b-c* interval is 15 map units, then the frequency of double crossing over would amount to the probability of the two single crossovers occurring simultaneously. This would be 10% of 15%, or 1.5%, *provided the two single crossovers are without influence on each other's occurrence,* that is, if they occur independently of each other.

Studies, however, have revealed that adjacent crossovers do not occur independently and that one occurring in, for example, the *a-b* interval would tend to suppress those occurring simultaneously in the *b-c* interval. The effect, therefore, would be to reduce the expected frequency of double crossing over. H. J. Muller, who first discovered the phenomenon, termed it *crossover interference,* and it may be defined as the tendency of one crossover to interfere with the occurrence of another in its vicinity. Cytologically, the term *chiasma interference* would be appropriate. Interestingly enough, interference does not generally extend across the centromere in a two-armed chromosome; each arm, therefore, acts independently in crossing over from the other.

Interference is more pronounced as the distance between successive genes becomes shorter; it decreases with increasing distance. An example will serve to illustrate its effect. The maize data discussed earlier illustrate partial interference. The map distances for the *bm-pr* and *pr-v* intervals are 22.3 and 43.4, respectively. The expected frequency of double crossing over, based on independent occurrence, would be 9.7%, that is, 22.3% of 43.4%. Only 7.8% doubles were recovered among the test-cross progeny. To express the degree of interference, Muller coined the term *coincidence,* which is an inverse measure of interference. The *coefficient of coincidence,* which is equal to one minus interference when the latter is given in terms of a decimal fraction, is calculated by dividing the observed number of double crossovers by the expected number. In the above experiment this would be $0.078/0.097 = 0.804 =$ coefficient of coincidence. Therefore, only 80.4% of the expected double crossovers were recovered, indicating a partial interference of 19.6%.

Analysis of Linkage in Human Beings

Although the principles governing gene order, theory of linkage and crossing over, and general strategy of linkage analysis established in experimental organisms are assumed to be transferable to human beings, the special circumstances encountered in the genetic analysis of the human species require that somewhat different strategies be used. Human families are small, the lifetime of the experimenter is that of the subject, and the investigator may not, for ethical reasons, manipulate the genetic organization of the subject in manner similar to that which can be done with experimental organisms. There have been developed, therefore, special, sometimes mathematical, approaches to problems of human genetic linkage. Nevertheless, the questions posed are the same: In what chromosomes are particular genes located? Are two loci, selected for intrinsic interest or convenience, located on the same chromosome (called *synteny*)? And if they are syntenic, what recombination distance separates them?

It is known, of course, that all genes whose transmission follows that of the X chromosome are syntenic; the X chromosome is thus a special case, but this information does not provide a knowledge of the map distance between any two X chromosome genes. The determination of map distances, or the strength of linkage, defined as some figure less than 50 map units, poses its own problems, even with the X chromosome, but with autosomal genes, the determination of both synteny and strength of linkage becomes more complicated. Considering the fact that the human haploid number is 23, it becomes immediately evident that the probability of two genes being *asyntenic,* i.e., not on the same chromosome, is considerably greater than that they are syntenic. Because of the differing lengths of the human chromosome, the odds that two loci *are not* syntenic are about 17.5 times the odds that the two loci are syntenic. Were all autosomes the same length, the odds would, of course, be 22:1.

The primary data from which analyses are accumulated come from pedigree investigations in which one can gather information on the segregation of allelic pairs of interest. Since human families are generally small, the number of opportunities to observe recombination in any given family is also small. Hence, there must be a pooling of many pedigrees in which the same genes are being followed. There is, however, an additional concern; this is the problem of the phase of linkage. Suppose that an assessment is being made of the possible recombinants related to two loci, A and B, each with two alleles A, A^1, B, B^1. In an experimental situation, the genotype of the parent can be adjusted at will, that is, it might be $\dfrac{A^1 \quad B}{A \quad B^1}$, in which case it is referred to as *repulsion*; or $\dfrac{A^1 \quad B^1}{A \quad B}$, in which case it is termed *coupling*. Experimental cytogenetics has demonstrated that the frequency of recombination between two loci is generally unaffected by the phase of linkage; also it is believed that, for any two loci, the heterozygotes will be found in coupling or in repulsion essentially at random. In many human families, on the other hand, only the heterozygosity of the parent will be known while the phase of linkage will not. As a result of this ignorance, it is difficult if not impossible to know which of the offspring is a recombinant. In the example given, $A^1 \quad B^1$ would be a recombinant from the parent in *repulsion*, while $A^1 \quad B^1$ would be a *non*-recombinant for the parent in *coupling*.

An additional complication in linkage computation is the fact that it is both necessary and desirable to use information from all possible matings, not only that from the most suitable, which may well be unavailable; the latter, of course, would be the double heterozygote by the double homozygous recessive, or $\dfrac{A \quad B}{A^1 \ B^1} \times \dfrac{A^1 \ B^1}{A^1 \ B^1}$. Some matings, of course, are more informative and provide better analytical possibilities than others, but the most commonly used technique which permits the use of as much pooled data as possible from multigenerational and bigenerational pedigrees was developed by Morton and is referred to as the *sequential* or *lods method*. The method, which is illustrated in Fig. 4.44, cannot provide yes-or-no answers to questions of linkage, but it does yield likelihood ratios, or odds, as to whether the data obtained from any particular pedigree are the consequence of recombination occurring at a certain frequency (θ) or the consequence of gene independence. By giving θ in the numerator of the equation a series of values greater than zero but less than 50, and by keeping the value of θ in the denominator at 50 (a value indicating no linkage), each family pedigree can be analyzed and a series of likelihood ratios can be obtained. By making use of the logarithms of the odds (hence the use of the acronym *lods*), the likelihood ratios of different families can be summed. The recombination fraction giving the highest *lod* value for any given family group is taken to be the maximum likelihood estimate (in Fig. 4.44 this would be when the θ value is 0.10, or 10% recombination). Such values of *lods* can be summated, and it has been determined that a *lod* value of +3.0 is very strong evidence of linkage while a value of –2.0 is equally strong evidence of gene independence. Any value between would provide only an equivocal estimate.

A note of caution should be added here. If the *lod* analysis suggests independence, this alone does not mean that the two loci are asyntenic. As the previous discussion has pointed out, two loci may be syntenic but sufficiently distant that the recombination value is around 50%, indicating independence; thus, the loci would not demonstrate conventional linkage.

Although the use of this technique for numbers of pairs of loci has proved useful in providing a map for many loci on the X chromosomes, and some clusters of genes elsewhere on the human chromosome, complementary approaches are required to identify genes and genes clusters with particular chromosomes. Even so, there are still "homeless" gene groups, i.e., loci known to be linked with good estimates of θ but without an identifiable chromosome with which they can be associated.

Techniques for the establishment of congruency between genetic loci (markers) and specific human autosomes have become

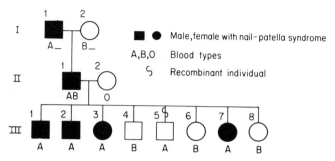

Fig. 4.44. Pedigree of the nail-patella syndrome (dominant gene, filled symbols) and the ABO blood groups. The genotype of the father in generation II is unambiguous, and shows that the nail-patella gene (NPa) and the A (I^A) blood group gene to be in a coupling phase, i.e., $\dfrac{I^A\ NPa}{I^B\ npa}$, since, with the exception of offspring #5, all others in generation III are either $\dfrac{I^A\ NPa}{I^O\ npa}$ or $\dfrac{I^B\ npa}{I^O\ npa}$. The mother in generation II must have been homozygous recessive for both alleles. Offspring #5 is a recombinant, and the recombinant fraction for this family grouping is 1/8. The lods method is not needed here to demonstrate linkage (the data make it clear that the genes are linked), but if the linkage phase were not known, then the method of lods might be applied as follows:

$$\text{Lod} = \log_{10} \times \frac{\text{likelihood of observed pedigree assuming } \theta = \theta}{\text{likelihood of observed pedigree assuming } \theta = .50}$$

For various values of θ (the recombination fraction) the likelihood of the pedigree is proportional to $\frac{1}{2}[\theta(1-\theta)^7] + [\theta^7(1-\theta)]$, the first element relating to the possibility that NPa and A are in coupling and the second to the possibility that they are in repulsion, these alternatives being assumed equally probable. For a sibship of 7 and 1, the lod values are:

θ	.05	.10	.20	.30	.40
lod	.650	.787	.730	.503	.193

Since the lod values can be summed, several additional family groupings providing equally good data, would put the lod values in the neighborhood of +3.0, lending strong support not only for the presence of linkage, but also for a map distance of about 10 units (Fig. 3 of McKusick and Ruddle (1977), copyright 1977 by the American Association for the Advancement of Science).

sufficiently sophisticated that linkage maps have been constructed for each of the human autosomes and on which relative gene locations have been identified. The data are being generated in numbers of laboratories, and as would be expected, the maps are changing rapidly with time. Certain of the approaches are cytogenetically familiar; others, particularly those using somatic cell hybridization, are novel.

A classic cytogenetic approach, first utilized by Donahue in 1968 to identify a given human gene with a specific autosome, is that of correlating the segregation of a morphologically marked chromosome with a particular genetic trait. In this instance, a variant chromosome 1, with abnormally extended centromeric heterochromatin, was shown to segregate regularly with the Duffy (*Fy*) blood group. This very important cytogenetic principle—the coordinate segregation of chromosomal and genetic phenotypes—has been refined still further by the localization of a specific genetic effect with a particular chromosomal region by means of deletions. Thus, the absence of an expected trait is correlated on a one-to-one basis with an observable deletion. Although less valuable than one might anticipate, this approach has assisted in the localization, for example, of the LDH-B locus to chromosome 12, and of the *Fy* locus to the immediate neighborhood of region 1*q*12 of chromosome 1 (see p. 149). By similar techniques, the locus determining the α-polypeptide of haptoglobin (Hpa) has been assigned to chromosome 16, the major histocompatibility complex (HLA) to chromosome 6, and the acid phosphatase locus to the distal end of the short arm of chromosome 2. Interestingly enough, the deletion method cannot be used for the assignment of X chromosome genes because the deleted chromosome becomes preferentially inactivated in somatic cells.

Once localized to a chromosome, markers may also be site-assigned by following the segregation of translocated chromosome pieces. For example, an X–14 translocation called KOP has a breakpoint on the X chromosome such that there are two pieces, short (X_p), and long (X_q). By analysis of cells derived from hybrid clones (see below) it has been shown that three X-chromosome markers (PGK, HGPRT, G6PD) are assignable to the long arm segment in that order, and with G6PD most distal in location.

The most recent and most fruitful approach to the establishment of gene assignment to particular chromosomes has made use of non-meiotic methods, principally those involving somatic cell hybridization. These so-called parasexual approaches take advantage of two critical phenomena. First, it is possible to produce true heterokaryons by the fusion of somatic cells from different species, even different genera. Second, as the hybrid cells are cultured, most of the chromosomes of one of the species are lost, leaving a more or less complete complement from one parental line with a few elements from the other. The principle again is the familiar cytogenetic strategy of correlating the presence of particular genetic markers (or, more properly, their phenotypic expression) with specific identifiable chromosomes. A number of requirements must be met for this approach to be effective: (1) genic expression must be evident at the cellular level for the cells being used, thus excluding, for example, any study of red blood cell antigens if cultured fibrocytes are being assayed; (2) the ready identification of each chromosome from the species in question must be made—a feature made possible in several mammalian groups by the several banding techniques available; (3) there must be a reasonably high efficiency of cell fusion; and (4) reasonably pure hybrid lines must be established from which the chromosomes are expected to segregate. In practice, cultured cells from two lines are co-cultivated as a monolayer in the presence of a strain of the Sendai virus 40 (SV40), which is inactivated prior to inclusion in the culture medium. More recently, polyethylene glycol has been used in place of the virus. These two agents appear to promote a weakening or partial degradation of the cell membranes as well as the formation of cytoplasmic bridges between cells, thus greatly facilitating the normally low rate of spontaneous cell fusion and permitting nuclear fusion. Once cells are fused, however, it is necessary to establish selective conditions which enable the hybrid cells to persist while, at the same time, eliminating the unfused parental cells. The approach makes use of two differently mutated cell lines which, when fused, would complement each other to the extent of covering genetic deficiencies. In the original method used by Littlefield, one strain was resistant to the drug 8-azaguanine, an analog of guanine, and it was deficient for the enzyme hypoxanthine guanine phosphoribosyl transferase (HGPRT$^-$), which prevented it from utilizing an exogenous source of either guanine or hypoxanthine. The other cell strain was resistant to the drug 5-bromodeoxyuridine, an analog of thymidine, and lacked the enzyme thymidine kinase (TK$^-$); it could not make use of an exogenous source of thymidine. The presence of the drugs in the respective culture media was necessary to prevent the revertants (HGPRT$^-$→HGPRT and TK$^-$→TK) from growing when they arise. Both strains, therefore, depended upon endogenous synthetic pathways to produce the appropriate monophosphates required for incorporation into DNA and/or RNA. The antimetabolite aminopterin was used to block both endogenous synthetic pathways in hybrid cells (Fig. 4.45).

To bring about cell fusion, the two strains are washed free of the 5-bromodeoxyuridine and 8-azaguanine and are co-cultured for several days, after which they are reinoculated into a medium that would favor fused, or hybrid, cells and discriminate against unfused parental cells. The components of this selective system consisted, therefore, of the parental cells, the hybrid cells, and a nonpermissive, or HAT, medium which contains hypoxanthine, thymidine, and aminopterin. Since the endogenous synthesis of the monophosphates was blocked by aminopterin, the parental cells, HGPRT$^-$/TK and HGPRT/TK$^-$, being unable to utilize the exogenously supplied hypo-

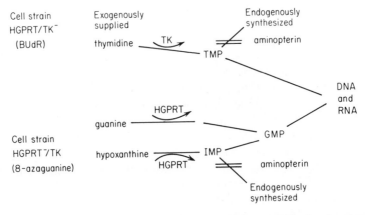

Fig. 4.45. Pathways of pyrimidine (guanine and hypoxanthine) and purine (thymidine) incorporation into DNA and RNA, with the sites of enzyme involvement and aminopterin blockage indicated. The drugs are indicated in parentheses. See text for explanation.

xanthine or thymidine, cannot multiply. The hybrid cells, with their complementing nuclei, possess both enzymes and can survive, even in the presence of aminopterin, since they could utilize the exogenously supplied metabolites. As these hybrid cells continue to divide, they tend to revert to a diploid condition, although the chromosomes carrying the appropriate genes must be present. Subsequent study of the hybrid cells revealed that their survival required the presence of the human chromosome 17 from the HGPRT⁻/TK strain, indicating that the gene for thymidine kinase could be assigned to that autosome. HGPRT was known, from other genetic evidence, to be assignable to the X chromosome, which in this experimental setup had to be derived from the HGPRT/TK⁻ strain.

Other selective systems can also be designed for particular kinds of cells or particular genetic situations. These may not require the presence of mutant cell lines, some of which may be inconvenient to use or difficult to acquire. For example, human cells are more sensitive to the drug ouabain than are rodent cells; an HAT-ouabain medium can be useful in selection in certain experiments. The sensitivity site for ouabain appears to be on chromosome 8. Other selective systems comparable to that which identified chromosome 17 as that bearing the thymidine kinase locus and chromosome 8 as that carrying the ouabain sensitivity site have identified chromosome 19 with a polio virus receptor and chromosome 10 with adenosine kinase activity.

The concomitant step is to determine which other chromosomes may be present or absent in hybrid cell cultures and which can be correlated with phenotypic markers present in the original human cell line. These would be loci not susceptible to selection by restrictive media. They, and the chromosomes on which they are located, would segregate more or less fortuitously from those required by the hybrid cell lines, but the association of specific chromosome and phenotypic marker would persist. The human chromosomes retained, other than those selected, will vary from line to line and from experiment to experiment, but the specific correlation will be unique, on a one-to-one basis, to the marker and the cell line.

A further refinement of this technique by Creagan and Ruddle has led to the development of a system which combines the features of somatic cell hybridization with that of marker chromosomes and cellular phenotypes. By establishing five clonal lines of cells, each with a different binary signature of chromosome and phenotype, any mutant gene with a detectable phenotype can be assigned to a specific chromosome with a minimum of difficulty ($2^5 = 32$, which provides more than enough possibilities). The success of this strategy is illustrated by the fact that several hundred human genes have been assigned not only to specific chromosomes but in many instances also to specific bands. A major constraint of this technique, however, is that only those loci whose products can be assayed from the cells used—fibrocytes and lymphocytes—can be readily localized. This may well eliminate the localization of the genes governing quantitative characters, for example, although such loci are difficult to handle by any method.

An additional procedure for the localization of particular loci to particular chromosomes is that of *in situ* DNA/DNA or DNA/RNA hybridization (see p. 124 for details of method). All that is needed is a sufficient amount of the appropriate DNA or RNA (rRNA, tRNA, or mRNA), properly labeled for identification, and the conditions needed for reannealing back onto the chromosome site from which it was produced or is complementary. By this means, the loci that code for human rRNA of the 18S and 28S kinds are located in the short arms of the satellited chromosomes 13, 14, 15, 21, and 22, while the locus coding for the 5S rRNA is in the long arm of chromosome 1.

The techniques described have indicated that the inherent difficulties involved in an expansion of our knowledge of human inheritance have in part been overcome by the development of a number of imaginative cytogenetic approaches, and it is to be expected that additional refinements of present procedures as well as new

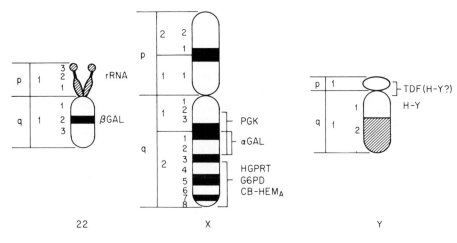

Fig. 4.46. Chromosome and gene maps of human chromosomes X, Y and 22 (compare with Figs. 3.58, 3.59 and 3.60). The chromosome maps are based on G-banding patterns, the gene maps by a variety of techniques. The Y chromosome has no identifiable genes located in the intensely fluorescent terminal portion of the long arm, but its male-determining influence is related to the presence of the histocompatibility antigen locus (*H-Y*), located within Yqll, and to *TDF* (testis-determining factor) located in the vicinity of the centromere. There is some question as to the separate identity of *H-Y* and *TDF* since H-Y has also been located in the short arm. The X chromosome has 102 known linked factors; among those located in the long arm are *PGK* (phosphoglycerate kinase); αGAL (αgalactosidase) associated with Fabry disease; *HGPRT* (hypoxanthine-guanine phosphoribosyl transferase (see Fig. 4.45)); *G6PD* (glucose-6-phosphate dehydrogenase; *CB* (deutron and protan color blindness; and *HEM$_A$* (classic hemophilia). In chromosome 22 βGAL (βgalactosidase) is located at 22q12, and rRNA (the nucleolar organizer region coding for rRNA) is at 22p12 (see McKusick and Ruddle, 1977 for details of other aspects of human linkage maps).

methods will lead to the kinds of linkage maps that have been produced for *Drosophila melanogaster,* corn, and tomato. Every human chromosome, including the Y chromosome, has at least one structural gene assigned to it (Fig. 4.46), and the X chromosome contains more than 100 known genes, although not all of them have yet to be assigned in proper order. Some govern morphological characters, but the majority relate to traits of a biochemical, viral, immunological, or medical nature. An interesting piece of comparative information to emerge from these studies is that the X chromosome of human beings and all other mammals is extraordinarily conservative in that any gene demonstrated in any mammal to be on the X chromosome would also be found to be present in human beings; the reverse is also valid.

A study of the events occurring in meiosis provides a clue to the physical basis of crossing over (a molecular consideration will be deferred until the next chapter). In prophase the homologous chromosomes synapse with each other in such a manner that by pachynema they are paired along their entire length. The synaptinemal complex, made visible through electron microscopical techniques (Fig. 4.15), reflects the intimacy of pairing. In diplonema the homologous chromosomes begin to separate from each other, the bivalents being held together only at certain points. These points of union are the chiasmata, and as pointed out earlier, it is generally agreed that they represent the points of prior genetical crossing over. Crossing over, therefore, must represent an actual exchange of chromatin material between homologous chromosomes. The homologues in diplonema are each longitudinally divided into two chromatids, and in clear preparations of meiotic cells it can be seen that a chiasma involves only two of the four chromatids of any bivalent (Figs. 4.22 and 4.23). Crossing over takes place between nonsister chromatids, not between whole chromosomes.

Figure 4.47 presents in diagrammatic fashion one interpretation of crossing over. It can be shown experimentally that chiasmata represent crossovers that have taken place between nonsister chromatids; those that might occur between sister chromatids, that is, the two chromatids of a single chromosome, would probably not be recognized as chiasmata since they would be indistinguishable cytologically from overlapping chromatids. In any event, in the absence of a special genetic constitution, such as a ring chromosome, sister-strand crossing over during meiosis could not be identified, although with radioactive thymidine or a nucleic acid analog it could be recognized in mitosis (Fig. 1.6). The sister chromatids remain associated on either side of the chiasma until anaphase, when the paired chromosomes separate. The chromatids not participating in the exchange will give rise to *noncrossover* offspring; the participating chromatids will give rise to the two complementary *crossover* types.

Fig. 4.47. Diagram of crossing over as it takes place in meiotic prophase through the breakage and reunion of nonsister chromatids at the four-strand stage. The precision of the process, occurring with no gain or loss of nucleotides in the affected chromatids, suggests that breakage is chemical (enzymatic?) rather than mechanical, but the nature of the process is unknown.

When more than one chiasma occurs between any two homologous chromosomes, the opportunities for rearrangement within linkage groups are increased, but it is obvious that a chiasma must occur between two gene loci in order for crossing over to be detected. The number of chiasmata between a pair of homologues does not vary greatly from cell to cell (for example, the X chromosome of *D. melanogaster* has from 0 to 3 chiasmata per oocyte) and is generally dependent upon the length of the chromosomes.

A specific relationship therefore exists between the cytologically visible chiasmata and the frequency of crossing over. A single chiasma within a bivalent leads to the formation of two noncrossover chromatids and two crossover chromatids; that is, recombinant frequency is one-half of the chiasma frequency. To state it otherwise: if, in 25 meiotic cells, a single chiasma always occurred between two gene loci, and if all the 100 chromatids were recovered from these cells, 50 would be noncrossovers and the other 50 would be crossovers. Consequently, 100% chiasma frequency leads to 50% crossing over, the maximum crossing-over frequency that can be observed between two genes. In the *pr-vg* linkage relationships discussed above, the 9.3% crossing over would result from a chiasma frequency of 18.6% in the region between the two genes. An approximation of the map length of any chromosome can, therefore, be determined by determining the mean number of chiasmata and multiplying by 50 (Table 4.2).

As the linear distance between two genes increases, the probability that two chiasmata may occur in the intervening region also increases. The occurrence of multiple chiasmata between two genes does not, however, alter the fact that the maximum *detectable* amount of crossing over between two genes, regardless of their distance apart, is still 50%. The chromosomal basis for this is illustrated in Fig. 4.48. Here the chiasma at the right (region I) is assumed to be constant in position, whereas those at the left (region II), designated *A, B, C,* and *D,* represent the four possible types that can each occur simultaneously with I. These types of double chiasmata, involving the four possible combinations of two nonsister chromatids, occur at random; in other words, there is no *chromatid interference* which promotes or diminishes the chances of any given pair of nonsister chromatids being involved in the second crossover. The recovered chromatids are indicated at the right of the figure. When the chromatids of the various types are summed [that is, the noncrossovers, the single crossovers in region I, the single crossovers in region II, and the double crossovers (regions I and II)], each is present four times. The crossover chromatids (those in which a chromatid exchange has

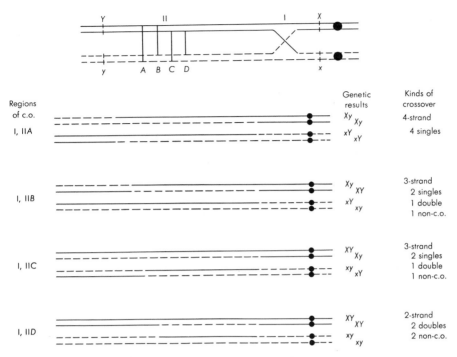

Fig. 4.48. Double crossing over between the linked genes X and Y and the resultant chromatids and genetic consequences. Crossing over in region I is considered as constant, while that in region II is at random with A to D designating the four possible types. A four-strand double crossover involves all four chromatids at some point; there are two three-strand doubles possible; a two-strand double involves the same two chromatids at both crossover points. The ratio of crossover to noncrossover chromatids is 4 noncrossover: 8 singles: 4 doubles. The genetic results add up to 50 percent crossing over; if a third crossover region were added between X and Y, the genetic results would not change.

actually taken place) are three times as frequent as the noncrossover chromatids. However, it is apparent that the two genes, X and Y, are not recombined in the two-strand double-crossover chromatids. Observationally, therefore, these would be indistinguishable from the noncrossover types, and the maximum frequency of crossing over between two genes would not exceed 50%. The same maximum frequency will also be obtained if there are 3 or more exchanges.

In most organisms, particularly in animals and the higher plants, it is not possible to examine directly all the products of a single meiotic sequence. It is thus necessary to sample gametes which are the products of many meiotic cells and to obtain an estimate of crossover frequency from the probability of recovering a recombinant or a nonrecombinant gamete. However, in the pink bread mold,

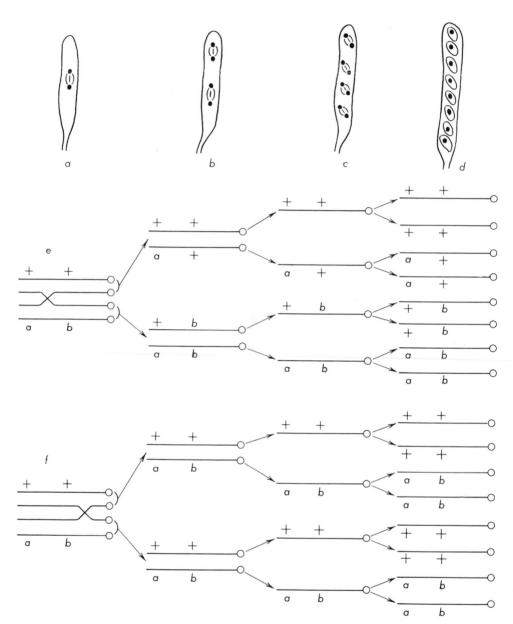

Fig. 4.49. Meiotic divisions and segregation in Neurospora. (a) Anaphase I; (b) anaphase II; (c) anaphase of the post-meiotic division; (d) mature ascus with eight ascospores; (e) behavior of one pair of homologous chromosomes and segregation of two pairs of genes (*a, b* and their wild type alleles) when the crossover is between the genes. The first division is equational for gene *a*, reductional for gene *b*, while the second division is the reverse. (f) Same as (e) except that the crossover is between gene *b* and the centromere, giving an equational first division and a reductional second division for both genes. The serial order of chromatids will be reflected in the serial order of the ascospores in the ascus.

Neurospora crassa, all four chromatids from any tetrad cannot only be recovered from a single spore sac or ascus, but they can also be recovered in serial order reflecting the two meiotic divisions. This provides information on the numbers and types of crossovers that occur. The manner of division and segregation is indicated in Fig. 4.49, and because there is good evidence that the ascospores do not, as a rule, slip past each other in the ascus during division, isolation of the ascospores in order, and the determination of their genotypes, can provide valuable evidence on cytological events taking place at the time of crossing over. Because homologous centromeres regularly segregate from each other in the first division, the map distance from the centromere to any gene can similarly be calculated by determining the proportion of asci in which alleles segregated at the second (vs. the first) meiotic division.

Chromosomal Evidence for Crossing Over. The basis of crossing over was discussed earlier, and it was pointed out that the chiasmata, visible initially in early diplonema, provided an appropriate mechanism for the reciprocal transfer of genes from one homologous chromosome to another. Through an exchange of chromatin material, all genes distal to the chiasma are reciprocally transferred to nonsister chromatids, and crossing over consequently appears to involve not just single genes but *blocks of genes.* This hypothesis was tested through the use of heteromorphic homologues that permitted the correlation of chromosome behavior with the disruption of linkage through crossing over. Stern, working with *D. melanogaster,* and Creighton and McClintock, working with maize, provided experimental evidence in 1931.

Stern's experiment is diagrammed in Fig. 4.50. The male parent in the experimental cross possessed a normal chromosomal set, with an X chromosome marked by the mutant eye-color gene *carnation (car)* and the normal allele $(+^B)$ of the gene *Bar (B),* which narrows the eye. The female parent was derived from a cross between two strains, one of which had a large portion of the Y chromosome attached (translocated) to the end of the X chromosome; this altered X chromosome carried the normal alleles of *carnation* $(+^{car})$ and *Bar* $(+^B)$. In the other strain the X chromosome had been broken into two separate parts; the piece bearing the X chromosome centromere carried the genes *car* and *B,* whereas the other fragment was translocated to the tiny IV chromosome. Both of these strains were viable, and when crossed they gave females that were heterozygous for the two genes in question. They also possessed two kinds of X chromosomes that could be morphologically distinguished from each

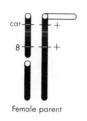

Female parent

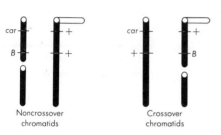

Noncrossover chromatids

Crossover chromatids

Fig. 4.50. Diagram of the chromosomes of *D. melanogaster* used by C. Stern to show that crossing over involves an exchange of chromatin between nonsister chromatids of homologous chromosomes. In the female parent, one X chromosome is represented as broken into two pieces; the top portion possesses its own centromere (open circle), while the remainder has a centromere derived from chromosome IV through translocation. The other chromosome in the female parent is distinguished by the fact that it possesses an arm derived from the Y chromosome (shown in outline). The crossover and noncrossover chromatids are indicated below, together with their genetic constitutions. The female was crossed to a male carrying *car* and the wild allele of *Bar* in its X chromosome. Any F_1 fly in which *car* and *Bar* were separated would reveal one of the crossover chromatids in its complement.

other as well as from the normal X chromosome, which, in the F_1 females, was derived from the father.

The parental heterozygous female can give crossover and noncrossover eggs, which, when fertilized by an X-bearing sperm containing the two genes *car*, and $+^B$, will give four types of female offspring (the males were disregarded). The two noncrossover types will be phenotypically *car-Bar* and wild-type, respectively. The former should, on cytological analysis, possess the two X chromosome fragments, and the latter should possess the X chromosome bearing the piece of the Y chromosome. Both, of course, will have a normal X chromosome in addition. The two crossover types will be either *carnation* with normal-shaped eyes or *Bar* with red (wild-type) eyes. The former should have two normal X chromosomes, and the latter should have a normal X chromosome plus the two X chromosome fragments, one of which will bear a portion of the Y chromosome. Stern made a cytological study of 364 crossover and noncrossover F_1 females, and there was good agreement between genetic and cytologic features.

Whether or not all chiasmata can be considered the cytological equivalent of genetic crossing over is a question of interest. The answer cannot be an unequivocal affirmative, because presumed chiasmata have been detected in *Drosophila* spermatocytes in which no crossing over occurs. Furthermore, the synaptinemal complex, found in some meiotic cells, indicative of homologous pairing, and a generally necessary precondition for crossing over, is absent in the spermatocytes of *Drosophila*. In a most exacting and exhaustive study made by Brown and Zohary in the meiotic cells of the lily, a 1:1 correspondence of chiasmata to chromatic exchange was found to hold true. The study in lily also demonstrated, as had earlier work, that crossing over at any single point along the chromosome involves only two of the four chromatids of a bivalent and that the exchange of chromatin is between nonsister chromatids. Ascospore segregation in *Neurospora* can be understood only if these conditions hold true (Fig. 4.49).

The relationship between chiasmata and crossing over in maize is illustrated in Table 4.2. With each chiasma being an exchange between only two of the four paired chromatids, each corresponds to 50 map units of genetical crossing over. An average of two chiasmata per bivalent would give a chromosome having a map length of 100 units. It is therefore possible to calculate rough map distances from chiasma frequencies. The observed crossover lengths for each of the 10 chromosomes (chromosome 1, being the longest, has been given an arbitrary pachytene length of 100, the other 9 being scaled to it) are lower than the lengths calculated from the chiasma frequency, but this is to be expected since there are some relatively unmapped regions of the chromosomes. The expected total crossover length is 1,350. In 1934, when Darlington first made his calculations, the observed total crossover length was 618, but by 1950 this had been in-

Table 4.2. Correlation of pachytene lengths, chiasma frequencies, and genetic maps of the ten haploid chromosomes of *Zea mays*. The longest chromosome, chromosome 1, is given an arbitrary length of 100 at pachytene, with the others being scaled to it in relative lengths. The three sets of observed values for the lengths of the genetic maps are given to indicate how map lengths change as the number of genes available for mapping increases.

				Length of Genetic Maps		
Chromosome Number	Pachytene Lengths	Chiasma Frequencies	Calculated	Observed 1934	Observed 1950	Observed 1976
1	100	3.65	187	102	156	161
2	86	3.25	163	58	128	155
3	78	3.00	150	92	121	128
4	76	2.95	148	80	111	143
5	76	2.95	148	44	72	87
6	53	2.20	110	52	64	68
7	61	2.45	123	50	96	112
8	61	2.45	123	20	28	28
9	53	2.20	110	52	71	138
10	44	1.95	98	68	57	99

Source of observed figures: C. D. Darlington (1934), M. M. Rhoades (1940), and M. G. Neuffer and E. H. Coe (1976).

creased to 904, and in 1970 to 1,119. Chromosome 8 shows the greatest departure between observed and calculated values, but it is the least studied of the maize chromosomes, possessing in 1976 only three definitely located genes. Chromosome 9 exceeds the calculated values, but since chiasma frequencies vary from one strain to another, the discrepancy is probably not unreasonable. There is, consequently, an approach to the 1:1 correspondence of chiasmata and crossovers as the chromosomes become better mapped, and this would hold true for the human as well as for any other organism.

Position and Frequency of Crossing Over

A precise determination of the position of crossing over in re-spect to chromosome distances can only be made after the genes have been located on a cytological map. The linear order of genes is the same for chromosomal and genetic maps, but physical and genetic map distances are not interchangeable units of measurement. The most convenient point of reference in all such problems is the centro-mere, although this can only be used in the chromosomes of higher organisms, particularly since there is good evidence that the centro-mere exerts a marked influence on the rate of crossing over in its neighborhood, even though a crossover on one side of the centromere is generally without influence on the frequency on the other side. A wealth of evidence indicates that the crossover frequency per unit of physical distance varies throughout the chromosome.

This can be demonstrated in chromosome 2 of *D. melanogaster* (Fig. 4.51). The left arm of chromosome II is approximately 55 map units long. The gene *purple (pr)*, 0.4 crossover units from the centro-mere, is one-quarter of the distance from the centromere to the end of the chromosome as judged by somatic chromosome lengths. *Black (b)*, 6.4 units from the centromere, occupies a position close to the middle of the arm. It is apparent, therefore, that the frequency of crossing over increases as one approaches the middle of the chromo-some arm and is greatest in the distal half of the chromosome, which has a map length of 48.2 units, or more than 85%. Other than to state that the centromere reduces crossing over in its neighborhood, the only other reason for this spatial variation would appear to be related to the compacted nature of the centric (constitutive) heterochro-matin.

Cytological maps can also be constructed from salivary chromo-somes, but they possess a disadvantage in that they do not give a true picture of the large heterochromatic portions adjacent to the centro-

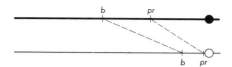

Fig. 4.51. Diagram of the cytological and genetical maps of the left arm of chromosome II of *D. melanogaster* to indicate that, while linear sequences are the same in both, the spatial distances are quite different. The heterochromatic portion of this arm occupies about one-fourth of the mitotic length and is a region in which crossing over is sharply reduced. Any gene such as *pr*, which lies close to the boundary between heterochromatin and euchromatin, would thus appear close to the centromere in the genetic map.

meres of the three major chromosomes in *D. melanogaster.* The detailed banded structure of euchromatin, however, permits an accurate mapping of the loci, and it also makes it feasible to determine whether local differences in crossover frequencies exist, a study manifestly impossible earlier with linearly undifferentiated somatic chromosomes but more feasible now with more recently developed banding techniques. That such local patterns of crossing over are found is indicated by a comparison of the cytological and genetical distances from *y* to *fa* in the left end of the X chromosome (see Fig. 4.43). Clearly, from these data, the *rst-fa* region is one of relatively high crossing over. The *w-rst-fa* region, located in four or possibly five detectable bands, has the same crossover frequency (1.5%) as the *y-pn-w* region, made up of 75 to 80 bands. It is not yet possible to explain such regional patterns in terms of the physical structure of the chromosome; we only recognize that they exist.

In all organisms tested, however, crossing over is infrequent in the vicinity of the centromere. This has been very nicely demonstrated in *Drosophila.* Use was made of a translocation involving the III and IV chromosomes, which was subsequently obtained in a homozygous state. The changes made were as follows, with the centromeres indicated by the vertical lines, points of breakage by arrows, and the numbers the locations of several genes:

Genes:					*Normal III*					*Normal IV*
ru	*h*	*th*	*st*		↓ *cu*	*sr*	*e*ˢ	*ca*	↓	
0	26.5	43.2	44.0		50.0	62.0	70.7	100.7		

Translocated Chromosomes

ru *h* *th* *st*| *ca* *e*ˢ *sr* *cu*

The genes *ru* to *st,* as a result of the translocation, remain unchanged relative to the centromere of chromosome III, whereas the remaining genes, in particular *cu,* are placed closer to the centromere of chromosome IV. The effect of this on crossing over is indicated with the percentages of crossing over indicated between the genes:

Normal	*ru*	*h*	*th*	*st*	*cu*	*sr*	e^s	*ca*
	27.5	21.3	0.4	6.2	13.3	11.6	35.0	

Homozygous Translocation	*ru*	*h*	*th*	*st*	*cu*	*sr*	e^s	*ca*
	29.1	21.2	0.8		0.9	2.8	32.2	

The crossover percentages in the successive regions from *ru* to *st* remain essentially unchanged. The displacement of *cu* to close proximity of the IV centromere markedly reduces the *cu-sr* and *sr-e*^{s} crossover frequencies. The e^s-*ca* region, being in the distal region of the arm, shows little variation. Clearly, proximity to the centromere alters the rate of crossing over, and it is apparent that the bunching of genes on the genetic maps of the X, II, and III chromosomes does not mean that they are physically closer to each other, but simply that crossing over is reduced in these areas. In this particular experiment, the reduction may be due to some property of the centromere itself; to the centric heterochromatin, which is probably compacted at the time of crossing over; or to some factor that has altered pairing relationships at synapsis.

Among the intrinsic factors affecting the frequency of crossing over are sex, age, and genotype; temperature is an obvious modifying extrinsic influence. It is, of course, well known that there is little or no crossing over in males of *D. melanogaster* except under unusual experimental conditions. The same holds true for the female of the silkworm, *Bombyx mori.* In organisms in which crossing over is regularly found in both sexes, the frequency of recombination between any two genes may be identical in the two sexes (or in the anthers and ovaries of dioecious plants), as in the garden pea; higher in females than in males, as in the mouse, the human, and the rat; or higher in males than in females, as in the pigeon. Where linkage differentials exist between the sexes, and where adequate data are available, it is the heterogametic sex in which crossing over is either lowered in frequency or is absent.

Discrepant Crossover Events. Any discussion of crossing over is usually based on the classical assumption that it is a reciprocal event taking place at the four-strand stage of meiosis and between nonsister chromatids. Because of chiasma interference, crossing over is also thought to be spaced such that adjacent crossovers are at some distance from each other; in *D. melanogaster,* for example, double crossovers within 10 map units are not recovered. It is obvious now that the classical view is in need of considerable revision in terms of location, frequency within limited areas, and reciprocity.

Somatic or Mitotic Crossing Over. We customarily think of crossing over as a phenomenon confined to meiotic cells, but it can take place in somatic cells as well, although with very much reduced frequency. The phenomenon was discovered in *D. melanogaster,* and it is detected by the occurrence of single or twin spots of body tissue displaying a recessive phenotype on an otherwise wild-type background.

Bridges had early postulated that the spots were due to the elimination of certain X chromosomes carrying wild-type genes, but Stern argued that the mosaicism was caused not by the elimination of chromosomes but by somatic crossing over followed by a normal mitotic segregation of chromatids. The genes *y (yellow body)* and *sn (singed bristles)* were used because, against the wild-type background of a heterozygous female, the *y* and *sn* phenotypes could be easily recognized. The basis of Stern's argument for somatic crossing over was the more frequent occurrence of twin spots, adjacent to each other, exhibiting *y* and *sn* phenotypes, than single spots of either *y* or *sn* tissue. Neither gene mutation nor chromosome loss explains these data.

In a typical experiment, involving the use of a *y +/+ sn* female, the mosaicism was expressed as twin spots of relatively equal size, one exhibiting the *y* phenotype and the other the *sn* phenotype. Less frequently only *y* spots were found, and still more rarely only *sn* spots. These results can be explained (Fig. 4.52) by assuming that the twin spots arose by a crossover after somatic pairing in the four-strand stage and between *sn* and the centromere, the *y* spots by a crossover between *y* and *sn,* and the *sn* spots by a double crossover between *sn* and the centromere and between *y* and *sn.* The greater frequency of twin spots results from a greater rate of crossing over in the area between the centromere and *sn* which covers 45 map units. The less frequent occurrence of single *y* spots is explicable on the basis of the relatively short map distance (21.0 units) between *y* and *sn.* The size of the spots would be dependent upon the time in development at which the crossover occurred, the larger ones being due to a crossover earlier in development.

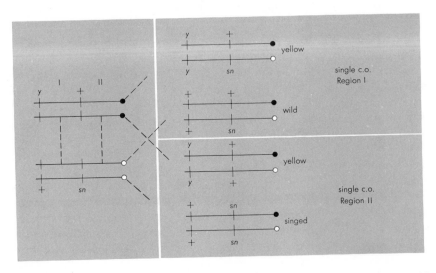

Fig. 4.52. Somatic, or mitotic, crossing over in the X chromosome of *D. melanogaster*. The chromatids segregate as in mitosis, not as in meiosis. A single crossover between *y* and *sn* (region I) will produce a yellow spot if the chromatids segregate as indicated; a single crossover between *sn* and the centromere (region II) will produce adjacent twin spots; single *sn* spots can result only through double crossing over and the proper segregation of chromatids and will be understandably infrequent. The size of the spots will be determined by the time during development when crossing over occurred. After C. Stern.

The relatively high frequency of somatic crossing over in *Drosophila*, which involves chromosomes II and III as well as the X, is due to the tendency of homologous chromosomes, abetted perhaps by the presence of centric heterochromatin, to synapse intimately, even in somatic cells. Both the frequency and position of such crossovers are influenced by the presence or absence of a group of loci called *Minutes*. How the latter factors exert their effect is not known, but sex-linked *Minutes* are more effective in enhancing somatic crossing over than are autosomal *Minutes*, and the *Minutes* in chromosome III are peculiar in limiting crossing over to the arm in which they are located. Interestingly enough, somatic crossing over occurs in males as well as females of *D. melanogaster*, although only females exhibit meiotic crossing over. The phenomenon, however, is not confined to *Drosophila*, and it has been found and studied in a number of fungal species. Known as the *parasexual cycle*, it contrasts with a normal sexual cycle which includes meiosis. There is some question whether the events in *Drosophila* and fungi are strictly comparable.

In *Aspergillus nidulans*, for example, the mycelium is ordinarily haploid, but diploid nuclei are occasionally formed when adjacent hyphal strands fuse and their nuclei mix and fuse in a common cytoplasm. When the haploid nuclei that fuse are of different genotypes, the diploid nucleus is consequently heterozygous for certain genes. Crossing over apparently occurs during the division cycle of these diploid nuclei. This by itself would not alter the total genetic character of a diploid nucleus because the gene content remains the same, but these nuclei then divide mitotically again to provide a segregation of genes in a manner similar to that depicted in Fig. 4.48. Also the diploid nuclei on occasion become haploid by a progressive loss of chromosomes, after which different gene groupings can be detected. In a fungal colony these would appear as sectors of different character.

The frequency of somatic crossing over is undoubtedly low in nature. It is, however, a source of variation, and in many fungal species that lack a sexual cycle the process may be quite important in evolution.

Negative Interference. Positive interference has been defined as the inhibiting influence of one crossover or chiasma on the occurrence of another one in its vicinity, the influence diminishing with distance between the crossover loci. In fact, beyond a certain evolutionary level the degree of positive interference is probably related in a meaningful way to the number of nucleotide pairs per average map unit (Table 4.3). There is little or no positive interference in the fungi; however, the molds, *Aspergillus nidulans, Neurospora crassa,* and a number of bacteriophages reveal a negative interference when intragenic rather than intergenic crossing over is being considered. This means that the occurrence of one crossover within a gene *en-*

Table 4.3. Genetic maps and map units viewed as a function of the amount of DNA in various organisms

	Total Linkage in Map Units	Total DNA in Nucleotide Pairs	Nucleotide Pairs per Average Map Unit
Lambda phage	20	4.5×10^4	2.3×10^3
T4 phage	2,500	2×10^5	0.8×10^2
E. coli	2,000	4.3×10^6	2.0×10^3
Aspergillus	660	4.0×10^7	7×10^4
Neurospora	800*	1.9×10^7	2.4×10^4
Drosophila	280	8×10^7	3×10^5
Mouse	1,954*	2.5×10^9	1.2×10^6
Maize	1,350*	7×10^9	8×10^6
Human	3,000*	2.9×10^9	1.0×10^6
Amphibian (*Batrachoseps*)	2,200*	6.8×10^{10}	3×10^7

*Estimated from chiasma frequencies.

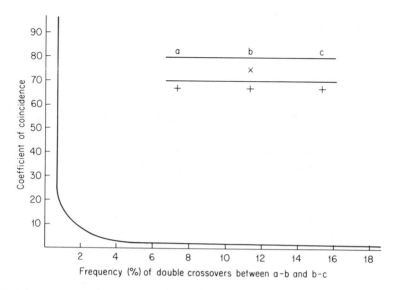

Fig. 4.53. Relation of the coefficient of coincidence and the frequency of double crossovers in a cross of *a b c* X *+ + +* in λ bacteriophage. The closer together the marker genes, the higher the frequency of double crossovers, indicating that the occurrence of one crossover promotes the occurrence of another in its immediate neighborhood (redrawn from Amati and Meselson, 1965).

hances the probability of another one in its neighborhood, the coefficient of coincidence will be greater than 1, and the frequency of double crossovers may be as high as a hundredfold greater than expected. For the bacteriophage λ, the coefficient is about 5 for almost any three genes closely linked and tested instead of an expected coincidence of less than 1, and any two marker genes separated by about one map unit or less will have many more double crossovers than would be expected (Fig. 4.53). The possible explanation of this phenomenon by the effective-pairing region hypothesis is discussed in the next chapter.

Gene Conversion. When a cross is made between a strain of yeast carrying the gene *A* and one carrying its allele *a,* the four haploid meiotic products would ordinarily show a phenotypic ratio of 2*A* :2*a. Neurospora,* with its eight ascospores, would exhibit comparable ratios of 4*A* :4*a*. Infrequently, but more often than can be accounted for on the basis of mutation rates, departures from those ratios have been found: 3*A* :1*a* or 1*A* :3*a* in yeast and 6:2 ratios in *Neurospora*. More infrequently still, odd numbered ratios, 5:3 and 7:1, have also been found in *Neurospora*. Clearly, these ratios are in violation of Mendelian and chromosomal expectations, and it appears as if one or more of the alleles had been converted to its alternate allelic form; hence the term *gene conversion* for the phenomenon.

Table 4.4. Four different asci obtained from a *Neurospora crassa* cross of + *pdxp* X *pdx* + and in which wild-type pyridoxine spores (+ +) appeared (Mitchell, 1955).

Spore Pairs	Asci			
	1	2	3	4
First	+ *pdxp*	*pdx* +	+ +	*pdx* +
Second	+ +	*pdx* +	+ *pdxp*	+ *pdxp*
Third	+ *pdxp*	+ +	*pdx* +	+ +
Fourth	*pdx* +	+ *pdxp*	*pdx* +	*pdx* +

Table 4.4 indicates the genetic analysis of four asci from *Neurospora* which revealed deviant ratios. They were derived from nearly a thousand dissected asci from a cross of + *pdxp* X *pdx* +. These are closely linked genes, both of which affect the synthesis of the nutritionally required pyridoxine. Only the genotype + + can survive in the absence of pyridoxine and the presence of the + + genotype should have been accompanied by the reciprocal product *pdx pdxp* if a normal crossover event had been involved. These, however, were missing, indicating that a typical crossover had not occurred. Subsequent studies in *Neurospora* and other organisms, using closely linked genes, flanked by outside gene markers, reveal that gene conversion is an intragenic event unaccompanied by the recombination of outside neighboring genes. The question as to why closely linked genetic markers within a gene can undergo nonreciprocal recombination, or gene *conversion,* while more distant alleles in different genes undergo reciprocal crossing over remains unanswered. Figure 4.54 is a diagram-

Fig. 4.54. The difference between reciprocal recombination (above) and gene conversion (below). In the latter situation, the section of chromosome at y^+ is copied three times while its allelic form y is copied only once. See Fig. 5.28 for a possible molecular explanation.

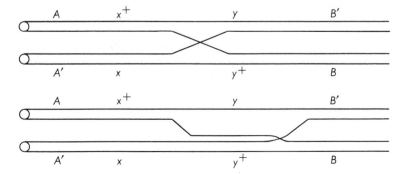

matic description of reciprocal recombination vs. gene conversion, but a possible molecular explanation will be deferred to the next chapter (Fig. 5.29).

Genetic Control of Meiosis

Meiosis was very probably derived from mitosis early in eukaryotic history by a series of mutational changes. We have no way of knowing whether this occurred only once and was subsequently retained by natural selection and then reinforced and varied in many small ways by additional mutations or whether it appeared repeatedly in the course of time. But like any chemically ordered process in the cell, meiosis is subject to genetic control. This is particularly evident in those processes which differentiate meiosis from mitosis and, as a consequence, among the meiotic products—eggs, sperm, zygotes, and/or spores—as opposed to the daughter cell resulting from mitosis. Mitosis is, of course, also under genetic control, but mutations which seriously disturb the mitotic process would be rigorously screened out because of their deleterious effect on the individual. By way of contrast, meiotic mutations, most of which are recessive in expression, can be tolerated and carried in a population, and as searches for them in maize, tomato, and *Drosophila* have shown, they are numerous and of varied character and they affect every process and structure of the meiotic cell from its initiation to the realization of its cellular products.

In this situation we need to recognize that we are dealing with both a controlling and a responding system. In cases in which it is the chromosome(s) alone which exhibits abnormal behavior or appearance, the chromosome is simultaneously the controlling and responding system, i.e., it is regulated by its own genes; where other parts of the meiotic cell or its cellular products are involved, the chromosome (or its gene) is the controlling system, the cell or its nonchromosomal parts the responding element. In most instances, the basic means of genetic control is but poorly understood, but, as with all cellular systems, they must be related to some transcriptional or translational system which, directly or indirectly, fails to function in a normal manner. A wild-type system, therefore, is one in which both controlling and responding systems are normal.

Genes which affect the meiotic process or its structures can be divided into three rather arbitrarily defined classes: (1) those that alter the structures or processes of meiosis without interfering with the general course of events necessary for the production of viable sexual gametes or asexual spores; (2) those that distort the structures or disrupt the process to bring about varying degrees of sterility or aberrant gametes or spores; and (3) those that are without obvious effect on the meiotic process itself but which, through one post-meiotic device or another, bring about a preferential survival or perpetuation of certain genetic combinations. The term "meiotic drive" has been coined to characterize the last group.

Examples of the first class of meiotic mutants have already been mentioned in other contexts. Heteropycnosis and precocious first anaphase segregation of the X chromosome in XO grasshopper males are obviously due to genes operating in one sex but not in the other; a similar explanation holds for the lack of crossing over in *D. melanogaster* males, accompanied by regular segregation of the chromosomes at anaphase; this is in contrast to the situation in females in which a lack of crossing over would often result in irregular disjunction. However, some animal groups, notably in the insects, are normally achiasmate, presumably undergo no crossing over but exhibit normal segregation at first anaphase. To these examples can be added a wide variety of selected inactivations and even eliminations of either single chromosomes or whole sets of chromosomes. Most of these examples are to be found among the insect groups, and some result in a pronounced form of meiotic drive (p. 276).

On a somewhat more restricted scale of change, the genetic circumstances in the perennial rye grass, *Lolium perenne,* are illustrative of the class of mutants governing chromosome size (Fig. 4.55). Here the chromosomes of various segregants differ in both size of chromosomes and the chiasma frequency per nucleus, suggesting a control that influences both the amount of crossing over and the amount of chromosomal material in the contracted bivalents. No other phenotypic effects are evident, and it would be of interest to know whether chromosomal size is determined by variations in the amount of DNA or non-DNA molecules; the latter is a more likely cause.

A large number of meiotic mutants belong in the second class mentioned; they are initially uncovered because they influence the degree of fertility or the production of aberrant offspring. This is particularly true in plants in which lack of fruit set, an obvious phenotypic expression, invites inspection of the meiotic process to determine the cause. Two examples involve the length of the chromosomes. The garden stock, *Matthiola incana,* normally possesses short, compact bivalents with chiasmata largely terminalized; a long-chromosome mutant has interstitially located chiasmata with frequent instances of nondisjunction. The mutant condition could be due either to a normal rate of contraction, but with an abbreviated

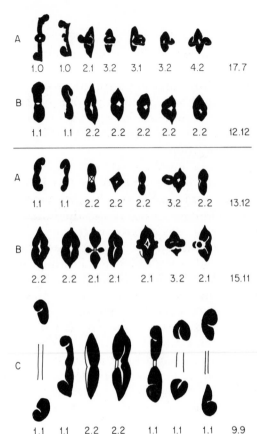

Fig. 4.55. Above, degree of terminalization in normal (A) and pollen sterile (B) plants of *Lathyrus odoratus*. Bottom, metaphase and early anaphase in meiosis in three plants of the perennial rye grass, *Lolium perenne*, showing differences in chromosome size and chiasma number and distribution; plant A was the parent of plants B and C.

period between the pachytene stage and metaphase, or to a normal prophase sequence in time coupled to a slower tempo of contraction. Either circumstance would leave the chromosomes longer and prevent terminalization of the chiasmata, but it would not account for the nondisjunction that takes place. A similar situation is found in the sweet pea, *Lathyrus odoratus* (Fig. 4.55), except that the normal condition is characterized by long chromosomes and interstitial chiasmata, the mutant type by short bivalents with terminalized chiasmata. The mutant is also male-sterile. In both species, and to a degree in *Lolium* as well, terminalization of chiasmata parallels the degree of contraction, indicating that in some way the latter process moves the chiasmata from the point of initial occurrence to the ends of the bivalents.

A large number of known mutants are believed to alter the synaptic process and, as a consequence, the frequency of crossing over. Those discovered in plants have largely involved microsporogenesis and are termed *asynaptic* because of the generally unpaired

nature of the pachytene chromosomes and the appearance of univalents at the metaphase stage. One might naturally assume that in
such instances the frequency of crossing over would be drastically
reduced, but in the case of asynapsis in maize crossing over had occurred, and, in fact, in two different tested regions in chromosomes
2 and 9 the frequency of double crossovers was nearly 25 times
higher than normal. The unpaired nature of meiotic homologues can
be due, therefore, either to a lack of pairing or to premature separation. There is no indication that all asynaptic genes in other organisms exert their influence similarly, but in the fungus, *Neurospora
crassa,* there is a gene which increases intra-allelic recombination at
the *histidine*-1 locus by a factor of 10 and without disturbing the
recombinational ratios of outside markers.

Mutant genes in *D. melanogaster* affecting crossing over have
their expression in the female sex only. They are not uncommon in
natural populations, and, as might be expected for a complex process
such as crossing over, the mutants differ from each other while sharing some similarities. A number of these can be mentioned: (1) the
C(3)G gene on chromosome 3 virtually abolishes recombination and
causes a good deal of nondisjunction at the first meiotic anaphase;
mei-S51 (mei = meiotic) reduces recombination to about one-half of
the normal level in all chromosomes, and it selectively brings about
nondisjunction of the X chromosome and chromosome 4, often in
the same cell; *mei-S282* strongly reduces crossing over at the ends of
the chromosome arms but leaves unaffected the frequency in proximal regions, suggesting that synapsis is initiated proximally and proceeds in a distal direction but is interrupted before completion.
Mei-S282 shows a strong similarity to *mei-S51* in that it also causes
the coincidental nondisjunction of the X chromosome and chromosome 4; *mei-332* differs from the others in that it increases recombination, with the increase greater proximally than distally. We shall see
later (p. 374) that inversions can also have a strong effect in increasing crossing over, an effect, however, that depends upon the particular rearrangement of chromatin rather than upon any specific gene.
Whether any of these effects relates to an altered structure or
behavior of the synaptinemal complex is unknown.

None of the meiotic mutants in *D. melanogaster* has an effect
on mitotic crossing over, the latter process having its own set of
genetic controls; that is, different strains exhibit different frequencies
of mitotic crossing over, and the *Minute* genes, of which there are
many, have their own highly individualized effect on the process.

Mutants of a quite different character are those that alter the
meiotic process by affecting the divisional apparatus of the cell. The

gene *divergent (dv)* in maize, when homozygous, causes the first division spindle to take on a divergent, flaring shape instead of a biacuminate one that converges at the poles. Several, widely spread, small nuclei rather than a single normal one form at telophase, and these at the second division have difficulty in bringing about regular segregation since each nucleus may form its own partial spindle in an uncoordinated way. A barley mutant leads to the absence of cell walls and possibly cell membranes among the pollen mother cells, with the result that these cells fuse with each other to produce polyploid nuclei and/or multinucleate cells. Instead of having 7 normal bivalents per cell at metaphase, the chromosomes may exist in any multiple of 14, and as many as 112 have been observed in a single cell. The chromosomes, which may exist as multivalents as well as bivalents and univalents, have difficulty being accommodated on a single metaphase plate because of crowding. The degree of sterility is obviously high. Another divisional mutant is the *polymitotic* gene found in maize. Acting in the microspore rather than in the microsporocyte, the gene induces a series of cytoplasmic divisions unrelated to the replication of the chromosomes. As a result, the microspore is segmented into a number of compartments, any one of which may or may not contain chromosomes. These microspores are, of course, nonfunctional.

The remaining mutants to be considered can be classified as being involved in the phenomenon of *meiotic drive.* In such cases, and from a state of either genetic or chromosomal heterozygosity, more than 50% of one kind of gamete is recovered. While the end result of meiotic drive is the distortion of Mendelian expectations in any breeding system, and the unequal perpetuation and increase of one kind of genome as opposed to another, the means to that end are quite varied.

The B chromosomes and the *abnormal 10 (K10)* chromosome of maize achieve their preferred retention through the process of preferential segregation. The situation with the B chromosome was mentioned in the previous chapter: It is preferentially distributed, as a result of nondisjunction, into one of the two sperm cells at the second microspore division (Fig. 3.60). The accessory chromosome of rye behaves similarly, except that nondisjunction occurs at the first microspore division, and with the generative cell becoming the preferred recipient. The B chromosomes of maize exhibit an additional preferential recovery in that any sperm possessing B chromosomes has a 70% (instead of a 50%) chance of fertilizing the egg instead of uniting with the polar nucleus, and they exert a further effect, also at the second microspore division, in that they cause the loss by elimination of any chromosomes of the A set which possess knobs.

Mendelian ratios of any genes found on these knobbed chromosomes are destined to be distorted by the B chromosome influence.

The normal chromosome 10 in maize exists as a knobless structure, whereas K10 possesses a large heterochromatic knob at the end of its long arm. When present in a heterozygous state (K10/10), the K10 chromosome is preferentially recovered up to 70% of the time on the female side because of the capacity of the K10 knob to act as a *neo-centromere*. As Fig. 3.13 illustrates, precocious movement of the knob, acting somewhat as a centromere, brings the knobbed chromosome into a position where it selectively enters the basal megaspore for inclusion into the egg nucleus. Crossing over between the normal centromere and the knob is an essential feature of preferential segregation. So far as can be determined the neo-centric capability is intrinsic to the knob itself rather than to any single gene. K10, whether in a homozygous or heterozygous state, will also influence the preferential recovery of any other knobbed chromosome when it is in a heterozygous state and when a crossover occurs between the centromere and the knob. K10, therefore, has a selective effect on heterochromatin but not on euchromatin. The circumstances are the same as those depicted in Fig. 4.50.

In addition to its distributional influence, K10 is also responsible for increased crossover frequencies in other chromosome pairs in the maize genome. The *Gl-Lg-A* region of chromosome 3 can serve as an example. *Gl* is close to the centromere; *A* is most distal. *Gl* and *gl* are recovered in nearly random numbers in the presence of K10 since few crossovers occur between this locus and the centromere. Substantial crossover increases occur in the *Gl-Lg* and *Lg-A* regions, largely because of an increased number of double crossovers, and the increases are enhanced still more so if chromosome 3 possesses a knob in the vicinity of the *Lg* locus. The reason for such increases remains obscure, although pachytene configurations suggest a more intimate pairing than is usual.

A gene in tomato called *Gamete eliminator (Ge)*, located in the centric heterochromatin of the long arm of chromosome 4, is responsible for a form of meiotic drive different from that induced by K10 of maize. In plants heterozygous for two alleles of the gene—Ge^c/Ge^p—the male and female gametes containing the Ge^c allele are effectively aborted, leading to the absence of expected genotypes and producing distorted ratios in the subsequent generation. The action of the Ge^c allele is essentially that of a "killer" gene functioning at the gametophytic level in both microspores and macrospores. Still another form of meiotic drive, different from that in maize or in tomato, is found in some species of *Oenothera,* but since a complex translocation system is involved a discussion will be delayed until Chapter 6.

Among *Drosophila* species, various examples of meiotic drive are known; they can affect either sex in a selective manner, and the genes involved act post-meiotically within populations of gametes. *Sex-ratio* males produce far more (over 90%) daughters than sons, regardless of the genetic character of the mother. The condition seems to depend upon an X chromosome gene whose action leads to the degeneration and dysfunction of most Y-bearing sperm. The *sex-ratio* influence is absent in females, and the *sex-ratio* X chromosome is passed on to sons in normal fashion.

Daughterless (da) and *abnormal oocyte (abo)* are chromosome 2 genes located in the left arm at about position 39 of the salivary chromosome map. The former has no effect on males, but the homozygous females *(da/da)* produce only male progeny, the females dying as a result of zygote mortality and/or developmental interference. Females homozygous for *abo* produce an excess of daughters. What is of considerable interest is that the effect of both *da* and *abo* can be lessened by the addition of X or Y heterochromatin to the genome of the mother, with *abo* more susceptible to heterochromatic influence than *da*. Neither *da* nor *abo* exhibits any other phenotypic effect, and it appears that both act as regulators of two different structural genes located in the heterochromatin of the X chromosome. The action of these structural genes and the nature of their products remain unknown, but they seem to be necessary for zygotic survival.

A final example of meiotic drive is that conditioned by the *segregation-distorter* complex in *D. melanogaster.* The situation is such that males heterozygous for an SD-bearing chromosome 2 and a normal (SD^+) homologue transmit the SD chromosome almost exclusively to the next generation. There is no sex preference in this instance, and failure to transmit the SD^+ chromosome is due to the failure of sperm bearing this chromosome to function properly. The SD^+ locus is located in the left arm of chromosome 2 in the region from *37D2-D7* to *38A6-B2,* not far from the location of *da* and *abo.* The SD locus does not act alone, however; like *da* and *abo, SD* interacts with another locus, *Responder (Rsp),* which is located in the proximal heterochromatin of the right arm of chromosome 2. The explanation that has been advanced is that SD produces a product (a protein presumably) which binds to the *Rsp* locus, inactivating it and thereby preventing normal spermiogenesis from taking place in those sperm containing the SD^+ chromosome. There is an allele of *Rsp* that is insensitive to the SD product, and it has been suggested that SD^+, an "allele" of *SD*, is simply the absence of *SD*. A third locus, *Enhancer of SD* or *E(SD)*, found in the heterochromatin of the left arm of chromosome 2, exaggerates the distortion brought on the *SD-Rsp* system.

The examples of meiotic drive that have been discussed pose a question. Are we dealing here, most probably in distorted form, with a phenomenon of differential gene expression that is a normal part of the pattern of growth and development? In many respects the action of *da, abo,* and the *SD-Rsp-E(SD)* complex is not too dissimilar from other patterns of inactivation and/or elimination: the inactivation of one of the X chromosomes in mammalian females; the elimination of the paternal X chromosome in marsupial females; the inactivation and eventual elimination of the whole set of paternal chromosomes in mealy bugs; and the inactivation of particular genes in differentiated cells. Whatever the final resolution of these problems that have been raised by the discovery of mutant forms, it now seems rather obvious that chromosomes are genetic entities carrying within themselves the means of their own regulation. It is when these systems go awry through mutations that we can gain some insight into the workings of the normal genome.

BIBLIOGRAPHY

Amati, P., and M. Meselsen, "Localized Negative Interference in Bacteriophage λ," *Genetics 51* (1965), 369–79.

Baker, B. S., et al., "The Genetic Control of Meiosis," *Ann. Rev. Genetics, 10* (1976), 53–134.

Boyd, J. B., et al., "The Mei-9^a Mutant of *Drosophila melanogaster* Increases Mutagen Sensitivity and Decreases Excision Repair," *Genetics, 84* (1976), 527–44.

Brown, S., and D. Zohary, "The Relationship of Chiasmata and Crossing Over in *Lilium formosanum,*" *Genetics, 40* (1955), 850–73.

Carpenter, A. T. C., "A Meiotic Mutant Defective in Distributive Disjunction in *Drosophila melanogaster,*" *Genetics, 73* (1973), 393–428.

Catcheside, D. G., *The Genetics of Recombination.* Baltimore: University Park Press, 1978.

Donnelly, G., and A. H. Sparrow, "Mitotic and Meiotic Chromosomes of Amphiuma," *Jour. Heredity 61* (1965), 91–98.

Flavell, R. B., and G. W. R. Walker, "The Occurrence and Role of DNA Synthesis during Meiosis in Wheat and Rye," *Exp. Cell Res. 77* (1973), 15–24.

Ganetsky, B., "On the Components of Segregation Distorter in *Drosophila melanogaster*," *Genetics, 86* (1977), 321-55.

Grell, R. F. (ed.), *Mechanisms in Recombination*. New York: Plenum, 1974.

——, et al., "Meiotic Exchange Without the Synaptinemal Complex," *Nature (New Biology), 240* (1972), 155-57.

Grzeschik, K. H., "Utilization of Somatic Cell Hybrids for Genetic Studies in Man," *Humangenetik, 19* (1973), 1-40.

Hasitschka, G., "Bildung von Chromosomenbündeln nach Art der Speicheldrüsenchromosomen under andere Strukturreigentumlichkeiten in den endopolyploiden Riesenkernen der Antipoden von *Papaver rhoeas*," *Chromosoma, 8* (1965), 87-133.

Henderson, S. A., "The Time and Place of Meiotic Crossing-Over," *Ann. Rev. Genetics, 4* (1970), 295-324.

Hotta, Y., and H. Stern, "The Appearance of DNA Synthesis During Meiotic Prophase in *Lilium*," *J. Mol. Biol., 55* (1971), 337-55.

Howell, S. H., and H. Stern, "The Appearance of DNA Breakage and Repair Activities in the Synchronous Meiotic Cycle of *Lilium*," *J. Mol. Biol., 55* (1971), 357-78.

Hughes-Schrader, S., "Meiosis Without Chiasmata in Diploid and Tetraploid Spermacytes of the Mantid *Callimantis antillarum* Saussare," *J. Morph., 73* (1943), 11-141.

Hultén, M., and J. Lindsten, "Cytogenetic Aspects of Human Male Meiosis," *Adv. Human Genetics, 4* (1973), 327-87.

Kattaneh, N. P., and D. L. Hartl, "Histone Transition During Spermiogenesis is Absent in Segregation Distorter Males of *Drosophila melanogaster*," *Science, 193* (1976), 1020-21.

Koo, G. C., et. al., "Mapping the Locus of the *H-Y* Gene on the Human Y Chromosome," *Science 198* (1977), 940-42.

Lindsley, D. L., et al., "Genetic Control of Recombination in *Drosophila*," in *Replication and Recombination of Genetic Material*, W. J. Peacock and R. D. Brock, eds. Canberra: Australian Academy Science, 1968.

McKusick, V. A., "The Mapping of Human Chromosomes," *Sci. Amer., 227* (1971), 104-13.

——, and F. H. Ruddle, "The Status of the Gene Map of the Human Chromosomes," *Science, 196* (1977), 390-405.

Meselsen, M. S., and C. M. Radding, "A General Model for Genetic Recombination," *Proc. Nat. Acad. Sci., 72* (1975), 358-61.

Miller, D. A., and O. J., Miller, "Chromosome Mapping in the Mouse," *Science, 178* (1972), 949-55.

Mitchell, M. B., "Aberrant Recombination of Pyridoxine Mutants of Neurospora," *Proc. Nat. Acad. Sci., 41* (1955), 215-20.

Moens, P. B., "The Onset of Meiosis," in *Cell Biology: A Comprehensive Treatise,* Vol. I, L. Goldstein and D. M. Prescott, eds., New York: Academic Press, 1978.

Neuffer, M. G., and E. H. Coe, Jr., "Linkage Map and Annotated List of Genetic Markers in Maize," in *Handbook of Biochemical and Molecular Biology,* 3rd ed., G. D. Pasman, ed. Cleveland: CRC Press, 1976.

Nicklas, B., "Chromosome Segregation Mechanisms," *Genetics, 78* (1974), 205-13.

Peacock, W. J., and R. D. Brock (eds.), *Replication and Recombination of Genetic Material.* Canberra: Australian Academy Science, 1968.

Perkins, D. D., and E. G. Barry, "The Cytogenetics of *Neurospora,*" *Adv. Genetics, 19* (1977), 134-286.

Radding, C. M., "Molecular Mechanisms in Genetic Recombination," *Ann. Rev. Genetics, 7* (1973), 87-112.

Rees, G., and R. N. Jones, *Chromosome Genetics.* Baltimore: University Park Press, 1978.

Renwick, J. H., "The Mapping of Human Chromosomes," *Ann. Rev. Genetics, 5* (1971), 81-120.

Rhoades, M. M., "Studies on the Cytological Basis of Crossing Over," in *Replication and Recombination of Genetic Material,* W. J. Peacock and R. D. Brock, eds. Canberra: Australian Academy Science, 1968.

Riley, R., "Cytogenetics of Chromosome Pairing in Wheat," *Genetics, 78* (1974), 193-203.

———, M. D. Bennet, and F. B. Flavell (organizers), "A Discussion on the Meiotic Process," *Phil. Trans., R. Soc. Lond. B., 277* (1977), 183-376. (Note: This is a collection of 16 papers dealing with all facets of the meiotic process.)

Ruddle, F. H., "Linkage Analysis in Man by Somatic Cell Genetics," *Nature, 242* (1973), 165-69.

Sandler, L., "On the Genetic Control of Genes Located in the Sex-Chromosome Heterochromatin of *Drosophila melanogaster,*" *Genetics, 70* (1972), 261-74.

———, and D. L. Lindsley, "Some Observations on the Study of Genetic Control of Meiosis of *Drosophila melanogaster,*" *Genetics, 78* (1974), 289-97.

———, and P. Szauter, "The Effect of Recombination—Defective Meiotic Mutants on Fourth-Chromosome Crossing Over in *Drosophila melanogaster,*" *Genetics, 90* (1978), 699-712.

Sears, E. R., "Genetic Control of Chromosome Pairing in Wheat," *Ann. Rev. Genetics, 10* (1976), 31-52.

Sheridan, W. F., and H. Stern, "Histones of Meiosis," *Exp. Cell Research, 45* (1967), 323-35.

Stadler, D. R., "The Mechanisms of Intragenic Recombination," *Ann. Rev. Genetics, 7* (1973), 113-28.

Stern, H., and Y. Hotta, "DNA Metabolism During Pachytene in Relation to Crossing Over," *Genetics, 78* (1974), 227-35.

Sybenga, J., *Meiotic Configurations: A source of Information for Estimating Genetic Parameters.* New York: Springer-Verlag, N.Y., 1975.

Ting, Y. C., "Synaptinemal Complex of Haploid Maize," *Cytologia, 38* (1973), 497-500.

Tschermak-Woess, E., "Notizen über die Riesenkerner und "Riesenchromosomen" in den Antipoden von Aconitum," *Chromosoma, 8* (1956), 114-34.

Van't Hof, J., and A. H. Sparrow, "A Relationship Between DNA Content, Nuclear Volume, and Minimum Mitotic Cycle Time," *Proc. Nat. Acad. Sci., 49* (1963), 897-902.

Wachtel, S. S., "H-Y Antigen and the Genetics of Sex Determination," *Science, 198* (1977), 797-799.

Walters, M. S., "Evidence on the Time of Chromosome Pairing from the Preleptotene Spiral State in *Lilium longiflorum* "Croft"", *Chromosoma, 29* (1970), 375-418.

————, "Variation in Preleptotene Chromosome Contraction among Three Cultivars of *Lilium longiflorum,*" *Chromosoma, 57* (1976), 51-80.

Watson, I. D., and H. G. Callan, "The Form of Bivalent Chromosomes in Newt Oocytes at First Metaphases of Meiosis," *Quart. J. Microscop. Sci., 104* (1963), 281-95.

Westergaard, M., and D. von Wettstein, "The Synaptinemal Complex," *Ann. Rev. Genetics, 6* (1972), 71-110.

5

Molecular Cytogenetics

The transmission and continuity of traits from one eukaryotic generation to the next can be followed in Mendelian fashion by observation and the determination of phenotypic ratios; the parallelism of genetic transmission with the behavior of the chromosomes in mitosis and meiosis provides a physical basis for inheritance as embodied in the chromosome theory of inheritance. Subsequently, with DNA identified as the substance of which genes are made, the subdisciplines of molecular genetics and molecular cytogenetics became possible, an attack on the nature and functioning of the gene and the chromosome was extended to molecular levels, and the distinctions between genetics, cytology, microbiology, and biochemistry became less distinct and often matters of technique and emphasis more than of kind. Through the techniques of chromatin isolation and fractionation down to the nucleotide and amino acid moieties; the labeling of chromosomes with nucleic acid analogs and the various radioactive and heavy isotopes; the newer techniques of fixation and staining; density gradient centrifugation; the use of restriction enzymes of exquisite precision for the cutting of DNA into specific segments; the cloning of these segments by means of bacterial plasmids; paper chromatography and gel electrophoresis; and DNA-DNA and DNA-RNA hybridization *in situ* and in solution: all of these techniques have shown that a subtle linear functional differentiation exists, and is detectable, within the seemingly monotonous molecular sequence of nucleotide pairs stretching from one end of a chromosome to the other. This differentiation, suggested earlier by genetic maps, chromomere patterns, and specific patterns of staining, now becomes amenable to analysis at more refined levels of organization, and it is possible to show that what was revealed genetically through phenotypic ratios can often be supported in molecular terms, with patterns of nucleotide sequences providing for patterns of behavioral or functional differentiation.

The unprecedented advances that have been made in this area of genetic and chromosomal biochemistry within the last several decades can be realized by contrasting the succeeding sections of this chapter with a statement made in the mid-1950s by L. J. Stadler, a leading geneticist of his day. In dealing with the nature of the gene, he stated that "The properties of the gene may be inferred only from the results of their action . . . our concept of the gene is entirely dependent upon the occurrence of gene mutations . . . any definition of a gene mutation presupposes a definition of the gene. . . . The gene cannot be defined as a single molecule, because we have no experimental operations that can be applied in actual cases to determine whether or not a given gene is a single molecule. It cannot be defined as an indivisible unit, because, although our definition provides that we will recognize as separate genes any determiners actually separated by crossing over or translocation, there is no experimental operation that can prove that further separation is impossible. For similar reasons, it cannot be defined as a unit of reproduction or the unit of action of the gene-string, nor can it be shown to be delimited from neighboring genes by definite boundaries."

The dilemmas of definition no longer exist in the same form as voiced by Stadler largely because DNA, RNA, and proteins, the triad of polymeric molecules underlying all genetic and cytogenetic phenomena, are identifiable, manageable, and manipulable *in vitro* and *in vivo*. Some aspects have already been discussed in our consideration of the molecular structure of prokaryotic and eukaryotic chromosomes. Here our concern centers not so much on structure as on the aspects of chromosomal behavior: replication, recombination, and repair, which, while differing as to end results, share common classes of enzymes and follow common pathways of change. Consideration will also be given to the involvement of the chromosome in transcription, to gene amplication or stepped-up activity as a means of increasing certain gene products at particular periods during the life cycles of some cells or organisms, and, lastly, to the nature of the gene as revealed at a molecular level or organization.

Replication

The replication of a eukaryotic chromosome is a lengthier and, probably a more complex process than is the replication of the smaller, relatively protein-free, naked chromosome of a prokaryote, but even among such seemingly simple and related chromosomes as the single-stranded DNA viruses of ϕX174, M13, and G4, the latter a variant form of ϕX174, replication does not follow the same pathway or require the same sequence of steps from initiation of the process, through elongation of the newly formed DNA strand, to the final termination. There are, however, basic similarities, whatever the organism, for the reason that the replication unit is a molecule of DNA or RNA. It will be to these similarities that most attention will be given.

In both prokaryotes and eukaryotes it has long been known that the genetic material is replicated in a *semi-conservative* manner. The essential details of the process are depicted in Fig. 5.1, together with other possible, but now disproven, methods of replication. The term *semi-conservative* is used to describe the process because the two strands of the original replicating molecule are preserved intact while the complementary polynucleotide strands are being assembled from the nuclear pool of nucleotides as the double helix opens up. If ^{3}H-thymidine is available for incorporation during the S period of a eukaryotic cell, the chromosomes of the subsequent metaphase will reveal both chromatids of each chromosome to be radioactively labeled. Such a distribution of label could result from any of the three possible modes of replication (Fig. 5.1); therefore, a further step is necessary to distinguish between them. A second division, carried out in the absence of labeled nucleotides, now shows the metaphase chromosomes to have one chromatid labeled and the other one free of radioactive label (Fig. 5.2). A comparison of Figs. 5.1 and 5.2 indicates that what happens at the level of the eukaryotic chromosome is paralleled by what took place at the molecular level of DNA.

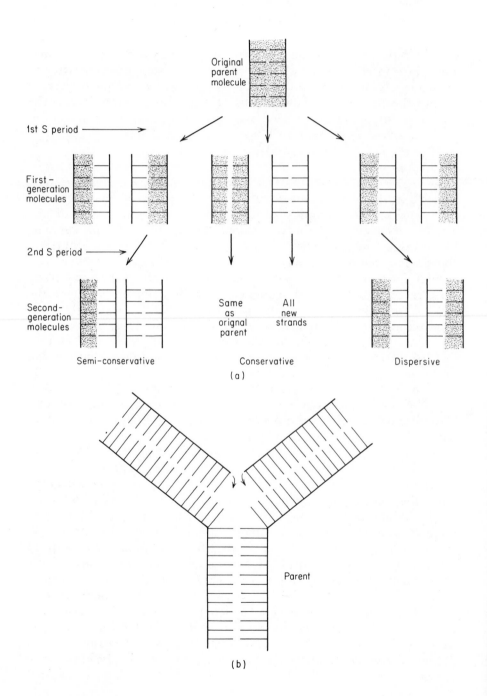

Original
parent
molecule

1st S period

First-
generation
molecules

2nd S period

Second-
generation
molecules

Semi-conservative

Same
as
orignal
parent

All
new
strands

Conservative

Dispersive

(a)

Parent

(b)

Fig. 5.1. (Opposite) Three possible modes of DNA replication. In *semi-conservative replication*. The double helix separates progressively into its two polynucleotide strands, and each one, acting as a template, forms a new double helix by the creation of a complementary strand. The new helices, or first generation molecules, consist of half parental, half newly synthesized strands as the completion of the first S period. As a result of a second S period, one helix will consist of half parental, half newly synthesized strands; the other helix will contain only newly synthesized strands. Each polynucleotide strand, therefore, is conserved as a unit, but the helix as whole is not. In *conservative replication*, the helix as a whole is conserved from one cell generation to another, and any new helices will consist only of newly synthesized strands. In *dispersive replication*, the helix is broken into small pieces, replicated either semiconservatively or conservatively, and then rejoined to form two double helices. Each helix, therefore, is a mixture of old and new material, with the original parental material diminishing in amount with each successive S period. B) semiconservative replication of a portion of a double helix to show the progressive opening up of the helix, and the formation of the complementary strands.

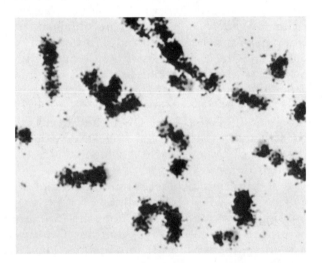

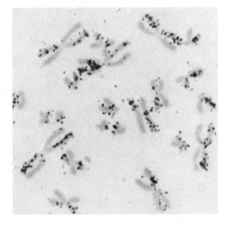

Fig. 5.2. Semiconservative replication as revealed through radioactive labelling. Top, the first metaphase after DNA synthesis in interphase, with synthesis occurring in the presence of radioactive thymidine (^{3}H-thymidine). Both chromatids of each chromosome show the presence of radioactivity, each dot representing a disintegration of a tritium atom. Right, the second metaphase following the incorporation of ^{3}H-thymidine, with its S period being completed in the absence of any radioactive materials. Only one chromatid of each chromosome is labelled radioactively. (Courtesy of Dr. T. C. Hsu).

Conformation of an identical process taking place in prokaryotes has also been obtained, in this instance using a heavy isotope of nitrogen, ^{15}N, instead of a radioactive nucleotide. DNA containing such heavy isotopes can be distinguished from that containing normal ^{14}N by density gradient centrifugation. By growing cells of *E. coli* for several generations in a medium in which the nitrogen source contains ^{15}N, a population of fully labeled DNA molecules is obtained (the ^{15}N will have been incorporated into the base portions of the nucleotides). DNA extracted from these cells would, because of its weight, sediment at a particular level on the density scale (Fig. 5.3). If now grown for a single cell generation in a medium containing only the normal isotope, all of the DNA, if replicated semiconservatively, would sediment at a density level intermediate between that expected for ^{15}N and that expected for ^{14}N. The same sedimentation picture would be obtained, however, if the process were dispersive or conservative instead of semi-conservative (Fig. 5.1). As with the eukaryotic system, additional cell divisions are needed to make a clear distinction among these options. Subsequent generations of cells also grown continuously in ^{14}N media would have the ^{15}N DNA constitute a progressively smaller and smaller proportion of the total DNA, but because DNA replicates semiconservatively, the original ^{15}N DNA would still be identifiable in the form of ^{14}N/^{15}N helices which would sediment at an intermediate density level.

Replication includes three sequential, but distinct steps, each involving its own set of conditions for successful completion of the process. These are: (1) initiation: the structure to be replicated must be primed for synthesis; (2) elongation of the strand during the main part of synthesis; and (3) termination. The latter step will be different for circular as opposed to linear chromosomes since a mechanism will be required for closing the gap in a circular form. The situation as worked out in several of the bacteriophages can serve as a basis of discussion and, when necessary, departure, but it seems evident from the analysis of replication in T4 and ϕX174 that it is a multi-enzyme process of considerable complexity. The majority of these enzymes have been isolated, and many of them have been characterized physically as well as in terms of biochemical action.

The complexity of replication as a biochemical event begins with the problems of initiation. Some 20 or more DNA polymerases have been isolated from a variety of sources—viral, bacterial, and higher forms—and while differences among them exist, all share in common an inability to initiate the formation of a polynucleotide chain by the linking together of mononucleotides. Such chain initia-

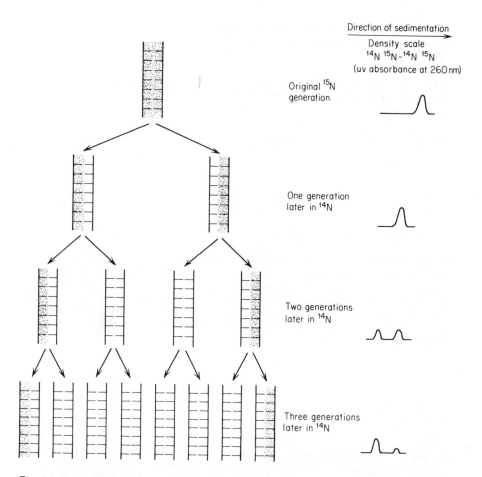

Fig. 5.3. Semiconservative replication in *E. coli* as demonstrated by Meselson and Stahl, and based on the fact that DNA formed in the presence of heavy nitrogen (^{15}N as opposed to ^{14}N) can be distinguished from that containing ^{14}N by centrifugation in a CsCl density gradient. The original heavy nitrogen polynucleotide strands are conserved, and although replication thereafter is only in the presence of ^{14}N, the ^{15}N/^{14}N helices can be distinguished from the ^{14}N/^{14}N helices. The sedimentation profiles are given on the right hand side of the figure.

tion is, of course, a requirement of replication. As a consequence, DNA polymerases are said to be "sequence blind" in that there are no nucleotide sequences specifically recognized by DNA polymerases, even though there are specific sites for the beginning of replication. Among the prokaryotes, each chromosome has a single point of origin. In the bacteriophage T7, this is an area close to one end of the linear chromosome (Fig. 5.4). In φX174 it is a stretch of 50 nucleotides within the *A* gene (Fig. 5.5), while in λ phage it is an

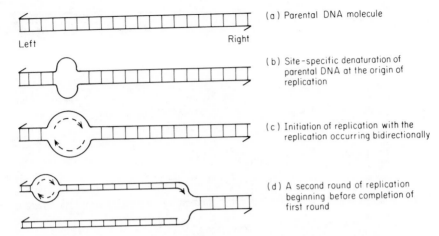

Fig. 5.4. Replication of the T7 chromosome is initiated at a point about 2 μm from the left end of the genome, and proceeds bidirectionally. Since the replication fork reaches the left end of the genome first, a Y-shaped structure forms, and in some instances a second round of replication may commence before the first is fully completed. The denaturation that occurs at the point of origin of replication, and opens the helix up, is due either to a DNA-protein interaction, or to an RNA primer molecule activating, or making accessible, the recognition site for the replicase.

(a) Parental DNA molecule

(b) Site-specific denaturation of parental DNA at the origin of replication

(c) Initiation of replication with the replication occurring bidirectionally

(d) A second round of replication beginning before completion of first round

Fig. 5.5. The replication origin of φX174 is a 50-nucleotide stretch located within the 1539-nucleotide long A gene. The function of that short sequence is to participate, along with host enzymes, in the replication of the double-stranded form of the virus. The gene begins at nucleotide #3981 and terminates at nucleotide #136, the numbering system being that used by Smith (1979) (see also Sanger, et. al., 1977 and Fiddes, 1977 for further details).

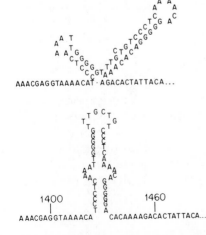

Fig. 5.6. A portion of the replication site in lambda (λ) phage which lies within the operator, or O gene. The exact size of the site is not known. O is a structural gene whose protein product, together with that of the promoter, or P gene, and with oop, a small RNA molecule encoded near the replication site, is required for the initiation of replication. Within the replication site, and between the nucleotides numbered 1409 and 1460, the sequence is such as to provide for the potential formation of the two kinds of hairpin loops illustrated. It is possible that these play a role in forming a replicase or a ribosome recognition site. However, comparable hairpin loops do not seem to be possible in the replication site of φX174 (Fig. 5.5) (Denniston-Thompson, et al., 1977, copyright 1977 by the American Association for the Advancement of Science).

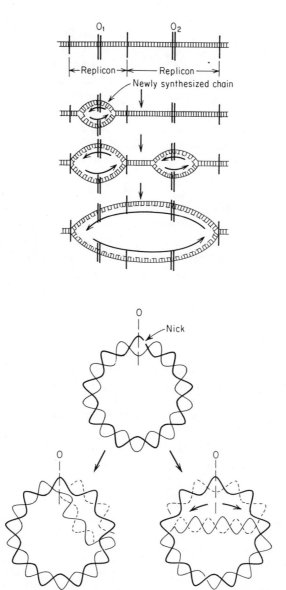

Fig. 5.7. Top, two adjacent replicons in a linear chromosome which, moving in both directions from their origins, O_1 and O_2 will eventually merge. The number of replicons in a chromosome can vary in number in one kind of cell as opposed to another, but how the control over numbers is exercised is not known. Bottom, replication in a circular chromosome, beginning at the origin, O, and proceeding in either a singular direction or bidirectionally. The nick in the circular chromosome is necessary in order that the newly formed chromosomes can be free of each other. Compare with Fig. 5.15, where the progress of replication is followed through the use of radioactive labelling.

area of somewhat greater length, within which two variations of hairpin loops are possible (Fig. 5.6). It is also within a structural gene *O* which codes for a protein required as an initiator of replication. The protein product of gene *P*, immediately adjacent to gene *O*, is also an absolute requirement for replication. Presumably, the initiator proteins interact in some manner with the hairpin loops(s) in a highly specific manner since the *O* and *P* proteins of closely related phages such as ϕ80 or P22 cannot serve as substitutes.

In any event, the actual process of replication seems to be "primed" by special RNA fragments of 50 to 100 nucleotides long— 81 nucleotides long in λ phage, 67 in ϕ80 phage. The fragments, which are complementary to the DNA to which they are attached, may be recycled tRNAs—a tRNA specific for trytophan is thought to be the primer when a reverse transcriptase converts the *Rous sarcoma virus* from an RNA to a DNA state—or they may possibly be especially prepared for priming purposes by an RNA polymerase. These primers, whatever their source, are necessary because the DNA polymerase III, the principal replication enzyme, must start from a base-paired position. Once initiated, however, the process proceeds bidirectionally in both linear and circular chromosomes by the establishment of replication forks (Fig. 5.7).

The forks, when once formed at the origin, are kept open and moving ahead of the growing polynucleotide chain by a DNA-binding protein, often referred to as an *unwindase*. In T4, the DNA-binding protein, a product of gene 32 (Fig. 5.8), has a monomeric molecular weight of 35,000 daltons and some 100 or 200 of these molecules are needed to keep each replication fork open. As the growing polynucleotide chain advances, the molecules of the DNA-binding protein are either released to become reattached at a more forward position

Fig. 5.8. The genes 32 and 43 in T4 phage produce, respectively, a DNA-binding protein which is responsible for the localized denaturation (opening up) of the double helix at the origin of the replication site, and a DNA polymerase which enters the denatured area to initiate replication in a $5' \rightarrow 3'$ direction, starting from the RNA primer (indicated as a dark sphere). Albert (1973; see Albert, et. al., in Bradbury and Javaherian, 1976) has also shown that the proteins coded for by at least four other genes participate in the replication process. The molecular weights of the two proteins are indicated in parentheses.

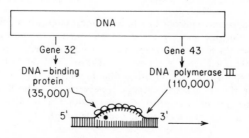

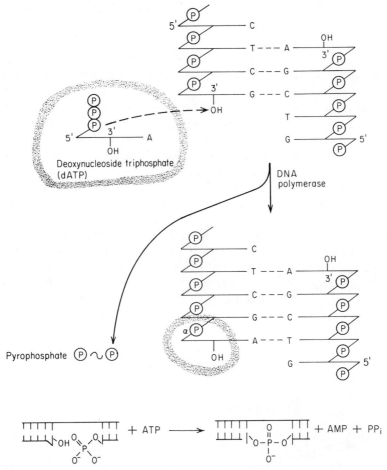

Fig. 5.9. Top, the linkage, by action of a DNA polymerase, of a triphosphorylated nucleotide to the 3′ OH end of a growing nucleotide strand, and bottom, the insertion of a triphosphate to close the gap in a DNA strand by action of the enzyme ligase. In both instances there is the release of a diphosphate.

or they slide forward to keep the fork advancing at a comparable pace by denaturing the double helix.

DNA polymerase III is a product of gene 43, has a molecular weight of 110,000 daltons, and promotes synthesis in a 5′ ⟶ 3′ direction. Each 5′ ⟶ 3′ polymerization event involves the attachment of a triphosphorylated nucleotide by its 5′-phosphate end to the 3′-OH end of the growing chain. During the process of incorporation, the two terminal phosphates are released (Fig. 5.9). The RNA primer used to initiate the process of replication must have been base-paired to form a short DNA·RNA hybrid chain, and there must have been a 3′-OH available for attachment purposes.

If at any time the base pairing is mismatched due to some accident of pairing or incorporation, DNA polymerase I functions as a $3' \longrightarrow 5'$ exonuclease to remove successive nucleotides until it reaches a properly matched base pair, after which further growth of the polynucleotide strand can be resumed. The fact that DNA polymerase I can, through its exonuclease function, correct mistakes of nucleotide incorporation, means that a cell or a virion possesses the means of minimizing the frequency of mutations that might arise during the process of replication.

Whether the chromosome is a circular one such as ϕX174 or a linear one such as T4, the initiating RNA fragment must be removed; this is done by the $5' \longrightarrow 3'$ exonuclease action of DNA polymerase I, leaving a gap that is then filled by the polymerase action of the same enzyme. A DNA ligase then links the two ends of the polynucleotide strand to close the circle. In *E. coli* and in some of the larger bacteriophages, another round, or even two more rounds, of replication may be initiated before the first is completed.

To summarize then, and using *E. coli* as an example, the replicative process (Figs. 5.10 and 5.11) requires a number of steps or processes, each with needed components for the completion of these processes, and with the components supplied by the action of specific genes: (1) initiation of replication, requiring the protein prod-

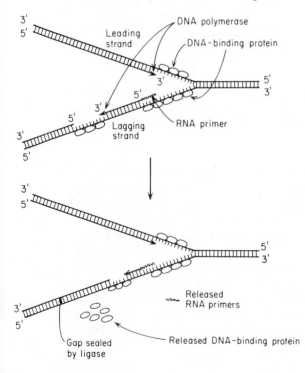

Fig. 5.10. Replication of both the leading and lagging strands of T4 phage (redrawn from Albert, 1973).

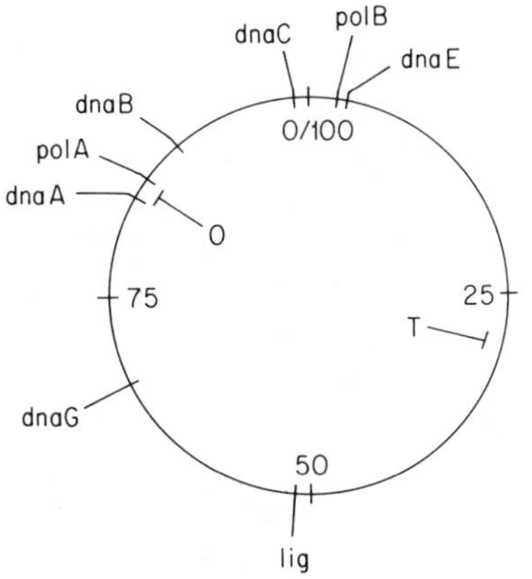

Fig. 5.11. The genetic map of the *E. coli* chromosome, showing the location of the genes involved in DNA replication, together with the point of origin of replication (O) and its termination (T). The termination would be 180 degrees from the origin since replication is bidirectional. The gene polB codes for DNA polymerase II, whose function is not known. See text for further explanation.

ucts of gene *dnaA,* followed by involvement of the products of gene *dnaC;* (2) unwinding, or opening up, of the double helix, brought on by a DNA-binding protein, a product of gene 32 in T4 phage, but of uncertain origin in *E. coli;* (3) RNA priming of the $5' \longrightarrow 3'$ leading polynucleotide strand, requiring an RNA polymerase, a product of gene *dnaG,* together with the products of genes *dnaB* and *dnaC;* the gene product of *dnaB* has an ATPase activity, stimulated by the presence of single-stranded DNA, and the products of both *dnaB* and *dnaC* require ATP for effective action; (4) DNA elongation of the leading strand requiring DNA polymerase III, a product of gene *dnaE,* plus elongation of the lagging strand by Okasaki fragments, which, on a continuous basis, requires the involvement of an RNA primer source and the DNA-binding proteins plus the products of genes *dnaB, dnaC, dnaE,* and *dnaG;* (5) RNA degradation, or removal of the RNA primer material which involves the exonuclease action of DNA polymerase I, a product of gene *polA;* (6) gap-filling and error-correcting, carried out by DNA polymerase I, acting both as a polymerase and as an exonuclease; and (7) closing of the linkages between Okasaki fragments by DNA ligase, coded for the gene *lig.* The exact role of all of the genes, particularly *dnaA, dnaB,* and *dnaC,* remains to be elucidated, but their required presence is indicated by the interference with replication brought on by their mutation.

In the phage ϕX174, the process described above for the leading strand is sufficient to transform a single-stranded molecule into its double-stranded replicative (RF) form. The Okasaki fragments require further explanation. These are short pieces of DNA which cause the replication of the lagging strand to elongate against the grain, that is, in a $5' \longrightarrow 3'$ instead of a $3' \longrightarrow 5'$ direction. Each fragment is started by being primed by a piece of RNA in the same manner as the continuous replication of the leading strand. The

Okasaki fragments, so named after their discoverer, are produced in prokaryotes at a rate of about 1 per sec, and they grow to a length of about 1,000 nucleotides, after which the RNA primers are enzymatically digested away by the exonuclease action of DNA polymerase I, the gaps are filled in by the same polymerase, and the fragment is covalently joined by the DNA ligase. The differences in replication between leading and lagging strands might suggest that the lagging strand (and hence its name) replicates at a slower rate, but if so, the rate is not so different as to leave wide gaps as the replication forks move along the double helix.

The existence of chromosomes such as those of T7 and ϕX174, and even those of *E. coli,* would suggest that while the pattern of replication just described is sufficient to account for the continued production of unique sets of chromosomes, it is inadequate to account for those that, like T4, are circularly permuted. The most adequate explanation of that process involves the *rolling circle* form of replication (Fig. 5.12). Since replicating bacteriophages seem to be associated with bacterial membranes, and since the chromosome of *E. coli* is membrane-associated by way of a mesosome, one of the nicked ends may be membrane-bound, with the other inner complementary, but unnicked, circle rolling away from the membrane to expose stretches of single-stranded DNA. Replication then begins in a $5' \longrightarrow 3'$ direction, with the inner circle acting as a template. The $5'$ end of the nicked strand is pushed out and grows in length as the inner circle continues to be copied in a complementary manner. The long tail can then be copied through the formation of discontinuous Okasaki fragments. Such long chromosomes have been found in the cytoplasm of T4-infected *E. coli.* To form a set of circularly permuted T4 chromosomes, the giant chromosome is pulled into the phage head and then clipped off when the head can hold no more DNA. This *headful hypothesis,* as it is known, is supported by the observation that if a T4 chromosome has lost a substantial portion of its genome through deletion, the chromosome size is still a normal 56 μm long, but the degree of terminal redundancy is increased sufficiently to make up for the previous loss.

A word more about the ϕX174 phage and its overall replicative process: It enters a cell of *E. coli* at a specific site on the bacterial membrane, and it is injected as a single-strand of DNA. This is the *plus* strand. It is immediately transformed into a double-stranded structure which is its replicative form. This phage chromosome, under the guidance of a protein coded for by gene *A* (Fig. 2.15), together with the host cell enzymes, replicates itself to give rise to about 20 double-stranded copies. The complementary, or *minus,*

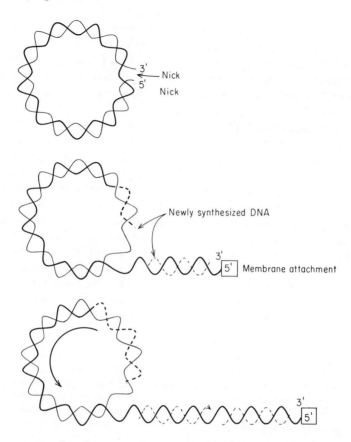

Fig. 5.12. Rolling circle method of replication in a circular chromosome (see Fig. 2.6 also). Following a nick induced by an endonuclease, the 5′ end attaches to a membrane while DNA synthesis
begins at the 3′ end and proceeds in a 5′ → 3′ direction. The inner
unbroken circle continues to roll as synthesis proceeds, pushing the
5′ tail further away; eventually the tail will also begin to act as a
template for synthesis, but since it is the "lagging" strand, synthesis
must be discontinuous, as in Fig. 5.10, with the fragments to be
linked together at a later time by a ligase action.

strand serves two purposes. Initially it serves as a transcribing template to produce the mRNAs which become translated into the viral
proteins. When an adequate amount of viral protein accumulates in
the host cell, the *minus* strand shifts its tactics to produce only viral
plus strands. About 200 *plus* strands are then packaged into capsids
for subsequent release from the cell during lysis. As Fig. 2.15 indicates, the latter steps are governed by the viral proteins.

As a relatively naked molecule of DNA, the prokaryotic chromosome is more or less ready to be replicated with a minimum of prior preparation; in fact, in a rich medium it can do so on a continuous basis and even to the point where two or more replicative cycles overlap with each other. The eukaryotic chromosome, however, has its chromatin organized in the form of nucleosomes, and it would seem that such highly compacted DNA would have difficulty in serving as a template for either the replicative or transcriptive enzymes. There is no suggestive evidence that the tightly bound histones are removed from the DNA preparatory to replication, but the structure of the nucleosomes gives a possible clue to what might be happening. The nucleosomes are made up of symmetrical half-nucleosomes, with each half unit containing one each of the four histones. By splitting into its halves, a partial uncoiling and lengthening of the DNA occurs, and replication is made possible (Fig. 5.13). Once replication is concluded, the nucleosome halves can be converted into full-sized nucleosomes by combining with newly formed histones, compaction of the DNA will take place, and the basic assemblage of the chromatin becomes reestablished. In this manner, each reconstituted nucleosome will contain one-half old and one-

Fig. 5.13. A diagrammatic representation of a possible method of eukaryotic replication with retention of nucleosome structure and the incorporation of newly synthesized histone. The uncoiling splits the nucleosomes into symmetrical halves (open and cross-hatched circles) and thereby lengthens the chromatin; replication takes place, followed by the addition of newly synthesized histone (solid circles). Reconstitution of the nucleosomes brings about a contraction of the chromatin. (after Albert, et. al., in Bradbury and Javaherian, 1976).

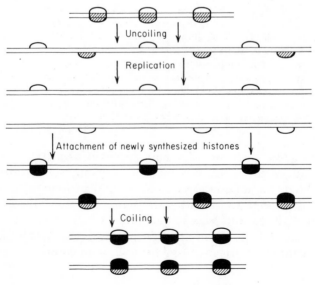

half newly synthesized, histone molecules. The entire process can go on with a minimum amount of DNA-histone and histone-histone linkages broken and reformed. However, experiments during which histone synthesis was inhibited while DNA synthesis was allowed to proceed, and consequently only preexisting histone was available for the reconstitution of the nucleosomes, showed that the preexistent histone octamers became complexed, or remained, with only one of the two daughter helices, while the newly formed histone was assembled onto the other helix in a nondispersive manner. This would suggest that the newly formed histones formed new octamers and that the old octamers did not separate into symmetrical halves. At the present time both hypotheses have some support.

Okasaki fragments are also involved in eukaryotic replication. Although there is evidence, in Chinese hamster cells, that the Okasaki fragments are about 1,200 nucleotides long, with 900 of the nucleotides being unique DNA and the remaining 300 being of repetitive DNA, Hand, in his review of eukaryotic replication, states that the eukaryotic Okasaki fragments are about 200 nucleotides long and that that length is related to the 200 nucleotide length of DNA involved in nucleosome structure. It is further suggested that replication begins at a recognition signal located in each of the linker areas between nucleosomes where histone 1 is found and that this separation of successive recognition signals accounts for the fixed length of the Okasaki fragments.

Rate of Replication. *E. coli* is probably representative of prokaryotes as to its rate of replication. It does so bidirectionally from a given point of origin (Fig. 5.11), and it proceeds at a rate of about 30 μm per min, completing a single replication in approximately 40 min at $37°C$. Since the chromosome has a length of about 1,100 μm and contains 3.9×10^6 bases, this means that about 48,000 bases are being incorporated per minute at each replication fork. It should also be remembered that it is a helical structure that is being replicated and that the parental double helix must rotate if the complementary strands are to separate. With one turn per ten nucleotide pairs in double-stranded DNA, and with the process completed in 40 min, the strands rotate at a rate of 4,800 rpm. An enzyme, DNA *gyrase,* is necessary, with energy derived from ATP. It is presumed that the rate is similar in the several kinds of viruses, although the single-stranded viruses such as ϕX174 must become double-stranded before becoming capable of replication, and some RNA viruses must, through the use of reverse transcriptase, produce DNA replicas before viral replication can commence (Fig. 5.14).

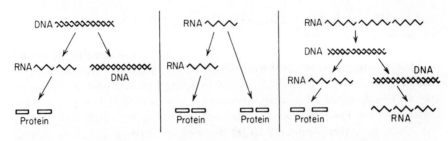

Fig. 5.14. Replication and information transfer systems found in the viruses. Left, a DNA virus that produces more copies of its DNA by direct replication, and its gene products— RNA and protein—by means of transcription and translation; the T4 phage would be an example of this system. Middle, an RNA virus that can, by replication procedures, produce more copies of itself directly, or, by translation, act as an mRNA to form protein, the polio-myelitis virus would be an example. Right, an RNA virus which has an intermediate form as a double stranded DNA produced by the action of a reverse transcriptase. This alternate DNA form can replicate to produce more intermediate DNA copies, and it can also produce copies of the original RNA virus. The intermediate DNA form can also produce mRNAs which are translatable into protein. An example of this kind of virus would be that which produces the Rous sarcoma in chickens; it is, therefore, implicated in the transformation of cells to a cancerous state which involves the integration of the intermediate DNA form into the host chromosome (after Temin 1972).

The rate in eukaryotic cells, however, is very much slower, pre-sumably because of the complex nature of eukaryotic chromatin as opposed to the naked DNA of the prokaryotes. For example, mam-malian cells in culture at 37°C show a fork displacement of 0.5 μm per min to a maximum of 2.5 μm per min. The rate seems to vary with the species, and with the temperature, but a mammalian average is around 1.0 μm per min. Embryonic cells of the frog at 20°C reveal a rate as low as 0.02 μm per min, while several kinds of cells in *Xenopus* and *Triturus* seem to have slightly higher rates. Pea and *Crepis capillaris* roottip cells would seem to fall in between mamma-lian and amphibian cells, having a rate of about 9.0 μm per hr at 23°C, while the average fork displacement in *Arabidopsis thaliana*, is 5.8 μm/hr. In *Helianthus annuus*, the displacement is 6.0 μm/hr at 10°C to 12 μm/hr at 30°. Therefore, when consideration is given to the large amount of DNA per eukaryotic chromosome and to the facts that not all chromosome replicate synchronously and that the S period of actively dividing plant and animal cells in culture is about 6 hr to 8 hr in duration, it becomes obvious that a eukaryotic chro-mosome cannot complete its replication within a single S period if it has but a single point of replication initiation. Thus, an average mam-malian chromosome conservatively contains an amount of DNA which measures around 20,000 μm to 30,000 μm in length; if there were but a single point from which replication begins, and if the rate of fork displacement of 1.0 μm per min it would take several weeks for the completion of replication instead of the known 6 hr to 8 hr. To make the point even more obvious, a *Tradescantia* chromatid consists of a single duplex molecule of DNA 1 meter, or 1,000,000 μm, long. With one point of replication proceeding bidirectionally and at an average rate of displacement of 9.0 μm/hr at 23°C, it would take over 230 days to pass through the S period.

The problem is further complicated by the fact that in any given organism the S period may vary widely depending upon the tissue. In *D. melanogaster*, for example, early cleavage nuclei divide every 9.8 min, with interphase occupying no more than 3.4 min. The S period cannot exceed interphase in duration. This is in contrast to an S period of 600 min for other somatic nuclei and a somewhat longer period of time for the polytene nuclei of salivary gland cells. Yet the rate of fork displacement, calculated from pulsed exposures to radioactive labels (Fig. 5.15), is approximately the same in all tissues, namely, an incorporation rate of 2,600 bases per minute per fork. A comparable situation has also been observed in *Triturus:* The S period in blastula nuclei is about 1.0 hr as compared to 200 hr for the S period in premeiotic spermatocytes; yet the rate of incorporation per displacement fork seems to be similar in the two kinds of cells.

The only way that the S period can be varied by means other than temperature is by varying the number of initiation points for replication. Since it is difficult to isolate a single intact chromosome and to determine the number of initiating points along its entire length, about the only way one can gain a measure of numbers is by determining the distance between successive initiation points along a single piece of DNA, and then dividing that distance into the total length of DNA in the haploid genome. Such a distance constitutes a *replicon* (Fig. 5.15). In the cleavage nuclei of *D. melanogaster*, the distance between adjacent points is 2.5 μm; it is 19.0 μm in somatic cells. In the somatic cells of the Chinese hamster, the rat, and the chicken at 37°C, the average distances are 50 μm, 30 μm, and 60 μm, respectively, giving in the hamster from 1.5 to 2.0 $\times$ 10^5 replicons per genome. There are some 43,000 replicons in the nuclei of pea roottip meristems, with the replicon sizes peaking at 36 μm to 38 μm, and at 54 μm. These lengths seem to be multiples of a more basic length of 18 μm, a size found in stelar and cortical cells which have been induced to divide. *Crepis capillaris* has approximately 60,000 replicons averaging 24 μm in length; each species, therefore, has its own number and size. It has also been shown that in *Helian-*

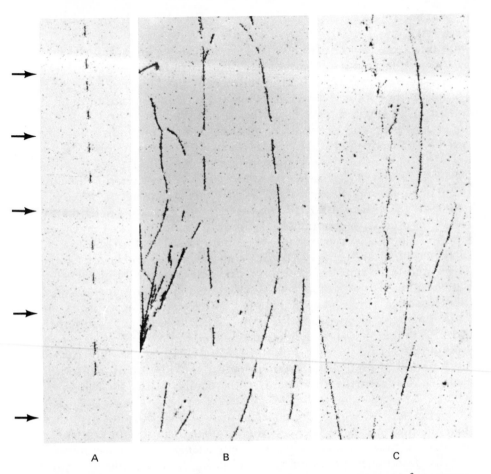

Fig. 5.15. Autoradiographs of replicons of *Helianthus annuus,* labelled with ^{3}H-thymidine during the S period. A; DNA isolated from seedling roots grown at 10°C, and given a single pulse of label toward the end of the replication period, hence there are no fading tails as in C. Arrows point to the location of the initiating points of replication. B; DNA from cultured roottips, given a single pulse of lable, with replication occurring throughout the pulse duration. C; DNA pulsed for 45 minutes with a label of high specific activity, followed by a 90-minute exposure to a pulse of low specific activity. This produces fading tails of silver grains, with the tails pointing in the direction of replication (courtesy of Jack Van't Hof).

thus annuus temperature seems to have no effect on replicon size, although it does on replication rate.

In *D. melanogaster* cleavage nuclei, all initiation points are activated within the space of 0.4 min, but in other more slowly synthesizing cells the initiation sites appear to be activated in some sequential manner. It is known, of course, that heterochromatin, whether constitutive or facultative, is synthesized late in the S period in somatic cells, but this is not so in cleavage nuclei. Furthermore, there is no evidence of heterochromatic compaction during interphase in these nuclei. In salivary gland cells, the heterochromatin of the polytene chromosomes is underreplicated and clustered in an ill-defined chromocenter, giving evidence of a close control exercised over the process both in terms of location as well as of time.

The average size of a replicon can be readily determined by autoradiography, but if one considers the length of the replicons and the rate of fork displacement, it is immediately apparent that the portion of DNA representing a replicon completes its replication in but a fraction of the S period. This means that not all replicons are activated simultaneously even though adjacent replicons, tandemly arranged, seem to replicate in unison (Fig. 5.15). As a result of such observations, the suggestion has been advanced that there are families of replicons, clustered along portions of the chromosome, which become activated in concert, followed by other families activated in a sequential manner. This would account for the late replication of heterochromatin and of sex chromosome in general, as well as the variation in replication timing observed among human autosomes. It has been postulated that mammalian genomes may have as many as twenty-five such families of replicons, but in *Arabidopsis thaliana,* with a low DNA value of 0.2 pg per haploid genome, only two families have been detected. Each replicon, averaging 24 μm in length, requires a bit more than two hours, or 74% of the S period, for the completion of replication, but the time of initiation of replication of the two families is separated by a 36-minute interval. Since DNA synthesis occurs throughout the S period of 2.8 hours, the two families of replicons must have an overlap period of replication of between 90 and 100 minutes.

Since the number of initiation sites for replication can vary by from one to two orders of magnitude from one tissue to another in the same species, a number of questions arise. What determines the number of sites? What determines their activation? Is an activation site a particular sequence of nucleotides recognized, say, by activating proteins? Little information is available, although in *D. melanogaster* the number of sites is thought to correspond to the number of chromomeres (bands) and that a *replicon* (i.e., that block of DNA being replicated between two initiation sites) averages about 26,000 base pairs long, a figure that corresponds roughly to band dimensions and to replication times. It seems almost certain, however, that there are no intrinsic termination points for replication other than the obvious ends of the eukaryotic chromosome. This lack is suggested by the fact that deletions, inversions, and so forth, do not interfere with the pattern of replication. The complementarity between RNA primers and repetitive sequences in the Okasaki fragments of the Chinese hamster suggest that initiation sites are fixed and that their frequency in any given cell might well be a function of the number and availability of initiating molecules.

Transcription

Transcription is similar to replication in that the chromosomal DNA acts as a template on which a polymerase builds a nucleic acid polymer complementary and antiparallel to the DNA strand that is being copied. The similarity, however, ends at this point, for in all other aspects the two processes differ as to function, mechanism, and product.

The function of replication is to produce exact replicas of the parental DNA, and thereby to conserve the coded information of inheritance for other viruses, other cells, or other generations. The entire genome of double-helical DNA is, therefore, copied with great exactitude, but other than to begin and end the process, the regulation actions are minimal in number. The purpose of transcription is to initiate the biochemical steps whereby individual genes, or clusters of genes, can express themselves phenotypically, through the formation of proteins, or, in the case of genes which produce rRNAs and tRNAs, to form molecules which participate in the process of phenotypic expression of other genes. Since all genes are not expressed in all cells, e.g., a hemoglobin gene would be functional in a precursor of a red blood cell but not in a brain cell, transcription is a highly selective process, copying only certain portions of the genome at any one time. The products of transcription are RNAs of several sorts which have a limited lifetime in the cell, after which they are degraded. Also, since genes rather than the genome are copied, there must be an appropriate number of initiation and termination points for the process; there is, consequently, no basis, except in *D. melanogaster,* for considering that a replicon and a transcribing stretch of DNA have one and the same identity.

From a genetic point of view transcription is the first step in a chain of biochemical events whereby a gene gains phenotypic expression. To the cytogeneticist, however, whose interest centers in the chromosome itself, the analysis of the products of transcription, i.e., the several kinds of RNA, being complementary to the DNA from which they were formed, provide an additional means by which the structure and function of the chromosome can be dissected and understood. This has already been pointed out for the rRNAs and tRNAs whose genes, through DNA/RNA hybridization procedures *in vitro* and *in situ,* can be located in the genome and shown to be redundant. Their transcription products have been shown to be larger than the final functional forms in that they possess substantial spacer DNA separating the structural segments and that these must be processed in the nucleus before being delivered to the cytoplasm. Here our attention is directed toward the character of the unique structural genes and to how their nature and behavior can be deduced from their immediate RNA products, the messenger RNAs.

Unlike replication, during which both polynucleotide strands of the double helix are copied in complementary form, transcription involves only one of the two strands for any given gene or at any given region of the chromosome (Fig. 5.16). This is the *sense* strand for that gene, but strand-switching can occur (Fig. 3.26), and the rRNA or the mRNA for a different gene may be copied from the other

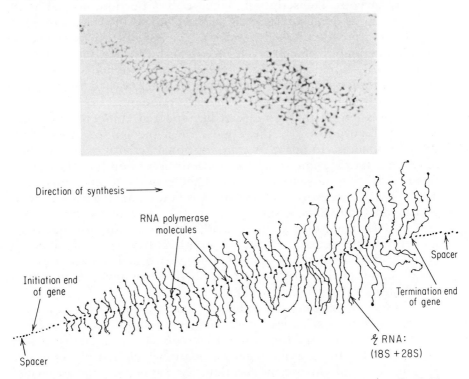

Fig. 5.16. Transcribing segment of DNA as seen in the electron microscope (see Fig. 3.26 also). Top, a gene from the nucleolar organizer region of the genome of *D. melanogaster,* giving rise to strands of rRNA which include both the 18S and 28S segments as well as the spacer between them. Bottom, an interpretation of the above figure; the spacer indicated here is between rRNA genes, not between 18S and 28S segments (courtesy of W. Y. Chooi).

strand. Transcription is carried out by an RNA polymerase which, unlike DNA polymerase, does not require a primer molecule with a free 3'-OH end to which an incoming nucleotide is to be attached. The RNA is transcribed, as in replication, in a 5' ⟶ 3' direction; this can be demonstrated by allowing the first half of a transcribing unit to do so in a nonradioactively labeled medium and the remainder to do so in a radioactively labeled medium. The radioactive label is found at the 3' end of the transcript. The DNA sense strand, therefore, being antiparallel, is being read in a 3' ⟶ 5' direction. Most of the aspects of transcription of both prokaryotes and eukaryotes can be carried out *in vitro,* and the differences between the two groups can probably be attributed to the simplicity or complexity of the genomes, together with differences arising during their separate evolutionary pathways.

Prokaryotic Transcription. When only a portion, instead of all, of a genome is being transcribed selectively, and the transcription of any given gene occurs only under a given set of circumstances, a minimum number of constraints must be built into the system to make it fully functional and responsive. If it is assumed that all other conditions are optimal, (1) there must be some way to repress those genes which are not being transcribed; (2) there must be a mechanism of derepression so that genes can be rendered competent for transcriptional purposes; (3) there must be a recognition site for the RNA polymerase to begin its transcriptional activities; and (4) there must be a termination signal for transcription to prevent the process from embracing unwanted regions of the genome. Constraints 3 and 4 are obvious necessities for any process limited as to place and time, while constraint 2 may simply be another way of expressing constraint 3 (it may, on the other hand, involve an additional step). A model for the control of these conditions has been presented by Jacob and Monod in the operon concept, worked out from experiments conducted on the constellation of genes involves in lactose metabolism in *E. coli* (Fig. 5.17).

Inducible enzymes are ordinarily absent from a cell, or present in very minute quantities, unless a specific substance, an *inducer,* catalyzes their prompt formation. In *E. coli,* the sugar lactose is such an inducer for the enzyme *B-galactosidase,* which hydrolyzes lactose to galactose and glucose. In addition, the enzymes *permease,* which facilitates the entry of lactose into the cell, an *acetylase,* whose function remains unknown, are simultaneously induced with B-galactosidase; the genes governing these three enzymes are adjacent to each other on the linkage map. Within 2 or 3 min after the introduction of lactose to the cell the B-galactosidase concentration in-

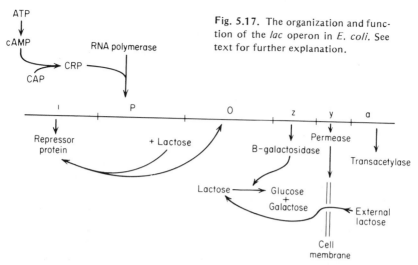

Fig. 5.17. The organization and function of the *lac* operon in *E. coli*. See text for further explanation.

creases until it is built up to several thousand molecules, a level maintained so long as lactose is still available. In the absence of any other information, it would have appeared that lactose somehow activated these genes as a transcriptional unit for the production of appropriate mRNAs. However, a group of mutants was discovered which suggested that this region of the *E. coli* chromosome had a more complex structure than simply the three stretches of DNA needed to code for the three enzymes. These mutants transformed the pattern of enzyme synthesis from an inducible to a *constitutive* process; i.e., the enzymes were now formed in substantial amounts in the absence of lactose, and hence at a time when they performed no essential cellular function. Mapped to the left of the loci of the B-galactosidase (z), permease (y), and acetylase (a) genes, the mutants, called i^-, suggested that the normal allele, i or i^+, was a regulator gene and that in a mutant condition it failed to regulate enzyme levels such that the enzymes and substrates were in reasonable balance, while at the same time it was without effect on the structure of the enzymes it regulated.

The regulator gene must, therefore, be a distinct genetic entity that is separate from the other three genes, but its mode of action was not made clear until a second group of mutants, different from those at the i locus, and mapping between the regulator and structural genes revealed that even in the presence of i^+, B-galactosidase and its companion enzymes could behave in a constitutive manner. These mutants revealed that an additional controlling step exists which governs transcription. Termed an *operator* locus, this gene could either contribute to the repression of transcription of the three structural genes when in a normal state (O^+) and in the presence of i^+ or prevent repression when in a mutant state (O^c) and independent of the state of the i locus. The term *operon* was coined by Jacob and

Monod, two French microbial geneticists, to designate this integrated unit of regulator, operator, and structural genes which in the absence of lactose was believed to function by the action of a represser substance, the product of the regulator gene i^+, which attached itself to the operator locus and blocked the attachment and, consequently, the action of an RNA polymerase. The diffusible nature of the repressor substance is indicated by the fact that transcription could take place in normal fashion in the presence of i^- if i^+ were present in the cell on a plasmid; that is, an i^- zya cell is inducible with i^+ on a plasmid. The O gene, on the other hand, can govern only the genes to which it is linked, and hence it functions as a site of action and not as a producer of a substance. In line with this interpretation, the mutant i^+ genes formed a faulty repressor that was incapable of binding properly to the operator, while the O^c mutants altered the operator in such a manner as to preclude the binding of a normal represser. Under normal conditions, the presence of lactose, as an inducer, led to the removal of the represser, thus freeing the operator so that attachment of the RNA polymerase could take place and transcription could begin.

Even this picture of an integrated set of genes, however involved, is still incomplete, for a number of strains of *E. coli* have been isolated which differ in the rates at which the three enzymes are transcribed, even in the presence of normal regulator, operator, and structural genes. The variable feature was tracked down to the site where the RNA polymerase attaches in order to commence transcription. Called the *promoter,* the character of this site determines the ease and frequency of attachment of RNA polymerase molecules, and thus the frequency with which the mRNAs are produced for ultimate translation into the three enzymes. The promoter, as it turns out, has a dual structure: a site adjacent to the operator for the attachment of an RNA polymerase and a second site adjacent to the regulator which binds another protein (the cyclic AMP-receptor protein, or CRP), whose presence is necessary for the proper binding of RNA polymerase. When the promoter is faulty through mutation or deletion, in either of its two components, transcription slows down or stops altogether.

The nature of the *lac-operon* is depicted in Fig. 5.17. The regulator-operator system, with the diffusible repressor, is an example of a *negative* control over transcription—the repressor must be removed for action to begin. The CRP system, however, represents a *positive* control; the cyclic AMP activates the protein CAP to form CRP which in turn must complex with the RNA polymerase before transcription begins. The repressor protein, coded for by the regulator, has been isolated and its amino acid sequence has been determined. The lac-operon has also been sequenced, and it, along with the lambda and φX174 systems to be described, can probably lay claim to being the most worked-over and best known pieces of DNA in the biological world. These particular stretches of DNA in the lac-operon, adding up to about 4,700 nucleotide pairs, functioning as regulator, promoter, operator, or structural genes, contain within them sites of "twofold symmetries" (Fig. 5.18), suggesting configurations of nucleotides to which specific proteins can be attached, possibly in the major grooves of the helical duplex DNA. In any event, the symmetry is too unusual to be a random phenomenon, and it is paralleled by symmetries found in other operons.

The organization of the lac-operon is corroborated through equally exquisite studies carried out on the repressor system in the lambda (λ) phage (Figs. 5.19 and 5.20). The phage, consisting of about 47,000 base pairs, can become integrated into the host chromosome of *E. coli* where it remains dormant as an episome, replicating along with, and inflicting no damage to, its host DNA or cell. Its genes are repressed by a protein, which exists as a dimer or tetramer made up of subunits of 26,000 molecular weight. The protein is the product of one of its own genes, *cI,* which binds to two sets of operators, O_R and O_L, one on either side of itself and separated from each other by about 2,000 base pairs which also include the *cI* and *rex* genes (Fig. 5.19), When the protein is bound to both operators, the immediately adjacent genes *N* and *cro* are repressed, as are the remaining 40 or more genes in the genome.

The size of the repressor protein is such that it covers between 15 to 30 base pairs, but as many as 100 pairs could be covered, indicating that there are several binding sites instead of only one (as in the lac-operon) for the repressor protein. What is striking, and in addition, what points to the significance of symmetry among the

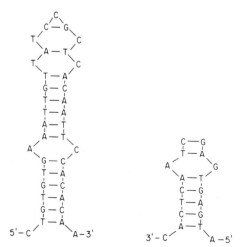

Fig. 5.18. The two-fold symmetry that exists in the sense strand of the entire operator (left) and in the promoter (right) of the *lac* operon of *E. coli.* As indicated, such symmetry makes possible the existence of hairpin loops such as those in Fig. 5.6, but the full significance of such loops remains to be clarified (after Dickson, et. al., 1975).

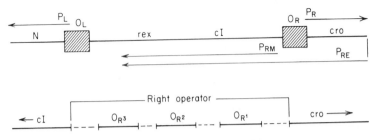

Fig. 5.19. Above, the structure of the operon of lambda phage, with its right and left operator and promoter components, together with the N and cro genes which are concerned with lytic growth, the cI gene which codes for a repressor protein which binds to the operators, and the rex gene whose function is less well known but which probably relates to a repression of some sort. Four promoters, or RNA polymerase binding sites, are present, and their general position is indicated by arrows, which also point out the direction of transcription. P_R is concerned with the transcription of the cro gene, and is located within or at the right side of O_R; P_L with the transcription of N; P_{RM} (promoter for repressor maintenance) and P_{RE} (promoter for repressor establishment) are both concerned with the transcription of cI, but P_{RE} is between 5 and 10 times more efficient in the production of the repressor protein coded for by cI, apparently because its transcriptional product carries a leader of RNA complementary to the end of the 16S rRNA, and thus binds the mRNA more effectively to the ribosome where it is more efficiently translated. Below, a more detailed picture of the structure of the right operator and its three repressor attachment sites. The cI gene codes for a protein having a molecular weight of about 26,000, and this, as a dimer or tetramer, binds preferentially to O_R1 at low concentrations; this attachment represses the cro gene, aids in the establishment of lysogeny, and stimulates the cI gene to greater activity. The repression is reinforced if additional repressor binds to O_R2. The repressor binds to O_R3 only when high levels are attained, at which time the repressor gene, cI, is itself repressed. The repressor protein, therefore, exerts a negative control over the function of the cro gene and, through auto-regulation of cI, governs its own level of concentration. The spacers between the operator segments seem to play a role in the binding of the polymerase during transcription since mutations within them (Fig. 5.20) interfere with the production of repressor protein. The left operator, O_L, is less well known in a functional sense, but is concerned with the regulation of the N gene. When both right and left operators are repressed, most of the genes of the phage are in an inactive state, largely because the product of N is needed for their functioning (see Ptashne, et. al., 1976, and Maniatis and Ptashne, 1976, for further details).

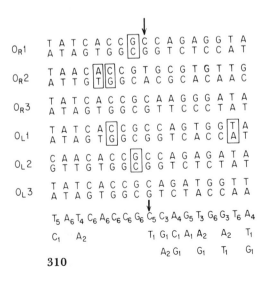

Fig. 5.20. The six repressor binding sites in the two lambda phage operators, O_R and O_L. Their sequence similarity is obvious; the figures at the bottom reveal the frequency with which a given base appears in each of the 17 positions. Each binding site shows a partial two-fold symmetry, with the arrows indicating the axis of symmetry. In this sense they are similar to those sequences indicated in Figs. 5.6 and 5.18, but their similarity and symmetry is more likely related to the fact that they all bind the same repressor protein, possibly in the major groove of the DNA duplex. The boxed areas are sites of base pairs which, if mutated, would have a decreased affinity for the repressor protein (Fig. 4 of Ptashne, et. al., 1976, copyright 1976 by the American Association for the Advancement of Science).

310

base pairs, is that there are six similar blocks of nucleotide pairs, each 17 pairs long, in the two operator regions of $O_R 1$ and $O_L 1$. When their nucleotide sequences are compared, they are more similar to each other than they are to other paired combinations; these are also the binding sites having the greatest affinity for the repressor protein. This protein can be readily removed by the exposure of *E. coli* to ultraviolet light, following which an RNA polymerase can replace the repressor and transcription can be initiated. The polymerase, which is substantially larger than the repressor protein, covers a 45-base pair sequence, the *promoter,* which includes most of one of the repressor binding site together with a number of base pairs at the beginning of the *N* or *cro* genes (Fig. 5.20).

Further study of the multiple binding sites of the two operators has revealed that maximum repression of either the *N* or *cro* genes requires attachment of the repressor protein to two binding sites within the respective operator, thus insuring a stricter control over transcription than would be possible with a single site. In addition, the right operator has two promoters, one related to the *cro* gene and the other to *cI*, in addition to a third (P_{RE}) which lies well to the right of O_R. By moving to the right or left, the RNA polymerase can transcribe in opposite directions, although on complementary strands of DNA. In addition, since *cI* codes for the repressor protein, it regulates its own level of activity by responding differentially to different levels of its own product. When large amounts of repressor protein are present in the cell, both *cro* and *cI* are repressed. Small amounts, bound to $O_R 1$, only repress *cro,* but they tend to stimulate the transcriptional activity of *cI*. There is, therefore, a sophisticated autoregulatory system in the λ phage that is not found in the *E. coli* lac-operon, a phenomenon which can possibly be accounted for by the existence of two very different stages in the life cycle of the phage: a genetically active lytic stage and a repressed lysogenic state.

To return to the *E. coli* lac-operon (Fig. 5.17), transcription will produce a polycistronic mRNA which will also include a transcribed portion of the operator. The RNA complementary to the operator will be degraded in the cytoplasm, while the mRNA will be translated into the three separate enzymes. The fact that B-galactosidase is formed about three times more frequently than are the other two enzymes indicates that the post-transcriptional controls are also operative in the cell, probably at the level of the ribosome.

The features of the operons just discussed provide examples of patterns of functional differentiation that can be built into the sequences of base pairs of DNA, patterns which are due in one way or another to protein-DNA interactions. This appears to be the typical structure of organization of the prokaryotic chromosome, whether phage or bacterium. What the situation is in blue-green algae remains

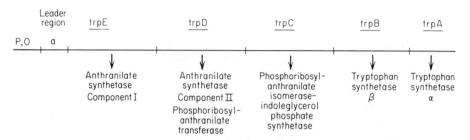

Fig. 5.21. The genetic structure of the *tryptophan* operon of *E. coli*. This differs from the *lac* operon in that the promoter (P) and operator (O) regions are followed by a leader region, a 160-nucleotide sequence within which is an "attenuator" governing the rate of mRNA formation from the five structural genes. Mutations within the attenuator prevent it from critically "sensing" the levels of tryptophan in the cell, with the result that the mRNAs for the five genes are transcribed in excessive amounts.

to be determined. The functional utility of the operon, with its binding sites for repressor proteins or RNA polymerase, is obvious; the regulation of gene expression is controlled with precision, and if polycistronic mRNAs are formed, these are for enzymes involved in a particular metabolic pathway, or in the case of phage chromosomes such as T7, the sequence of events needed for phage replication. In *E. coli,* the structure of the leucine, tryptophan, arabinose, galactose, and histidine operons are comparable to the lac-operon, even though variations of both structure and control exist. The tryptophan operon, for example, includes an *attenuator* region of about 160 base pairs between the promoter-operator region and the structural genes, which function as an optional termination signal, thus permitting the operon to gear its rate of transcription to the level of the product it codes for (Fig. 5.21). The histidine operon (Fig. 5.22) is a repressible rather than an inducible system; that is, histidine governs its own level in the cell by acting as a co-repressor, combining with, and activating, an inactive repressor which can then bind to specific sites on the operator.

Finally, one can inquire as to whether *E. coli* operons, for example, are self-regulating in a manner comparable to the *cI* gene in λ phage. As Fig. 5.16 suggests, mutual rather than self-regulation of chromosomal activity is probably more characteristic of the *E. coli* operons, but what is clear is that the naked DNA of the prokaryote chromosome, through a variety of positive and negative feedbacks, is governed with exquisite sensitivity by DNA-protein interactions. These same interactions are also the basis of all eukaryotic transcriptional control.

If we now carry an examination of transcription to the molecular level of DNA, the circumstances in ϕX174 are the most illuminating that we have to date. As Fig. 2.15 indicates, the ϕX174

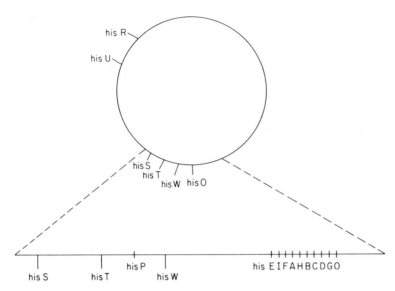

Fig. 5.22. The *histidine* operon in *Salmonella typhimurium*, together with other genes concerned with histidine production. The sequence of structural genes in the operon is not
the same as the sequence of metabolic steps which transforms phosphoribosylpyrphosphate
+ ATP into 1-histidine. The metabolic steps are as follows: $G \rightarrow E \rightarrow I \rightarrow A \rightarrow H \rightarrow F \rightarrow B \rightarrow C$
$\rightarrow B \rightarrow D$, but mutations in any structural gene can be offset by the addition of 1-histidine to
the medium, indicating that the sequence of reactions is linear and non-branching. The genes
hisR, hisS, hisT, hisU and *hisW* are all involved in the formation of a functional tRNA^{his},
but are not under the control of the operator (O) itself. *HisP* is concerned with histidine
transport (after Ames and Hartman, 1974).

chromosome contains ten genes, but among them, as will be shown
later in Fig. 5.36, there are separate genes, overlapping genes, and
genes within genes to code for its ten necessary proteins. In addition,
there are four noncoding areas. The nucleotide structure of the
A-gene contains a promoter of six nucleotides (Fig. 5.23); this is a
recognition site for the RNA polymerase occurring immediately
before the hairpin loop. Transcription begins part way up the left
side of the loop, includes a block of five nucleotides which makes up
the ribosome recognition site (and which is probably complementary
to some portion of one of the rRNAs in the ribosomes), passes
through an mRNA termination site, and continues transcription to
form an mRNA of uncertain length. Two other initiation sites for
transcription are also known, one within gene *A* and the other within
gene *C*. Three termination sites have also been identified, one at the
beginning of gene *A* and the other two between genes *D* and *F* and
between genes *F* and *G*. All are associated with hairpin loops, but
whether there is a 1:1 relation between the three mRNAs that form
and the three termination sites is not known.

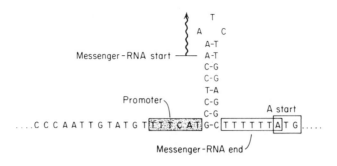

Fig. 5.23. The nucleotide sequence of a region of the ϕX174 genome which includes a 6-nucleotide promoter region which serves as a recognition site for RNA polymerase which will transcribe the-*A* gene; a hairpin loop which contains a 5-nucleotide mRNA initiation site; an mRNA termination sequence for the transcription of the *H* gene (located to the left) and the block of non-genetic nucleotide between the *H* and *A* genes; and the -ATG- triplet code that marks the beginning of the *A* gene (from Sanger, 1977, Fiddes, 1977, and Smith, 1979).

Eukaryotic Transcriptional Systems. The success of the pro-karyotic operon as a model for explaining a wide variety of interrelated genetic and molecular phenomena prompted a search for a similar system among the eukaryotes. None has been found. Clustered genes involved in the sequential events of a single metabolic pathway, and inducible enzyme systems are present in such eukaryotes as yeast and *Neurospora,* but a thorough analysis of their mode of operation and a search for regulator or operator mutants have revealed the absence of a prokaryotic-like operon. For example, the *arom* (aromatic) locus in *Neurospora* (Fig. 5.24) consists of five linked genes concerned with the production of chorismic acid, a precursor of tryptophan, but at least four other unlinked genes are also concerned with the same pathway of reactions, and there is no evidence of operator, promoter, or regulator genes located adjacent to the locus. In addition, the five linked genes, on the basis of mutational studies, code for the five enzymatic activities, but these are all expressed by a single protein of 230,000 molecular weight made up of two identical units of 115,000 molecular weight. However, a comparable cluster of four linked genes in the *qa* (quinic acid) cluster of *Neurospora* codes for separate enzymes, but again, no obvious operon-like control system is to be found.

There must, of course, be several kinds of acceptance, or recognition, sites for the RNA polymerases, a set of circumstances for making this site competent for the initiation of transcription (i.e., to derepress it), and a termination point for the process. The finding that *in vitro* DNA freed of histone is capable of a substantially greater degree of transcription than is isolated chromatin, has led to the belief that the histones are the general repressors of eukaryotic DNA, while the nonhistone chromosomal proteins play the role of derepressors, thus promoting the transcriptional process. The conservatism of histone sequences of amino acids, their ubiquitous presence in all nuclei, and their relation to nucleosome formation and maintenance preclude their playing any highly specific or selective role in DNA transcription. The histones, however, can undergo phosphorylation, methylation, and acetylation to alter their structure and, undoubtedly, their actions. It is now clear that actively transcribing genes, which, as Fig. 5.13 suggests, are partially uncoiled, and consequently more exposed to the RNA polymerases, are more susceptible to degradation by deoxyribonucleases than are inactive genes, i.e., DNA sequences which code for globin proteins are preferentially degraded in erythroid cells but not in non-erythroid cells. The histones, therefore, protect as well as repress DNA.

The fact that *Neurospora* seems to possess but two of the known major histones (H2a and H2b) associated with its DNA might suggest that it would retain some semblance of a prokaryotic operon as a regulatory system, but while the clustered genes in the *arom* and *qa* locus may be remnants of evolutionary history, their control is more typically eukaryotic, which is equivalent to stating that the mode of

Fig. 5.24. The *arom* (aromatic) region of *Neurospora crassa*. A cluser of five genes involved in the synthesis of chorismic acid is found in linkage group II, and is transcribed into a single large HnRNA that becomes translated into a single protein possessing five different enzymatic functions. Also involved in the same synthesis is another gene, *arom-3*, located distantly in the same linkage group, and three genes in other linkage groups. As in the histidine operon of *Salmonella*, the sequence of genes in the cluster is not the same as the sequence of metabolic reactions, but *arom-3* and the unlinked genes function at the beginning and the end of the reaction pathway. No evidence of the presence of a promoter or an operator has been uncovered, indicating that the control of the formation of chorismic acid must be by other mechanisms (after Chase and Giles, 1971).

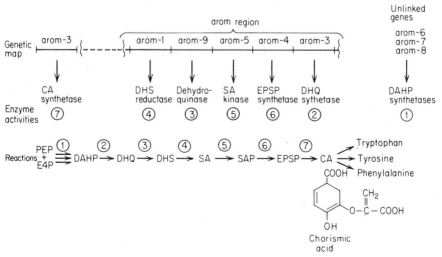

control is known only to the extent that it does not conform to a prokaryotic protocol.

The role of the nonhistone proteins as stimulators of DNA transcription is suggested by a number of findings. Their great diversity is indicative of diverse functions, and their presence in one differentiated cell as opposed to their absence in others, coupled with a concomitant increase in RNA synthesis, suggests that they are concerned with the processing of genetic information at the chromosomal level. Thus, when rat liver cells are stimulated with cortisol, there is a correlated rise in the level of a single nonhistone protein and increased RNA synthesis. A nonhistone protein of somewhat similar molecular weight (43,000 compared to 41,000) makes its selective appearance in *Drosophila* salivary gland cells when ecdysone is added as the stimulating agent.

Further studies of rat liver chromatin by Bonner and his group have provided additional insight into the nature of active vs. inactive DNA. Partial hepatectomy leads to regenerative mitotic activities until the liver once again reaches its original size. DNA synthesis and mitosis in the remaining tissue commence in about 18 hr to 24 hr after the operation, but much earlier and within a few hours there is a sharp rise in transcribing DNA until about 40% of the DNA is involved (Fig. 5.25). Accompanying this rise in active DNA, a parallel disappearance of histone brought on by the action of a histone protease has been reported, which degrades the histone during or after its separation from DNA. The uncovered DNA is then assumed to unite with a nonhistone chromosomal protein, in which state it is transcriptionally competent. It is clear that this interpretation is not fully in accord with that presented earlier (Fig. 5.13), so questions remain to be answered fully.

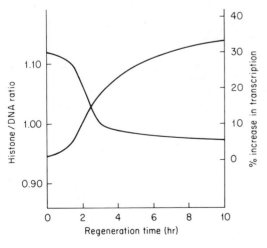

Fig. 5.25. Relations of regeneration time following a partial hepatectomy in the rat to both transcribing activity and to the histone/DNA ratio, suggesting that a rise in transcription is related to histone changes (after Bonner, 1974).

Since histones bind to the phosphate groups of any DNA, and hence are nucleotide "blind," the transformation of inactive to active DNA has been viewed as a positive act brought on by the action of the nonhistone chromosomal proteins. Up to 115 different nonhistone proteins of this sort have been identified from rat nuclei, and they are present in numbers of molecules per genome from 4,000 to over a million copies. Some of these proteins are the enzymes relating to replication, transcription, and translation, some bind immediately to newly formed RNA to make ribonucleoprotein particles, possibly ribosomes, and some bind specifically to the middle-repetitive (150 to 200 base pairs long) DNA of liver chromatin, suggesting (but not proving) that this portion, about 2%, of the protein is a controlling element for transcription. RNA complementary to middle-repetitive DNA sequences binds to this same protein, again implying some kind of protein-nucleic acid specificity paralleling transcriptional activity.

In line with these findings is the fact that estrogen, which stimulates the transcriptional activity of DNA of cells of the uterus and oviduct of the chick, binds to a protein specific for this hormone. This complex in turn binds to repetitive DNA sequences located in the transcribing portion of the chromatin.

The technique of DNA/RNA hybridization *in vitro* permits one to trace the primary transcriptional products found in the nucleus back to the complementary DNA from which it was derived. A comparison of these transcripts with the mRNAs of the cytoplasm permits one to determine how a primary transcriptional product is processed in order to make it capable of being "read" as a message. It was pointed out earlier that the two principal 18S and 28S rRNAs are derived from a common precursor molecule from which the spacer material is removed during processing and that the precursor molecule for 28S rRNA also has a nontranslatable insertion removed before the two ends of the rRNA are spliced together. The nuclear precursors of the tRNAs are also larger in size than their ultimate mature state; a 15-base sequence in the nuclear transcript is not present in the mature tRNAs of yeast, and cell extracts contain substances, presumably enzymes, which can remove these sequences and bring about the RNA-RNA splicing of the ends. It is possible, then, to determine the character of unique sequences of DNA by an examination of their initial transcriptional products.

The bulk of extractable nuclear RNA consists of a heterogeneous RNA of variable size (HnRNA) and a larger amount of low molecular weight RNAs. The function of the latter remains unknown, and may be, at least in part, the degraded remnants of previously processed RNAs, but a portion of the HnRNAs appear to be related to the unique gene sequences, and hence to be the precursors of cytoplasmic mRNAs. Euchromatin, as indicated in Chapter 3, consists of unique, single-copy sequences of DNA interspersed with repetitive segments, the latter averaging around 150 to 200 nucleotide pairs in length in the rat, about 300 pairs in *Xenopus,* and much longer in *Drosophila.* Unlike rRNA, which is derived from, and will hybridize with, only a very restricted portion of the genome, HnRNA has a nucleotide composition characteristic of the total DNA of the species from which it is derived, and it will hybridize generally with all parts of the genome. The size range of HnRNA in the higher eukaryotes is great—from 4,000 to 50,000 nucleotides—while the lower forms such as the slime molds and amoebae yield molecules rarely larger than 6,000 nucleotides. The repetitive sequences, which in the rat often are doubly repeated, make up about 10% to 15% of the larger pieces of HnRNA, that is, those longer than 15,000 nucleotides in length.

The HnRNA molecules do not function as such in translation; they have to be further processed before being transported to the cytoplasm (Fig. 5.26). Much of the HnRNA is degraded in the nucleus and apparently never used for coding purposes, but some are selected for transport to the cytoplasm and are provided with a "cap" at the 5′ end, while a stretch of about 200 adenosines are attached to the 3′ end of the transcript. The "cap" is a 7-methylguanosine linked to the penultimate nucleotide by a triphosphate bond to give an inverted 5′ ⟶ 5′ structure that is resistant to further degradation. The role of the Poly(A) attached to the 3′ end of the mRNA is not known, but it may act as a termination signal in translation or it may function to stabilize the mRNAs in their attachment to the ribosomes. Also, if Poly(A) attachment is blocked, the mRNAs, with the exception of the histone mRNAs and the rRNAs, do not make a cytoplasmic appearance; in addition, mRNAs lacking Poly(A) have a briefer half-life than do those with Poly(A) termination—10 min to 15 min as compared to 180 min. The addition of Poly(A) to the processed mRNAs is, therefore, a post-transcriptional, nuclear event and not a transcriptional tail from a thymine-rich region of a coding section of DNA. There is, however, some evidence from the slime mold *Dictyostelium* to suggest that its structural genes may be unique in being flanked at its 5′ ends (which would be complementary to the 3′ end of the mRNAs) by a series of thymines which could be transcribed into a Poly(A) stretch. The thymine-rich 5′ ends, however, seem to be shorter than the 200-nucleotide Poly(A) region, so that some post-transcriptional addition also takes place.

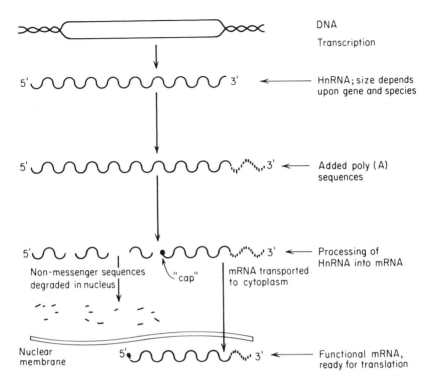

Fig. 5.26. Sequence of events which produces and then transforms a piece of HnRNA within the nucleus into a functional mRNA ready to be translated into protein in the cytoplasm. Degradation of the non-coding sequences of HnRNA takes place in the nucleus, and from the 5' end. The "cap", which is a 7-methylguanosine triphophate bonded with a 5' $\longrightarrow$ 5' linkage to the penultimate base, is indicated as being added in the nucleus along with the Poly(A) sequences of about 200 nucleotides, but whether this is the time of adding the "cap" is not known. The mRNAs destined to be translated into histone would lack the Poly (A) sequences, and presumably the "cap" as well.

HnRNA is transcribed in a 5' $\longrightarrow$ 3' direction, with the DNA being transcribed by the RNA polymerase in a 3' $\longrightarrow$ 5' direction. Consequently, transcriptional and replicational directions are the same. The recognition site for the RNA polymerase is, therefore, on the 3' side of the structural strand being read (Fig. 5.24). The nature of the eukaryotic recognition site, as well as the recognition site(s) for the processing of HnRNA into mRNA, remain unknown, but both within and outside the coding regions of HnRNAs there are palindromic sections of nucleotides of various lengths—that is, reverse repeats of nucleotide sequences—which can form hairpin loops and stems, and it may well be, as in the prokaryotes (Fig. 5.6 and 5.24), that these can serve that purpose.

The enzymatic requirements of various kinds of cells and the selective nature of gene activation would predict that the extractable HnRMAs and mRNAs would vary in size, character, and amount from one differentiated cell to another. This, indeed, is what is found. The average coding sequence of mRNA is about 1,000 nucleotides, to yield, when translated, a protein of 300 to 500 amino acid residues, but the mRNA that translates into fibroin in the silkworm contains 14,000 nucleotides in its coding length and has a total length of 16,000 nucleotides. In terms of variability, *Drosophila* embryos have about 15% of their total DNA transcribing during early development; adults transcribe only 10% in any given kind of cell, a late instar larva may produce only a few species of HnRNA in its salivary gland cells, while during pupal metamorphosis, when the whole anatomy of the fly is undergoing change and restructuring, 30% is active. The latter figure is exceptionally high, and it must involve hundreds, perhaps thousands, of individual transcriptional segments of RNA. By way of further contrast, *E. coli* may have as much as 40% to 50% of its DNA actively transcribing when it is growing in a rich medium, while among humans, liver averages 4% to 5% and brain as high as 22%. It has been estimated that mouse brain cells have as many as 6,000 active units even if each of the HnRNA molecules is as large as 50,000 nucleotides. In this connection, it must be remembered that only one of the two strands of the double helix transcribes at any given locus, indicating that 80% to 100% of the *E. coli* DNA and 44% of that in the human brain is transcriptionally active.

The process of transcription as well as the transcribed product of a specific gene or locus can be examined to advantage in a number of different kinds of cells: the amphibian oocyte, with its lampbrush chromosomes, and the stretches of rDNA which are producing 18S and 28S rRNAs (Fig. 5.16); the spermatocytes of *Drosophila hydei* with its active Y chromosomes (Fig. 3.41); and the salivary gland cells of dipteran flies with their polytene chromosomes (Fig. 3.73). Polytene puffing, which is a visible manifestation of transcription, is cell-, developmental stage- and species-specific, as well as being amenable to easy experimental manipulation.

Puffing occurs when specific bands of the polytene chromosomes open up into a diffuse state. RNA, identifiable as specific sequences of HnRNA, accumulates rapidly, as evidenced by the ready incorporation of radioactive uridine at the site of puffing, and is then quickly complexed with protein to form a ribonucleoprotein. Actinomycin D, which specifically inhibits DNA-directed RNA synthesis, inhibits puffing as well.

Mention has already been made of the fact that the molting hormone, β-ecdysone, has a pronounced effect on puffing when added *in vitro* to isolated salivary glands of third instar larvae of *D. melanogaster*. Bands which are puffed at the time of dissection, e.g., bands 25AC and 68C, and that are actively transcribing, gradually regress in size and cease their activity. The addition of ecdysone will not affect these bands, but it will cause about six other bands to begin to puff within 5 min to 10 mins; these will reach maximum synthetic capability within 1 hr to 4 hr, after which they slowly regress. For example, band 23E is at maximum size within 1 hr to 2 hr, bands 74EF and 75B by 4 hr. Some 3 hr later, and without further hormonal application, another set of puffs arises, reaching maximal activity between 5 hr and 10 hr. These include bands 22C, 62E, 78D, and 82F. Late puffing is dependent upon the activity and, presumably, the products of the earlier puffed regions, indicating that the genes involved in molting are responsive to hormonal control and are brought into play in a regulated and sequential manner. A comparable system of specific hormonal control of ovalbumin production in the chick oviduct by estrogen and progesterone is also known, but it cannot be similarly visualized.

In another dipteran fly, *Chironomus tentans*, a particular band known as Balbiani ring 2 (BR2) in chromosome 4, produces, when puffed, an HnRNA molecule of unusual size. It has a 75S sedimentation value and a molecular weight of from 15 to 35×10^6 daltons. After having Poly(A) affixed to its 3'OH end, it passes to the cytoplasm relatively unchanged in size. Here it is believed to code for a large polypeptide containing over 4,000 amino acid residues and having a molecular weight of 200,000 daltons. DNA/RNA hybridization studies reveal that BR2 contains repeated sequences within the chromomere, but whether the huge HnRNA contains but a single, long, unique coding sequence, with shorter units of repetitive, but noninformational, DNA making up the remainder of the HnRNA, or whether the unique sequence itself is repeated a number of times, is not known.

Genetic Recombination and Repair

Gene recombination frequencies and the construction of genetic maps in both eukaryotes and prokaryotes are determined by means of phenotypic ratios. Chromosomal and genetic homologies are, of course, a necessary precondition for the regular occurrence of recombination just as marker genes (mutants) are necessary for its detection, and it is now quite clear from a variety of studies that the concept of homologies can be extended to the molecular level as well. Furthermore, the formation of chiasmata in eukaryotic meiosis and gene recombination in prokaryotes are both associated with the processes of breakage and reunion—of chromatids in the former and of double-stranded DNA helices in the latter. Since there is virtually nothing in the cell that takes place without enzymatic involvement, these facts place recombination and its cytological counterpart— chiasma formation—in the same general molecular picture with replication, episomal integration and excision, and recombinant DNA phenomena, as well as chromosomal breakage and repair. The latter process should not be thought of as an abnormal event for it occurs regularly in normal and abnormal cells (Fig. 1.6). Enzymes exist which are capable of breaking and then of rejoining DNA strands without a net loss or gain of nucleotides and, hence, of genetic information, and it also now seems to be that, in spite of variations that each process will exhibit, the same general classes of enzymes are involved in all of these events.

As will be pointed out later, chromosomal repair, like replication, may involve only a single DNA helix; in all of the other phenomena, and in some aspects of breakage and reunion of chromatids, at least two double-stranded helices take part, and since homologies are at the basis of all recombinant events, some mode of recognition of homologies is needed to bring together the interacting strands. In meiosis, it is possible, although uncertain, that the synaptinemal complex aids in bringing homologous regions of chromosomes into appropriate juxtaposition, but the mode, or the initiation, of recognition of double-stranded prokaryotic helices has not been recognized.

The involvement of enzymes in recombinational processes is indicated by the presence of particular enzymes at characteristic periods of the cell cycle, and their absence at other stages, and through the use of enzyme mutants which reveal a correlation between interference with the recombinational process and enzyme modification or absence. In the pollen mother cells of the lily, for example, an endonuclease, a polynucleotide kinase, and a polynucleotide ligase are present during zygotene and pachytene stages when chiasma formation and DNA incorporation occur, but they are absent at other stages of meiotic prophase. Also at these stages there is present a DNA-binding protein comparable to that produced by gene 32 of the T4 phage. The actual role of this protein in recombination is only conjectural at this time, but it is possible that it facilitates the opening up of the helices and that it promotes the complementary base pairing of single strands from two different but homologous helices.

Four exonucleases have been identified and characterized in *E. coli* and lambda phage, and their involvement with recombination has been established through the use of mutants. Similar mutants exist in *Neurospora*. Mutants in T4, deficient in polynucleotide ligase and DNA polymerase and incapable of effective recombination, have been particularly useful in that recombination in these mutants can proceed up to an intermediate point and then stop, leaving unresolved hybrid or heteroduplex DNA which forms early in the process.

A number of schemes have been proposed to explain recombination at the molecular level; all involve heteroduplex DNA, but the origin of such DNA remains obscure. Figure 5.27 suggests that as homologous chromatids, or the DNA strands of prokaryotes, recognize each other in some fashion at the level of nucleotide homologies, "nicks" or breaks arise in the single polynucleotide strands as a result of enzymatic action of endonucleases, or possibly as unfilled gaps which remain at the ends of replicons at the conclusion of the S period. Opening up of the helices occurs, possibly by the action of a DNA-binding protein, followed by the pairing of complementary strands. Since the "nicks" need not be exactly opposite each other, the subsequent formation of hydrogen bonding may leave both gaps and tails, the latter recognized as heteroduplex DNA. The tails are removed by the action of an exonuclease and the gaps are filled in by a DNA polymerase and are closed by a ligase.

The involvement of a DNA-binding protein is suggested by the fact that the gene-32 protein is necessary for recombination in T4, and a similar one has been found in the pollen mother cells of the lily, but not in vegetative cells. The gene-32 protein preferentially binds to A-T rich areas of DNA, and if this same preference is exhibited by the comparable protein in eukaryotes, the initiation of recombination may well have its origin in the A-T rich areas of repetitive sequences.

Figures 5.27 and 5.28 illustrate the sequence of events taking place during the process of crossing over. The polarity of the polynucleotide strands in the nonsister chromatids is not given in Fig. 5.27 since conflicting theories differ as to whether the strands are of like or of opposite polarity, but in Fig. 5.28 strands of like polarity are indicated, a situation that is now generally accepted. The same enzymes and proteins are believed to function in either instance. Where heteroduplex DNA is formed, however, there is a strong likelihood of mismatched base pairs being formed; this is particularly the case where the mutants used to detect recombination are base substitution mutants (for example, a G-C pair instead of an A-T pair). Enzymes, not yet identified, but probably similar to the prokaryotic DNA polymerase I with both an exonuclease and a polymerase

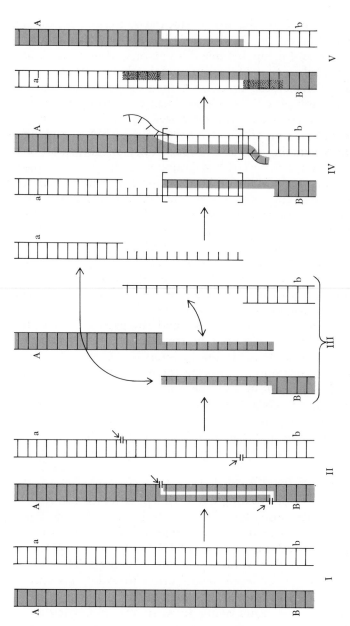

Fig. 5.27. Crossing over at the molecular level. (I) The two nonsister, but homologous, chromatids that will undergo crossing over be-tween the two linked genes (the other two uninvolved chromatids are not included in this illustration). (II) the "nicks" are induced by endonucleases in the polynucleotide strands of the two double helices. (III) The unravelling of the strands (here exaggerated for purposes of illustration) which results from the rupture of hydrogen bonds between paired bases, and which precedes the rejoining to form new genetic combinations. (IV) The chromatids rejoined in complementary fashion (the areas within brackets), with gaps remaining to be filled in by enzymatic action, and with loose and excess polynucleotide strands remaining to be excised by appropriate exonucleases. (V) The end result of crossing over, with the gaps filled (dark areas), the loose ends excised, and the breaks in the strands repaired by ligation. All of the events—nicking, breaking of hydrogen bonds, gap filling and excision—are enzymatic, and probably occur in all chro-mosomes (see Fig. 1.6), but which in meiotic cells at the time of synapsis leads to crossing over.

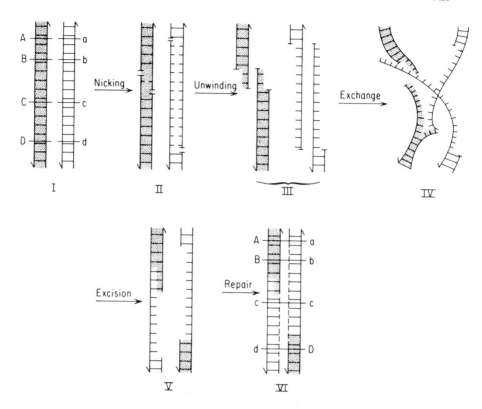

Fig. 5.28. A possible explanation of gene conversion $(C \longrightarrow c)$ during crossing over and as a result of the excision of mismatched sections of a polynucleotide strand from one of the nonsister chromatids, and a later filling in of base pairs to reconstitute the double helix, giving c instead of C in the corrected area. In all other respects, the process is the same as indicated in Fig. 5.27.

activity, can correct these mismatched pairs, giving rise to gene conversions associated with recombination, a not uncommon situation in the ascomycetes.

It should be emphasized, however, that heteroduplex DNA has not been demonstrated conclusively in any eukaryote undergoing recombination, so that the molecular details of recombination in different species may take a variety of forms, even when identical end results of genetic exchange are achieved. Figure 5.29, for example, illustrates how pieces of DNA can be taken up by a bacterial cell in the process of transformation and incorporated, on a nonreciprocal basis, into the bacterial genome. There is no evidence that such phenomena can occur in eukaryotic cells on a regular basis since recombination is a reciprocal affair.

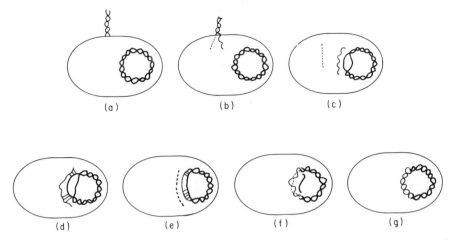

Fig. 5.29. A diagrammatic representation of the sequence of events by which a piece of externally derived DNA may be incorporated into a bacterial genome. a) An external piece of DNA attaches itself to the bacterial membrane prior to entry. b) As the DNA enters, one of the polynucleotide strands begins to disintegrate while the other is attracted to the bacterial genome (c) at a denatured region. Base pairing of the entering strand with the bacterial genome (d) is followed by excision of the unpaired strand (e), and integration of the entering strand into the genome. Once integrated, the entering section undergoes replication (f), with a release of strand to which it was paired. The released strand disintegrates (g) while the entering stand, by ligation of its ends, becomes an incorporated portion of the bacterial genome, the transformation is completed with no gain or loss of nucleotides.

Episomal Integration and Excision. Some bacterial viruses, as pointed out in an earlier chapter, can enter a host chromosome and become integrated in such a way that they are replicated along with the genome of the host. In this *prophage* stage their presence is not harmful to the host cell. A virus genome, however, can be excised from the host chromosome, multiplied in the host cytoplasm, encapsulated in viral protein, and eventually lyse and thus kill the host cell. Lambda phage is such an episome. As explained previously in Chapter 2, it possesses cohesive, single-stranded ends which are necessary for infection and which circularize after the phage enters the cell to produce a structure which undergoes a single recombinational event to enter (integrate with) the host chromosome. Recombination takes place at a specific *att* site under the guidance of specific enzymes, some of viral and others of bacterial origin. The first step after recognition and positioning of the strands is the induction of a pair of staggered cuts made in by restriction enzyme within a 15-nucleotide area common, and hence homologous, to both phage and bacterial chromosomes (Fig. 2.13). The cuts, induced by an endonuclease, permit the opening up of the two helices, followed by complementary and hybrid pairing, and with the cuts then closed by a DNA ligase. A reversal of the process would excise the phage chromosome and permit it to multiply within the host cell.

A somewhat comparable situation exists in the case of *recombinant DNA*. Recombinant DNA studies deal with the introduction of foreign DNA into a bacterial cell by means of a vector which can be either a plasmid or a bacteriophage such as lambda phage. If lambda phage is used, there is also the possibility of having the foreign DNA inserted into the host chromosome as well; the plasmids, if used as vectors, would exist and multiply within the bacterial cytoplasm. The plasmids, of which there are many kinds (Table 2.1), including the *F* (sex) factor as well as those carrying genes for antibiotic resistances (*R agents*) or for the synthesis of the protective protein colicin (*colicinogenic agent*), are double-stranded circles of DNA capable of independent multiplication in the host cell, and, in the case of *F* (sex) factors, also of integration at the site of *insertion sequences* (Fig. 5.30). They consequently provide bacterial cells with additional genes not found in the normal genome, but since they are dispensable and are not coordinated in replication with the host genome, they are not considered to be a second chromosome.

The plasmids, although variable in size, are about one one-thousandth that of an *E. coli* chromosome. They enter the cell through pili, and they can increase in numbers up to 1,000 per cell under special circumstances. Any piece of DNA, even that of quite unrelated organisms, that can be inserted into these plasmids can, therefore, be effectively cloned, provided, of course, that the introduced DNA does not kill the bacterial cell or interfere with its metabolism, and if the phenotypic expression of the introduced DNA can be recognized in a bacterial culture.

The introduction of foreign DNA into a bacterial cell by means of a plasmid is a rare event requiring a number of interrelated steps. It is probable that the events to be described have occurred time and again under natural conditions, but in the laboratory the foreign DNA, whatever its source, is inserted into the plasmid by means of a group of endonucleases called *restriction enzymes* (Table 5.1). These enzymes, some 50 or more of which have been discovered, isolated, and characterized as to cleavage properties, occur in the cytoplasm of bacterial cells and have the normal function of protecting the cell by cleaving and ultimately destroying any foreign DNA that might enter. Their specific usefulness in recombinant DNA studies, as well as in DNA sequencing, is that each restriction enzyme recognizes and cleaves single polynucleotide strands of DNA within a highly specific "restriction" sequence of 4 to 8 nucleotides. Furthermore, these restriction sequences have been found to possess a 180°-rotational,

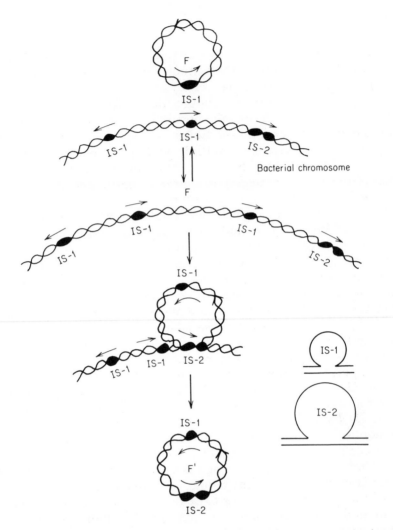

Fig. 5.30. The integration of an F (sex) factor into an *E. coli* chromosome at the site of an insertion sequence (IS). The normal chromosome has 8 copies of insertion sequence 1 (IS-1), each of which is about 800 nucleotides long, and 5 copies of IS-2, each of which is 1500 bases long. These are found in the *coli* genome at a number of regions, are capable of moving around in the genome, and can be inserted in a clockwise or counter clockwise direction. The were first recognized because one became inserted into the *Gal*⁺ locus where it induced a mutant phenotype. When a section of the chromosome containing the mutant *Gal*⁺ was hybridized in vitro to a normal *Gal*⁺ gene, a loop was evident in electron micrographs (lower right); depending upon the particular *Gal*⁺ mutant, the loop had a length of either 800 or 1500 bases. When the ISs are found in the F factors, a region of homology is provided from which they could be integrated or excised.

Table 5.1. Some restriction enzymes used in recombinant DNA studies.

Enzyme	Bacterial Source	Recognition Sequences	Results of Cleavage
Eco RI	*E. coli*	G$\downarrow$AATTC	Cohesive termini
Hae III	*Haemophilus aegyptius*	GG$\downarrow$CC	Flush ends
Hha I	*Haemophilus haemolyticus*	GCG$\downarrow$C	3'-dinucleotide extension
Hpa I	*Haemophilus parainfluenzae*	GTT$\downarrow$AAC	Flush ends
Hpa II	*Haemophilus parainfluenzae*	C$\downarrow$CGG	5'-dinucleotide extension
Mbo I	*Moraxella bovis*	$\downarrow$GATC	5'-tetranucleotide extension
Mbo II	*Moraxella bovis*	GAAGA-8bp*	—
Bam I	*Bacillus amyloliquefaciens*	G$\downarrow$GATCC	5'-tetranucleotide extension
Hind III	*Haemophilus influenzae*	A$\downarrow$AGCTT	Cohesive termini

*Mbo II does not cleave within the recognition sequence but 8 base pairs distant. Those with extensions also have cohesive ends, but they are not known in detail. See Roberts, 1976; Beers and Bassett, 1977; and Helinski, 1977 for additional details.

Fig. 5.31. Cleavage by restriction enzymes, and the fusion to unite donor and vector. Cleavage by HhaI produced a 3' extension while that by MboI a 5' extension when a circular vector (or a lambda phage) is nicked. The donor and vector must have complementary cohesive ends in order for them to be integrated with each other, so the same restriction enzyme is used to nick them both. Flush ends, produced by a restriction enyzme such as HaeIII, can first be acted upon by terminal transferases to give poly(A) or poly(T) extensions for cohesiveness, after which the ends can be annealled and then ligated by a DNA ligase.

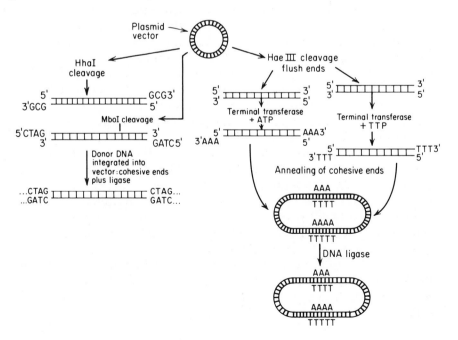

or palindromic, symmetry; that is, the sequence in one strand and a comparable sequence in the opposite strand are identical if read from opposite directions, but with similar chemical polarities; the two sequences, however, may be separated by nonsymmetrical nucleotide pairs (Fig. 5.31). The restriction enzymes, for example, the widely used EcoR1, cut these symmetrical but antiparallel sequences near their ends, producing the same kind of staggered cuts required for the integration of lambda phage into a bacterial genome, and thereby generating cohesive ends which can unite with other similarly induced cohesive ends. The restriction enzyme, HaeIII, however, cuts both strands at the same site, thereby producing flush ends.

Since there are only four kinds of base pairs, and since the restriction sequences capable of being recognized are short, any large piece of DNA, including the plasmid itself, will by chance possess at least one sequence available for cleavage. Lambda phage, for example, has five restriction sequences recognized by Eco R1, while the plasmid pSC101, which is resistant to the antibiotic tetracycline (Table 2.1), has only one. When foreign donor DNA is being inserted into a plasmid vector, the same restriction enzyme is, therefore, used to cleave both so that both will have the same cohesive ends. The complementary and cohesive ends can then reanneal under appropriate conditions, and a DNA ligase can then covalently link the foreign DNA into the plasmid circle. When flush ends are generated by a restriction enzyme, the ends of both vector and donor must be joined by a terminal deoxynucleotidyl transferase, as indicated in Fig. 5.32.

The frequency with which this occurs is low, as is the frequency with which plasmids enter a bacterial cell. The latter frequency, however, can be markedly increased by heating the bacterial cells in the presence of calcium chloride. Once a plasmid has entered a cell it can replicate itself independently of the host chromosome. Under normal conditions about 25 copies of the ColEl plasmid, which possesses genes for colicin, are present in the cytoplasm of an *E. coli* cell, whereas only 6 to 8 of the pSC101 plasmids are normally found. When the antibiotic chloramphenicol is added to the culture medium, the replication of the *E. coli* chromosome is inhibited at the same time that the ColEl plasmid can increase its number up to 1,000 per cell. Any piece of DNA attached to the plasmid can, as a consequence, be most effectively cloned.

In order to provide an advantage to, and thus to isolate, only those bacterial cells containing the plasmid plus its appropriate piece of foreign DNA, a selective method of some sort has to be used. In such a selective system plasmids containing genes for resistance to one or more antibiotics are useful since the nonplasmid-containing cells can be eliminated by exposure to the appropriate antibiotic(s). Again, if the integrated piece of foreign DNA being sought codes for a protein with a known function—e.g., an enzyme catalyzing the final step in the synthesis of an amino acid—then a bacterial cell lacking the ability to form that particular amino acid serves as an appropriate host. Thus, in a minimal medium containing the antibiotics and lacking the appropriate amino acid, only those cells containing the plasmid plus its selected foreign DNA will be able to survive and replicate. The selective procedure, therefore, depends upon the problem being investigated, with plasmid, restriction enzyme, foreign DNA, and host cell all playing crucial roles. The introduced DNA must, of course, be capable of phenotypic expression; otherwise its presence would not be detected. Lambda phage is sometimes used as a plasmid, but if the restriction enzyme EcoR1 is also used, care must be exercised since of the five sites cleaved by EcoR1, two lie within essential genes required for phage growth. Mutant forms of lambda are known which lack these two sites, and it has been possible, as an example, to clone a his^+ (histidine$^+$) gene of yeast in such a system and in an *E. coli* that was his^-.

The source of foreign DNA for insertion into plasmids can be as varied as the problems under investigation. In addition to the strictures mentioned above, it is only necessary that the foreign DNA does not interfere with the replication of the plasmid or the host. Ribosomal and transfer RNAs, as well as mRNAs from eukaryotic cells producing a particular kind of protein—hemoglobin in the mammalian red blood cell precursors or ovalbumin from the oviduct of the chick—are useful in that the genes producing these mRNAs can be obtained with reverse transcriptase. Once obtained in sufficient quantity and then freed from their plasmid carrier by reversing the processes of incorporation, these genes can be analyzed for nucleotide structure or for location in the genome by *in situ* hybridization methods.

One of the practical hopes of the recombinant DNA studies is that by inclusion of appropriate genes into a plasmid, and its successful entry and propagation in a bacterial cell, the bacterial cell can become a miniature factory for the production of important protein substances, for example, the hormone insulin for use in the correction or alleviation of diabetes or the small hormone, somatostatin, which is of mammalian hypothalamic origin and of medical significance in that it inhibits the secretion of a number of other hormones. Such production would require not only the cloning of the genes in question but also their transcription into an mRNA, followed by translation into a protein. Obviously, it is easier to clone foreign DNA than to bring about subsequent transcription and translation as well, but a 500-fold increase in *E. coli* DNA ligase has been achieved, and a significant amount of somatostatin has been formed, both within *E. coli,* the former by means of a lambda phage vector and the latter through the plasmid pBR322. Bacterial genes from related strains inserted into plasmids seem capable of effective transcription and translation, but whether all eukaryotic genes will function similarly remains debatable. Segments of DNA from such varied eukaryotic sources as the human, chick, frog, rabbit, yeast, *Euglena,* and sea urchin have been cloned in bacterial cells, but whether these genes can successfully surmount the specificities of transcription and translation—i.e., specificities of initiation, ribosome binding, termination, etc.—involve questions for future research.

On the other hand, recombinant studies are not confined to bacterial cells. One of the early bright prospects of recombinant DNA research was the possibility of insertion of nitrogen-fixing genes from a bacterial species into the genome of a eukaryotic higher plant, a prospect made more alluring by the realization, in this energy-short world, that the production of ammonium fertilizers requires, on a worldwide basis, energy equivalent to 2×10^6 barrels of oil per day. There is no certainty that this can be done; in *Klebsiella pneumoniae,* a nitrogen-fixing bacterium, eight different genes in two separate clusters are involved, and their collective length seems too great for a plasmid to handle. However, *Klebsiella* is very close, evolutionarily speaking, to *E. coli,* and it has been possible to transfer the nitrogen-fixing genes to *E. coli,* thus providing another source of nitrogen fixation. It is possible, however, by inserting nitrogen-fixing genes into plant viruses or into the DNA of mitochondria or chloroplasts, that such genes can be made part of genomes of important crop plants.

There is always the danger that the manipulation of genetic material, particularly into the human colon bacterium, *E. coli,* could transform a ubiquitous and necessary intestinal form into one capable of epidemic damage. One can easily imagine a strain of *E. coli* having devastating properties as well as one which might be equally beneficial. The prospects have generated much public concern, discussion, and controversy, with the result that recombinant DNA studies must now be carried out under strict federal guidelines in order to minimize the probabilities of creating a bacterial monster.

Two additional facets of recombinant DNA studies can be mentioned. First, plasmid or phage vectors are not necessarily needed for this research, highly useful though they are. If a particular mRNA can be isolated in reasonable amounts and one desires to analyze the gene producing that mRNA, it is possible to have the mRNA transcribe back to the gene in question through the use of a reverse transcriptase; subsequently, Poly(A) residues can be attached to the 3' end of one polynucleotide strand through the use of terminal transferases and Poly(T) residues can be attached to the complementary strand at its 3' end. This then provides cohesive ends which permit the gene units to circularize and to act as an intact plasmid without additional vector material, assuming, of course, that this artificial plasmid can enter and replicate within a host cell.

Second, the restriction enzymes present in bacterial cells function as a defense mechanism to protect the cell against invasive and foreign DNA. These enzymes are generally without effect on the host DNA even though the host genome, because of its size, is very likely to possess numerous recognition sites which could be attacked by its own restriction enzymes. These sites, however, are protected by another group of enzymes—*modification enzymes*—which alter the nucleotides in the recognition sequences, generally by methylation, and thus make them unavailable to the restriction enzymes. The relative nakedness of the prokaryotic genome is, therefore, illusory, in that it possesses its own means of stabilization, analogous, possibly, to the chromosomal proteins of eukaryotic chromatin.

DNA Repair Mechanisms. The exactness of transmission of DNA, generation after cell generation, and the generally low frequency of mutations and chromosomal changes in normal individuals, would suggest that the DNA of both prokaryotes and eukaryotes possesses an extraordinary degree of stability, a stability that might have its primary origin in the fidelity of the replication process as well as in the continued integrity of the DNA molecule itself. It is a matter of observation that in most individuals or cells the end product of replication is a copy of a molecule having a very low level of copy-error (or mutation), but it is now apparent that the faithfulness of the replication process and the stability of the DNA molecule are due, in varying degrees, to the existence of repair systems that all cells possess in one form or another. These repair systems performing "SOS functions" are not infallible, however, and may indeed be error-prone and unable to correct fully the DNA mistakes.

Reference has already been made (p. 294) to the fact that some DNA polymerases, through their exonuclease actions, can correct mismatched nucleotide pairs during the replicative process, thus reducing errors of nucleotide incorporation to very low frequencies of occurrence. It has been estimated that the incorporation of incorrect nucleotides occurs at a rate of one in 10,000 to 50,000. Polymerases can "sense" the presence of these errors and bring about their correction to minimize the possible genetic damage that might ensue from altered DNA sequences. A similar process of correction does not exist for any errors of transcription, but since the mRNAs have a relatively brief existence in the cell, such errors of incorporation are not generally serious. Even if a long-lived, but faulty, mRNA were to be translated into an enzyme or structural protein capable of causing damage, the effect would be likely to be confined to a single cell or a unicellular organism, and long-range effects would be absent.

The process of recombination is based upon an orderly system of breakage and the repair of broken strands of DNA, generally without gain or loss of nucleotides. A somewhat similar process of recombination must also take place between sister chromatids in somatic cells (Fig. 1.6), implying not only the occurrence of frequent, spontaneous breaks in the helices of DNA but also the existence of a means for bringing these broken ends together for rejoining. Enzymes, of course, are required for this purpose. Undoubtedly, most of the breaks are rejoined without disturbing the original alignment of nucleotides, and hence they escape detection; it has been estimated that 95% of all X-ray induced breaks rejoin in this way, so it is only when exchanges such as those in Fig. 1.6 occur that their existence becomes visibly evident. It is, in fact, believed that the repair enzymes patrol the genome in much the same way that repair trucks patrol our major highways and correct breakdowns as they take place.

Every organism presumably has its own rate of spontaneously occurring chromatin breaks, a rate that is influenced by genotype, degree of hybridity, rate of cell division, nutrition and, among seeds, length and conditions of storage. Thus, the frequency of breaks increases with length of storage, but it tends to be diminished as the growing roottips lengthen. When genetic circumstances are involved it is more than likely that faulty enzymes are the cause, and in some instances serious levels of breakage and damage can occur. In human populations, for example, individuals with the inherited skin disease *xeroderma pigmentosum* lack one or another of the nucleases involved in repairing damage to DNA brought on by exposure to ultraviolet light; for each of the several expressions of the disease, a different enzyme is either defective or absent. Patients suffering from *progeria,* which causes premature aging, lack a ligase for uniting broken polynucleotide strands; cells of those suffering from *Bloom's disease* exhibit high levels of chromosome breaks and rearrangements, although in this instance the basic enzymatic relations remain unresolved. In normal cells, therefore, the propensity of DNA to undergo constant breakage is masked by the efficiency of the repair processes.

Damage to DNA, revealed phenotypically as mutations, lethality, or chromosomal changes, can occur in a variety of ways. It can arise spontaneously or it can be induced by a wide spectrum of radiation and chemical agents. Damage may include alteration of base pairs, breaks in both single and double nucleotide strands, and cross linking of DNA with DNA, or DNA with protein, all of them interfering with the proper functioning of DNA or the structural integrity of the chromosome. Evidence for the existence of repair systems came from the observation that strains of bacteria or cells vary significantly in their response to mutagenic agents such as X rays or ultraviolet light, and, most importantly, from the observation that the survival rate of bacterial cells exposed to ultraviolet light depended upon whether, following such exposure, they were kept in the dark or exposed to visible light for varying lengths of time (Fig. 5.32).

The repair system best understood is that which corrects the *dimerization* process caused by exposure to ultraviolet light. Thus, the absorption of ultraviolet light, with a wavelength of around 2,600 Å, causes adjacent thymines to bond together to form a *dimer*, distorting the double helix as the result of a lack of hydrogen bonding

Fig. 5.32. Survival curves of the cells of two strains of *E. coli*, B/r and B$_{s-1}$. Curve I, B/r cells surviving after exposure to ultrraviolet light and recovery in the dark; Curve II, B/r cells surviving after exposure to ultraviolet light, followed by maximum recovery in visible light; Curve III, B$_{s-1}$ cells after exposure to ultraviolet light and recovery in the dark. The two strains obviously differ in their sensitivity (or resistance) to the lethal effects of ultraviolet light, and it is probable that B/r cells possess a far more efficient DNA repair system than do those of B$_{s-1}$.

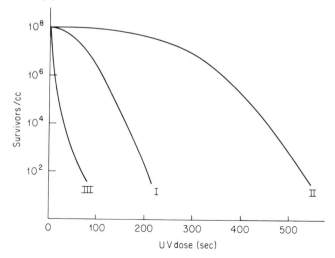

between the complementary nucleotides (Fig. 5.33). These dimers can be removed and the regularity of the helix can be restored by one of several repair mechanisms. The first is known as *photoreactivation,* a process existing in virtually all organisms with the exception of the placental mammals. It is a process in which a specific repair enzyme is activated by the blue portion of the visible spectrum, which, when activated, can recognize and break the bonds between adjacent thymines, convert them back to their previous linear arrangement, and permit them to reestablish normal nucleotide pairing. It is this process which leads to an increased survival of UV-irradiated cells after exposure to visible light (Fig. 5.32).

Fig. 5.33. The formation of thymine dimers by ultraviolet light, generally generated from germicidal lamps emitting at 2537 A, and their removal by the process of photoreactivation (exposure to visible light) and by a "dark repair" system. In the latter system the number of nucleotides removed and restored may number in the hundreds.

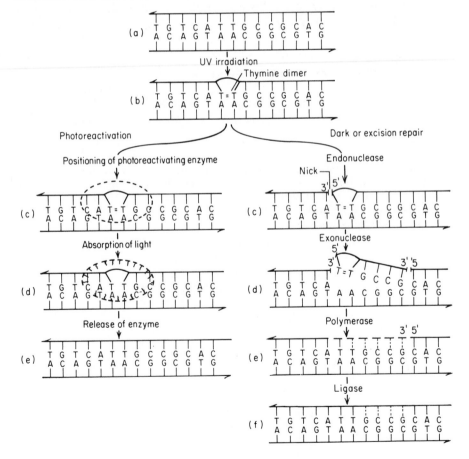

The removal of thymine dimers (other dimers can also form but these are most prevalent) can also occur in the dark by a process known as *dark* or *excision repair*. It is a process bearing a good deal of resemblance to normal recombination. In this instance, enzymatic recognition of the dimer in bacterial cells is followed by a single-stranded break immediately adjacent to the dimer by a specific endonuclease, leaving an exposed 5' end. This step is a requirement for subsequent repair, and it is the one that is either missing or is present to a low degree in some *xeroderma* patients. Another specific enzyme, in this case an exonuclease, then excises the damaged region. From 150 to 600 nucleotides are replaced in each repaired region of bacterial genomes, and up to 5,000 may be replaced in UV-irradiated *Tetrahymena*. The gap is then filled in by a special DNA polymerase and closed by the action of a ligase (Fig. 5.34). The enzymes involved in this "cut and patch" system are similar to, but in some species different from, those involved in meiotic recombination. Thus, the *rec⁻* (lacking or depressed in recombination) mutants of *E. coli* are deficient in both repair and recombination, a situation also encountered in *D. melanogaster* with the *mei-9* mutant. However, other *D. melanogaster* meiotic mutants may exhibit only a recombinational response while the repair system remains fully functional.

X rays are more likely to induce single- or double-strand polynucleotide breaks instead of generating dimer formation, the effect being brought on by the active radicals resulting from the ionization process which then interact with DNA. Single-strand breaks are readily repaired, with 1 to 5 nucleotides being inserted per induced single-strand break, but the enzymatic involvement in this instance remains somewhat uncertain, although the "cut and patch" procedure is very likely to be involved.

Gene Amplification

As a primary gene product, the HnRNAs provide the investigator with an access route back to the structure of that part of the chromosome that produced them, that is, through the use of prokaryotic reverse transcriptases the DNA complementary to the HnRNAs can be synthesized *in vitro* and in sufficient quantity to permit subsequent analyses of nucleotide sequence. What is not revealed by such procedures is the rate at which these chromosomal sequences transcribe and how this rate is controlled. During the course of eukaryotic development and adult life, it is clear that some proteins are needed more at one time than at another, more in one cell or tissue type than in another. As pointed out in the discussion of transcription, this involves a selective activity among the genes, but it also implies selective rates of transcription and translation. For example, the production of hemoglobin by the cells of the hemopoietic system must take place not only constantly to keep up with replacement rates but even at an increased rate should a greater demand be placed on the system. The production of fibroin in the silk glands of the silk moth, *Bombyx mori*, is another example. It has been estimated that approximately 10^{10} molecules of fibroin must be made in 4 days by each silk gland cell to produce the material for the pupal case, a phenomenal rate when it is considered that the DNA/RNA hybridization studies have revealed only one or, at most, two fibroin genes per haploid complement. On the basis of two or four genes per diploid complement, this averages out, respectively, to approximately 7,000 or 14,000 molecules of fibroin per gene per sec. Since the fibroin molecule requires a coding length of 14,000 nucleotides to form a protein of approximately 4,650 amino acids, and since an average rate of amino acid incorporation is around 15 per sec, some form of gene amplification is necessary to achieve the level of RNAs and/or proteins needed by the organism.

There are several obvious mechanisms of enhancement which can bring about increased levels of gene products, several of which have already been mentioned in other contexts. The first is the kind of *preexistent enhancement* exemplified by the genes producing ribosomal RNAs, tRNAs, and histones. These redundant genes, readily identified by DNA/RNA hybridization techniques, are involved in the expression of the genome as a whole rather than of its selective parts; that is, the histones play a general role in chromatin organization and possibly gene suppression, the ribosomal RNAs in forming, along with ribosomal proteins the sites for translation, and the tRNAs in bringing about translation. These molecules, therefore, are an integral part of the working machinery of the cell at a basic level of function, and they are needed in large numbers whenever the cell is actively synthesizing.

Only very rarely are genes coding for unique protein found to be redundant in the genome. An example of such genes, however, are those coding for keratin, the chick feather protein. In the 21-day chick embryo there are between 19 and 25 distinct keratin molecules discernible by gel electrophoresis, suggesting at least this many different nonallelic, but clearly related genes, while DNA hybridization studies reveal from 100 to 300 loci. On average, therefore, each nonallelic gene would be represented from 4 to 12 times in the genome. The gene for the variable portion of the antibody molecule has been estimated to be represented by some 200 to 1,000 copies, but there is no certainty that they are identical with each other.

Another kind of enhancement, also discussed previously, is that represented by the polytene chromosomes of the dipteran flies, or the kind of redundancy found in the macronuclei of the hypotrich protozoa. The polytene chromosomes, occurring in selected somatic cells, consist principally of multiply replicated unique sequences, with the heterochromatin which contains the rRNA loci under-replicated, whereas the protozoan macronucleus contains multiple copies of both unique as well as the usually reiterated ribosomal genes. Such *replicative enhancement,* however, is characteristic only of differentiated cells which will no longer be divided or of nuclei which will not contribute genetically to the next generation.

Where gene amplication is not involved, but amplication of gene product is, as in the case of the fibroin gene, the process must involve either one or both of two situations: an increased rate of transcription in order to increase the number of fibroin mRNAs and/or a stable mRNA which can persist a long time and be read repeatedly without destruction. The fibroin gene is believed to form up to 10^5 mRNAs, with each one presumably lasting through the 4 days of fibroin formation. Whether the rate of incorporation of amino acids under these circumstances can be enhanced is not known.

The ribosomal genes, however, have evolved an additional mode of *replicative enhancement,* found principally in the maturing oocytes of amphibian species, although they are also present in some worms, beetles, and mollusks. A mature amphibian oocyte contains about 10^{12} ribosomes which have been accumulated during the several months of maturation and stored as monosomes for subsequent use during early embryonic development. An average somatic cell can produce up to 4×10^4 ribosomes per day, and as Tartof has pointed out, without some form of amplification, it would take over 600 years for an oocyte to complete its ribosome production. To meet this contingency, up to a 1,000-fold replicative enhancement of rDNA occurs, permitting the oocyte to accomplish its necessary formation of ribosomes within a reasonable length of time. This is done by a selective and repeated replication of the rDNA region of the chromosome, with each replicated piece being eventually released into the nucleoplasm where it forms an unattached nucleolus which comes to lie against the nuclear membrane. The piece of rDNA can be detected on one side of these nucleoli by Feulgen staining or by the incorporation of ^{3}H-thymidine. It can also be released from its nucleolar site and shown to be actively synthesizing rRNA (Fig. 5.16), but it differs from the chromatin extractable from the total genome in that it lacks the H1 histone which is normally found in the linker DNA between nucleosomes.

How the replication of the rDNA is accomplished remains unknown. On isolation, many of the rDNA molecules appear as circles ranging in molecular weight from 87 to 140 $\times$ 10^6 daltons, or from about 10,000 to about 46,000 bases long; the smallest of them are equivalent to the single rDNA unit from which they are derived and the largest ones correspond to about 15 rDNA copies. The presence of circles and rolling circles as well as linear molecules suggests that the rolling circle mode of replication is involved during at least a part of the replicative enhancement process, and it suggests that this occurs after the release of the first replicative form from the parental chromosome itself, but it is not known how the first replicate is formed and released. A reverse transcriptase has been postulated, but none has been positively identified.

The rate of rDNA formation, and of the associated nucleoli, is under some kind of cellular control since a mutant *Xenopus*, heterozygous for the anucleolate genes, has as many extra chromosomal nucleoli as a homozygous individual, even though it has only one-half as much rDNA. Furthermore, the rate of formation of 18S and 28S rRNA is coordinated with the production of 5S rRNA, even though the 5S is not amplified through replication enhancement. This fact may account for the need for a far greater redundancy of the 5S genes. This kind of control is also expressed in anucleolate mutants: They are incapable of forming 18S and 28S rRNA after exhaustion of those present in the oocyte, but neither is 5S RNA formed to any significant extent even though the genes are present.

The amplification described above in *Xenopus* oocytes brings about an increase in rDNA 2,500 times greater than that found in the chromosomes: it is not, however, the only amplification that occurs. In the primordial germ cells, when only 9 to 16 germ cells are present, a wave of amplification raises the level of ribosomal gene number 10-fold to 40-fold, a level that is maintained until meiosis. An analysis of the primordial germ cell rDNA reveals that a good deal of spacer DNA heterogeneity exists from one oocyte to another, indicating that not all rDNA loci are being amplified uniformly and that some kind of mechanism preferentially amplifies some loci while leaving others underamplified. The rDNA resulting from the first wave of amplification is subsequently lost in all meiocytes. The spermatocytes do not regain this lost amount, but the maturing oocytes then go through a second and major amplification, thus providing the oocytes with a massive amount of transcribing rDNA. The rRNA so produced will last, without further addition, until the gastrular stage, when the rDNA in the nucleolar organizer regions begins to function again.

A different kind of amplification of nucleolar material has been described in human oocytes and in plant meiocytes. There is no evidence for the *Xenopus*-type amplification of rDNA loci in human oocytes, but at the diplotene stage numerous micronucleoli, ranging in number from 15 to 40, and containing RNA (and possibly rRNA) make their appearance. Their diameter averages 1.6 μm as contrasted to the 6.9 μm diameter of the two or three normal nucleoli attached to the satellited chromosomes, and while they are generally dispersed in the nucleoplasm, they are associated, when they are attached, with centric heterochromatin and, in particular, with the secondary constriction in chromosome 9 heterochromatin, and possibly, with the secondary constrictions in chromosomes 1 and 16. Whether these micronucleoli participate in ribosome formation is not known.

The situation in higher plant meiocytes of both sex organs is indicated in Fig. 5.34. The cytoplasm, except for isolated pockets formed by concentric circles of the endoplasmic reticulum, is scrubbed clean of basophilia in meiotic prophase; a dramatic drop in ribosome number takes place from pre-leptotene stages to diakinesis in male meiocytes and between leptotene and zygotene stages in female megasporocytes. An analysis indicates that the loss of baso-philia is due to a loss of rRNA since whatever RNA that remains has a nucleotide content characteristic of that of chromosomal origin. A rise in ribosome number at the end of the meiotic prophase coincides with, and is made possible by, the appearance of supernumerary nucleoli which given evidence of RNA synthesis in pachytene and diplotene stages when the ribosome population was at its lowest. These supernumerary nucleoli form at the nucleolar organizer regions as chains of small bodies which are released by late diplotene, or they appear at times as accessory nucleoli attached to, or budded from, the surface of the original nucleolus which has not yet been dis-

Fig. 5.34. Changes taking place in the number of ribosomes per unit volume during the early meiotic stages in the microsporocytes of *Lilium henryi*. PL = preleptotene; L = leptotene; etc. (redrawn from Dickinson and Heslop-Harrison, 1977).

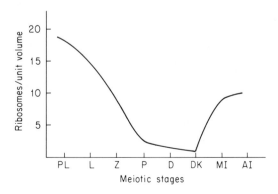

persed. The supernumerary nucleoli disappear in very late prophase, and it is apparent that their contents are directly related to the restoration of the ribosome population as the cells complete the meiotic cycle. The plastids and mitochondria undergo a similar, though less extreme, reorganization of their ribosome population and internal membrane structure. The apparent reason for the organization that takes place in the meiocytes is that it is a preparation for the onset of the gametophytic generation. Since the cells cannot selectively rid themselves of sporophytic mRNAs, they scrub themselves clean of all RNAs, after which they repopulate the cytoplasm with ribosomes for the attachment of gametophytic mRNAs.

The several kinds of amplification tactics engaged in by genes and cells, the ability of the 18S and 28S rRNA genes of *D. melanogaster* to be enhanced or diminished in number in order to achieve a wild-type level of capability (p. 99), and the elimination of parts of chromosomes, whole chromosomes, and even whole sets of chromosomes (p. 117), indicate that the genomes of various species have acquired sensitive and flexible ways of controlling and responding to developmental and metabolic demands. As pointed out in the next section, it may well be that the large amounts of eukaryotic DNA, so puzzling to cytogeneticists, may be related to a variety of complicated regulatory mechanisms, including those just described.

The Nature of a Gene

We began this chapter by raising some of the problems of definition encountered by anyone trying to clarify the nature of a gene in the mid-1950s. The interim period, however, has witnessed a phenomenal growth of genetic information, largely of a molecular nature made possible by new techniques and equipment, more appropriate experimental organisms, and new questions to be answered, with the result that the nature of the gene can be stated in far more definitive terms. Genetically, genes can still be identified by their phenotypic effects, separated from adjacent loci by *cis-trans* relationships, modified by the genomic background in which they are found, and mapped through recombinational procedures. Biochemically, a gene has definite molecular boundaries, a beginning, an end, and a particular sequence of nucleotides in between, and, serving as a template, it can replicate itself with high precision as well as through transcriptional processes and can produce several kinds of complementary RNAs. The task of definition becomes more complicated when the gene is viewed as a cytogenetic unit; as the amount of DNA per genome increases as one goes from the prokaryotes to the eukaryotes (Fig. 1.1), one of the questions that arises is whether the gene identifiable by mutational, recombinational, or functional criteria is the same as that associated with given amounts of chromatin. This problem can best be approached by examining the gene as a functional unit in such well-known experimental organisms as the phage ϕX174, *E. coli,* and *D. melanogaster.*

A gene in its simplest form can be seen in ϕX174, even though a number of genetic surprises are revealed by an examination of its nucleotide sequences. As pointed out earlier, the single-stranded viral genome consists of 5,375 nucleotides, comprising 10 genes and 4 short nongenetic, but possibly functional, regions (compare Fig. 5.35 with Fig. 2.15). On entry into a bacterial cell, the virus genome becomes double-stranded, and the newly formed complementary strand serves as the template for mRNA transcription. Examination of a single gene will serve to illustrate its molecular and cytogenetic nature.

The *A* gene proper (i.e., the translatable portion of the gene) consists of 1,539 nucleotides, beginning with the triplet codon A-T-G, which specifies methionine, and ending just before the termination codon T-G-A (Fig. 5.5). (*Note:* The codons are given in terms of DNA instead of the conventional RNA listings and in terms of the entering strand; to put the codons in the usual RNA frame of reference, simply replace any T with a U). As Fig. 5.23 indicates, however, gene *A* possesses a promoter region, followed by a hairpin loop which contains a recognition site for RNA polymerase where mRNA formation begins, and a block of five nucleotides which serves as a ribosome recognition site (the latter site is possibly complementary to a portion of an rRNA molecule in the bacterial ribosomes). If any of these sites is faulty, the gene is not transcribable.

Between the hairpin loop and overlapping the initiating codon of gene *A* is a termination signal for the mRNA related to the preceding *H* gene. This signal functions apparently only if the whole hairpin loop has been transcribed; it is nonfunctional so far as the *A* gene is concerned because the mRNA for the *A* gene does not involve the entire loop and because a different reading frame is involved. The *A* gene also includes a block of 50 nucleotides representing the site where DNA replication is initiated (Fig. 5.5), and which therefore must include a recognition site for an RNA polymerase to form an RNA primer molecule.

It is apparent, therefore, that gene *A* is more than a block of nucleotides capable of being transcribed into a particular mRNA; its nucleotide sequence is engaged in multiple activities, and it involves other genes as well. Included within it is the complete nucleotide

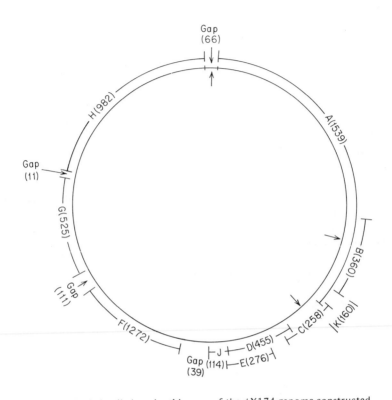

Fig. 5.35. A detailed nucleotide map of the φX174 genome constructed from data derived from DNA sequencing techniques (compare with Fig. 2.15 which was constructed earlier from genetic data alone). The genes are lettered with the number of nucleotides per gene in parentheses. The inside arrows point to the three sites where mRNA transcription can begin; each such site possesses a recognition site for RNA polymerase attachment. Gene B is included entirely within the limits of gene A, as gene E is within the limits of gene D; their reading frames, however, are different from those within which they are located, and hence the included genes encode for quite different sequences of amino acids. Gene C overlaps gene B at its origin, and with gene D at its termination; gene D also overlaps gene J at its termination. The existence of gene K was suspected because of its known existence in a closely related bacteriphage G4. It was eventually shown to overlap genes A and C. The gaps are non-coding sequences between some of the genes, although they may be transcribed. The total number of nucleotides in the coding sequences, obtained by adding the nucleotides of each gene (which means counting some nucleotides more than once), amount to 5941, with 227 in the non-coding gaps. This is substantial more than the 5375 in the genome itself, and is an example of possibly unique genetic economy (see Fig. 5.33) for the details of one particular area). Each gene also possesses a ribosome recognition signal which lies in the sequence of the gene immediately preceding it (clockwise) (after Sanger, 1977; Fiddes, 1977, and Smith, 1979).

sequence of gene *B*, together with its ribosome and RNA polymerase recognition sites; furthermore, the termination codon of gene *A* overlaps with the initiation codon of gene *C* (Fig. 5.35). Such overlaps also occur between genes *C* and *D* and between genes *D* and *J*. Gene *E*, like gene *B*, is an included gene, being wholly within gene *D*. Such inclusions and overlaps are possible, and, of course, are features of remarkable genetic economy, only because the included or overlapped genes have different reading frames and, consequently, code for different kinds of proteins. Even before these circumstances were fully understood, it was apparent that something of the sort must exist because there was not enough DNA in the genome to code for all of the known proteins, if it was assumed that each gene was separate and distinct.

This kind of genetic economy is exemplified by the block of nucleotides where genes *D, E,* and *J* overlap (Fig. 5.36). The ribosome recognition site, *AAGGAG*, not only serves this purpose for gene *J*, but it also codes, by different reading frames, for amino acids specified by the codons of genes *D* and *E*. The stop signal for gene *E* is part of two codons in gene *D*, and it is in a noncoding area preceding the initiation codon of gene *J*. Genes *D* and *J* are overlapping in their respective termination and initiation codons, made possible again because the reading frames are different.

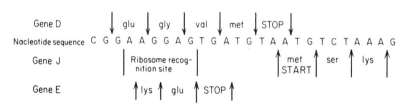

Fig. 5.36. A stretch of nucleotides in the φX174 genome and the functions which they serve for genes D, E and J. Genes D and J overlap each other, and gene E is included within gene D (after Fiddes 1977).

How widespread the phenomenon of such gene arrangements is conjectural, but the SV40 virus also contains overlapping and included genes, while *QB*, an RNA virus of *E. coli*, has an interesting example of an included gene sequence. Figure 5.37 illustrates the closed, double-stranded genome of SV40. The *VP* genes code for structural proteins of different size and in different amounts; they are obviously related to each other in nucleotide sequence since both *VP2* and *VP3* overlap with *VP1* and since *VP2* is included within *VP3*. *VP2* and *VP3* have the same reading frame with *VP3*, having an

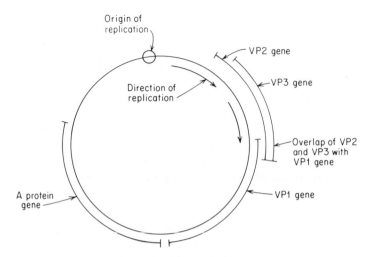

Fig. 5.37. The circular genome of simian virus 40 (SV40). SV40 replicates itself in the kidney cells of the green monkey. The A protein gene transcription occurs early, and is followed by the VP genes which code for viral proteins. VP2 and VP3 have the same reading frame and the same termination point, but VP3 is entirely within VP2; the initiator codon for VP3 is possibly a methionine codon (AUG) located near the beginning of VP2. Both VP2 and VP3 overlap the coding sequence of VP1. The unmarked region of the genome contains other genes which are not indicated (redrawn from Thimmappaya, et. al., 1977).

initiator codon, possibly one coding for an internal methionine, within *VP2*, and both terminating similarly. In *QB*, the two genes in question have a common initiation site, and hence a common reading frame, but while the shorter gene, which codes for coat protein, terminates about 400 nucleotides away, the longer gene, which aids in the formation of infectious viral particles, terminates about 800 nucleotides farther on. The shorter has one termination codon which is skipped, or read through, about 3% of the time; the longer gene ends in a pair of termination codons. Both proteins are necessary, but the longer one only in limited quantities. Regulation in amounts of the two kinds of proteins is therefore determined by the frequency with which the first termination codon is ignored during translation, but how this regulation is determined is unknown.

The existence of overlapping or included genes must, inevitably, exercise an extreme dampening effect on permissible mutations, since a single nucleotide change could conceivably affect the character of two proteins simultaneously and adversely. The coordinate evolution of two genes would be a very slow process.

The lactose operon of *E. coli* is representative of bacterial genetic systems. The genome of *E. coli* is enormously longer than that of ϕX174—over 3,000,000 nucleotide pairs as compared to 5,375 nucleotides—and a greater degree of complexity per functional locus is evident. All of the elements of the viral system—ribosome and RNA polymerase recognition sites, initiating and terminating codons, and promoter—are presumably present, but so too is the repressor-operator system, thereby instituting additional mechanisms of control; genes now govern other genes, and an increasingly larger portion of the genome takes on a regulatory rather than a structural function. (*Note:* The relative simplicity of the viral system as regards control or regulation may be deceptive; a virus is a parasite, and host regulatory systems may be usurped by the virus to govern its own genome.)

With an increasing amount of DNA in *E. coli,* and with regulatory functions quite obvious, it could be argued that a need for genetic economy, achieved through overlapping or included genes, would be relaxed and seem less pressing than in the viruses. Several examples of overlapping genes have been suspected in *E. coli,* but the prevalence of this kind of genomal complexity remains undetermined at present and awaits a more detailed analysis of the bacterial genome.

If we turn now to *D. melanogaster,* we find that the increasing amount of DNA per genome is paralleled by an increasing amount of what appears to be regulatory DNA. The haploid genome of *D. melanogaster* contains 1.4×10^8 nucleotide pairs, 80% of which are included in, or related to, unique or single-copy DNA. Assuming that an average protein could be coded for by a gene of about 1,000 nucleotides, there is sufficient DNA for about 100,000 structural genes. There is good reason, however, for believing that the number of structural genes in this species is less than 10,000, and an excellent case can be made for the hypothesis that each chromomere seen in a polytene chromosome of a salivary gland cell represents a single gene.

The total number of bands or chromomeres in the X chromosome of *D. melanogaster* is 1,024. If each of these bands corresponds to a single gene, then all mutations, spontaneous or induced, that appear in this chromosome should fall into 1,024 groups that are separable from each other by the *cis-trans* test for complementarity. An early experiment indicated that some 968 genes on the X chromosome were capable of mutating to a lethal state, a reasonable approximation to the number of bands. Chromosome 4 in the same organism has about 50 bands, and again an extensive study has indicated that approximately 50 identifiable genetic loci are present. A more feasible approach, however, has been to examine a selected region of the polytene map, to plot all mutations which fall within this region, and to assign such mutations to their appropriate complementary groups.

This has been done with the *3A1-3C2* region of the X chromosome, a region which contains 18 identifiable bands (*3A1*, 10; *3B1*, 6; *3C1*, 2) (Fig. 4.43). A large number of mutations of various kinds—lethals as well as those having morphological, sterility, or delayed emergence effects—have been mapped in the area, and they fall into a minimum of 19 complementary groups. For example, the *3B* region has 6 bands, but 9 complementary groups have been described. These data, together with those from chromosome 4, and an extensive series of X-ray induced rearrangements in the X chromosome, argue that the one gene-one chromomere hypothesis is a reasonable one, but it is an approximation, not an exact correlation. It must be recognized that the limits of a chromomere are difficult to define in a cytogenetic sense and that the screening procedures used may not have detected all possible mutational events.

Keeping these qualifications in mind, we now have the question of how much of the DNA of a chromomere is transcribed and how much of the transcribed RNA is mRNA for translation into appropriate polypeptides. It has been estimated that the average *D. melanogaster* chromomere consists of 25,000 to 30,000 nucleotide pairs per chromatid (the range is 5,000 to 100,000), enough DNA for up to 30 individual genes. Mutational data, however, show that each chromomere has but a single genetic function, and in keeping with this, produces only one kind of mRNA. Evidence from a variety of sources, including that derived from a study of the Balbiani ring 2 of *Chironomus,* suggests that the entire chromomere is transcribed into a single, large HnRNA containing, on average, 25,000 to 30,000 nucleotides, and with this transcribed HnRNA containing but a single copy of a specific mRNA sequence. This would seem to be an extravagant use of DNA, and it raises the question about the purpose of the remainder of the HnRNA as well as its DNA source, particularly since DNA/RNA hybridization experiments show that mRNA sequences in *D. melanogaster* are many times smaller than the original transcript, containing only 500 to 4,000 nucleotides. Does the remainder of the HnRNA serve any purpose? Clearly, the DNA of a chromomere is far more than simply a code specifying the amino acid sequence of a particular polypeptide. If it is assumed that evolutionary selection pressures would eventually rid a species of unnecessary DNA and that what is retained has a function, the remainder of the HnRNA probably consists of a regulatory system that governs the processing and translation of its own mRNA, and all parts of this apparatus must function properly if the mRNA is to perform its essential role. Mutations could occur in the structural or the regulatory portion of the chromomere, but the effects would be referable to a single gene.

The emerging picture of the eukaryotic gene and its transcriptional RNA product is further complicated by the recent discovery that a number of genes, perhaps most eukaryotic genes, are interrupted; that is, their coding sequences are made up of two or more segments separated from each other by noncoding stretches of spacer DNA. How general this property is remains unclear at the moment, but such genes have also been found in viruses (adenovirus-2 and SV40) and eukaryotes, but not in bacteria. (It might be argued that this is a point in favor of the view that the viruses such as Ad-2 and SV40 are eukaryotic derivatives which have attained a parasitic existence.) First described in Ad-2, these interrupted genes include those of the mouse and rabbit that code for the B-globin of hemoglobin; a mouse gene that codes for an immunoglobin; an ovalbumin gene in the chick; a *D. melanogaster* gene coding for the 28S type of rRNA; and at least four yeast genes, each of which codes for a different tRNA. The mouse gene coding for B-globin, for example, consists of 3 coding units separated by two spacers, one spacer of 116 nucleotides located between codons 30 and 31, and a very much longer spacer of 646 nucleotides between codons 104 and 105 (Fig. 5.38). A region for capping the mRNA precedes the initiator codon on the left, and a site for the addition of Poly(A) follows the termination codon. The entire structure is made up of 1,567 base pairs. Within the structure of each of the 4 tRNA genes of yeast is an identical 15-nucleotide spacer sequence adjacent to the anticodon, but the finished and functional tRNAs show no evidence of these spacer nucleotides.

The mouse immunoglobulin gene presents still another facet of gene structure, promoting Broker and his colleagues to describe the situation as an example of "baroque molecular architecture," baroque in this instance meaning "fantastically overdecorated." The mouse immunoglobulin λ chain is coded for by an embryonic gene (also a germline gene since it is transmitted in this form) that is in three parts (Fig. 5.39): an *L* region coding for the leader peptides of the mRNA, an insertion of 93 base pairs (I_1), an amino terminal half coding for the variable (*V*) portion of the light chain, and a region of unknown length followed by a 39-base pair region (*J*), a 1,250-base pair insertion (I_2), and then a region which codes for the constant (*C*) portion of the immunoglobulin chain. A somatic recombination must take place at some earlier stage of development to bring the coding units closer together in the myeloma cells, a process which

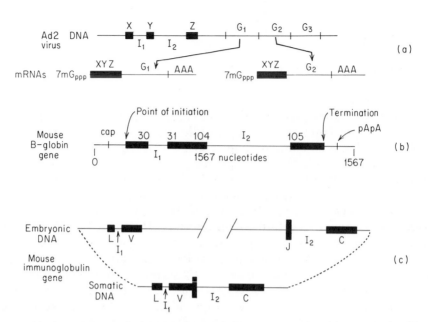

Fig. 5.38. Diagrammatic representation of three interrupted coding sequences. A) A section of the Ad2 viral genome. Segments X, Y and Z, separated from each other by the intervening segments I_1 and I_2, are transcribed, and the segments I_1 and I_2 excised post-transcriptionally, after which the XYZ block is attached as a leader to the several mRNAs that will be transcribed from genes G_1, G_2, etc., to the right. The 7mGppp cap and the Poly(A) tails will be added post-transcriptionally. B) The mouse B-globin gene, with its three coding segments (black), the point at which the cap is added, the initiation and termination codons, the point at which the Poly(A) is added, and the intervening segments I_1 and I_2, containing respectively, 116 and 646 nucleotides. The numbers 30, 31, 104 and 105 refer to the codons between which the intervening segments are located. The functional mRNA will be spliced post-transcriptionally. C) The mouse immunoglobulin gene as found in embryonic and germ line cells, and in myeloma cells the soma. L codes for a hydrophobic leader, V for the variable portion of the light immunoglobulin chain, C for the constant portion of the chain, I_1 and I_2 are intervening segments of 93 and 1250 bases respectively, and J is a region concerned with somatic recombination and RNA splicing. The large segment between the end of V and the beginning of J is "removed" to bring V and J in close proximity to each other. (A after Broker, et. al., 1977; B after Konkel, et. al., 1978; C after Brack, et. al., 1978).

presumably removes a block of nucleotides lying between *V* and *J*. The *J* region itself has a dual function, its left half being involved in the process of somatic recombination and its right half somehow aiding in the RNA-RNA splicing events which must occur post-transcriptionally to remove the included insertions, I_1 and I_2. If this is the correct interpretation of the gene structure, its several functions, and its subsequent alteration, it suggests a possible correlation of cellular differentiation with chromosome structure, with internal chromatin elimination part of the process of change.

Let us examine more fully the possible ways in which the spacer sequences are handled during transcription and/or translation. The functional yeast tRNAs, lacking their earlier spacer nucleotides, could be formed by one of two different processes: Either (1) the entire tRNA gene is transcribed, with the spacer sequences subsequently excised and the coding segments spliced by a post-transcription mechanism to produce a functional tRNA, or (2) the two coding sections of the gene, but not the spacer, are transcribed, and these are then later linked together to form a functional tRNA. When the transcriptional product is destined to function as an mRNA, three possibilities present themselves: the two described above for the tRNAs, followed by a direct translation into a polypeptide, plus a third, namely, the formation of a precursor mRNA that includes the appropriate codons plus the spacer, but with only the codons being translated, followed by a linking up of the several smaller peptides to form the full-sized polypeptide. The best available evidence is that the first process is correct, that is, the entire gene with its spacer(s) is transcribed to form a large HnRNA from which the spacer sequences are excised and the coding sections are spliced together before translation begins. Presumably, the Poly(A) stretches are added to the 3′ end and the 5′ end is capped at about the same time. Excision of spacer material must obviously be done with exquisite precision, for a gain or loss of even a single nucleotide would alter the reading frame and, consequently, the resultant polypeptide. Crude extracts of yeast cells are known to contain the enzymes required for these events. On the other hand, imprecise

Fig. 5.39. A portion of the genome of the SV40 virus that is transcribed early in the viral life cycle; the genome, arbitrarily numbered from 70 on the left to 10 on the right, is divided into units of equal length (the entire genome would be equal to 100 units). The three vertical lines between units 60 and 50 are translation termination signals which, in B and C, are included in the regions excised during maturation of the mRNAs. B; the mRNA which will code for the *t* antigen after the excised region has been removed (dotted line), and the 630 and 1900 nucleotide sections have been spliced; C; the mRNA which codes for the *T* antigen; it has the same 1900 nucleotides at its right end as does the mRNA for the *t* antigen, but a shorter piece at its left end due to a larger portion of the original transcription product being excised (redrawn from Flint, 1979).

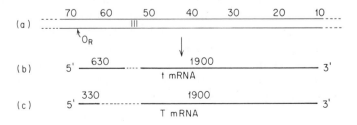

excision and splicing could yield new kinds of mRNAs, and these, on translation, could produce new proteins with new functions and still leave the original gene intact. Further, imprecision could, by selection, become regularized if the newly translated product were of significant importance to the cell or the virus. For example, the *t* and *T* tumor antigens of the SV40 virus share certain qualities in common: their ends are similar, as are their antigenic determinants, but *t* possesses some methionines not present in *T*. Because of size considerations, an overlap of the coding sequences has been postulated, but because the reading frame is the same and since both seem to be derived from RNAs produced during the early phase of viral activity, it is believed that they differ by an internal segment of RNA which is present in one mRNA but not in the other (Fig. 5.39).

Gilbert, indeed, has argued that novel proteins from parts of old ones, particularly in proteins larger than monomers, provides a rapid means of evolution. This topic will be dealt with at greater length in the final chapter, but the basic argument is that interrupted genes are primitive genes, and that the tightly-bound organization of prokaryotic genes is a more recent development brought on by the elimination of spacer sequences. As a result, the bacteria, on a nucleotide basis, are more efficient and more effective coding units than are the eukaryotic species, but only at the sacrifice of morphological and functional flexibility and complexity. Higher forms, according to Doolittle, are higher not because of the acquisition of increased genetic plasticity, but because they have retained the plasticity of primitive and more generalized cells.

It is not possible to draw firm conclusions as to the significance of genic interruption, and as yet there is little information as to its distribution in eukaryotic genomes. Those genes within which spacers have been identified do not constitute unique groups of genetic determinants. The β-globin genes of mice and rabbits, the immunoglobin light chain gene of mice, and the chick ovalbumin gene code for specialized products in highly differentiated cells, but the 28S rRNA genes of the sea urchin and *D. melanogaster,* and the tRNA genes of yeast are functional in every cell engaged in metabolic activity. The histone genes, on the other hand, conform to the classic view of a gene as an uninterrupted sequence of nucleotides, each one of which is a part of a particular codon. Nevertheless, the existence of interrupted genes, and the fact that their mRNAs are brought to a functional state by post-transcriptional processes, indicate that diversity, in either an evolutionary or a developmental sense, can be introduced at several levels, and not only at the level of nucleotide change in DNA.

BIBLIOGRAPHY

Abelson, J., "Recombinant DNA: Examples of Present-Day Research," *Science,* *196* (1977), 159-60. (Note: This article is a prelude to an additional 24 reports, all of which deal with recombinant studies.)

Albert, B., "Studies on the Replication of DNA," in *Molecular Cytogenetics,* B. A., Hamkalo and J. Papaconstantinou, eds. New York: Plenum, 1973.

Ames, B. N., and P. E. Hartman, "The Histidine Operon of *Salmonella Typhimurium,*" in *Handbook of Genetics,* Vol. I, R. C. King, ed. New York: Plenum, 1974.

Beers, R. F., Jr., and E. G. Bassett, *Recombinant Molecules: Impact on Science and Society.* New York: Raven Press, 1977.

Brack, C., et al., "A Complete Immunoglobulin Gene Is Created by Somatic Recombination," *Cell, 15* (1978), 1-14.

Bradbury, E. M., and K. Javaherian (eds.), *The Organization and Expression of the Eukaryotic Genome.* New York: Academic Press, 1976.

Broker, T. R., et al., "Adenovirus-2 Messengers: An Example of Baroque Molecular Architecture," *Cold Spring Harbor Symp. Quant. Biol., 42* (1977), 531-54.

Callan, H. G., "DNA Replication in the Chromosome of Eukaryotes," *Cold Spring Harbor Symp. Quant. Biol., 38* (1973), 195-204.

Case, M. E., and N. H. Giles, "Partial Enzyme Aggregates Formed by Pleiotropic Mutants in the *arom* Cluster of *Neurospora Crassa,*" *Proc. Nat. Acad. Sci., 68* (1971), 58-62.

Catcheside, D. G., *The Genetics of Recombination.* Baltimore: University Park Press, 1977.

Chooi, W. Y., "RNA Transcription and Ribosomal Protein Assembly in *Drosophila Melanogaster,*" in *Handbook of Genetics,* Vol. 5, R. C. King, ed. New York: Plenum, 1976.

Cleaver, J. E., et al., "Human Diseases with Genetically Altered DNA Repair Processes," *Genetics, 79* (1975), 215-25.

Cohen, S. N., "The Manipulation of Genes," *Sci. Amer., 233* (1975), 24-33.

Darnell, J. E., "Implications of RNA-RNA Splicing in the Evolution of Eukaryotic Cells," *Science, 202* (1978), 1257-60.

Denniston-Thompson, K., et al., "Physical Structure of the Replication Origin of Bacteriophage *Lambda,*" *Science, 198* (December 9, 1977), 1051-56.

Dickinson, H. G., and J. Heslop-Harrison, "Ribosomes, Membranes and Organelles During Meiosis in Angiosperms," *Phil. Trans. R. Soc. Lond. B., 277* (1977), 327-42.

Dickson, R. C., et al., "Genetic Regulation: The *Lac* Control Region," *Science, 187*(1975), 27-35.

Doolittle, W. F., "Genes in Pieces: Were They Ever Together?" *Nature, 272* (1978), 581–82.

Fiddes, J. C., "The Nucleotide Sequence of a Viral DNA," *Sci. Amer., 237* (1977), 54–67.

Flint, S. J., "Spliced Viral Messenger RNA," *Amer. Sci. 67* (1979), 300–311.

Furth, M. E., et al., "Genetic Structure of the Replication Origin of Bacteriophage *Lambda*," *Science, 198* (1977), 1046–51.

Gilbert, W., "Why Genes in Pieces?" *Nature 271* (1978), 501.

Goulian, M., and P. C. Hanawalt (eds.), *DNA Synthesis and Its Regulation.* Menlo Park: W. A. Benjamin, 1975.

Hanawalt, P. C., "Molecular Mechanisms Involved in DNA Repair," *Genetics, 79* (1975), 179–97.

Hand, R., "Eucaryotic DNA: Organization of the Genome for Replication," *Cell, 15* (1978), 317–25.

Helinski, D., "Plasmids as Vectors for Gene Cloning," in *Genetic Engineering for Nitrogen Fixation,* A. Hollaender, et al., eds. New York: Plenum, 1977.

Hockman, B., "Analysis of a Whole Chromosome in *Drosophila,*" *Cold Spring Harbor Symp. Quant. Biol., 38* (1973), 581–90.

Hollaender, A., et al. (eds.), *Genetic Engineering for Nitrogen Fixation.* New York: Plenum, 1977.

Holliday, R., "Molecular Aspects of Genetic Exchange and Gene Conversion," *Genetics, 78* (1974), 273–87.

Hotchkiss, R. D., "Molecular Basis for Genetic Recombination," *Genetics, 78* (1974), 247–57.

Huberman, J. A., and H. Horwitz, "Discontinuous DNA Synthesis in Mammalian Cells," *Cold Spring Harbor Symp. Quant. Biol., 38* (1973), 233–38.

Judd, B. H., and M. W. Young, "An Examination of the One Cistron-One Chromomere Concept," *Cold Spring Harbor Symp. Quant. Biol., 38* (1973), 573–80.

Konkel, D. A., et al., "The Sequence of the Chromosomal Mouse β-globin Major Gene: Homologies in Capping, Splicing and Poly(A) Sites," *Cell, 15* (1978), 1125–32.

Landy, A., and W. Ross, "Viral Integration and Excision: Structure of the *Lambda att* Sites," *Science, 197* (1977), 1147–60.

Lefevre, G., Jr., "The One Band-One Gene Hypothesis: Evidence from a Cytogenetic Analysis of Mutant and Nonmutant Breakpoints in *Drosophila melanogaster,*" *Cold Spring Harbor Symp. Quant. Biol., 38* (1973), 591–600.

Lehamn, I. R., "DNA Ligase: Structure, Mechanism and Function," *Science, 186* (1974), 790–97.

Maniatis, T., and M. Ptashne, "A DNA Operator-Repressor System," *Sci. Amer.,* *234* (1974), 64–76.

Miller, O. L., Jr., "The Visualization of Genes in Action," *Sci. Amer., 228* (1973), 34–42.

Moore, D. E., et al., "Construction of Chimeric Phages and Plasmids Containing the Origin of Replication of Bacteriophage *Lambda*," *Science, 198* (1977), 1041–46.

Painter, R. B., "DNA Damage and Repair in Eukaryotic Cells," *Genetics, 78* (1974), 139–48.

Paul, J., "The Transcriptional Unit in Eukaryotes," *Genetics, 79* (1975), 151–58.

Ptashne, M., et. al., "Autoregulation and Function of a Repressor in Bacteriophage Lambda," *Science, 194* (October 8, 1976), 156–161.

Roberts, R. J., "Restriction Enzymes," *CRC Crit. Rev. Biochem., 4* (1976), 123–64.

Sanger, F., et. al., "Nucleotide Sequence of Bacteriophage ϕX174 DNA," *Nature 265* (1977), 687–695.

Smith, M., "The First Complete Nucleotide Sequencing of an Organism's DNA," *Amer. Sci. 67* (1979), 57–67.

Stadler, L. J., "The Gene," *Science, 120* (1954), 811–19.

Stern, H., and Y. Hotta, "DNA Metabolism During Pachytene in Relation to Crossing Over," *Genetics, 78* (1974), 227–35.

Strickberger, M. W., *Genetics,* 2nd ed. New York: Macmillan, 1976.

Temin, H. M., "RNA-Directed DNA Synthesis," *Sci. Amer., 226* (1972), 24–33.

Thimmappaya, B., et al., "The Structure of Genes, Intergenic Sequences, and mRNA from SV40 Virus," *Cold Spring Harbor Symp. Quant. Biol., 42* (1977), 449–57.

Van't Hof, J., "DNA Fiber Replication of Chromosomes of Pea Roots Cells terminating S," *Exptl. Cell Res., 99* (1976), 47–56.

Van't Hof, J., et. al., "Replicon Properties of Chromosomal DNA Fibers and the Duration of DNA Synthesis of Sunflower Root Tip Meristem Cells at Different Temperatures," *Chromosoma 63* (1978), 161–191.

Van't Hof, J., and C. A. Bjerknes, "Chromosomal DNA Replication in Higher Plants," *BioScience 29* (1979), 18–22.

Weintraub, A., "The Assembly of Newly Replicated DNA into Chromatin," *Cold Spring Harbor Symp Quant. Biol., 38* (1973), 247–56.

Whitehouse, H. L. K., "Advances in Recombination Research," *Genetics, 78* (1974), 237–45.

Young, M. W., and B. H. Judd, "Nonessential Sequences, Genes, and the Polytene Chromosome Bands of *Drosophila melanogaster*," *Genetics, 88* (1978), 723–42.

6

Variation: Nature and Consequences of Altered Chromosomal Structure

Wherever adequate genetic and cytological data have been obtained, the general features of eukaryotic inheritance in both plants and animals have been found to be remarkably constant. This fact has, of course, enormously simplified the study of inheritance, because the basic principles derived from experiments on favorable organisms can then be generally extended to the great majority of species which do not readily lend themselves to laboratory study. To be sure, differences exist and must be taken into account, but in the main the physical features of meiosis—synapsis, crossing over, reduction in chromosome number, and gene segregation—are sufficiently constant to permit generalizations to be made from the cytological fact to the genetical interpretation, and vice versa. Indeed, modern cytogenetic interpretation is a composite derived from the observations of many eukaryotic species. Our concern here will be largely with eukaryotic systems, since a good deal has already been stated about the prokaryotic chromosome, but additional prokaryotic information will be dealt with when it serves a purpose.

Cell division, whether it be mitotic or meiotic, is a series of events coordinated in both time and space to give a reasonably predictable result: two genetically identical daughter cells if mitotic, a variety of gametes or asexual spores which may differ in genotype if meiotic. Thus an organism heterozygous for a given gene produces gametes carrying each of the alleles with a frequency of 50%. Mendelian ratios demonstrate this equality, as does the constancy of allelic frequencies from one generation to the next in normal diploid populations. Both, of course, depend upon the regularity of the meiotic and post-meiotic processes, and in the absence of any form of meiotic drive.

Mitosis, at both morphological and molecular levels, provides a mechanism for the maintenance of precise quantitative and qualitative genetic continuity; that is, the daughter cells are the genetic equivalent of the mother cell from which they arose. Meiosis, on the other hand, while providing for genetic continuity, also injects variation into patterns of inheritance through the segregation of alleles, the disruption of linkage groups by crossing over, and the union of gametes of dissimilar genotype. When meiosis is regular, however, these variations occur in predictable ratios, as we have earlier discussed, but we find also that additional variation can be introduced by other means. These departures from a regular inheritance have evolutionary significance.

The constancy of the chromosome as a structural entity lies in its capacity to reproduce itself at each cell division with extraordinary precision. However, chromosomes can undergo change spontaneously even as genes mutate, and the newly constructed chromosome, like its original counterpart, is replicated exactly at each cell division thereafter. Under natural conditions, such changes in chromosomal structure are rare events; they can, on the other hand, be induced with relative ease by ionizing radiations such as X rays and by chemicals. Certain genotypes, such as Bloom's syndrome in humans, can also increase significantly the frequency of structural changes.

Chromosomal aberrations leading to alterations in the linear order of genes may be grouped into four classes: (1) deletions or deficiencies, (2) duplications, (3) inversions, and (4) translocations. The first three, as a general rule, affect only single chromosomes, whereas translocations may involve one, two, or more chromosomes. Their detection can be made both cytologically and genetically in favorable material; in less favorable material, certain aberrations can be inferred from the chromosomal configurations found at metaphase and anaphase of meiosis I.

A deficiency involves the detachment and loss of a block of chromatin from the remainder of the chromosome. The deleted portion will not survive in higher organisms if it lacks a centromere, because it will have no capacity for movement in anaphase. The portion of the chromosome carrying the centromere functions as a genetically deficient chromosome, its movement being relatively unimpaired even if its genetic potential is altered.

Deficiencies can be either terminal or interstitial. The former can arise by a single break in a chromosome followed by a "healing" of the broken end; the latter results from two breaks followed by the reunion of broken ends (Fig. 6.1). Each type, if large enough, can be recognized in pachytene or salivary gland chromosomes by the manner in which pairing takes place with a normal homologue. The location of deficiencies can be determined with considerable exactitude in the salivary gland chromosomes of *Drosophila* by comparing the band structures of the deficient and normal chromosomes (Fig. 6.2), or possibly in lampbrush chromosomes by an absence of specific loops, but such a procedure is not possible in plants or in animals lacking giant chromosomes.

There is some question about the nature of terminal deficiencies. Their formation results in the loss of the normal end of the chromosome, or telomere, and the transformation of a bipolar segment of the chromosome into a unipolar structure; that is, the broken end must heal into a stable condition to prevent elimination or further change resulting from the fusion of chromatid ends when the chromosome divides. Undoubtedly, terminal deficiencies occur, as depicted in Figure 6.1, but they are certainly much less frequent than are interstitial types.

A deficiency represents a quantitative change in the genotype in that it involves the loss of genic material; so it would be expected that deficiencies would have deleterious effects on an organism, the effect depending upon the amount of genic material and its specific function. Homozygous viable deficiencies would be expected to be rare, although viable deficiencies do exist in phage and bacteria, which of course are haploid. A mutant lambda phage has been described, for example, which has lost 23% of its chromosome through deletion, but special nutrient conditions are required for successful propagation. In the great majority of higher plants, deficiencies act as gametophytic lethals, because they cause pollen abortion. Those that are transmitted through gametophytic generation presumably include loci that do not affect pollen functions. In

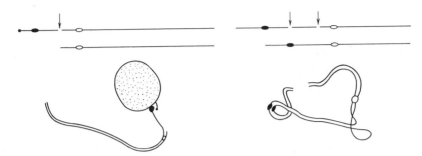

Fig. 6.1. Diagram to illustrate the origin of terminal (left) and interstitial (right) deletions and their appearance at pachynema in the microsporocytes of maize when paired in the heterozygous state. Left, chromosome 6 with one of the homologues missing part of the short arm, including the nucleolar organizer and terminal knob. Right, pairing is relatively complete, with the normal chromosome forming an unpaired loop corresponding in size to the missing piece in the other chromosome. Redrawn from B. McClintock.

Fig. 6.2. Portion of the right arm of chromosome 2 in *D. melanogaster*, showing the position of the deleted segment from 44*F* to 45*D* (above) and the pairing relations that have been observed when the deletion is present in the heterozygous state (below). The arrows indicate the position of the deletion in the deficient chromosome and the region present in the normal homologue.

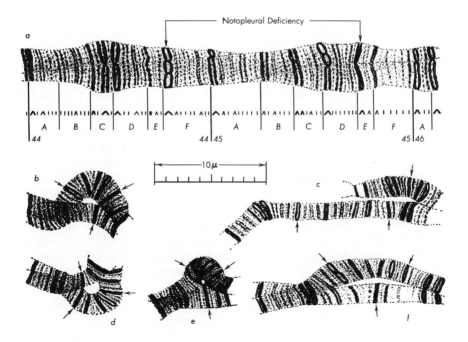

Drosophila, homozygous losses involving the tip of the X chromosome are viable provided they are very tiny. Apparently, loss of the genes *yellow, achaete, white,* and *roughest* can be withstood in either a hemizygous (single, as in a male) or a homozygous condition, but these genes constitute a very small portion of the entire chromosome. Cytologically visible loss of bands 3C2, 3, and 5 in the X can also be tolerated when homozygous. However, even deficiencies that are viable in a heterozygous condition have their size limits. In *D. melanogaster* a loss of more than 80 bands (8% of the chromosome) is generally considered to be severely detrimental if not lethal even when the homologous chromosome is intact. When one considers that the X chromosome alone possesses approximately 1,000 bands, it must mean that physiological balance can readily be upset or that certain genes are developmentally crucial and that under natural conditions elimination of large deficiencies would soon take place. Deficient gametes, however, can survive to take part in fertilization in animals, suggesting that minimal gene actions occur in these cells. Deficient chromosomes can survive more readily through the female side in higher plants, perhaps because of better nutritional circumstances in the ovule, but on the male side, if they are not entirely eliminated by the formation of nonviable pollen grains, the pollen grains that contain them cannot successfully compete with normal grains because of their slower rate of growth.

Because deficiencies result in the loss of genes, it is not surprising that they give rise to recognizable genetic consequences and that many act as recessive lethals. In addition, and again not surprisingly, they frequently produce detectable morphological changes that are inherited as dominant characters. The following *Drosophila* characters are all associated with deficiencies, and all are dominants: *Blond, Pale, Beaded, Carved, Snipped,* and *Plexate.* In addition, most of the *Notch* and *Minute* mutations are known to be deficiencies even though some do not appear to involve a discernible loss of chromatin. This suggests considerable genetic heterogeneity within the phenotype. It is of interest that the *Minute* mutations, which are usually deficiencies, are located in many regions of the chromosomes, yet the phenotype of delayed development, small size, and small bristles is constant. Evidently, many different loci, when deficient, may determine the same phenotype, suggesting that a common phenomenon, e.g., altered or diminished protein synthesis, may be involved in each one. The hypothesis has been advanced that *Minutes* may be the result of a loss of the loci coding for transfer RNA, or alternatively, those coding for ribosomal proteins.

In humans a number of congenital abnormalities have been traced to chromosomal deficiencies in the heterozygous state.

Chronic myeloid leukemia is associated with the Philadelphia (Ph′) chromosome, identified earlier as chromosome 22 minus a substantial portion of its long arm, but now known, as a result of banding studies, to be related to the translocation of the distal portion of 22*q* to the end of 9*q*. The *cri du chat* (cat cry) syndrome, so named because of the characteristic mewing cry of the affected child in infancy, but also including severe mental retardation and other physical abnormalities, results from a loss of the short arm of chromosome 5 (5*q*14–5). Loss of the short arm of chromosome 4 (4*p*-) leads to abnormalities somewhat similar to those associated with the *cri du chat* syndrome, but without the characteristic cry; loss of a region of chromosome 13 (13*q*14-) is associated with retinoblastoma. Losses of parts of other chromosomes, including the X and Y chromosomes, have also been identified, and, as would be expected, each is associated with a more or less characteristic phenotype. Of particular interest are several cases of mosaicism in which the Y chromosome shows a reduced chromatin content, and it is assumed to be deleted for a substantial portion of its long arm, and particularly that portion which fluoresces brilliantly. The individuals, all females, are 45, X/46, XY mosaics. The effect of the 45, X cell line is evident in that several of the individuals show short stature, abnormal genitalia, and other features of Turner's syndrome (p. 408), but others appear normal except for the absence of germ cells. The presumed deletion of the Y chromosome has reduced its male-determining effect to the extent where any aspects of maleness are virtually absent. The localization of the histocompatibility antigen, with its possible testis-determining influence, in the short arm of the Y casts doubt on the above interpretation.

Deficiencies have been used to locate genes on the salivary gland chromosomes of *Drosophila*. The cytogenetic principle is a simple one. A correlation, as illustrated in Fig. 6.3, is made between the absence of a particular band in the chromosome and the presence of a particular morphological phenotype. The deletions labeled *285–14, 264–31,* and *N–8 Mohr* are of particular interest in that they share in common the loss of the band *3C1.5,* and most importantly, all show the *white* phenotype when heterozygous for the *w* allele. Hence, the deficiency of band *3C1.5* removed the $+^w$ allele, so that *w* is acting hemizygously. It appears reasonably certain that the *white* locus is at or very near the *3C1.5* band. The method therefore provides a valuable landmark for mapping the chromosome and for correlating the genetic and cytological maps. The *white* locus is more complex than suggested here; on genetical grounds, it appears to consist of at least three separate genes, two essential and one nonessential, associated with the *3C2-3* doublet. Other genes have been similarly

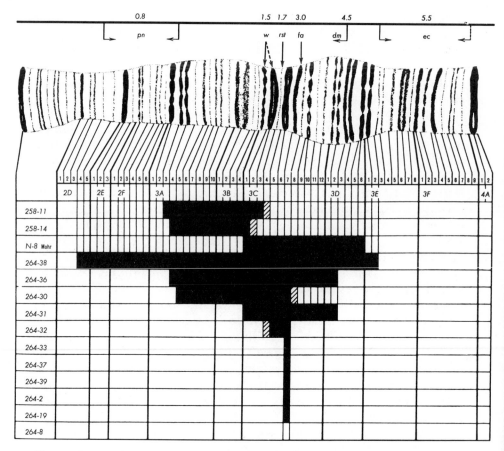

Fig. 6.3. Portion of the left end of the X chromosome of *D. melanogaster,* showing the genetic map, the banded structure, and a group of deletions. The black segments below indicate the bands removed by each deletion. Through the use of overlapping deletions, it is possible to correlate the absence of a particular band with the appearance of a particular morphological phenotype. See text for details.

located through the use of overlapping deletions; still others, owing to the absence of the requisite deletions, have been located only in relation to certain groups of bands.

In some instances in *Drosophila* there is a clear correlation between the phenotypic effect and the bands deleted. Thus, the mutants *lozenge, forked, cut,* and *singed* in the X chromosome are the only phenotypes induced when their respective bands are deleted or interrupted by chromosome breaks. Also, alterations in band *1B3* always induce a *scute* effect, and *3F1* an *echinus* phenotype, suggesting that a given band is genetically homogeneous and that each region acts as a single functional unit. Certain deletions in the *vestigial* (reduced wing size) regions behave as dominant characters, and the greater the loss of chromatin the more severe the phenotype in its departure from normality. The *Notch* effect, on the other hand, appears whenever either the distal or proximal edge of the *3C7 (facet)* band is lost; thus the magnitude of phenotypic effect bears no relation to the number of bands missing (Fig. 6.3). Indeed, *Notch* may appear with no detectable loss. Again, as indicated above, the *Minute* phenotype can be associated with changes at a large number of loci throughout the complement.

McClintock has provided evidence that viable morphological variations can be produced in maize by tiny homozygous deficiencies. The method of producing deficiencies (chromosome 5) was through the use of small ring chromosomes that included a functional centromere plus the proximal chromomeres of the short arm (Fig. 6.4). The aberrant behavior of the ring chromosomes led to progressive losses of chromatin, and it could be demonstrated that simple mutants (affecting a single character) resulted from a particular minute segment of chromatin, whereas compound mutants, affecting several

Fig. 6.4. McClintock's method of producing deficiencies through the use of ring chromosomes. Left, diagrams of chromosome 5 of maize, with the normal 5 at the left, Def. 1 with its small ring chromosome in the center, and Def. 2 with its somewhat larger ring at the left. The arrows indicate the break points that produced the rings. Upper right, pachytene configuration of the normal 5, Def. 2, and the larger ring synapsed with each other; middle right, normal 5, Def. 1, and the small ring free, and with the loop formed by the normal 5; lower left, normal 5 and Def. 1 without the ring; the unpaired loop has a tendency to slide along the arm and does not always give a true indication of the position of the deficiency. (McClintock, 1938).

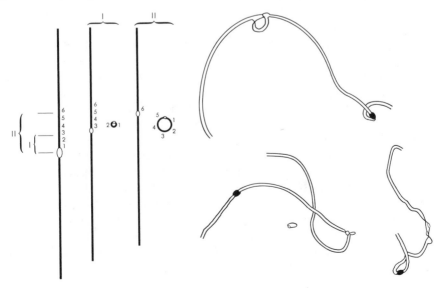

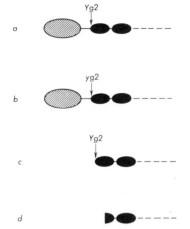

Phenotype appearing when homozygous	Phenotype appearing following combinations
a + a green seedling	*a + b, c,* or *d* green seedling
b + b yellow-green seedling	*b + c* green seedling
c + c pale-yellow seedling	*b + d* yellow-green seedling
d + d white seedling	*c + d* pale-yellow seedling

Fig. 6.5. Diagrammatic representation of the chromomeres at the end of the short arm of chromosome 9 in maize, together with the correlation of the extent of the deletions and the phenotypes resulting from homozygous and heterozygous combinations. (McClintock, 1944).

characters, involved larger pieces of the ring. These ring chromosomes were always used in association with a deficient rod chromosome from which the ring was derived, so loss of chromatin from the ring would lead to the total absence of that chromatin segment, and the mutant phenotype, which in some instances duplicated the appearance of known mutant genes, would be expressed. A comparable study of the end of the short arm of chromosome 9 led to similar conclusions: that loss of chromatin produced mutant phenotypes when in a homozygous condition, suggesting that the usual mutant phenotype was due to an effectively absent gene product. In this instance, terminal rather than proximal chromomeres were involved (Fig. 6.5).

The use of deficiencies, or deletions, of genetic material for purposes of determining the structure and limits of a gene has been carried to an even further point of refinement in prokaryotes. Benzer, for example, making use of overlapping deletions, has produced a *fine-structure map* of the *rII* locus of the T4 bacteriophage. Deletions in this material are recognized by the fact that the mutant phenotype does not revert to wild type, and mutants in *rII*, whether point mutations or phenotypic effects resulting from deletions, are further characterized by the fact that they cannot conduct a lytic cycle in strain K of *E. coli* lysogenic for lambda, and that their growth in strain B of *E. coli* is uncontrolled (rapid lysis and the production of an abnormal plaque on bacterial plates). When tested, the mutant *rII* phages fell into two classes which would complement each other; that is, in a mixed infection, strains of class A could complement (cover) the functions missing in class B, and vice versa, so that lysis could proceed in the absence of any recombination between the two classes. Hence, the rII locus was made up of two genes, *rIIA* and *rIIB*.

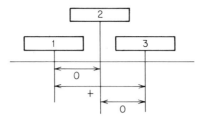

Fig. 6.6. Benzer's (1955) method of using overlapping deletions to map gene positions in T4 bacteriophage. The three deletions are shown in relation to each other on the chromosome. Mixed infection with strains possessing deletions 1 and 2, or deletions 2 and 3, cannot yield wild type progeny following recombination since the progeny would still possess a deletion whose length would be related to the degree of overlap. Only mixed infection with strains possessing deletions 1 and 3 could yield wild type progeny, with the frequency of such progeny a function of the distance between deletions. The presence (+) or absence (o) of wild type progeny is indicated under the appropriate arrows designating the crosses.

Benzer hypothesized that if two mutants carrying deletions overlapped with each other, they could never, in a mixed infection, recombine to form a wild-type phage, whereas if they did not overlap, wild-type phage would occasionally appear and be recognized by the characteristics of the plaque which would form (Fig. 6.6).

Fig. 6.7. The deletions used by Benzer to construct a fine-structure map of genes rIIA and rIIB in T4. The large deletions at the top are used for a rough location of the mutation in question, after which the shorter deletions can be employed to more narrowly locate the positions. Where the limits of a particular deletion are known, vertical lines are drawn to segment the genes into identifiable areas. See text for explanation.

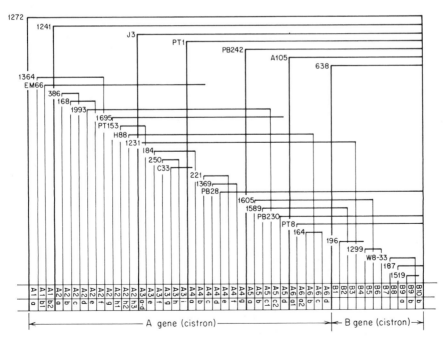

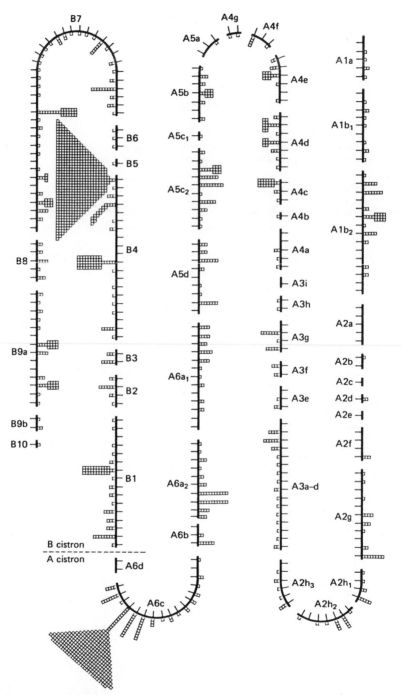

Fig. 6.8. A fine-structure map of genes rIIA and rIIB in T4. Each square reveals a single area where mutations have occurred; some loci mutate more frequently in some areas than in others, although for reasons unknown (Benzer, 1961).

Some 40 or more deletion mutants were uncovered which terminated at different regions in genes *A* and *B*; 7 of the deletions were extensive and extended from gene *A* into or beyond gene *B* (Fig. 6.7). Once one constructed a map of the area based on deletions, any point mutation could be readily located by first crossing it with each of the 7 strains possessing the large deletions. If, for example, no wild-type recombinants were formed in mixed infections with deletions *1272, 1241, J3,* and *PT1*, but were with others to the right of *PT1*, then the mutant must lie in the region bracketed by *PT1* and *PB242*. The mutant can then be tested in mixed infections with the shorter deletions *221, 1368,* and *PB28*. If no wild-type revertants are obtained with *221*, but are *1368* and *PB28*, it can then be assumed that the mutant lies in the area designated *A4d*.

As a result of such endeavors, Benzer has constructed a detailed map of genes *rIIA* and *rIIB* (Fig. 6.8). In this figure, each mutant is indicated by a square, and it is apparent that in the area between *A6c* and *A6d*, and between *B4* and *B5*, two highly mutable "spots" occur. The bases of this genic instability is not known, but clearly some parts of the chromosome of T4 behave quite differently from others in terms of mutability.

It would seem unlikely that deficiencies would exert a profound evolutionary influence since they diminish metabolic controls. However, we do know that in many organisms—annual herbs as opposed to their perennial relatives—a reduced chromosome number and comparably reduced DNA values accompany specialization, but where this process has been examined in detail, as it has in closely related species of *Crepis*, the loss appears to be restricted to centromeres and adjacent heterochromatin. When one deals with more distantly related forms, the nature of the loss is not so evident, but we know too little of the properties of heterochromatin to comment on the evolutionary aspects of its gain or loss. It is instructive, however, to recognize that segments of nucleic acid, replicating *in vitro,* can reduce their nucleotide content significantly when subjected to pressures which favor such reductions.

Duplications

The existence of multiple copies of genes such as those coding for the rRNAs and for histones, and of satellite or simple-sequence DNA which may be represented a million or more times in a eukaryotic genome, both aspects of which have been previously discussed, injects a note of caution into any discussion of duplications. Here the term "duplication" will refer to the grosser, and generally cytologic-ally visible, aspects of quantitative chromosomal increases occurring in euchromatin, leaving the term "repeat" to cover the repetitive DNA sequences detectable through molecular techniques. There is, on the other hand, a probable continuous spectrum of change from the smaller heptanucleotide repeats in *Drosophila virilis* to the larger duplications of the eukaryotic genome, so the distinction made here is, in all probability, an arbitrary one.

An extra piece of a chromosome of the normal eukaryotic com-plement, whether attached in some manner to one of the members of the regular complement or existing as a fragment chromosome, is known as a duplication. When attached to a chromosome in the form of an added section, the duplication may be in tandem, in reverse tandem, or as a displaced piece. Thus if the duplicated piece is represented by the letters *def,* a tandem duplication would be *abcdefdefghi;* a reverse tandem *abcdeffedghi;* and *rstdefuvw* or *rstfeduvw* a displaced duplication.

Figure 6.9 illustrates the tandem duplication associated with the well-known *Bar* phenotype in *D. melanogaster.* Four or five bands in the *16A* region are involved, and the salivary gland picture shows that the normal males have the *16A* region represented once, *Bar* males twice, and *Bar-double* males three times. In females, the wild-type B^+/B^+ has nearly 800 facets in each of its eyes; in the hetero-zygote B/B^+, the number is reduced to about 360, while in the homozygous B/B, the eye is reduced to about 70 facets, arranged into a long, narrow structure. As shown by Sturtevant, *Bar* has a tendency to segregate altered phenotypes, and the frequencies of reversions of *Bar* to normal and to a more exaggerated *Bar-double* are similar, each altered phenotype occurring about once in 2,000 individuals. Their occurrence can be accounted for by the phenom-enon of unequal crossing over. The manner in which this takes place is indicated at the bottom of Fig. 6.9. By the same process, the number of *16A* regions in a single X chromosome can be increased beyond three, and individuals have been obtained with as many as eight regions in tandem sequence. With each addition, the number of facets in the eye is reduced. It has also been shown that the pheno-typic effect of the *Bar* duplication can be partially prevented by feeding larvae appropriate nucleic acid precursors.

As might be expected, duplications are observed more fre-quently in nature and are less likely to be lethal to the individual than are deficiencies. Duplications can be more readily detected cytologically in species such as *Drosophila* because of the detailed structure of salivary chromosomes, but even in plants, studies of

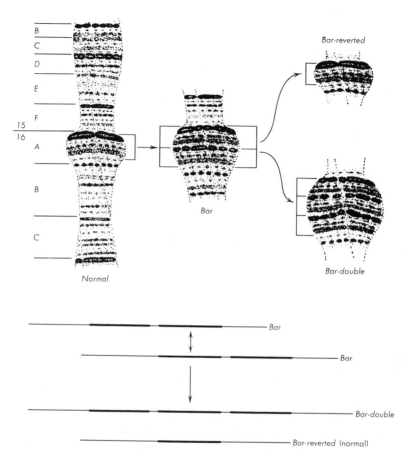

Fig. 6.9. The *Bar* phenotype in *D. melanogaster* is due to a duplication of the *16A* region, as indicated above, while *Bar-double* is due to triplication of the same region of the X chromosome. Below, diagram to illustrate how unequal crossing over in *Bar* females leads to *Bar-double* and to normal phenotypes. T. H. Morgan, C. B. Bridges, and J. Schultz, *Carnegie Inst. Wash. Yrbk., 31* (1932), 303–07.

meiosis in haploid individuals, where no pairing would be expected, reveal some unexpected synapsis, which would suggest the presence of duplications. It would appear, indeed, that certain environmental circumstances can favor the retention of duplications for better adaptation, a situation somewhat similar to the loss of nucleotides under rigorous selection. Thus, in yeast, selection for increased acid phosphatase activity, resulting from a diminished supply of an essential nutrilite, led to the recovery of five independent duplications of the structural gene coding for the enzyme. Two of the duplications arose spontaneously; the other three were induced by ultraviolet light; all were transposition duplications.

Just as deficiencies can be uncovered genetically by the absence of an expected dominant phenotype in heterozygotes, so duplications can be recognized by the absence of recessive phenotypes. For example, fertilization of an attached-X *Drosophila* female, multiply recessive for X-chromosome genes, by treated sperm from a male homozygous dominant for the genes in question will yield occasional females exhibiting unexpected dominant phenotypes. The duplicated section covering the recessive genes, derived from the male chromosome, but now located in the male-derived Y chromosome or autosomes, can then be located by cytological examination of the salivary chromosomes.

Structural and regulatory genes do not arise *de novo* in organisms today, although the "primordial" gene or genes must have evolved from some simpler biochemical system in the primeval past. Because all organisms do not have the same number of genes (their variable DNA content would suggest this), a difference in gene number must have risen through a variety of mechanisms during the course of evolution. Abrupt loss of genetic material and, consequently, gene product through deficiency, is an unlikely evolutionary event, for reasons given previously. Duplications, however, do not possess such evolutionary limitations, and in fact the duplication of loci would appear to provide a feasible method for the acquisition of new genes, and hence new physiological functions.

Inherent in this hypothesis is the further assumption that two genes, contained in the same genome and identical as to structure and function, can diverge through mutation to such an extent that they, at a later stage in evolution, may control entirely different separate functions. In well-adapted organisms, mutations in general lead to the loss or impairment of function and are consequently likely to be selected against in a population because of their adaptive disadvantage. Should the mutated gene be present as a duplication along with the normally functioning gene, the possibilities of its retention and continued mutation, possibly in new directions, become considerably enhanced, particularly if novel substrates would favor the existence and functioning of a mutant enzyme.

To prove that such an event is anything more than a theoretical possibility, three stringent criteria must be satisfied: (1) the existence of two or more identical but genetically separable units must be demonstrated; (2) the duplicate origin of these loci from an original single locus must be proved; and (3) the qualitatively different functions governed by these once-similar but now diverged loci must be established. It has been suggested that in such circumstances one copy of the structural gene in question would be genetically "silent," with its product inactive because of an improper folding into a tertiary shape. Mutation and recombination, taking place without the constraints of natural selection, would alter the "silent" gene, and, by chance, produce a protein which could fold in such a way as to act on a different substrate. The similarity of the tertiary structures of a number of proteins such as trypsin and chymotrypsin suggest that they had their origin from duplicated genes.

There is a considerable body of evidence that points to the establishment of duplications in natural populations of *Drosophila*. The most frequent type appears to be that which involves adjacent gene repetitions, called *repeats*. The "doublet" or "capsule" structure of the bands in salivary gland chromosomes has been interpreted as repeated band duplications, often in inverted order but not necessarily so. These may be seen in the salivary map of *Drosophila melanogaster*, and confirmatory genetic evidence has come from studies of crossing over in the pertinent region and aberrations involving doublets. Thus the two doublets in *89E* of the right arm of chromosome 3 include the bithorax series of five pseudoalleles, that is, genes which behave as alleles when tested against each other but which can be separated by crossing over (Fig. 6.10). The two pseudo-alleles of the sex-linked *vermilion* locus are apparently located in the $10A_{1-2}$ double, with one in the left half and the other in the right (Fig. 6.11). Such repeat sections, common to the genus *Drosophila*, have also been found in the fungus gnat, *Sciara*, but they, of course, would not be cytologically detectable in any other type of chromosome. However, genetic tests have revealed examples of pseudoallelism in maize, *Aspergillus*, *Neurospora*, bacteria, and bacteriophages, while independently segregating alleles for phosphoglucoisomerases have been identified in the plant genus *Clarkia*.

Most of these mutant genes, generally adjacent to each other as tandem duplications, were first classified as examples of multiple allelism until crossing over between them, or their independent loss, established their separateness. Thus, the two garnet genes (g^1 and g^2) in chromosome 2 of *D. melanogaster* at 44.4, can yield new wild-type chromosomes from a *trans* configuration in about one per 30,000 individuals. In some instances each of the separate genes can be shown to have its own group of true alleles, similar to and paralleling the action of, the alleles of the neighbor gene. The action of suppressor genes, which affect one part of a duplicate locus but not the other, also supports the concept of their separate and distinct nature (Fig. 6.11). On the basis that the gene cannot, because of its

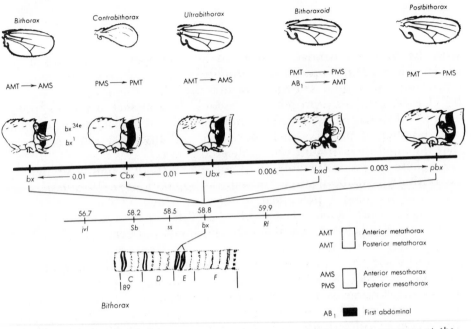

Fig. 6.10. Phenotypes, linkage relations, genetic map, and chromosomal structure at the *bithorax* locus in chromosome 3 of *D. melanogaster.* E. A. Carlson, *Quart. Rev. Biol., 34* (1959), 33-67.

Fig. 6.11. Phenotypes, linkage relations, genetic map, and chromosomal structure at the *vermilion* locus in the X chromosome of *D. melanogaster.* E. A. Carlson, *Quart. Rev. Biol., 34* (1959), 33-67.

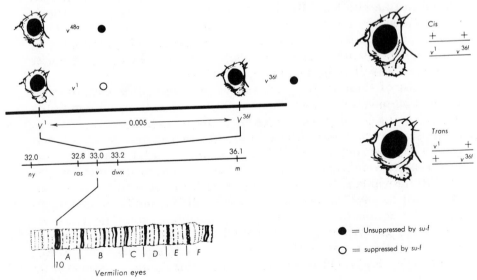

molecular structure, provide an unlimited source of variation, such parallelism of alleles is expected and understandable. Their similar patterns of mutation, together with their association with doublet structures, support the conclusion that the duplicate or triplicate loci stemmed from an original locus.

Satisfaction of the third criterion—that the duplicated locus can carry out a qualitatively different function—and thus provide a source of new genes and of additional variation—is somewhat more difficult to prove. It would be necessary to ascertain that one of the genes is acting in an altered manner, but with lingering evidence of their true origin from a former single locus and yet with sufficient diversity of action to justify the assumption that further divergence would qualify them as having totally different modes of action. As might be imagined, such proof would not be easy to obtain by genetic means alone but suggestive evidence from *Drosophila* (Figs. 6.10 and 6.11) and the human has indicated that transitional states do exist, for example, the amino acid sequences of related enzymes point strongly to their close relationship of separate loci through duplications.

The situation in human beings is represented by the variants in hemoglobin and bears on the question of the origin and evolution of new genes and, consequently, of new proteins. Hemoglobin A (Hb-A), the form most commonly found, is a tetramer composed of four polypeptide chains, two alpha and two beta. Hb-A$_2$ differs in having the beta chains replaced by delta chains, and is, therefore, alpha$_2$delta$_2$. The beta and delta loci are closely linked and are thought to be duplicate loci, comparable perhaps to garnet and *vermilion* in *D. melanogaster* (Fig. 6.11). An abnormal hemoglobin (Hb-Lepore), studied by Baglioni, possesses the two normal alpha chains, but the other two monomers have amino acid sequences characteristic of both the beta and delta chains (Fig. 6.12). Baglioni suggests that unequal crossing over has taken place between the beta and delta loci, giving a deleted segment in one chromosome and a duplicated segment in the other. The deleted chromosome would have parts of the beta and delta loci directly adjacent to each other rather than being spatially distinct, and the whole segment would be activated when producing hemoglobin rather than each one independently. By transcription and translation, this segment would produce the Hb-Lepore variant. A similar mechanism had been previously proposed by Smithies to explain the altered amino acid sequence of one of the polypeptide chains in human haptoglobin, but other than these two circumstances, duplications in humans are either rare or difficult to detect.

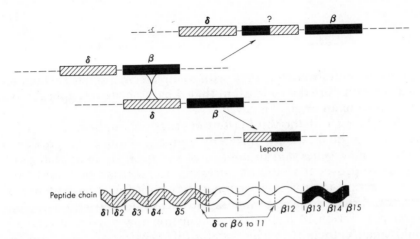

Fig. 6.12. Diagram of the circumstances giving rise to the composite locus responsible for the formation of a portion of the protein in Hb-Lepore. Top, unequal crossing over between the beta (β) and delta (δ) loci gives rise to a triplication, which has not been found, and a composite Hb-Lepore locus. Below, the polypeptide chain found in Hb-Lepore, which is delta-like at its left end and beta-like at its right end; the middle region, (6–11) is common to both chains and is strong indication that these are duplicate loci. After V. M. Ingram, *The Biosynthesis of Macromolecules* (New York: W. A. Benjamin, 1965).

Prokaryotic duplications are virtually absent in natural populations, yet they arise as frequently as gene mutations in laboratory cultures. Their ease of formation might well be related to the relative ease with which external DNA can be introduced into a genome through transformation and/or transduction. The tandem duplication of the histidine operon in *Salmonella typhimurium* was discovered following transduction. The rapid excision of duplications from a genome, however, indicated that their presence is generally detrimental and that the organism possesses the means for their prompt removal. In *E. coli,* for example, a duplication of a genomic segment including the *trpB* cistron (related to the synthesis of the B protein of tryptophan synthetase) was translocated to another part of the genome, brought under the control of an operon not regulated by tryptophan levels, and hence not subject, like the normal operon, to tryptophan repression. Such a lack of control would be obviously detrimental to the organism.

Inversions

The structural aberration most frequently encountered in wild populations of higher organisms, and probably the one most useful to the geneticist in a wide variety of experimental designs, is the inversion. Since Sturtevant first detected them in *Drosophila* through the altered order of genes in linkage groups, inversions have been found in a wide variety of species; in such plant genera as *Paris* and the North American species of *Tradescantia*, few individuals seem to be free of them. Many species of *Drosophila*, and in particular *D. willistoni* and *D. pseudoobscura*, show a wealth of inversions which can be studied in precise detail in the salivary chromosomes. Some groups of animals, on the other hand, seem to be relatively free of inversions, although small inversions are difficult to detect cytologically and genetically. *D. meridiana* falls into this group along with the anopheline mosquitoes, the urodele Amphibia, and many species of grasshoppers.

An inverted chromosome is one in which a portion of the gene sequence, evidently as a consequence of chromosome breakage, has become rearranged in reverse order. There are two general types, distinguished from each other as to whether or not the centromere is included. The more commonly encountered form is confined to a single arm of a chromosome, and is termed *paracentric*. No visible change in chromosome arm ratio in somatic cells is produced by this type of aberration. The *pericentric* inversion, however, includes the centromere, and it may markedly alter chromosome morphology. For example, it is evident that an acrocentric chromosome can be converted to a metacentric type by an appropriately placed pericentric inversion. This aspect will be considered further below. There are, in addition, significant differences in the genetic consequences of the two types of inversion, these differences stemming from crossovers that can occur within the inverted segments of chromatin.

Paracentric Inversions. If, for example, the sequence of a normal chromosome is represented as *ABCDEFGH*, an inverted homologue might be *ABFEDCGH*, with the *CDEF* section being shifted in its relation to neighboring loci. If the *CDEF* region contained known genes whose linkage to other genes outside the inversion were altered, such an alteration could be detected, in inversion homozygotes, by genetic tests. The process is laborious, however, and would be employed today only when other, simpler means would not suffice. In pachytene and salivary chromosomes, an inversion in a heterozygous condition can be recognized by the inversion "loop" which is formed when all portions of the two chromosomes synapse in an homologous fashion (Fig. 6.13). In anaphase of meiosis, the commonly observed inversion bridge, with its accompanying acentric fragment, results when a crossover takes place within the inverted section. The frequency with which such inversion crossing over takes place depends

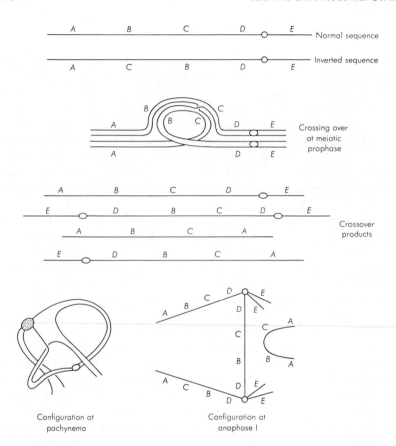

Fig. 6.13. Diagram of a pair of homologues, one having a normal, the other an inverted, sequence of genes, the synaptic configuration at pachynema with a single crossover occurring within the inverted segment, the four chromatid products from the crossover event, configuration as seen in a pachytene stage of maize, and the configuration found at anaphase I.

upon the length of the inverted segment, its location in the chromosome, and the crossover characteristics of the individual. It is evident that the resultant chromatids will be abnormal, owing to the formation of a dicentric and an acentric. The acentric fragment will be lost, having no capacity for movement at anaphase, and the bridge will be broken either by the strain of anaphase movement or by the cell wall that will cut across it. Occasionally neither event will happen, and the dicentric bridge will be simply suspended between the two polar groups of chromosomes, to be eventually lost by failure to enter one or the other of the telophase nuclei. Breakage of the dicentric bridge and loss of the acentric fragment would lead to inviability in the haploid cells arising from meiosis or death of the resultant embryo.

In *Drosophila* no serious reduction in gamete viability is encountered as the result of inversion heterozygosity. The situation in males is readily explicable because crossing over is generally absent, but in females it must be assumed that the failure of inversions to reduce egg viability results either from a lack of crossing over within an inversion or from the exclusion of the dicentric bridge from the egg nucleus. The latter hypothesis has been shown to be correct. The dicentric chromatid, resulting from a single crossover within an inversion, passes into the polar bodies, and the innermost chromatid, which is not involved in crossing over, remains to be included in the functional egg nucleus (Fig. 6.14).

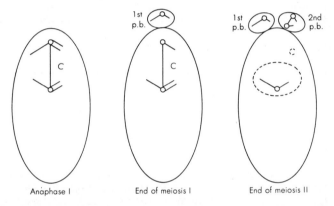

Fig. 6.14. Schematic representation of the fate of an anaphase I bridge resulting from inversion crossing over in the oocyte of *D. melanogaster*. Meiosis occurs near the top of the oocyte, and the chromatid nearest the surface passes into the first polar body. It is presumed that the bridge passes into the second polar body, while the bottom chromatid, which has not been involved in crossing over, passes into the functional egg nucleus. The acentric fragment probably disintegrates in the cytoplasm of the egg.

Whether a similar situation holds in the embryo sacs of the higher plants is known for certain only in maize, but survival would depend upon the mode of embryo sac formation. In maize, where the basal megaspore of a linear tetrad develops into the embryo sac, the noncrossover chromatid may be regularly included in the basal cell, but since the inversion bridge tends to be broken at anaphase I, the deficient chromosome can be included in the basal megaspore and thus contribute to zygotic inviability.

When two crossovers are formed within an inverted segment, the results will depend upon the number of chromatids involved. From the diagrams in Fig. 6.15, it can be seen that two-strand double crossing over will yield four normal chromatids, two of which were

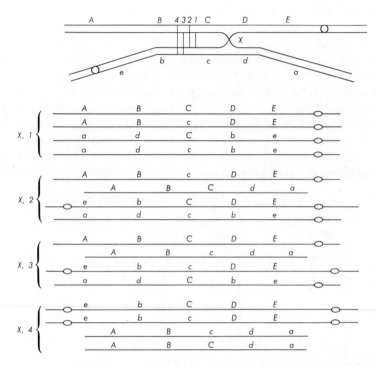

Fig. 6.15. The chromatid and genetic consequences of double crossing over within a paracentric inversion. The crossover at X is assumed to be constant, with the second crossover being at the *1*, *2*, *3*, or *4* position. *X,1* would be a two-strand double, *X,2* and *X,3* would be three-strand doubles, and *X,4* would be a four-strand double.

involved in crossing over, and the other two were not. The presence of appropriate genetic markers within the region of crossing over would be necessary to detect such an event. Three-strand crossing over yields one noncrossover chromatid, one crossover chromatid, and a dicentric bridge with its acentric fragment; four-strand crossing over would yield two dicentric chromatids and two acentric fragments; and duplication and deficiency would presumably lead to death of the four products of meiosis, or of the zygote if the gamete were involved in fertilization.

Clearly, therefore, inversions seriously interfere with the recovery of chromatids that have been involved in crossing over, but since multiple crossing over occurs infrequently under such circumstances of structural heterozygosity, use has been made of inversions whenever experimental design demands freedom from crossing over in a particular region of a chromosome. The use of multiple inversions reinforces the situation, of course, and the recovery of crossover chromatids can be prevented. The further incorporation of a lethal gene within the limits of an inversion also serves as a means for indefinitely preserving structural heterozygosity. The now famous *ClB* stock of *D. melanogaster* developed by Muller for his early radiation studies owes its usefulness to the *C* component, which, in the form of an inversion, acts as a crossover suppressor; the *l* component, a recessive lethal, which prevents homozygosity for the *ClB* chromosome; and the *Bar (B)* phenotype, which permits identification of individuals carrying the *ClB* chromosome in a heterozygous state. Other balanced lethal stocks make use of a similar system of inversions and lethal genes, but as stated above, the usefulness of inversions in this manner is not that crossing over is totally suppressed, but rather that they result in the loss of crossover chromatids.

Although paracentric inversions have a drastic effect on the recovery of crossovers that have occurred within their limits, and an equally severe reduction on crossing over in the regions immediately adjacent to the inversion limits, their effect on the frequency of crossing over in other nonhomologous chromosome pairs within the genome is generally to bring about increases. Crossing over, or its lack, in one pair of chromosomes is, therefore, not independent of that in other nonhomologous pairs, a phenomenon recognized by Sturtevant early in the period of *Drosophila* genetics. It would appear as if a meiotic nucleus possessed the potential for a specific frequency of crossovers; if that potential was not drawn upon by one pair of chromosomes, owing to the presence of an inversion, other chromosome pairs could utilize this potential to increase their own frequency of crossing over. However, since an inversion does not necessarily reduce crossing over within its limits, but rather influences the recovery of the crossover chromatids, the increase in crossing over is an actual increase and not a redistribution of crossovers. The uncertain nature of the phenomenon is made even more evident by the fact that the magnitude of the effect is not correlated with the size of the inversion being tested, the positions of inversion limits, or the amount of crossing over that takes place within an inversion. Also, the inversion may affect two nonhomologous pairs of chromosomes to different degrees.

Table 6.1 and Fig. 6.16 indicate the effects that can be achieved. Each inversion exerts its one specific influence on crossing over (some inversions, when tested, exert no effect), and the effect when present varies along the length of the chromosome, with the most pronounced effect in those regions adjacent to the centromere or to the centric heterochromatin.

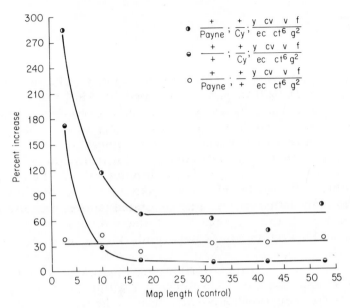

$$\frac{+}{Payne} ; \frac{+}{Cy} ; \frac{y \quad cv \quad v \quad f}{ec \quad ct^6 \quad g^2}$$

$$\frac{+}{+} ; \frac{+}{Cy} ; \frac{y \quad cv \quad v \quad f}{ec \quad ct^6 \quad g^2}$$

$$\frac{+}{Payne} ; \frac{+}{+} ; \frac{y \quad cv \quad v \quad f}{ec \quad ct^6 \quad g^2}$$

Map length (control)

Percent increase

Fig. 6.16. Effect of the *Payne* and *Curly* inversions on crossing over in the X-chromosome of *D. melanogaster*. Abscissa represents the control map length in individuals lacking the inversions; the ordinate represents the percent increase obtained by the three inversion combinations indicated in the square (Steinberg and Frazer 1944).

Table 6.1. Crossing over in chromosome 2 of *Drosophila melanogaster* as modified by the presence of CIB inversion in the X chromosome and the Payne inversion in chromosome 3. The centromere lies between the genes *pr* and *c*

Cultures	Number of Flies	al − ap	ap − b	b − pr	pr − c	c − px	px − sp	Total C.O.
			Regions Tested (C.O. in Percent)					
Control	2894	10.3	27.7	3.5	16.2	20.0	3.5	80.2
CIB	1776	12.4	29.0	9.3	21.7	23.5	5.9	101.8
Payne	2519	14.2	29.7	5.8	18.9	25.3	6.0	99.9
CIB + Payne	1652	18.5	37.3	12.4	31.7	29.9	7.8	137.6

Source: T. H. Morgan et al. *Carnegie Inst. Yearbook, 31* (1932), 303–07.

Pericentric Inversions. If the two breaks involved in formation of a pericentric inversion were equidistant from the centromere, the chromosome would appear unchanged morphologically. Should they occur at different distances from the centromere, a shift in the position of the centromere would obviously take place. To what extent this type of aberration has altered the shapes of chromosomes in natural populations cannot be estimated at present, although it is customary to postulate the existence of pericentric inversions to account for centromeric shifts from acrocentric to metacentric, or vice versa. However, supportive evidence in the form of meiotic configurations or of shifts in genetic position of recognized loci is difficult to obtain. The salivary gland chromosomes of *Drosophila,* on the other hand, have permitted the identification of at least 32 pericentric inversions, and in members of the Orthoptera, where centromeric shifts are common, there is reason to believe that pericentric inversions have been influential in producing new karyotypes.

Crossing over within a pericentric inversion, when heterozygous, also produces characteristic chromatid products. As Fig. 6.17 indicates, a single crossover within the inversion loop does not give rise to a dicentric chromatid and an acentric fragment, but instead produces two new chromatids, each of whose ends are identical in genic content. Duplication and deficiency obviously result, and sterility or reduced fecundity would ensue because of gene loss or gain. The only type of crossovers that would not result in duplication-deficient chromatids would be the 2-strand doubles, and these would have their frequency determined by the size of the inversion loop. This would argue for the establishment only of short pericentric inversions, and there is evidence from both grasshopper and rodent groups that centromere shifts have been more common in short than in long chromosomes.

Fig. 6.17. A pericentric inversion, with the break points equidistant from the centromere, the pairing relations at pachynema with a single crossover within the inversion loop, and the chromatid products resulting from such a crossover.

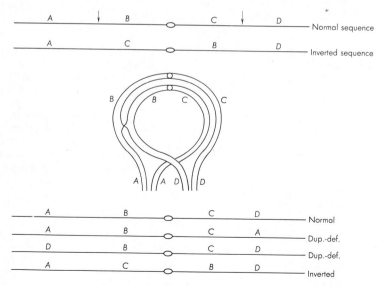

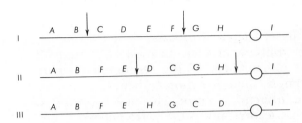

Fig. 6-19. Three chromosomes assumed to be sequentially related to each other through overlapping inversions. If they are related to each other in origin, the order of origin must be I→II→III, III→II→I, or I←II→III. Arrows indicate the position of breaks needed to form the inversions.

Inversions and Evolution. Interest in the evolutionary significance of inversions centers largely around the paracentric type. In an organism possessing a number of inversions in the same arm and in which the salivary gland chromosomes can be studied, their family history, as well as that of the species, can be reconstructed through the use of overlapping inversions. This has been done in the third chromosome of *D. pseudoobscura* by Dobzhansky. Figure 6.18 illustrates three chromosomes that have come to differ from each other by successive inversions that ovelap each other. Their order of origin, sequentially, must be I → II → III, III → II → I, or I ← II → III.

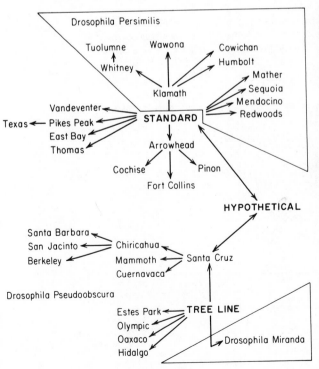

Fig. 6.19. Family tree of the various inversion sequences found in the left arm of chromosome 3 of *D. pseudoobscura* and the related species *D. persimilis*. The relationships have been established through the use of overlapping inversions, and sequences connected by a single arrow differ by a single paracentric inversion. The names assigned to the inversions indicate the place where they were first discovered. Those below Standard occur only in *D. pseudoobscura*, those above only in *D. persimilis*, with Standard being common to both species. (Dobzhansky, 1970.)

It would be unlikely, because of the breaks required to form them, that I gave rise directly to III, particularly if II were known to exist. A number of inversions in *pseudoobscura* and the related species *persimilis* can be shown to be related to each other, as Fig. 6.19 indicates. The Standard (ST) arrangement is the only one common to the two species.

Many of these inversions are present in the same natural population. Although no population contains all of the inversions common to a species, in *D. pseudoobscura* as many as eight have been found in a single test area. When relative frequency is plotted on a regional or seasonal basis, certain patterns emerge. For example, ST is more frequent at lower altitude in the Sierra Nevada mountains than is Arrowhead (AR), but a reversal in frequency takes place at high altitudes. Chiricahua (CH), which is present in the same population, shows no altitudinal variations. On other mountain ranges the pattern may be different. Furthermore, ST, at any given altitude, is higher in frequency when the temperature is warm than when it is cool, whereas CH tends to increase in frequency as cool weather comes on. This observation can be tested in the laboratory in population cages. A sample population of *D. pseudoobscura* consisting of 11% ST and 89% CH individuals was allowed to breed at random for 9 months at a temperature of 25°C. Within 4 months, or four generations, ST had trebled its numbers relative to CH, and at the end of the period equilibrium had been established with the population consisting of 70% ST and 30% CH chromosomes. In an equivalent population maintained at 16°C, no shift in frequency occurred. The fact that ST does not displace CH at 25°C and the presumed stability of the population at 70% ST and 30% CH argues for the positive maintenance of a chromosomal polymorphism and for the superiority of the heterozygous over the homozygous individuals. In fact, if the fitness of the several chromosomal combinations is computed— with fitness being based on preadult viability, developmental rate, fecundity, longevity, and rapidity of mating after emergence—and if the fitness of ST/CH is set at 1.0, then that of ST/ST is 0.89 and that of CH/CH is 0.41. At 16°C the fitness values would, of course, be altered. The homozygotes, ST/ST and CH/CH, would permit crossing over to take place within the inversion limits and lead to the breakup of possible favorable gene combinations; this would have an effect on their fitness values.

Not all species of *Drosophila* are similar to *D. pesudoobscura.* Those closely associated with the human—*simulans, virilis,* and *hydei*—show little or no inversion polymorphism. Where inversions are common, as in *pseudoobscura, persimilis, nebulosa,* and *busckii,* the inversions tend to accumulate in a single chromosome or chromosome arm, for reasons not immediately evident. *D. willistoni* illustrates another facet of inversion polymorphism. About 50 inversions have been identified in populations that range from Florida to Argentina, and in central Brazil, which seems to be the central point of dispersal and polymorphism, the mean number of inversions per individual is nine. From the center to the periphery of the range, the number decreases. This could result from the fact that in the center of distribution, time has permitted the establishment of variation and flexibility, and hence the occupation of a variety of ecological niches. The few individuals that make it to the periphery are basically experimentalists, carrying selected genotypes and inversions with them, and needing close inbreeding and homozygosity to establish genotypes capable of existing in a few limited niches initially available.

These observations permit certain conclusions. Inversions are not only tolerated but may actually be selected for in natural populations, presumably because each block of chromatin, which is kept relatively intact because of the paucity of crossing over, will tend to become allelically different from other comparable blocks of chromatin because of random mutations that may accumulate in the absence of recombination. Because selection operates not on single genes but on the total genotype, these partial gene sequences or supergenes, which vary in size according to the length of the inversion, can be selected for or against more effectively than can gene sequences which are constantly being reshuffled through crossing over. The size of the inversion is, therefore, of importance; it must not be too long or crossing over within it will alter the allelic sequence, but it must be long enough to accumulate a genetic difference of sufficient magnitude to allow for a differential response to environmental variations. The seasonal and regional studies show that such inversions have definite selective values, which in *Drosophila* at least are sufficient to cause the population to respond, in terms of frequency of individuals, to the changing character of the local environment. Whether the same type and size of inversions in a population having a longer generation time would act similarly is not known, but in *Drosophila* a population heterozygous for inversions has a genetic structure which, at any given time and place, is a function of the local environment. It is equally probable that inversions can serve as foci for species divergence, given sufficient time to accumulate genetic differences and the erection of barriers to prevent breeding with the parent population.

It is useful, at this point, to consider what is involved in the formation of an inversion when viewed at the level of the DNA molecule. Presumably, the unit involved is the DNA double helix. The antiparallel nature of the polynucleotide strands of a helix raises the following problem: Not only is the gene order inverted in the inversion segment, but the polynucleotide segments must switch strands, as indicated in Fig. 6.20. Since only one of the two poly-nucleotide strands appears to be read off in the transcriptional process, the message, in the form of mRNA, will be correspondingly altered. Many inversions lead to inviability when present in a homo-zygous state, and we may infer from this that inviability results from alteration in protein synthesis or structure. Some inversions, how-ever, produce no phenotypic effect in either a heterozygous or homozygous state. This fact suggests that the breakpoints for an innocuous inversion lie in the nonessential portions of the chromo-some, that the altered polarity of transcription is of no consequence, and, in addition, that there are numerous points along the chromo-some for the initiation of transcription, thus minimizing the possible effects of a change in polarity. On the other hand, some inversions show a dominant effect in the heterozygous state—the *Curly* inver-sion in *D. melanogaster* is an example—but it is uncertain whether this is due to the breakpoints being within essential portions of the

Fig. 6.20. The consequences of an inversion in a DNA helix. (I) A segment of DNA indicat-ing the antiparallel nature of the complementary nucleotide strands, and with arrows designating the break points. (II) The consequences of an inversion if the inverted segment is simply switched end-for-end; the antiparallel nature of the nucleotide strands would not permit such an arrangement. (III) The consequences if the inverted segment is not only switched end-for-end but also rotated 180 degrees before reinsertion. The information code would obviously be different, but the antiparallelism would be retained.

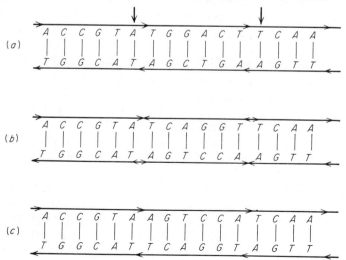

chromosome, to a loss of chromatin when the inversion occurred, or to genes being brought into new linkage arrangements.

The only inversions of note in the prokaryotes are those involving the *lac* region of the *E. coli* chromosome. These have been produced by infecting a bacterium deficient for the *lac* region with an *F* (sex) factor carrying a functional *lac* segment. Some of these segments become inserted in an inverted way as judged from the polarity of the genes during chromosomal transfer. The inverted insertion of the *lac* segment does not seem to interfere with its normal function.

Most, if not all, of the inversions detected in the human are of the pericentric type, possibly because the shift in centromere position makes them more easily identifiable. These have been found in the Y and in chromosomes 9 and 21. A particular inversion in the C group of chromosomes has appeared in four families, and it seems to have been passed on to the offspring at about the same frequency as the normal homologue. Although most of the inversions have been found in individuals with some kind of congenital abnormality, there is no certainty that there is a positive relation between the abnormality and the presence of the inversion.

Translocations

The process through which two chromosomes are broken and then mutually exchange blocks of chromatin is known as a *reciprocal translation* (Fig. 6.21). The two new chromosomes will function normally in division if each possesses a single centromere. Should the reunion be such as to produce chromosomes with two centromeres (dicentric) or with no centromeres (acentric), these will generally be eliminated, because they would fail to segregate properly at anaphase.

Translocations, like inversions, may exist in either a homozygous or a heterozygous state, provided the aberration in question is not associated with a lethal condition. Homozygous translocations generally behave as do the normal chromosomes from which they arose, except that new linkage groups become established. If they persist in nature they can give rise to a new chromosomal race. These are found sporadically in a great many species but are common to some genera of plants, such as *Oenothera, Paeonia,* and *Datura,* and to scorpions and roaches in the animal kingdom. Translocation heterozygotes are readily recognized by the characteristic pairing configurations seen during prophase and metaphase of the first meiotic division (Fig. 6.21). Homologies determine synapsis, so complete

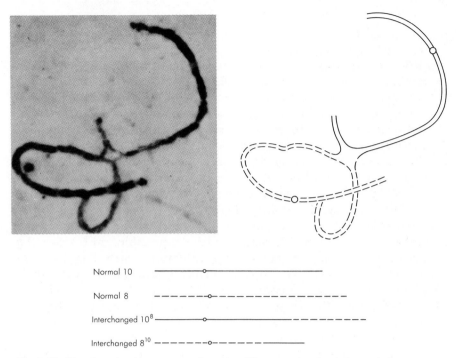

Normal 10 ————————○————————————

Normal 8 ——————○——————————————

Interchanged 10⁸ ————————○—————— ——————————

Interchanged 8¹⁰ ——————○———— ————————

Fig. 6.21. Translocation heterozygote in maize. The normal and translocated chromosomes are indicated below; above, pachytene configuration (photograph and diagram) characteristic of a translocation heterozygote. M. M. Rhoades, in *Corn and Corn Improvement* (New York: Academic Press, 1955), pp. 123–219.

Fig. 6.22. Translocation heterozygote in Sciara involving the X chromosome and chromosome II. In the diagram: c, centromere ends of the two chromosomes; f, the free ends. Courtesy of Dr. Helen V. Crouse.

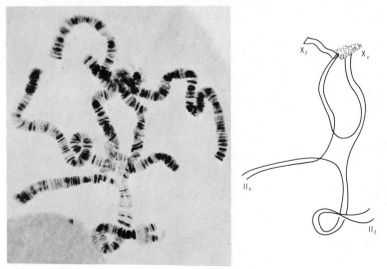

pairing requires that the chromosomes form the typical cross-like figure at pachynema, with the pairing partners being changed in the vicinity of the translocation breaks. This relationship permits a determination of the position of the translocation break, provided the synaptic behavior is exact. If chiasmata are formed in each of the paired arms, a ring of four chromosomes would result at metaphase; should one arm fail to form a chiasma, the result would be a chain of four. The degree of terminalization of chiasmata would determine the appearance of the ring or chain at diakinesis or metaphase. In dipteran salivary gland chromosomes, translocations are easily recognized, and the banded structure permits the exact points of breakage to be determined with certainty (Fig. 6.22).

When a translocation ring of four chromosomes reaches the metaphase plate, several arrangements are possible, and, since the arrangements determine anaphase disjunction, each has its own genetic consequences. The three possible arrangements, with their anaphase results, are illustrated in Fig. 6.23. In adjacent-1 and adjacent-2 types, the rings are so oriented that adjacent chromosomes go to the same anaphase pole, and the gametes formed, although different from each other, are both deficient and duplicated for certain regions of the chromosomes. The remaining arrangement differs in that alternate chromosomes go to the same pole, and the gametes formed are of two kinds, one having a normal set of chromosomes, the other a translocated set. Neither gamete is deficient or duplicated in any way, each having a full complement of the genes represented by the two chromosomes.

If the three types of orientation occurred at random, it would be expected that a translocation would lead to inviability in approximately two-thirds of the gametes. In plants, however, and particu-

Fig. 6.23. Diagram of alternate (left), adjacent-1 (middle), and adjacent-2 (right) arrangement of a translocation ring of 4 chromosomes. Segregation from these positions would lead to viable gametes from the alternate position, duplication and deficiency from adjacent-1, with homologous centromeres going to the same pole, and duplication and deficiency from adjacent-2, with nonhomologous centromeres going to the same pole.

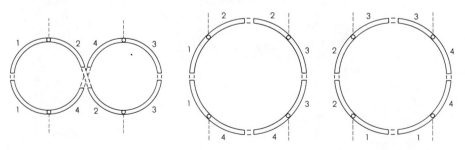

larly in maize, where the phenomenon has been studied extensively, sterility, as determined by the frequency of inviable seeds or defective pollen, is nearer to 50%. This can only mean that in maize the zigzag (alternate) metaphase arrangement, leading to movement of alternate chromosomes to the same pole, occurs with a frequency approaching 50%.

A number of factors govern the formation of a ring of four chromosomes and its orientation on the metaphase plate. The chromosomes have to be sufficiently long so that chiasmata will form in each arm; if the frequency of chiasmata per arm is less than one, chains of four or less are likely and distribution is irregular. If the chromosomes are too long, the chiasmata may be too numerous to permit complete terminalization, and the ring too flexible or too big to orient properly on the spindle. It follows that the pieces of chromatin exchanged must be of sufficient length to permit regular chiasma formation. The pieces of chromatin exchanged must also be equal in length to form symmetrical rings of four; asymmetrical rings have difficulty in orientation and segregation. The species having translocation complexes and a high degree of fertility and fecundity have metacentric chromosomes of similar length, with the breakpoints adjacent to the centromeres and presumably in the centric heterochromatin. A high degree of terminalization, or chiasma formation restricted to the ends of the chromosome arms, is also essential for proper orientation and adjacent disjunction and segregation at anaphase. Chiasmata localized at the ends of the chromosome arms means that crossing over in the proximal regions is minimized and linkage is greatly enhanced.

The translocations discussed have involved only two nonhomologous chromosomes. However, should the arm of one of the translocated chromosomes be involved in a second interchange with a third nonhomologous chromosome, a ring or chain of 6 would form at metaphase. A third interchange would give a ring of 8. The process can go on until the entire complement of chromosomes is involved and produces a translocation complex. This situation is encountered in *Rhoeo discolor,* a monotypic genus in the family Commelinaceae, in which a ring of 12 chromosomes can be formed (Fig. 6.24). Surprisingly enough, the plant is quite fertile. *Oenothera* and *Paeonia* represent the various stages that may be found. In *Paeonia californica,* which has 5 pairs of chromosomes, plants have been found with a ring of four ④ and three bivalents (3_{II}), and ⑥ and 2_{II}, a ⑥ and a ④, a ⑧ and 1_{II}, and a ⑩. *Oenothera* exhibits a similar tendency, which varies with the species. *O. hookeri* has seven normal bivalents, whereas other species may have all 14 chromosomes linked in a ring

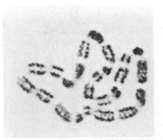

Fig. 6.24. Meiotic metaphase in *Rhoeo discolor.* Left, a ring of twelve chromosomes; right, an open chain of twelve chromosomes. Courtesy of Dr. K. Sax.

at meiosis. The disposition of these three genera to form translocations cannot be explained by any similarity of chromosome morphology. *Rhoeo* and *Oenothera* have medianly placed centromeres in their chromosomes, but the chromosomes of *Rhoeo* are large, those of *Oenothera* relatively small. *Paeonia* has large chromosomes, but one pair is acrocentric and the break points vary in position rather than being localized at the centromere, as in the other two genera.

A particularly interesting kind of translocation complex has been reported in two species of the East African mistletoe, *Viscum fischeri* and *V. engleri*. In *fischeri,* the male plants exhibit seven bivalents and a translocation chain of 9 chromosomes at meiosis ($2n = 23$). Alternate segregation from the complex is the rule so that gametes of 11 and 12 chromosomes are regularly formed. Since the female is $2n = 22$, and further since the translocation complex involves the sex-determining chromosomes, male plants of *fischeri* are characterized by a genome consisting of 14 autosomes plus $X_1 X_2 X_3 X_4 - Y_1 Y_2 Y_3 Y_4 Y_5$. The female, therefore, must be $X_1 X_1 X_2 X_2 X_3 X_3 X_4 X_4$, be homozygous for all chromosomes, and form 11 bivalents at meiosis. Whatever adaptive advantage there is to this type of permanent translocation heterozygosity is consequently male-restricted. How much of the *fischeri* complex is actually involved in sex determination is difficult to determine, *V. engleri* ($2n = 28$) is more variable, exhibiting a ring of 6 chromosomes plus 11 bivalents in one plant and 2 rings of 4 plus 10 bivalents in another; whether the sex chromosomes are involved in these translocation rings is probable but not certain.

Translocation complexes of the sort described for *Rhoeo* or *Oenothera* do not exist in the animal kingdom, although translocations are prevalent in many groups. A number of insect families— roaches, scorpions, grasshoppers, weevils, and beetles—are highly heterozygous for translocations, and White estimates that in the Australian grasshopper, *Keyacris (Moraba) scurra,* a newly formed translocation is found in one out of every 1,000 individuals, a rather high rate of chromosome alteration. However, a similar rate is also characteristic of the human, with many of the changes appearing among abortuses.

Although Down's syndrome is generally associated with trisomy of chromosome 21 in the human (3 chromosome 21s are present instead of the normal 2), and consequently with a somatic number of 47 chromosomes (see Chapter 7), the same phenotype is occasionally encountered in some individuals with only 46 chromosomes. Analysis reveals that at least two kinds of translocations are involved, but in both chromosome 21 participates. In the first, a translocation between chromosomes 14 (but also chromosome 15 as well) and 21 gives a karyotype that includes a normal chromosome 14, a translocated 14–21 chromosome, and a pair of 21 chromosomes. The segment of chromosome 21 represented in triplicate is assumed to be responsible for the syndrome, and the individuals correspond phenotypically to the chromosome 21 trisomies. In the other, a translocation presumed to have taken place between the two chromosome 21s has produced a new chromosome with both of its arms roughly similar in genetic content. When present in an individual also possessing a normal chromosome 21, the phenotype is again realized because of a triplicated region of this chromosome even though the somatic number is 46 instead of 47.

Virtually all combinations of human chromosomes have been shown to be involved in translocations, most of them of the reciprocal type [the Philadelphia (Ph') chromosome, involving a *9q22q* translocation would be an example of a nonreciprocal type]. The distribution of translocations, however, is not a function of the length of the chromosome or of a chromosome arm; a survey of breakpoints in a group of translocations uncovered in 75 different families indicates that the short arm of chromosome 2 is involved in these events far more frequently than expected. The same is also indicated for chromosomes in the B, D, and G groups, again with the short arm of each, as a rule, being more frequently involved than the long arm. Whether these regions are more susceptible to spontaneous breakage and reunion to form translocations than they are to restitution which would reestablish the original alignment of chromatin, or whether these regions are slower to heal when once broken, and hence are more available for illegitimate reunion, is difficult to establish. A distribution of breakpoints induced by quinacrine mustard and by X rays does not follow the same pattern, although the results are equally nonrandom in character; for example, the breaks induced by both agents are almost exclusively located in the pale bands following Giemsa staining, with two bands, one in Xq and the other in $9q$ near the centromere, being particularly sensitive to breakage. Only a single dark band in $2p$ has more than its share of induced breaks.

Individuals heterozygous for translocations, like those possessing inversions, show no disposition toward congenital abnormalities, indicating that rearrangement of the genes has no significant phenotypic consequences. However, the effects of irregular segregation from translocation complexes at meiosis are severe on the subsequent generation. In the 75 families surveyed, and in which 157 parents heterozygous for translocations were the parents of 571 offspring, 270 children, or 47.3%, received a balanced chromosomal complement resulting from alternate segregation (100 had a normal complement, 170 were balanced but heterozygous), 79, or 16.5%, received unbalanced genomes, and 22% of the pregnancies were stillborn or aborted. This is a significant amount of fetal wastage, which can be attributed to the fact that the amount of chromatin exchanged was unequal in length, was not confined to centric heterochromatin, and led to chains of 4 rather than to rings of 4, with irregular segregation being frequent. It is of interest to find that a heterozygous mother was three times more likely to be responsible for the unbalanced complement of an offspring than was a heterozygous father; the reasons remain unknown.

Another type of translocation that is prevalent among animal groups is that known as the *centric,* or *Robertsonian, fusion.* This occurs when two acrocentric chromosomes undergo a fusion or reciprocal exchange to form a metacentric chromosome. Since this leads to a reduction in chromosome number, the discussion will be deferred to the next chapter.

Translocations and Evolution. The evolutionary consequences of translocations are intriguing, with *Oenothera* and *Datura* providing most of the available information. It is clear that in a translocation heterozygote, the chromosomes involved in a ring of four are not inherited independently; the two normal chromosomes are inherited as a group, as are the two translocated ones, if a full complement of genes is to be transmitted. What were initially two independent linkage groups are now united into one, despite the fact that the chromosomes exist as independent entities and segregate independently of each other in mitosis. As the translocation complex increases in size, the number of independent linkage groups decreases until, as in many *Oenothera* species, the seven haploid chromosomes behave as one large linkage group. This has the aggregate effect of sharply reducing the component of genetic variability due to independent assortment of chromosomes. The genes and their alleles at the ends of the arms, where chiasma formation regularly takes place, undergo Mendelian segregation even though bound up in translocation complexes. It is only when a rare crossover takes place in the proximal regions of the arms that significant variability is released, pointing up the fact that these regions behave similarly to the "supergenes" that become isolated within inversion limits. As such, they represent foci for future divergence and evolution.

To appreciate the chromosomal situation in *Oenothera* it is necessary to understand the nature of the *Renner complex* and the *Renner effect*, so-named after their discoverer. *Oenothera* has a haploid complement of seven chromosomes with 14 pairing ends. If it is assumed that any particular end can be reciprocally translocated to any other end, then 91 different end combinations are possible (1-2, 1-3, 1-4, 1-5, 1-6, 1-7; 2-3 . . . 2-7; . . . 6-7) (Fig. 6.25). Most of these have been found in natural populations, a fact that itself indicates either that there is a strong propensity for *Oenothera* chromosomes to undergo translocation or that this is a structural alteration possessing considerable survival value.

Fig. 6.25. Chromosome pairing at meiotic metaphase in *Oenothera lamarckiana* which possesses a ring of 12 chromosomes and one normal bivalent. The haploid complements contain two different sets of genes, are referred to as the *gaudens* and *velans* sets, and when outcrossed to another form, yield quite different offspring. The ends of the chromosomes in each set have been identified; the end arrangement in the *velans* set is 1-2, 3-4, 5-8, 7-6, 9-10, 11-12, and 13-14. The *gaudens* set is 1-2, 3-12, 5-6, 7-11, 9-4, 8-14, 13-10. Only the chromosomes with ends 1 and 2 can pair to form a bivalent; the others can only pair at their ends. In a ring, therefore, the pairing would be 3-4, 4-9, 9-10, 10-13, 13-14, 14-8, 8-5, 5-6, 6-7, 7-11, 11-12, and 12-3, with the two 3 ends pairing to complete the ring.

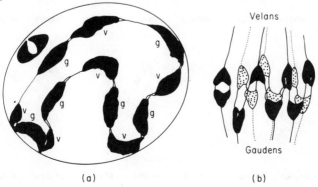

(a)   (b)

Of the known end combinations, seven in particular outnumber the others in frequency and distribution, and for a variety of reasons these 7 chromosomes are believed to constitute the ancestral arrangement from which others arose. A single translocation would give rise to a ring of 4 chromosomes in meiosis. By successive translocations, with each involving the previously existing ring and a nonhomologous pair, rings of 14 are built up. These would be unstable, however, in a population, for structural homozygosity would result through inbreeding, and individuals with rings of variable size and numbers of bivalents would be expected. Structural heterozygosity becomes enforced in one of two ways: (1) when lethal genes are included in the ring of chromosomes, with the end point being a ring of 14 chromosomes with different lethals in each of the two haploid sets of 7, or (2) when embryo sac or pollen competition is such that only certain combinations of chromosomes are found in the offspring. The species of *Oenothera* differ in the ways that they achieve heterozygosity.

Only alternate segregation from the ring leads to the formation of viable gametes; so each group of 7 chromosomes becomes, in essence, a single large linkage group with recombination of genes being confined to the pairing ends of each chromosome (Fig. 6.26). These linkage groups, consisting of 7 separate but collectively inherited chromosomes, are called *Renner complexes,* and each individual possessing balanced rings of chromosomes is a dual entity in

Fig. 6.26. The pattern of inheritance in a complex translocation heterozygote such as *Oenothera strigosa,* indicating that the haploid chromosomes form a linked group that are passed intact from one generation to the next and that there is no independent assortment of the chromosomes. The chain of chromosomes (center) customarily forms a ring although, as Fig. 6.24 shows, chains are also possible.

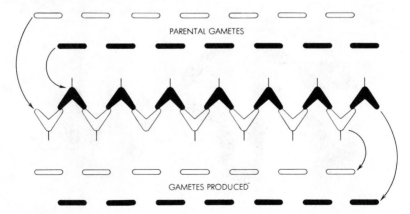

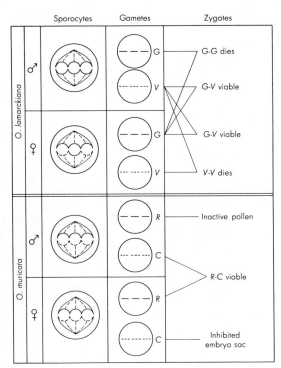

Fig. 6.27. Operation of the Renner effect in two species of *Oenothera*. In *O. lamarckiana*, zygotic lethality removes the homozygotes from the population; in *O. muricata*, homozygosis is prevented through gametic lethality.

the sense that each contains two complexes that, because of mutations and lack of proximal recombination, may differ appreciably from each other in genic content. For example, in *O. lamarckiana*, which has a ring of 12 plus a single bivalent rather than a ring of 14, the two complexes, called *gaudens* and *velans,* yield two very different species hybrids when outcrossed to other forms (Fig. 6.25). The individual plants of *O. lamarckiana*, which are generally self-pollinated, breed true, however, because a system of zygotic lethals prevents the formation of homozygotes and the breakdown of the complexes (Fig. 6.27).

The *Renner effect* is a further refinement of the system in that it reinforces the continued persistence of translocation heterozygosity. Commonly explained as a system of balanced recessive lethal genes which prevent homozygosity by the elimination of zygotes—a system which probably does function in *O. lamarckiana*—the Renner effect is actually based on gametic competition, coupled at times with selective gametic lethality and with a chemotropic recognition system. The circumstances in *O. muricata,* a European species, serve to elucidate the effect. The diploid complement of *muricata* consists of two complexes, *rigens* and *curvans,* the former promoting embryo

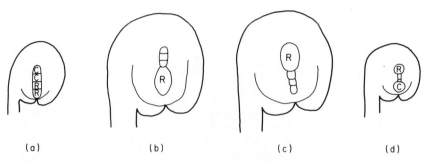

Fig. 6.28. The Renner effect as it operates in the ovules of *O. muricata*. A) The linear tetrad of megaspores resulting from meiosis in the megasporocyte, with the *rigens* (R) and *curvans* (C) complexes randomly located within the tetrad. B) The *rigens* complex in the micropylar position, where it is favored both by its position as well as by its competitive strength. C) The *rigens* complex in the chalazal position where its competitive strength is such as to generally overcome the *curvans* complex even when it is in a micropylar position. D) Rarely the *curvans* complex in the micropylar position can compete successfully, and form viable seeds, but these are always heterozygous, suggesting that while *curvans* can function in an egg nucleus it probably contains lethals which prevent homozygosity. The ability of one complex to compete with another is a matter of degree, and varies with the species of Oenothera and with the mixing of complexes in interspecific hybrids (after Cleland, 1972).

sac development and the latter found in active pollen grains. *Rigens* pollen is shrunken and has round pollen grains; *curvans* pollen is plump with spindle-shaped grains. *O muricata* is, therefore, heterogamous as regards its pollen, whereas *O. lamarckiana* is isogamous because all pollen grains are viable and active. *Muricata* lethals function at the gametic level in pollen but only at the zygotic level in *O. lamarckiana.*

On the female side of *O. muricata,* the *rigens* complex is the favored one, but on the basis of competition, not lethality. Embryo sac development in the *Onagraceae,* to which the *Oenotheras* belong, is unusual in two respects: Unlike the situation in most higher plants, the embryo sac develops from the micropylar megaspore rather than that at the chalazal end of the linear tetrad (Fig. 6.28), and it contains only four nuclei: that of the egg, two synergids, and a polar nucleus. There are no antipodals, and the endosperm, obviously, would be diploid instead of triploid. When the *rigens* complex finds itself in the micropylar megaspore, it is favored both by position and by the nature of its genome. When in the chalazal position, however, it can still outcompete the *curvans* complex to become the functional embryo sac, although in a low percentage of instances the *curvans* complex can succeed and form a normal embryo sac. The general inability of the *curvans* complex to compete with *rigens* in an ovular environment is, therefore, not due to a lethal gene, as such genes are generally viewed, but to a slower developmental tempo. Pollination of a *rigens* embryo sac by a *curvans* pollen tube will consequently preserve the heterozygosity.

The aspects of gametic competition, when both complexes function in pollen and embryo sacs, can be further illustrated by two examples. When the complexes *albicans* and *rubens* of *O. biennis* of de Vries are competing with each other, the former is found only in the embryo sacs and the latter is found in pollen, but when *rubens* is combined in a hybrid with the complex *tingens* from *O. rubicaulis*, *rubens* is found in the embryo sac and *tingens* is found in pollen. In *O. bertriana*, its *A* and *B* complexes function in both pollen and embryo sacs, but no homozygotes are recovered. Schwemmle, for example, found, on self-pollination, 4,300 seeds, all heterozygous, but in the ovaries 45% of the ovules were sterile. It might be supposed that the sterile ovules were the aborted homozygous ones, but when *O. bertriana* is crossed to another species with *C* and *D* complexes (*AB* X *CD*), the viable seeds were *AC, AD, BC,* and *BD* in equal numbers, and the same percentage of inviable ovules was observed. The homozygotes, therefore, were never formed and then

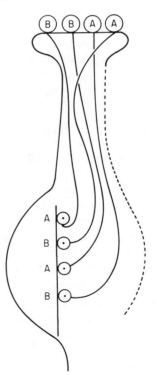

Fig. 6.29. Diagrammatic representation of a longitudinal half of the pistil of *O. bertriana*, indicating that A pollen selectively seeks out B ovules, and B pollen seeks out only A ovules, thus maintaining the heterozygosity of the complex. The stimulus recognized by the incoming pollen tubes is a chemical one emanating from the synergids in the embryo sac.

eliminated on the basis of lethality. Evidence now indicates that, on self-pollination in *O. bertriana,* the *A* and *B* pollen tubes grow equally well in the *AB* style but that pollen tube *A* will selectively enter embryo sac *B*, and pollen tube *B* will seek out its opposite number, embryo sac *A*. Such a recognition system (Fig. 6.29) can only be due to a chemical response, and it now seems evident that chemotropic signals emanate from the haploid synergids, diffuse into the style, and are there sensed, in a discriminating way, by the incoming pollen tubes.

Another example of competition among embryo sacs has been described in *Paeonia* by J. Walters. Each ovule contains several megasporocytes, each of which forms an embryo sac. An incoming pollen tube has, therefore, the opportunity for the selection of an appropriate genotype. This system, of course, differs from that in *Oenothera,* but the consequences are similar.

Alternate segregation from a ring of 14 is highly regular in *Oenothera,* and nondisjunction of chromosomes is not frequent enough to affect fertility in a marked manner. Regularity of segregation can be attributed to the fact that all of the chromosomes, despite many translocations, have median centromeres, a feature that permits greater maneuverability of chromosomes on the metaphase plate. The surviving translocations have been not only reciprocal, with the breakpoints in centric heterochromatin, but have also been approximately equal in length.

The North American *Oenotheras,* which include a number of Central American genera as well as the more widely dispersed species in the United States, provide an overall picture of the manner by which evolution has proceeded via the translocation-complex route. The advantages as well as the limitations of both the system and their present status are fairly evident; indeed, it might be said that the *Oenotheras* have achieved their present evolutionary position and diversity by means of a unique unorthodoxy. As Fig. 6.30 indicates, *O. hookeri* is an orthodox form, regularly forming 7 pairs of bivalents and occupying the western regions of the United States. *O. strigosa* is a translocation form adapted to the drier regions of the Rocky Mountains and the Great Plains; its two complexes are inherited through the egg (*alpha*) or the pollen (*beta*). *O. biennis I,* a mesic species, shares the *beta* of *strigosa,* together with its own *alpha* complex; *biennis II* is made up of the *alpha* complex of *strigosa* together with a *biennis beta* complex; while *biennis III* seems to have arisen from a cross between *biennis I* and *II*. *O. argillicola* and *O. grandiflora* are normal, bivalent-forming species which have given rise to *O. parviflora* through hybridizing.

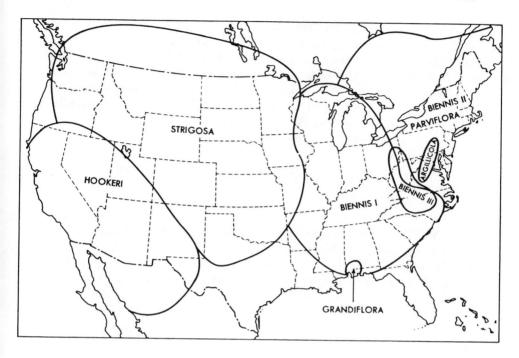

Fig. 6.30. The distribution of some species of Oenothera in the United States. *Oe. hookeri,*
grandiflora and *argillicola* are normal bivalent-forming species; all the others possess translocation complexes. See text for further explanation.

The *Oenotheras* have consequently utilized a number of genetic
strategies, some of them deleterious to a certain degree, and have
combined them into a unique set of systems that functions very well.
These devices are as follows: (1) reciprocal translocations, which
ordinarily lead to a reduction in fertility because of irregular segregation, but which, because of the maneuverability of the ring complex,
favors alternate segregation and high fertility; (2) a linkage system
which drastically reduced genetic recombination; (3) gametic and
zygotic lethals, coupled with gametic competition and selective
recognition patterns based on a chemotropic mechanism; and (4)
self-pollination, which in itself leads to inbreeding and a general lack
of vigor. In an open pollinated species, any one of these would be
genetically disadvantageous; yet the translocations lead to the
development of diverse linkage groups (the complexes); the lethals,
gametic competition, and the recognition system enforce structural
heterozygosity; and self-pollination prevents outcrossing, which
would tend to break up the complexes. On this basis, the course of
evolution in the *Oenotheras* is first the gradual formation of ring

complexes, followed by the incorporation of lethals and/or the establishment of competitive and recognition mechanisms, and then the shift from outcrossing to self-pollination. It is only in this order of occurrence that the entire system makes evolutionary sense when it is viewed in terms of its adaptive success: despite the complexity of the heterozygotes, their enforced uniformity is such as to make them into virtual clones. As a consequence, they share both the advantages and disadvantages of the asexually reproducing apomicts (see Chap. 8.).

Isochromosomes

Certain variant metacentric chromosomes have been found whose synaptic behavior would suggest that the two arms are of an identical genetic and structural nature. Such a chromosome is, in essence, a reverse duplication, with the centromere separating the two arms; the attached-X chromosome of *D. melanogaster* is a classical example. Its genetic content can be represented as *ABCDE · EDCBA;* its origin is uncertain, but it has proven to be a most useful experimental chromosome for studying patterns of crossing over.

The meiotic behavior of isochromosomes has been studied in maize, wheat, and rye; these have also been found in grasshoppers,

Fig. 6.31. Chromosome 5 in maize. Top, normal chromosome with centromere represented by circle; middle, fragmentation of chromosome through the centromere; bottom, isochromosome formed through duplication of the short arm to give a reverse tandem duplication. (*a*) Pairing of the short arm with a normal pair of homologues; synapsis is two-by-two although the centromeres may form an association of all three, (*b*) the isochromosome folding back to pair with itself at pachynema. After M. M. Rhoades, *Genetics, 25* (1940), 483–520.

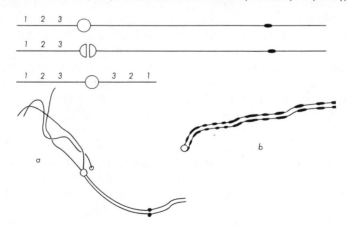

rodents, and humans. In maize an original chromosome 5 gave rise spontaneously to a telocentric chromosome consisting only of the short arm. This chromosome, like most telocentrics, was unstable, and it gave rise to an isochromosome (Fig. 6.31); when existing singly in a meiotic cell, it can fold back upon itself to permit the two arms to pair with each other, thus providing clear evidence of synaptic homology. Isochromosomes can arise from meiotic univalents which have a propensity to "misdivide" on the spindle, the plane of division being across the centromere instead of along the length of the chromosome; this creates telocentrics, and the two chromatids of a telocentric can then fuse at their centric ends to give an isochromosome. Isochromosomes can also form from the fusion of two acrocentric homologues, a circumstance which seems to have occurred in the rodent, *Sigmodon minimus.* Isochromosomes involving both the short and long arms of the human X and Y have been identified—they would be $Xp \cdot Xp$, $Xq \cdot Xq$, $Yp \cdot Yp$, or $Yq \cdot Yq$— and for some as yet unknown reason, even the X isochromosomes appear to have been paternally derived. An isochromosome involving the long arm of chromosome 21 has also been identified, and in the presence of a normal chromosome 21, it is responsible for the appearance of the Down's phenotype; another isochromosome, *13q · 13q,* is correlated with Patau's syndrome with its very severe mental and physical manifestations, a condition also induced by addition of a normal chromosome 13 to the diploid genome (i.e., trisomy-13). The supernumerary B chromosomes in rye and grasshoppers have been shown to give rise rather readily to isochromosomes; the ones in rye are of interest in that the one which involves the long arm reveals that the directed segregation of the B chromosomes in the first microspore division is dependent upon a locus in that long arm. An isochromosome made up of the short arm has no effect on such divisional processes.

Ring Chromosomes

Ring chromosomes have been described in a number of organisms from viruses to humans, but it is now evident that as a group they are similar only in a superficial, morphological sense; some are naturally circular, others arise as spontaneously formed aberrations, while still others can be induced by X rays, ultraviolet light, or a variety of chemicals. As previously discussed and illustrated in Fig. 2.12, the circularity of the lambda bacteriophage chromosome is due to the complementary redundancy of single polynucleotide sequences which terminate the chromosome. The chromosome of *E. coli* is similarly circular, as apparently are those of all bacteria and many of the viruses. The genetic maps of such chromosomes would, of course, be circular as well, for the topography of the map and the physical nature of the chromosome must be consistent with each other. However, as also pointed out in Chapter 2, such consistency is not found in the T4 bacteriophage: the map is circular, but the chromosome is a linear structure. The reason for this, as indicated in Fig. 2.10, is that the T4 chromosomes are a circularly permuted group of molecules, with each possessing a terminal redundancy amounting to about 2% of the total length. Partial enzymatic digestion followed by annealing produces circular chromosomes, providing proof of the terminal redundancy.

The ring chromosomes of higher organisms, however, must be viewed as aberrant types. They have been extensively studied in *D. melanogaster,* maize, and humans, and although they can be perpetuated in certain experimental stocks, they would, in the long run, tend to be eliminated in any situation where they compete with their normal linear homologues. This fact is evident from their behavior, for although they can replicate in such a manner as to give two unentangled rings, they also give rise to interlocked rings and to double-sized, dicentric single rings (Fig. 6.32). The fact that cleanly separating rings can be formed at all is, by itself, surprising since the DNA molecule is a helical structure with a complete twist for every 34 nm of length, and it replicates semi-conservatively. In maize, however, small ring chromosomes separate freely a fair percentage of the time; larger rings have a greater tendency to be interlocked or to form double-sized dicentric rings. Breaks in the polynucleotide strands must therefore occur during replication to allow the helical molecules to disentangle enough to be free to separate as chromatids during anaphase.

Ring chromosomes in the human have involved the X chromosome, chromosomes 1 (46, XX, 1r) and 3 (46, XXp$^+$, 3r), several members of the 6–12 group, and chromosomes 13, 17, and 18. When the X is involved, it is frequently lost since cells can exist in an XO condition, but when the major autosomes are involved, most of the cells show the presence of rings, their absence in all probability leading to lethality. All individuals possessing ring chromosomes exhibit phenotypic abnormalities, with microcephaly common when the larger autosomes are ring-shaped. It seems reasonable to suppose that the loss of chromatin accompanying the formation of the rings and the irregularity of transmission of these chromosome are responsible for the observed physical and mental deviations.

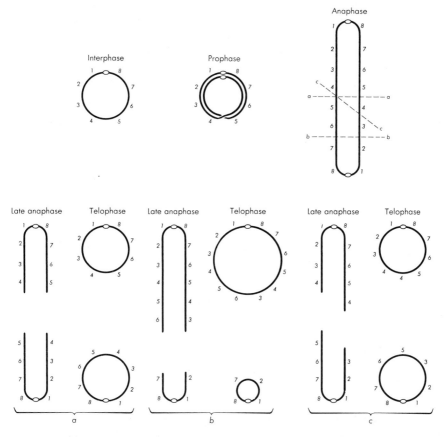

Fig. 6.32. Diagram illustrating how a ring chromosome, as a result of replication, can be transformed into a double-sized, dicentric ring. The ring would either be hung between the poles at anaphase or, as is illustrated, fragment into two pieces, each with a centromere. The brokens ends can re-fuse in telophase, to repeat the same cycle in the next cell division. Some ring chromosomes—particularly small ones—can separate cleanly at anaphase without forming a dicentric ring or becoming interlocked (not illustrated), but how this is accomplished is not known. B. McClintock, *Genetics, 26* (1941), 542-71.

BIBLIOGRAPHY

Anderson, R. P., et al., "Tandem Duplications of the Histidine Operon Observed Following Transduction in *Salmonella typhimurium,*" *J. Mol. Biol., 105* (1976), 201-18.

Benzer, S., "The Fine Structure of a Genetic Region in Bacteriophage," *Proc. Nat. Acad. Sci., 41* (1955), 344-54.

————, "On the Topography of the Genetic Fine Structure," *Proc. Nat. Acad. Sci., 47* (1961), 403-15.

Bloom, A. D., "Induced Chromosomal Aberrations in Man," *Adv. Human Genetics, 3* (1972), 95–192.

Carlson, E. A., "Comparative Genetics of Complex Loci," *Quart. Rev. Biol., 34* (1959), 33–67.

Carr, D. H., "Genetic Basis of Abortion," *Ann. Rev. Genetics, 5* (1971), 65–80.

Cleland, R. E., *Oenothera: Cytogenetics and Evolution.* New York: Academic Press, 1972.

Darlington, C. D., *The Evolution of Genetic Systems,* 2nd ed. New York: Basic Books, 1958.

Dobzhansky, Th., *Genetics and the Evolutionary Process.* New York: Columbia University Press, 1970.

Grant, V., *Plant Speciation.* New York: Columbia University Press, 1971.

Grant, V., *The Genetics of Flowering Plants.* New York: Columbia University Press, 1975.

Hansche, P. E., "Gene Duplication as a Mechanism of Genetic Adaptation in *Sacchromyces cerevisiae,*" *Genetics, 79* (1975), 661–74.

——, et al., "Gene Duplication in *Sacchromyces cerevisiae,*" *Genetics, 88* (1978), 673–87.

Ingram, V. M., *Biosynthesis of Macromolecules.* Menlo Park: W. A. Benjamin, 1965.

Jackson, E. N., and C. Yanofsky, "Duplication–Translocation of Tryptophan Operon Genes in *E. coli,*" *J. Bact., 116* (1973), 33–40.

McClintock, B., "The Production of Homozygous Deficient Tissues with Mutant Characteristics by Means of the Aberrant Mitotic Behavior of Ring-Shaped Chromosomes," *Genetics, 23* (1938), 315–76.

——, "The Stability of Broken Ends of Chromosomes in *Zea mays,*" *Genetics, 26* (1941), 234–82.

——, "The Relation of Homozygous Deficiencies to Mutations and Allelic Series in Maize," *Genetics, 29* (1944), 478–502.

Morgan, T. H., et al., "Constitution of the Germinal Material in Relation to Heredity," *Carnegie Inst. Wash. Yearbook, 31* (1932), 303–07.

Ohno, S., *Evolution by Gene Duplication.* New York: Springer-Verlag, N.Y., 1970.

Rigby, P. W., et al., "Gene Duplication in Experimental Evolution," *Nature, 251* (1974), 200–04.

Stebbins, G. L., *Chromosomal Evolution in Higher Plants.* Reading, Mass.: Addison-Wesley, 1971.

Steinberg, A. G., and F. C. Fraser, "Relations Between Chromosome Size and the Effects of Inversions on Crossing Over in the Third Chromosome of *Drosophila melanogaster,*" *Genetics, 29* (1944), 83–101.

Sturtevant, A. H., and G. W. Beadle, "The Relation of Inversions in the X-Chromosome of *Drosophila melanogaster* to Crossing Over and Disjunction," *Genetics, 21* (1936), 554–604.

Wiens, D., and B. A. Barlow, "Permanent Translocation Heterozygosity and Sex Determination in East African Mistletoes," *Science, 187* (1975), 1208–09.

Young, M. W., and B. H. Judd, "Nonessential Sequences, Genes, and the Polytene Chromosome Bands of *Drosophila melanogaster*," *Genetics, 88* (1978), 723–42.

7

Variation: Sources and Consequences Involving Chromosome Numbers

I f the chromosome numbers of a randomly selected group of individuals of a particular species were determined, in all likelihood they would be the same. This situation is to be expected, for species are reasonably constant biological entities, and it is not difficult to appreciate that this stability is related to a constancy in the numbers and kinds of genes and chromosomes. Any departure from this constancy is likely to be selected against since the imbalance brought on by the gain or loss of genes would most probably have a detrimental effect. The fact that a variety of neoplastic cells, lacking the regulatory control of normal cells, are often characterized by deviant chromosome numbers bears out the idea that genetic and developmental stability is a delicately maintained state, dependent upon a specific constellation of genes properly apportioned among the chromosomes of the genome. The chromosome number of a species is a significant biological datum in the definition of a species. But, even as the genes mutate or change in number through loss or addition, so also do chromosomes. The process is sporadic, for cell and chromosome divisions are remarkably regular phenomena, but variations do occur, and they are sometimes perpetuated to give rise to new chromosomal races and to new species.

Variation in chromosome number produces two types of individuals or cells: (1) those whose somatic complements are exact multiples of the basic haploid number characteristic of the species and (2) those whose somatic complements are irregular multiples of the basic number.

Individuals or cells of the first type are termed *euploid*. They may be haploid (monoploid), diploid, triploid, tetraploid, and so on, with the higher multiple members above the diploids being referred to collectively as *polyploids*. A tetraploid plant, for example, could produce diploid gametes and gametophytes in much the same manner as a diploid plant produces haploid gametes and gametophytes. However, an irregular meiotic distribution of chromosomes in some polyploids due to irregular synapsis and metaphase orientation often leads to comparable irregularities in chromosome number in the gametes and resultant offspring. In a polyploid series, therefore, the initial point of reference is the basic haploid complement of chromosomes, and the term diploidy, for example, implies that each chromosome of the haploid set is represented twice, even though homologous chromosomes may differ from each other in genic content or arrangement as a result of deficiencies, duplications, or inversions.

Individuals or cells having irregular chromosome numbers are *aneuploid*. A diploid organism that has gained an extra chromosome of the A set (B chromosomes, when present, are usually not thought of as contributing to aneuploidy) is designated as $2n + 1$; a comparable loss would be indicated as $2n - 1$. The terms *hyperdiploid* and *hypodiploid* have also been used to designate these two chromosomal conditions, but their use is not recommended, because they can be used with equal propriety to refer to $2n$ diploids possessing duplications and deficiencies, respectively. A combination of polyploidy and aneuploidy can lead to variable numbers of chromosomes in any single cell, organism, or race of organisms.

Aneuploidy

In general, the two members of a pair of homologous chromosomes regularly segregate during meiosis in a normal diploid to give a haploid set of chromosomes in a gamete or a spore; in mitosis, two

Table 7.1. Abnormal zygotes resulting from nondisjunction and loss of sex chromosomes in meiosis

Ovum	Sperm	Normal Meiosis		Nondisjunction in Meiosis I		Nondisjunction in Meiosis II		
		X	Y	XY	O	XX	YY	O
Normal Meiosis	X	XX	XY	XXY	XO	XXX	XYY	XO
Nondisjunction in Meiosis I or II	XX	XXX	XXY	XXXY	XX	XXXX	XXYY	XX
	O	XO	YO*	XY	OO*	XX	YY*	OO*

*The OO, YO, and YY combinations are probably nonviable, and do not appear among live births.

Table 7.2. Types, sex syndrome, and frequency of some aneuploid chromosome constitution in humans (see Table 7.1 and 7.4 for further information on sex-chromosome variants)

Chromosome Constitution	Sex	Syndrome	Frequency in Abortuses	Frequency at Birth
XO	F	Turner's syndrome: short stature, webbed neck, streak gonads, sterile, underdeveloped breasts, little or no mental retardation	1/18	1/3,500
XXY	M	Klinefelter's syndrome: underdeveloped testes, infertile, some breast enlargement, varying degrees of mental retardation	0	1/500
XYY	M	Tall stature, long limbs, fertile, aberrant social behavior debatable	?	1/2,000
XXX, XXXX, XXXXX	F	Mental retardation increases with increasing numbers of X chromosomes, normal phenotype	0	1/1,400 for XXX, less so for others
Trisomy-21	M or F	Down's syndrome: peculiarly folded eyelids and facial characteristics, stubby hands and feet, mental retardation, cardiac malformations, peculiar pigmentation of the eye, average survival of 16 years	1/40	1/1,000 to 1/600, increasing age of mother
Trisomy-18	M or F	Edward's syndrome; severe abnormalities, average survival of 6 months	1/200	1/4,500
Trisomy-13	M or F	Patau's syndrome: hairlip, cleft palate, cerebral, ocular, and cardiac defects, average survival of 3 months	1/33	1/14.500

cells of like chromosomal constitution are the usual consequence of somatic divisions. Exceptions occur, however, to give cells or organisms deficient or duplicated for a particular chromosome. A variety of mechanisms are implicated.

The phenomenon of *nondisjunction,* described by C. B. Bridges (Fig. 4.40) is a recognized source of aneuploid gametes (Tables 7.1 and 7.2). In his study of sex-linkage in *Drosophila,* Bridges recognized that the X chromosomes that failed to disjoin properly, giving $n + 1$ or $n - 1$ eggs, had not previously undergone crossing over. Failure to synapse and cross over results in the particular homologues arriving at the metaphase plate in an unpaired, or univalent, state. Univalents, as a rule, exhibit difficulty in achieving proper orientation between the poles, and they may fail to segregate, may pass randomly to one pole or the other, or may divide, as in mitosis, into their two chromatids. If the latter situation occurs, they cannot again divide and segregate properly at the second meiotic division. Univalency can also occur through precocious terminalization and loss of chiasmata prior to meiotic metaphase, but the frequency of this type of segregational error is probably rather low.

In mitotic cells, anaphase lag leads to aneuploidy, primarily through elimination. In cells that regularly eliminate certain chromosomes from the soma (see later), the chromosomes behave normally up to early anaphase, but the chromatids fail to pass to the poles as

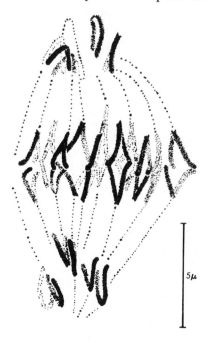

Fig. 7.1. A cleavage division in a cecidomyiid, *Mycophila speyeri,* at which a number of the chromosomes are eliminated. It appears that a selective anaphase tension fails to develop, and certain of the chromosomes do not migrate to the poles to be included in the telophase nuclei. B. Nicklas, *Chromosoma, 11* (1960), 402-18.

5μ

a result of insufficient mid-anaphase tension. This suggests a selective action of the spindle fibers connecting the chromatids to the poles (Fig. 7.1), or it may be due to some form of incomplete replication, but the mechanism of selectivity remains unknown. It seems possible, however, that the same mechanism might be responsible for the high degree of aneuploidy encountered in cultured cells, in diseased cells as found in both malignant and nonmalignant growths, and occasionally in the leucocytes of older human beings.

Aneuploid numbers in natural populations, involving chromosomes of the A set as distinct from the heterochromatic B chromosomes, are rather uncommon, but the spring beauty, *Claytonia virginica,* regularly has individuals with diploid numbers of 12, 14, and 16. Whether the haploid numbers of 7 and 8 represent legitimate examples of aneuploidy, with 6 being the basic number, is not known; it may well be that the genetic complement is the same in all of the diploid individuals, but somehow distributed among a different number of chromosomes in different individuals. If these are true aneuploids, their origin derives in all probability from triploid or tetraploid members of the population, these producing some unbalanced gametes which manage to survive and participate in the reproductive processes. The garden hyacinth, *H. orientalis,* is also tolerant of aneuploid numbers; the diploid number is believed to be 16, but numbers of 17, 19, 20, 21, and 23 have also been found. Aneuploids in animals are relatively rare, and such individuals tend to be quite abnormal; they can be readily produced in amphibia by heat shock, but their abnormalities are such that they could not survive under natural conditions.

Complementary gametes, that is, $n + 1$ and $n - 1$, will, on union with a normal gamete, produce individuals that are $2n + 1$ and $2n - 1$. These are commonly referred to as *trisomics* or *monosomics,* respectively; the particular chromosome in question is represented in triplicate or singly instead of in the usual disomic number. In trisomics, the extra chromosome is generally assumed to be a normal member of the basic haploid set unless proved otherwise by cytological or genetical means.

Trisomic Types. $2n + 1$ types are found occasionally among the offspring of diploid organisms, as described above, but they can more readily be obtained for study by crossing a diploid with a triploid. The irregular distribution of chromosomes in a triploid, because pairing is 2 by 2 at any given region along the length of the chromosomes, is such that unbalanced gametes are frequently formed. The possibilities are somewhat greater for producing unbalanced offspring in plants if the female is a triploid; the egg seems to be better able to survive genic imbalance than does an aneuploid pollen grain, presumably because the egg exists in a more or less protected environment and does not have to compete in the manner of a pollen grain. Aneuploid sperm are readily transmitted in animals without loss since there is no gametophytic generation to act as a gametic screen.

Trisomics have been studied extensively in *Datura,* maize, tomato, wheat, tobacco, *Drosophila,* and the human. They generally produce characteristic phenotypes, depending upon the genic content of the added chromosome; in humans, autosomal trisomics in particular lead to severe abnormalities while those involving the X and Y chromosomes, which are far more frequent, have a lesser, but nonetheless, noticeable effect (Tables 7.1 and 7.2). In *Drosophila,* the small chromosome 4 is the only autosome that can exist in a trisomic state to give a functional breeding fly, and the genetic ratios obtained among the offspring of triplo-4 females are in agreement with what would be expected on the basis of chromosome segregation and of nonelimination of unbalanced gametes. Thus, if a + + *ey* (*ey* = *eyeless*) triplo-4 female is crossed to an *ey ey* male, a 5:1 ratio of normal to *eyeless* flies is obtained. This results from the segregation of the three chromosome 4s to give four types of gametes in a ratio of 2+:2+*ey*:1++:1*ey*. The ratio also shows complete dominance of + since all flies with even one + allele have normal eyes despite the presence of two *ey* alleles. In plants, however, the unbalanced pollen grains either fail to mature properly, or they compete only with difficulty with normal pollen grains in pollen tube growth and hence in fertilization. In maize, for example, between 1 and 2 percent transmission of $n + 1$ pollen grains occurs, but in the embryo sacs the transmission is from 25 to 50 percent in a $2n + 1$ individual, the percentage being determined by which chromosome exists in a trisomic state. The percentage of transmission is also likely to be lower than expected on a random basis because the extra chromosome often fails to be included into a telophase nucleus as a result of synaptic upset or anaphase lag.

In maize and tomato the trisomics have been used effectively as a device for locating genes in particular linkage maps. For example, if an *mm* (where *m* is a mutant gene) plant is crossed to a trisomic with *m* in a heterozygous state, a distortion of the basic 1:1 backcross toward a 4:1 or a 5:1 will indicate that the chromosome in a trisomic state carries the gene in question.

The study of trisomics made in the Jimsom weed, *Datura stramonium,* has been particularly illuminating. *Datura* has a haploid number of 12 chromosomes, with the consequence that 12 different

Normal (2n)

| 2n+1·2 | 2n+3·4 | 2n+5·6 | 2n+7·8 |

| 2n+9·10 | 2n+11·12 | 2n+13·14 | 2n+15·16 |

| 2n+17·18 | 2n+19·20 | 2n+21·22 | 2n+23·24 |

Fig. 7.2. Seed pods, or capsules, of *Datura stramonium* as modified by the presence of an extra chromosome (2n + 1). Each chromosome, when in a trisomic state, produces a characteristic phenotype that can be recognized and includes a number of other features of morphology in addition to those depicted here. A. G. Avery, S. Satina, and J. Rietsema, *Blakeslee: The Genus Datura* (New York: Ronald Press, 1959).

trisomics are possible. All have been identified, and each produces certain characteristic phenotypic aspects that distinguish it from the others. This is to be expected, because we may suppose that the gene content of each chromosome is different, although how the extra chromosomal material determines the phenotype is unknown. Figure 7.2 illustrates the 12 possible types as they affect the morphology of the seed pod.

Each of the 12 normal chromosomes has been designated by labeling their ends. Thus the chromosomes become 1·2, 3·4, 5·6, . . . , 23·24. The trisomic Rolled, for example, has the 1·2 chromosome in triplicate. Such trisomics, in which an unmodified chromosome is present in triplicate, are referred to as *primary trisomics*.

The frequency of transmission of $n + 1$ gametes through the egg in *Datura* can be determined by selfing the trisomic forms. The average transmission rate is about 20%, but for particular chromosomes, the rate varies from 2.73% to over 31%, a variation that is unrelated to the size of the added chromosome but is undoubtedly determined by genic content and its effect on pollen viability and transmission. Most of the trisomics in *Datura* are not transmitted through the pollen.

Secondary trisomics have also been found and described in *Datura*. A primary trisomic, defined as one with the added chromo-

412

some being a normal member of a normal complement, can form a chain trivalent in meiosis, a "frying pan" configuration, or a bivalent and a univalent (Fig. 7.3). Secondary trisomics, however, are identifiable by the presence of rings of three chromosomes, although chains of three, or a bivalent and a univalent, are possible as well as is the "frying pan" arrangement. The extra chromosome, therefore, is an isochromosome, at least so far as its pairing ends are concerned. If there are 12 possible primary trisomics in *Datura*, the haploid number being 12, there obviously can be 24 secondary trisomics. All have been discovered and identified. *Tertiary trisomics* are also known, the extra chromosome in this instance being one which, as a result of a translocation, possesses the ends of two nonhomologous chromosomes. Chains of five chromosomes, or a double "frying pan" configuration, are therefore possible and are uniquely characteristic, with various combinations of chains, bivalents, univalents and "frying pans" produced in one meiotic cell or another.

With the discovery of chromosomal abnormalities in humans, and the relative ease with which these may be identified, it is understandable that a great deal of interest has arisen and that a vast amount of information has been generated. Much of the interest has been focused on spontaneous abortions. It has been estimated that 15% of all conceptions terminate as abortions, but this figure is undoubtedly low since it does not include preimplantation losses which are known to occur but are difficult to quantify. It is known that there is about a 30% zygotic loss in laboratory and domestic animals, and it would appear that the human is no exception in this matter.

Fig. 7.3. Types of meiotic configurations found in the three kinds of trisomics at metaphase. The chain of 3 and the so-called "frying pan" configuration are typical, but not unique, in a primary trisomic. A ring of 3 positively identifies a secondary trisomic, although chains of 3, a "frying pan", or a bivalent and a univalent, are also possible as well. In the tertiary trisomic, a chain of five and the double "frying pan" are identifying characteristics, although various combinations of chains, bivalents, univalents and "frying pans" are also possible.

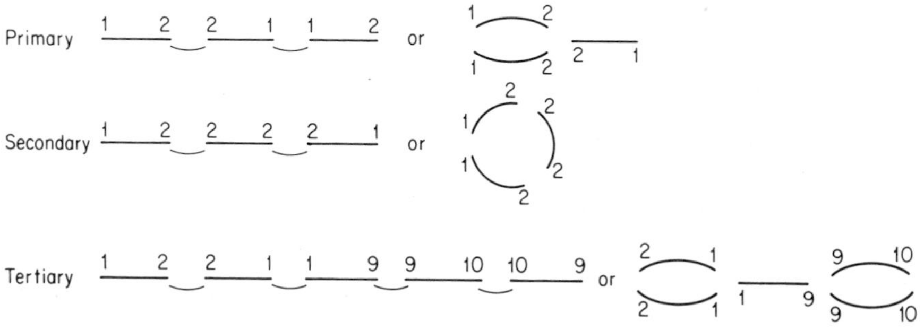

The process of abortion is, therefore, a highly effective and selective screen for the removal of abnormal phenotypes from the human population; depending upon the chromosome(s) involved, the abortion rate for abnormal genomes is between 60% and 100% effective.

The most common group of chromosome anomalies are the trisomics. Among abortuses, every chromosome has been found to be in a trisomic condition, with those of the D, E, and G groups being most common. These are also the most common autosomal trisomics among live births as well. Collectively, the autosomal trisomics comprise 47.8% of all abnormal fetuses, and 11.9% of all abortions, but those involving the A, B, C (exclusive of the X), and F groups of chromosomes, as well as chromosome 16, never come to full term and have been found only in early fetuses, suggesting that they induce such drastic genetic imbalance that survival is out of the question. Of the E group trisomics, which make up about 15% of the chromosomally abnormal fetuses, chromosome 16 is by far the most common, but only chromosome 18 trisomics can be carried to full term (Table 7.2).

Trisomy-21, which is responsible for the Down's phenotype, along with chromosomes 22, 13, and 15, make up the largest group of human autosomal trisomics. These chromosomes have in common the fact that each possesses a nucleolar organizer. Since the female meiocyte remains in a more or less suspended state of meiosis (in the diplotene, or dictyotene, state) from a late prenatal period until the egg is shed some 12 to 50 years later, and further since nucleoli tend to aggregate with each other, it may well be that the satellited chromosomes experience more difficulty in segregating at anaphase I than do the nonsatellited homologues, and hence undergo nondisjunction more frequently. This suggests, of course, that the source of meiotic error leading to trisomy-21 is maternal, but more recent studies, utilizing the banding techniques so that the source of each chromosome can be ascertained, reveals that about one-quarter of the trisomy-21s may be related to paternal nondisjunction, and hence unrelated to any kind of nucleolar involvement.

On the other hand, a definite increase in the frequency of trisomy-21 births is correlated with an increase in the age of the mother. This would suggest that there is an increased frequency of nondisjunction of chromosome 21 as the mothers age, a pattern that has also been observed with aging mice, where an increasing frequency of univalents is age correlated. The cause is not known, but the fact remains that in humans the chance of a trisomy-21 birth to a 45- to 59-year-old woman is 50 times greater than it is to a mother aged 15 to 19 (Fig. 7.4). The frequencies of other trisomies similarly

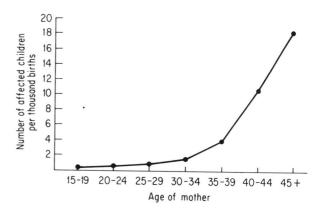

Fig. 7.4. Incidence of Down's syndrome in the children of mothers of different ages.

increase with an increasing age of the mother, but not in such a dramatic fashion.

A number of partial trisomies have been identified. One of these has previously been mentioned (p. 391): This involves the long arm of chromosome 22 and is correlated with the incidence of chronic myeloid leukemia. Others involve the short arms of chromosome 4, 5, 9, and 10, while the long arm of chromosome 21, translocated to some other chromosome, and existing in the same nucleus with a pair of normal chromosome 21s, brings on Down's syndrome. About 5% of individuals with this phenotype result from this kind of partial trisomy, and it locates the source of imbalance in the long arm.

It is of interest to point out that trisomy in the human, in addition to determining more-or-less diagnostic phenotypes, also seems to be correlated with the retention of fetal or embryonic characteristics. The excess autosomal material appears, in some manner, to have blocked or slowed down the normal progress of differentiation and growth required for the attainment of adult form and behavior, and particularly that related to sexual and mental maturation. It is generally assumed that differentiation is directed by only a fraction of the coded information present in the genome, the remainder being transcriptionally suppressed. The extra chromosomal material, therefore, interferes with the programmed information necessary for normal growth and development, and fetal properties tend to persist.

Monosomics and Nullisomics. The loss of a chromosome in a diploid organism is, as expected, an even more deleterious genetic change than is the addition of an extra chromosome. In fact, monosomic individuals ($2n - 1$) among diploids are relatively rare unless the chromosome involved is very small or lacking in significant genetic content. Haplo-4 individuals in *Drosophila melanogaster* are monosomic for the very small chromosome 4. Even in such species as tomato and corn, where monoploid individuals are often found in field plots, monosomics are very infrequent. The fact that the loss of an entire genome is less deleterious than the loss of a single chromosome from a diploid genome points to the importance of genetic balance in determining the viability and survival of the diploid organism. Among haploid organisms, loss of a chromosome from the basic genome would undoubtedly be a lethal condition, but the haploid generation of diploid plants can transmit deficient gametes with differential efficiency; as with $n + 1$ gametes, $n - 1$ gametes are poorly transmitted through the male side, but with a somewhat higher, although variable, efficiency through the embryo sac.

The only viable monosomics in man are the XO individuals, with their viability due to the existence of a dosage compensation mechanism which permits a modified and subnormal development accompanied by complete sterility (see next section). Autosomal monosomy is therefore an inviable state in man, but that it does occur, probably as a result of nondisjunction, can be judged from the fact that it is not an uncommon state among spontaneous abortuses.

Again, as might be expected on the basis that a polyploid can tolerate the loss of a chromosome with greater ease than can a diploid, monosomics are much more commonly found in polyploid species. All 24 monosomic in the tetraploid tobacco, *Nicotiana tabacum,* and all 21 in the hexaploid wheat, *Triticum aestivum,* have been identified and extensively studied. These plants generally are not as robust as those having a normal complement of chromosomes, and they show a variable phenotype according to which chromosome is missing, but they can survive and reproduce with fair success. Only a few $n - 1$ pollen grains are viable, and when viable succeed only rarely in affecting fertilization, but monosomic eggs in tobacco and wheat average about a 50 percent efficiency as compared to euploid eggs, the particular frequency of success being dependent upon the genetic content of the missing chromosome. Their usefulness to the plant breeder is comparable to that of trisomics in diploid species: they make it relatively easy to assign genes to particular linkage groups. Thus, in wheat for example, a newly arisen mutation can be located with little difficulty by crossing the mutant plant (as a male) to all of the 21 monosomics, and determining the patterns of segregation that deviate from the expected.

The *nullisomic* $(2n - 2)$ is a particular kind of double monosomic in that it lacks both homologues of a given chromosome pair.

Just as there are 21 possible monosomics in wheat, so are there 21 possible nullisomics, most of which have been identified. These are obtained by selfing a monosomic: $2n - 1 \times 2n - 1$ yields an occasional $2n - 2$. The usefulness of such nullisomics is indicated by the fact that a nullisomic strain of wheat which lacked a pair of chromosome 5's from the B genome permitted the initial identification of a gene which governs the diploidization of a hexaploid species (see p. 437).

Aneuploidy of Sex Chromosomes. The sex of an individual is the most obvious structural and behavioral phenotypic trait among bisexual organisms; thus a substantial amount of attention is focused on the chromosomes associated with sex. Gain or loss of X and/or Y chromosomes has been frequently observed, and the aneuploid types permit an interpretation of the roles of the X, Y, and autosomes in sex determination. In general, and apart from the effect of the X and Y on the reproductive system, the developmental anomalies caused by variation in these two chromosomes is less severe than those induced by autosomal trisomies or monosomies, with the anatomy and physiology of the affected individual being more compatible with normal life.

As Table 7.3 indicates, a variable number of Xs and Ys allows us to state that the X chromosome in *Drosophila* is female-determining, while the autosomes possess a male-determining influence. The particular sexual phenotype depends, therefore, upon a ratio between the Xs and the autosomes. The Y chromosome is without influence in this regard, although it does control male fertility; that is, an XO *Drosophila* is a normal appearing male, but it is totally sterile. In humans, however, as in the plant *Melandrium*, the Y chromosome is clearly male-determining (Table 7.4). Thus testes (even though they may be nonfunctional) are found in individuals with XY, XXY, XYY, XXXY, and XXXXY karyotypes, but they are lacking in those with XO, XX, XXX, and XXXX karyotypes. The influence of the Y chromosome is probably at the time of gonadal differentiation as well as during gamete formation (Fig. 3.41). In mice, for example, removal of the undifferentiated gonadal tissue by embryonic castration leads to female sex expression even in those individuals possessing an XY karyotype, while loss of parts of the Y chromosome results in incomplete male determination and defective testicular development. The mammalian situation appears to be similar to that found in the plant *Melandrium*, where the Y chromosome also exerts a positive male-determining influence even in the presence of multiple X chromosomes (Fig. 7.5).

Studies conducted on some XYY individuals among humans have led to a further clarification of the function of the Y in the development of the reproductive system. Phenotypically, these are generally tall, long-limbed males whose social behavior, said by some to be aggressively antisocial, is still being debated; they are fertile, but they do not produce viable XY or YY sperm, as indicated by the fact that their male offspring are normal. A question about the influence of the Y arises since some humans with no detectable Y chromosome present are 46,XX true hermaphrodites with some testicular tissue; others are 46,XX and show the Klinefelter's syndrome. There seems to be little doubt, however, that the histocompatibility Y antigen (H-Y antigen) is a manifestation of the testis-determining gene and is male restricted. XYY males have significantly more antigen than normal XY males, and through the use of isochromosomes of the $Y-Yq \cdot Yq$ or $Yp \cdot Yp$—the gene has been localized in

Table 7.3. Chromosome constitution in *Drosophila* and *Melandrium*, indicating that sex determination in the former genus is a matter of balance between the X chromosomes and the autosomes, whereas in the latter it is a balance between X chromosomes and Y chromosomes

Drosophila			Melandrium		
Chromosome Constitution	Sex	X/A Ratio	Chromosome Constitution	Sex	X/Y Ratio
2 A XXX	Superfemale	1.5	2 A XX	Female	0.00
			2 A XYY	Male	0.5
2 A XX 2 A XXY 3 A XXX 4 A XXXX	Female	1.0	2 A XY 3 A XY 4 A XY 4 A XXYY	Male	1.0
3 A XX 3 A XXY	Intersex	0.67	4 A XXXYY	Male	1.5
4 A XXX	Intersex	0.75			
2 A X 2 A XY 2 A XYY 4 A XX	Male	0.50	2 A XXY 3 A XXY 4 A XXY 4 A XXXXYY	Male (occasional ♀ blossom)	2.0
3 A X	Supermale	0.33	3 A XXXY 4 A XXXY	Male (occasional ♀ blossom)	3.0
			4 A XXXXY	Hermaphrodite (occasional ♂ blossom)	4.0

Table 7.4. Some sex chromosome variants in human beings*

Chromosome Number	Sex-chromosome Constitution	Number of Sex Chromatin Bodies (Barr)	Fertility	Remarks
		Male Habitus		
46	XY	0	+	Normal male
47	XYY	0	+(−)	Male
47	XXY	1	−	Klinefelter's syndrome; mentally retarded
48	XXXY	2	−	Like Klinefelter's
49	XXXXY	3	−	Severely retarded
49	XXXYY	2	−	Like Klinefelter's
48	XXYY	1	−	Like Klinefelter's
		Female Habitus		
46	XX	1	+	Normal female
45	XO	0	−	Turner's syndrome
47	XXX	2	±	Triple-X syndrome; mentally retarded
48	XXXX	3	?	Mentally retarded
49	XXXXX	4	?	Mentally retarded
		Mosaics		
45/46	XO/XY	0	−	♂ or ♀ appearance: ♂ pseudohermaphrodite
45/46	XO/XX	0/1	−	Turner's
46	XX/XY	0/1	−	True hermaphrodite
46/47	XX/XXY	1	−	Klinefelter's
46/47	XX/XXX	1/2	−	Some are true hermaphrodite
45/46/47	XO/XX/XXX	0/1/2	−	Like Turner's
		Structural Changes		
45 + fragment	Xy	0	−	"Deleted Y"; may have ♂ or ♀ habitus
45 + fragment	Xx	1 (small)	−	"Deleted X"; no severe mental defect; ♀
46	XX	1 (large)	−	"Isochromosome X"; like Turner's

*See O. J. Miller, *Am J. Obstet. Gynecol.*, **90** (1964), 1078–139, for further examples.

the middle of the short arm. It seems probable, from studies on other mammals, that while the Y may be concerned with testicular development, it does not control the entire male phenotype.

The mammalian X chromosome also plays a positive role in sex determination, although with varying degrees of influence in different species. In mice, XO individuals are phenotypically normal and fertile females, but they are less vigorous than their XX mates. The XO phenotype in humans, on the other hand, while still female, is

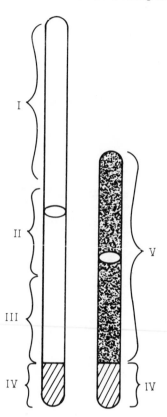

Fig. 7.5. Diagram of the X (right) and Y (left) chromosomes of Melandrium. I, female suppressor region in the Y which when absent leads to bisexual development; II, essential male promoting region which when absent leads to female development; III, essential male fertility region which when absent causes anthers to abort; IV, pairing region common to both chromosomes; V, differential portion of the X which has no homology with any portion of the Y. Whether the differential portion of the X has female promoting properties is uncertain, just as it is in humans.

highly aberrant, exhibiting Turner's syndrome which is characterized generally by short stature, infantilism, streak gonads which are incomplete and sterile, multiple pigmented nevi (birthmarks), and a webbing of the neck, in addition to other less obvious defects. Mental retardation, however, is not a significant aspect of the syndrome, and it does not appear unless the Xs rise above the normal female number. Study of a number of individuals possessing aberrant X chromosomes, and who also exhibit a modified Turner's syndrome, suggests that the syndrome is due to monosomy of genetic loci in the short arm of the X, which may be homologous with loci in the Y chromosome. This would explain why there is a range of phenotypic expression from normal female to Turner's syndrome, and from normal male to Turner's syndrome, depending upon the degree of sex-chromosome deficiency intermediate between XX and XO and between XY and XO karyotypes, respectively.

X chromosome aneuploidy is about twice as frequent in male newborns as in females; it is very probable that this discrepancy is due to the fact that about 98% of all XO individuals are incapable of completing gestation. The XO condition is the second most common cytogenetic abnormality found in abortuses, comprising about 25% of the total. In 75% of XO individuals, it is the paternally derived chromosomes that are missing. It is still unclear why XO individuals are so frequently inviable, since one X is destined to become inactivated (lyonized) early in embryonic development, but it seems likely that inactivation does not occur until one or more critical developmental processes are completed and that these processes are vital for further embryogeny.

Centric Fusion and Fission

Aneuploidy is considered to be a numerical departure from the normal diploid complement of chromosomes, with a gain or loss of whole chromosomes of the A set. Variable chromosome numbers can occur, however, without apparent gain or loss of critical genetic material by the processes of *centric fusion* or *centric fission*. As the terms imply, the centromeres are involved in each instance. Centric fusion is a special but fairly common type of rearrangement of chromatin in which two acrocentric chromosomes fuse to form one larger chromosome with a more or less medianly placed centromere (Fig. 7.6). Centric fission is the reverse of fusion; a two-armed chromosome is transformed into two acrocentrics. The fact that both processes can occur, and have been documented in a wide variety of both plants and animals, leads one to assume that terminal centromeres can unite with, or dissociate from, each other without impairment of centromeric function, and that a functional centromere, constructed as it is of several chromomeres (Fig. 3.11), can cleave in two, and with both parts retaining the ability of anaphase movement. The latter situation is borne out by McClintock's study of the ring chromosome resulting from breaks in chromosome 9 of maize, with one of the breaks being through the centromere (Fig. 6.4); both the depleted chromosome 9 and the ring chromosome were fully capable of anaphase movement. In this instance, however, the centromeres were flanked on either side by additional chromatin, thus conveying stability to the diminished centromeres; a cleaved centromere, on the other hand, is unstable when it exists in a terminal position (Fig. 6.31). Stable terminal centromeres, nevertheless, have been described in both natural populations of grasshoppers, and in Chinese hamster cells *in vitro*.

Fig. 7.6. Possible ways by which a centric fusion or fission can take place with a minimum, or no, loss of essential genic material. Possibilities I, II and III assume that the acrocentric chromosomes actually have a short stretch of chromatin as a second arm, and which can be lost without serious phenotypic effect; this need not be so since stable truly telocentric chromosomes have been described. Possibility IV assumes that the centromeres are truly terminal, and that there is no second short arm; whether the centromeres can fuse with, or dissociate from, each other without actual breaks through the centromeric regions is an unlikely, although theoretically possible, event. If centric fissions occur from the metacentrics in I, II and III, the lost fragments could not, of course, be recovered. Possibility II is the most unlikely means of producing a stable metacentric; such dicentrics are rarely stable entities since their stability and continued persistence would depend upon both centromeres always being oriented to the same pole; even so, these dicentrics have been described in man, wheat, grasshoppers and several species of blackflies.

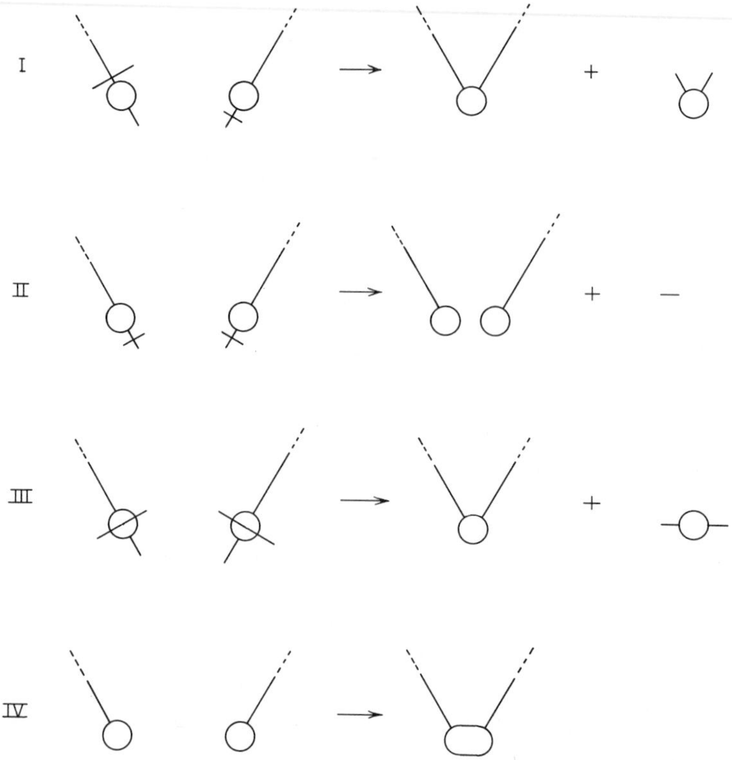

While there is little question that both centric fusions and fissions occur, the mechanisms which lead to these rearrangements are not fully understood. Fig. 7.6 diagrams the several possibilities of occurrence. What is involved is the organization of the centromere and the extent to which it can be broken into functional parts. With the exception of possibility IV in Fig. 7.6, which involves fusion or fission without actual breaks, and which offers the least likely route of change since the ends of chromosomes (telocentromeres?) are stable entities, the others are known to have occurred. Possibility I has been recently described in a cultivar of celery (*Apium graveolens* var. *dulce*); the small fragment arising from the fusion process has not been lost, but rather has persisted in the genome where it sometimes associates with the derived metacentric in meiosis. Possibilities II and III differ only in the placement of the breaks; both lead to fused centric regions, with II resulting in a true dicentric, the two centromeres being separated by a segment of chromatin, and III resulting in a centromere that could be smaller, the same size, or larger than a normal centromere, depending upon where within the centric regions the breaks occurred. Possibility III has been described in the Australian grasshopper, *Percassa rugifrons,* by John and Freeman, while Moens has analyzed a particularly interesting case in the grasshopper, *Neopodismopsis abdominalis,* which may be an example of either possibility II or III. In pachytene stages, centromeres of the acrocentric chromosomes lie against (or are attached to?) the nuclear membrane, where they appear as deep staining knobs, and from which the synaptinemal complex arises. A metacentric formed from the fusion of two acrocentrics shows two such knobs, or a knob about twice the size of a single one. In addition, and reinforcing the notion that the metacentric possesses either two centromeres or a centromere twice the size, it has been shown (by Moens) that at meiotic telophase the acrocentrics of approximately the length of the arms of the metacentric are connected to the poles by 30 to 40 microtubules, while the metacentric chromosomes are drawn to the poles by about 70 to 80 microtubules, a number expected if both centromeres were intact, or nearly so. In addition, Moens has also shown that the smallest of the acrocentrics has only half the number of microtubules connecting it to the pole as that characterizing the larger members of the genome; whether this is due to a smaller centromere, or is related to length of the chromosome is not clear.

The two processes of fusion and fission, therefore, reduce or increase the chromosome number without a change in the number of major arms of chromatin, giving rise, within a species, to what has been termed a Robertsonian type of chromosomal polymorphism; Robertson's law states that the chromosome number may vary but the number of chromosome arms remains constant. Care must be taken in order that centric fusions and pericentric inversions not be confused with each other. In the latter case only one chromosome is involved, two in the former, but unless one is certain of the number of basic arms, and of homologies (which are relatively easy to determine in the banded polytene chromosomes of the Diptera) confusion can readily arise. In those species lacking polytene chromosomes, the techniques of banding through Giemsa staining offer the means of discrimination.

Robertsonian fusions are far more common in animals than in plants. In the common English shrew, the total number of autosomal arms is 36, but from individual to individual the diploid number of chromosomes ranges from 22 to 25 in females and 22 to 27 in males. The karyotypes are depicted in Fig. 7.7, and it is evident that the smaller autosomes can exist as acrocentrics or metacentrics. In the marine snail *Thais lapillus*, the monoploid number ranges from 13 to 18. In the 18-chromosome form, all of the chromosomes are acrocentrics, whereas the 13-chromosome form possesses 5 metacentrics and 8 acrocentrics (Fig. 7.8). Intermediate forms with 1, 2, 3, or 4 metacentrics have also been found in the same population, leading, of course, to a local chromosomal polymorphism. An interesting ecological preference is also exhibited by these chromosomal races, the 18-chromosome form being found in the relatively constant environment below low tide-level, and the 13-chromosome form in the more variable high-tide zone. The intermediate numbered individuals occur in the intervening tidal areas. Just how centric fusions

Fig. 7.7. Chromosome polymorphism in the common shrew (*Sorex araneus*). The two individual chromosome complements depicted here have a somatic number of 24, but it shows that the smaller chromosomes may appear as acrocentrics or fuse to form metacentrics. After C. E. Ford, J. L. Hamerton, and G. B. Sharman, *Nature, 180* (1957), 392-93.

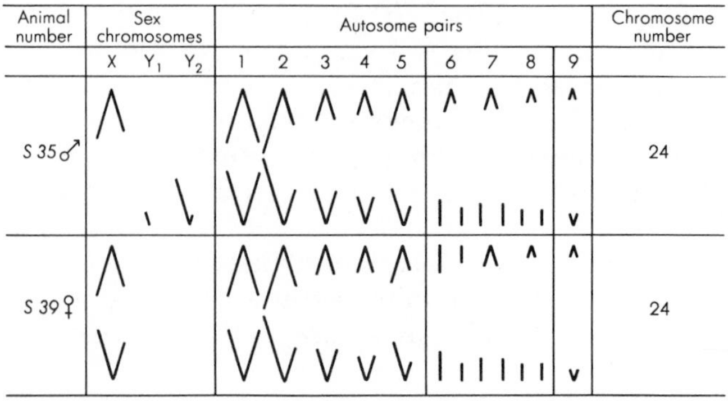

Animal number	Sex chromosomes			Autosome pairs									Chromosome number
	X	Y₁	Y₂	1	2	3	4	5	6	7	8	9	
S 35 ♂													24
S 39 ♀													24

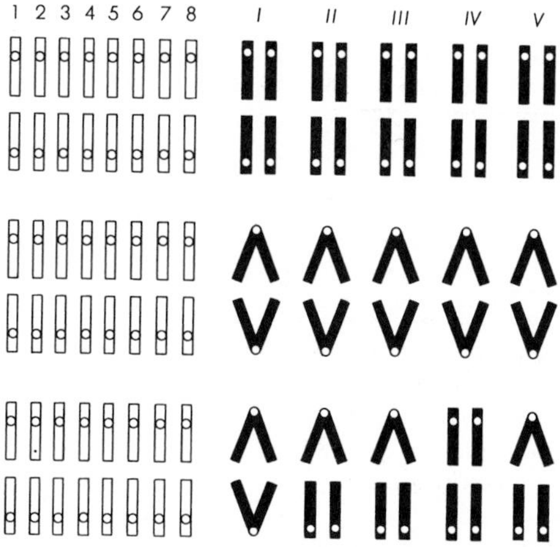

Fig. 7.8. Chromosome polymorphism in the marine snail *Thais lapillus*. The 18-chromosome form (top) occurs primarily at low-tide levels and the 13-chromosome form (middle) at high-tide levels, while the forms with intermediate numbers of chromosomes (bottom) are found between the tidal zones. All chromosome numbers between 13 and 18 have been found. After Hans Staiger.

are correlated with ecological preference in an adaptive way remains unknown, but a somewhat parallel case has been discovered in the mole rat (*Spalex ehrenbergi*) in Israel. Robertsonian polymorphism has led to the formation of chromosomal races which possess diploid numbers of 52, 54, 58 and 60, with a distinct distribution along a north–south axis. The 52 and 54 numbers are found in the humid north, the 60 chromosome race in the arid south, a trend that is also accompanied by a decreasing basal metabolic among the individuals.

In the group of primates most closely related to humans, that is, the great apes, there is marked variation in the number of acrocentric chromosomes, although the diploid number remains constant (Table 7.5). The number of arms, however, ranges from 72 in the orangutan to 82 in the pygmy chimpanzee; the number of human chromosome arms is 78. The evolutionary significance of this remains obscure, particularly since some groups such as the Felidae (cats) show remarkably little karyological variation. A rather unusual case of variation has been found in the mouse, *Mus musculus*. The usual karyotype consists of 40 acrocentrics as the diploid number, but a 22-chromosome population has been found in the Central Apennines of Italy and a 26-chromosome population in more northern valleys of the Alps. It takes 9 centric fusions to transform the diploid number from 40 to 22, and seven fusions to attain the $2n$ number of 26. The polymorphism is even more pronounced, however, since among the 22-chromosome individuals not all show the same fusion patterns. Where the two chromosomal races of 22 and 40 overlap, hybrid populations exist, and diploid numbers of 34, 36, 37, 38, and 39 have been found. Laboratory-produced hybrids with 31 chromosomes all have a low fertility, with about 70% of the gametes containing unbalanced karyotypes.

Table 7.5. Karyotype morphology of the primates most closely related to humans (data on humans included for comparative purposes)

		Karyotype Morphology				
			Autosomes		Sex Chromosomes	
	Diploid Number	Meta Centric	Acrocentric			
Species			Large	Small	X	Y
Pongo pygmaeus (orangutan)	48	26	16	4	M	M
Gorilla gorilla gorilla (lowland)	48	30	12	4	M	M
Gorilla gorilla beringei (highland)	48	30	12	4	M	?
Pan troglodytes troglodytes (chimpanzee)	48	34	8	4	M	A
Pan troglodytes paniscus (pygmy chimp.)	48	36	8	2	M	A
Homo sapiens	46	34	6	4	M	A

SOURCE: J. L. Hamerton et al., (1963).

Centric fusion can also unite acrocentric X chromosomes to autosomes. When this occurs, there is a shift from autosomal to sex linkage for the genes involved, and the creation of a new or neo-Y chromosome (Fig. 7.9). This has taken place in at least 12 species of *Drosophila,* in certain phasmids (walking sticks), in the beetles *Tribolium confusum* and *Agrilus anxius,* and on at least 7 different occasions in the Morabinae, an Australian group of wingless grasshoppers.

The chromosomal situation in the rainbow trout, *Salmo irideus,* indicates that centric fusion and dissociation are responsible for the polymorphism existing not only among individuals of the species but also in the tissues of a single individual. The trout has a diploid number of 104 chromosome arms, and $2n$ numbers ranging from 58 to 104 have been reported. The 58-chromosome form has 12 acrocentrics and 45 metacentrics; the 104-chromosome form would, presumably, possess only acrocentric chromosomes. Within a given individual, the chromosome number in different tissues can vary, with each tissue having a predominant, but not an absolute, number of chromosomes. This is indicated in Table 7.6. In all instances of fusion, nonhomologous chromosomes are involved, and genetic imbalance in somatic cells is not a problem. Fusion of homologous chromosomes would, of course, lead to unbalanced complements in gametes if present in gonadal tissue, and would be selected against in evolu-

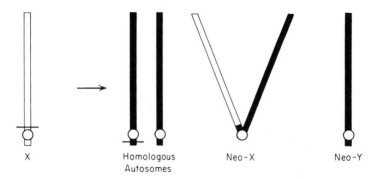

Fig. 7.9. The creation in an XX-XO species of an XX-XY sex-determining mechanisms through the centric fusion of the original acrocentric X and an acrocentric autosome. In meiosis the neo-Y would segregate from the neo-X, and since it would now be confined to the male line, it is considered a Y chromosome regardless of its genetic potentialities.

tion. What evolutionary advantage there is to fusion and dissociation of nonhomologous chromosomes is, on the other hand, not immediately obvious, although chromosomal polymorphism may be a prelude to further karyological change, and hence to evolutionary divergence within the species. Offsetting this possible evolutionary advantage is the fact that gonadal cells undergoing meiosis, and containing variable numbers of acrocentrics and metacentrics, would form multivalents, paving the way for irregularities of segregation. In *Salmo*, however, the species seems to be adjusted to this form of chromosome variability, for few unbalanced gametes are produced.

Table 7.6. Modal chromosome number variation in tissues and development of *Salmo iriaeus* (trout) (Ohno, 1965).*

Diploid Number:	58	59	60	61	62	63	64	65
Acrocentrics:	12	14	16	18	20	22	24	26
Early embryo		6	40	8	10	2	5	
1 month								
Spleen	2	14	7					
Kidney	3	14				1		
Liver			3	19	22	3	6	1
8 months								
Spleen		13	15	3				1
Kidney	10	10	4					
Ovary	3	1	6	5	9	2		
or								
Testis			19					
18 months								
Spleen		19						
Testis		4	16		5		3	

*Figures refer to number of cells containing the indicated number of chromosomes. Note that, as the diploid number increases by one, the number of acrocentrics increases by two.

A haploid organism contains a single genome, or set of chromosomes. A diploid organism, however, may have its two haploid sets similar or dissimilar. Members of a diploid species would generally have similar genomes; an organism resulting from a cross between two diploid species would have dissimilar genomes and is generally referred to as a *diploid hybrid*. Where more than two genomes are involved, that is, in triploids, tetraploids, and so on, organisms possessing similar genomes are known as *autopolyploids,* those with dissimilar genomes as *allopolyploids.* Allopolyploids must obviously arise through hybridization. By definition, therefore, a clearcut distinction can be made between auto- and allopolyploidy; in practice, however, the distinction breaks down, as may be seen from the diagram in Fig. 7.10.

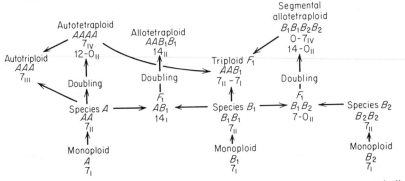

Fig. 7.10. Relationships between autopolyploidy, allopolyploidy, and segmental allopolyploidy in a hypothetical group of related species having a base number of seven chromosomes. Under each form is given the number of univalents (I), bivalents (II), trivalents (III), and quadrivalents (IV) that might be expected to form in meiosis. The chromosome structure of species B_1 and B_2 are assumed to be more similar to each other than either is to species A, and the segmental allotetraploid would be expected to show more multivalency than a normal allotetraploid, but not so much as an autotetraploid.

Monoploidy. Although usually considered abnormal in diploid organisms, monoploidy may be a normal state. The gametophytic stages of lower plants and the males of certain insects, such as the bees, wasps, and other hymenopteran forms, are regularly monoploid. Production of gametes proceeds in normal fashion because special mechanisms of meiosis have arisen through evolution. In the meiotic cells of the male honeybee, for example, synapsis cannot occur, and during the first meiotic division all chromosomes pass into one cell, none into a cytoplasmic bud, which is pinched off. The second meiotic division, which is mitotic-like and equational, is quite regular except that the cytoplasmic division is uneven, producing daughter cells of different size. Only the larger is transformed into a functional sperm. There is consequently no reduction in the number of chromosomes during meiosis. Monoploid organisms that are normally diploid must of necessity pass through a highly irregular meiosis. The chromosomes, which have no homologues with which to pair, pass as univalents onto the metaphase plate, where they segregate at random (some may divide equationally at times). Because this leads to deficient gametes or spores, the sterility of such monoploids is very high. Occasionally, however, monoploids produce functional gametes, as would be expected on a random basis of segregation. Thus if the probability of any particular chromosome passing into a certain anaphase nucleus is $1/2$, then the probability of obtaining a functional gamete is $1/2^n$ where n equals the number of chromosomes in the monoploid genome. Clearly, the higher the monoploid number, the less likely will be the probability of viable gamete formation.

Monohaploids are of cytogenetic interest for two reasons. Once they are obtained in plants, the chromosome number can readily be doubled by colchicine. These individuals would be completely homozygous for all genes and would achieve in one step a condition that would require many generations to approach through close inbreeding. The importance of this in plant breeding, where isogenicity is an important factor in experimental design, can readily be appreciated. Such monoploids may rise spontaneously at a slow rate, and may be recognized initially by their smaller stature and high sterility, but a technique for their production has been developed in the last decade. Some members of the Solanaceae—*Datura* and *Nicotiana,* in particular—and to a more limited extent, the grasses, provide favorable material. Microspores are taken from anthers at the time of the first microspore division and are cultured on artificial media. Some of the microspores will continue through division to produce the typical generative and tube cells, and others will succumb to the treatment, but a small percentage of the cells will undergo a somatic type division to yield monoploid embryos. After a period of growth, these can be treated with colchicine to produce homozygous diploid plants. An analogous technique has been developed by C. L. Markert for use in animals. Fertilized eggs of mice, removed from the uterus before the male and female pronuclei have fused to form a diploid nucleus, can have one of the pronuclei removed surgically without damage to the egg. A brief subsequent treatment of the egg in cycloheximide prevents cell division without inhibiting the division of the chromosomes of the remaining pronucleus. Removal of the egg from the effects of the inhibitor then allows division to proceed, but by now the chromatids have separated and the nucleus has doubled its chromosome number. Reinsertion of the egg back into a receptive uterus permits implantation to take place, and a homozygous female results. No males are possible by this technique. Which pronucleus is removed surgically is a random affair since one cannot be distinguished from the other, but if the Y-bearing sperm pronucleus were retained, the resultant diploid would be YY, a lethal condition.

Haploids from diploid barley (*Hordeum vulgare*) are readily induced when reciprocal crosses are made with diploid *H. bulbosum*. Seed set from such crosses is about 50% of normal, but approximately 10 days after pollination seed abortion sets in. A fair percentage of the embryos can be salvaged, however, by removing them at 8 to 10 days of age, and culturing them on an artificial medium. All of the surviving plants—about 11% of those cultured—are haploids ($n = 7$), and all of the chromosomes are from the *H. vulgare* genome. These can be doubled by exposure to colchicine to produce homozygous diploids. Similar crosses of hexaploid wheat (*Triticum aestivum*) with either diploid or tetraploid *Hordeum bulbosum* also yields haploids ($n = 21$), with only the wheat chromosomes surviving. Seed set is lower than it is in the *H. vulgare* crosses, and seed abortion begins between the 14 to 18 days after pollination. The embryos can be rescued by the same culturing technique. The cause of seed abortion is not clear; in both the barley and wheat crosses, pollination is successful, the embryos begin to develop with both genomes intact, but then the *H. bulbosum* chromosomes are progressively eliminated from the cells by failure to undergo appropriate anaphase separation and segregation.

A somewhat similar method can be used to produce haploids in the potato, although the culturing of embryos is not a necessary step. In *Solanum tuberosum* ($2n = 48$), for example, dihaploids ($n = 24$) are readily induced through female parthenogenesis, with *S. phureja* being the pollinating stimulator of egg development. The dihaploids are highly fertile, and if these are crossed again with *S. phureja*, monohaploids ($n = 12$) arise parthenogenetically with low frequency (0.49 monohaploids/1000 seeds). These can be doubled through the use of colchicine.

A study of monoploid meiosis also provides a clue to the amount of homology within the basic chromosome complement. In haploids of *Sorghum*, for example, which has 10 instead of the usual 20 chromosomes, most cells have 10 univalents. Some, however have an occasional bivalent, suggesting either that 10 is not the basic number or that duplications are present, and hence that each chromosome is not genetically unique. A similar situation has been found in the pepper ($n = 12$), in which as many as 6_{II} have been found, leading to the suggestion that the supposed diploid is really a polyploid in which many, if not all, of the genes are duplicated. In economic plants, this becomes a factor of importance when the genetics of a particular character is being determined.

Autopolyploidy. If a diploid species has its two similar genomes designed *AA,* then an autotriploid becomes *AAA,* and an autotetraploid *AAAA* (Fig. 7.10). The latter would have its origin directly from the diploid by doubling of its chromosomal number either by somatic doubling or by the union of two diploid (unreduced) gametes, whereas the former could arise as an offspring of a tetraploid and a diploid parent or from diploid parents by the union of an unreduced and a reduced gamete. Once thought to be quite common in plant species, autopolyploids are now believed to be relatively rare in natural populations. Through the use of colchicine, autotetraploids are relatively easy to produce artificially.

Autopolyploids are, in general, larger than their related diploids as a result of an increase in cell size, just as haploids are smaller. There is, in general, an increase in size of various plant parts, particularly in such organs as flowers and seeds which have a determinate type of growth, a delay in growth and in flowering, and often an increase in the darkness and thickness of the foliage as one goes from haploid, to diploid, to tetraploid plants of the same genetic stock. As one goes beyond the tetraploid level, however, increases in chromosome number often result in abnormalities such as dwarfing, wrinkled foliage, and weak plants. These factors generally contribute to the relative inferiority of the autopolyploids as compared to their diploid progenitors, and they provide a possible explanation for their rarity in natural populations. The degree of polyploidy at which unbalance sets in will evidently depend upon the species, or even the individual, in question.

Cytologically, autopolyploids are characterized and identified by the presence of multivalents formed at metaphase of meiosis I. In autotriploids, the three homologous chromosomes can pair with one another (although 2 by 2 at any given point along the chromosome) to give trivalents Fig. 7.11; in autotetraploids, quadrivalents would result. The number is not constant from cell to cell and will depend upon the degree of synapsis and chiasma formation taking place in meiotic prophase. Figures 7.12 and 7.13 illustrate several types of multivalent formation in autopolyploids. An interesting

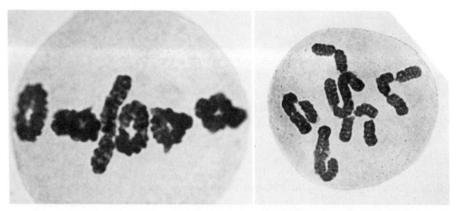

Fig. 7.11. Microsporocytes of Tradescantia at metaphase I: left, diploid; right, triploid. The chains of three are evident in the triploid; the relative difficulty of 2-by-2 pairing keeps pairing and chiasma formation confined to the ends of the chromosome arms.

exception to the 2 by 2 pairing in meoisis is the complete synapsis found in the polytene chromosomes in triploid individuals of *Drosophila melanogaster*.

The number of certain autopolyploids existing in nature is meager. In both plant and animal groups it is possible that some species perpetuating themselves through parthenogenetic means (see Chapter 8) are genuinely autopolyploid, although in the absence of a detailed structural analysis of the chromosomes this would be difficult to establish. The tiger lily, *Lilium tigrinum*, is thus thought

Fig. 7.12. Pachytene pairing and metaphase I configurations that are possible in autoploids: left, triploid; right, tetraploid. Univalents are depicted in outline; centromeres are assumed to centrally located in all chromosomes. Univalents are randomly distributed and sometimes lost.

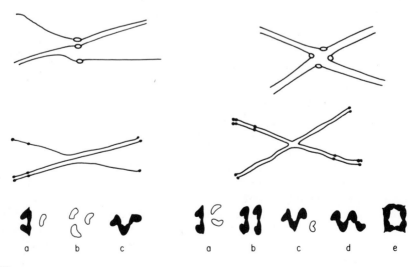

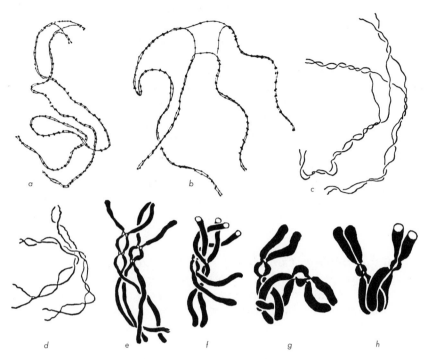

Fig. 7.13. Quadrivalent formation in the tetraploid *Allium porrum* at pachynema (*a* and *b*), diplonema (*c* and *d*), and at diakinesis (*e* to *h*). The chiasmata are strongly localized around the centromeres, but the quadrivalents remain held together until metaphase I by some form of residual attraction. A. Levan, *Hereditas, XXVI* (1940), 454–62.

to be an autotriploid, and it reproduces vegetatively by means of axial bulbils. Some triploid species of Fritillaria are believed to be of a similar nature. Diploid-tetraploid populations of an autopolyploid character have been reported for several frog genera in East Africa, but a lack of information on meiotic behavior again raises the question of uncertainty. In the plant kingdom, *Galax aphylla* and *Achlys triphylla,* both remnant members of a Tertiary circumpolar flora, have 2*n* and 4*n* types, and with no known close relatives. Several of the North American species of *Tradescantia,* for example, *T. virginiana,* have both 2*n* and 4*n* members which, except for minor size differences, are virtually indistinguishable from each other morphologically; their meiotic behavior regularly leads to the formation of quadrivalents. *T. rosea* (also known as *Cuthbertia graminea*) possesses 2*n* and 4*n* members of similar morphology. The potato, *Solanum tubersum,* on a variety of grounds, is assumed to be an autotriploid, as are some subspecies populations of the orchard grass, *Dactylis glomerata.* No polyploid higher in chromosome number than a tetraploid has ever been assumed to be an autopolyploid. As compared to allopolyploids, however, autopolyploidy had had no evident evolutionary impact, although it is possible that transitory autopolyploids, capable of outcrossing interspecifically more readily than their diploid ancestors, may have served to foster subsequent species crosses and then to have disappeared. Horticulturally, the induction of larger, hardier, and later and longer blooming autopolyploids is of some importance, particularly if the form, when once induced, can be perpetuated and increased through vegetative means. In addition to the potato, such domesticated species as the celandine, *Ranunculus ficaria*, the apple, *Malus pumila*, and the hyacinth, *H. orientalis*, are apparent natural autoploids, while tetraploid snapdragons and marigolds have been produced artificially for the horticultural trade.

Allopolyploidy. Considering for the moment only allotetraploids, the genome constitution can be represented as *AABB* (Fig. 7.10), with the individual having arisen by a doubling of the chromosome number of an F_1 hybrid between species *A* and species *B*. If the genomes are sufficiently dissimilar structurally, no synapsis will occur in the diploid hybrid, and high sterility will ensue as the result of the random segregation of unpaired chromosomes. Doubling of the chromosome number to give the allotetraploid *AABB* will, however, provide for regular synapsis and segregation. Genome *A* will pair with genome *A*, genome *B* with genome *B*. The gametes will be *AB*, and if sterility in the F_1 hybrid is due only to irregular chromosome distribution, then the allotetraploid can be expected to have a high degree of fertility. However, when hybridity sterility is due to genic as well as to chromosomal imbalance, doubling of the chromosome number will not restore fertility. Thus, Stebbins has shown that some sterile diploid hybrids of grass species, when subsequently brought to a tetraploid state, did not achieve a fertile state.

The classical intergeneric cross between *Raphanus* (radish) and *Brassica* (cabbage), made by the Russian cytogeneticist Karpechenko, illustrates the point just made (Fig. 7.14). The 9 haploid *Raphanus* chromosomes are distinctly different, probably in a structural way, from the 9 haploid *Brassica* chromosomes, although they must be fairly similar to each other genically in order to have permitted hybridization in the first place. No pairing occurred in the F_1 species hybrid, and a high degree of sterility ensued because of the irregular distribution of univalents. A number of unreduced (diploid) gametes were formed, however, and several tetraploid individuals were recovered, their origin stemming presumably from an unreduced egg being fertilized by a sperm produced by an unreduced pollen grain.

If *R* and *B* represent the *Raphanus* and *Brassica* genomes, respectively, then the genome of the tetraploids is *RRBB,* and each individual is an allotetraploid (Fig. 7.10), fully fertile and with normal meiosis. The only gametes produced are diploid and *RB* because, in meiosis, the members of the *R* genome pair only with their homologous members in the other *R* genome, and the same with the members of the two *B* genomes, thus providing for homologous pairing, metaphase orientation, and regular anaphase I segregation. The allotetraploids, therefore, behave in all respects similarly to normal diploids, and provide a clear example of how two successive events—hybridization between two species, followed by a doubling of the chromosome number—can cataclysmally bring about the creation of a new species. *Raphanobrassica,* to be sure, is an artifically created alloploid species, but it meets all of the criteria required of a legitimate species encountered in nature—a distinct and constant

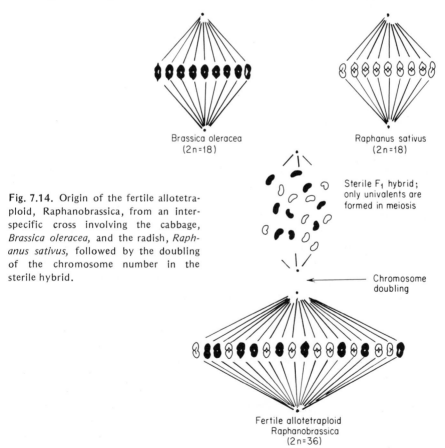

Brassica oleracea
(2n=18)

Raphanus sativus
(2n=18)

Sterile F₁ hybrid;
only univalents are
formed in meiosis

Fig. 7.14. Origin of the fertile allotetraploid, Raphanobrassica, from an interspecific cross involving the cabbage, *Brassica oleracea,* and the radish, *Raphanus sativus,* followed by the doubling of the chromosome number in the sterile hybrid.

Chromosome
doubling

Fertile allotetraploid
Raphanobrassica
(2n=36)

morphology, and the ability to reproduce its own kind as the result of regular meiotic processes giving rise to viable gametes. *Primula kewensis* is another example of an allotetraploid artifically created from a sterile hybrid previously resulting from a cross being two diploid species, but a number of allotetraploids are known to have occurred in nature: *Nicotiana tabacum, Gossypium hirsutum* (upland cotton), the maritime grass, *Spartina townshendii* and *Galeopsis tetrahit* are examples whose diploid ancestors are known, and from which the allotetraploids have been resynthesized. It is, of course, possible that each arose in a one-step process, that is, by the fusion of two unreduced gametes, but it is far more probable that each had its origin via the two stages known to have brought about the creation of *Raphanobrassica* and *Primula kewensis.*

Allopolyploids of a more complex nature than that represented by the allotetraploids can arise, and probably have arisen many times in the evolution of plant species. It is also clear that not all diploid species hybrids show the lack of pairing characterizing the initial hybrid between *Raphanus* and *Brassica;* the tetraploids derived from such crosses will exhibit partial quadrivalent formation at meiosis. This occurs when the two genomes are not totally dissimilar, and partial synapsis and crossing over can take place between the chromosomes of different genomes. Stebbins has referred to such individuals as *segmental allopolyploids.* Clearly then, auto- and allopolyploids are the extremes of a spectrum of forms in which there is a range from complete homology between all of the sets of chromosomes (auto-) to a complete lack of homology (allo-) between the two kinds of genomes. The segmental allopolyploids fall into between the autoploids and the alloploids, and it is believed that they are more common in nature than either of the two extreme types.

Inheritance and a high degree of fertility in polyploids is possible only when pairing between homologues leads to bivalent formation, and anaphase segregation is highly regular. This is what happens in *Raphanobrassica,* but if the *R* chromosomes also paired with members of the *B* genome, this would be considered *homoeologous pairing* to distinguish it from homologous pairing; multivalent formation and irregular segregation would sometimes occur, and *Raphanobrassica* would then be judged a segmental polyploid rather than a strict allopolyploid. The situation uncovered in the bread wheats, *Triticum aestivum* (sometimes called *T. vulgare*), indicates that the distinction between these two kinds of polyploids may be a matter of genetic control rather than of structural or genic dissimilarities.

T. aestivum is a hexaploid ($2n = 42$), and its three genomes, labelled *A, B,* and *D,* have been traced to specific wild relatives: the

A genome to *T. monococcum;* the *B* genome to *T. searsii,* although *T. (Aegilops) speltoides* is a possible source; and the *D* genome to *T. tauschii,* with *Aegilops squarrosa* also suggested as the origin (Fig. 7.15). Only bivalent formation occurs, with no homoeologous pairing evident; thus, gametes with 21 chromosomes are regularly produced. The 21-chromosome *aestivum* genome, therefore, consists of three homoeologous groups, each of which has its chromosomes identified, respectively, as *A1* to *A7, B1* to *B7,* and *D1* to *D7. A1* is homologous with the other *A1,* but homoeologous with *B1* and *D1,* and the same for the other six sets of chromosomes. The close genic similarity of these three genomes to each other is indicated by the fact that various chromosomes can be substituted without serious effects. For example, a strain nullisomic for *A5* but tetrasomic for *B5* maintains a somatic number of 42 chromosomes, and presents a fairly normal phenotype.

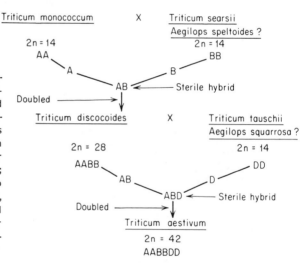

Fig. 7.15. Origin of the allohexaploid bread wheat (*Triticum aestivum*) from three different diploid grass species, coupled with chromosome doubling. *T. searsii* has recently been cited as the origin of the *B* genome, although earlier *A. speltoides* was given that honor; also *T. tauschii* is now thought to be the source of the *D* genome, although *A. squarrosa* once had that distinction. See text for further discussion of these interrelations.

The fact that the hexaploid wheat behaves meiotically as a diploid has been traced to the influence of a single gene *Ph (Pairing homoeologous)* located in the long arm of the B5 chromosome; it has also been identified in natural populations of *T. searsii*. Its presence was made evident when a strain nullisomic for *B5* was produced by selfing a *B5* monosomic; at meiosis a relaxation of pairing constraints led to a shift from strict homologous pairing to a mixture of homologous and homoeologous pairing, and as a consequence, from bivalent to multivalent formation at meiotic metaphase. The locus exhibits normal mendeliam behavior in transmission, trisomic and monosomic ratios are observed, and through of use telocentric chromosomes consisting only of the short or the long arm of *B5*, it could be demonstrated that the effect was independent of centromere involvement.

When the *B5* chromosome is in a monosomic state, meiotic pairing is reasonably normal, indicating that the effect of a single dose of the *Ph* gene, mediated probably by a single protein, is sufficient to maintain the diploid-like behavior of *T. aestivum* meiosis, but in individuals trisomic and tetrasomic for B5, the depressant action of the gene extends to homologous pairing as well, as evidence by a progressive reduction in chiasma frequency. Chromosome *B5* is not alone in exerting this kind of influence, for a similar but less marked effect is observed in plants nullisomic for *A5, B3*, an arm of *D3* and the short arm of *B5*. Also, in certain interspecific hybrids supernumerary *B* chromosomes can compensate for the absence of a *B5*-like chromosome by preventing or reducing homoeologous pairing. Thus, in the interspecific hybrid between *T. boeticum* X *Aegilops mutica* ($2n = 14$), where homoeologous pairing occurs with fair frequency, the presence of one supernumerary *B* chromosome drops the chiasma frequency in individual cells from an average of about 4.5 to well below 1.0. A similar effect has been found as well in interspecific hybrids of *Lolium*. How the influence of the *Ph* gene is exerted is problematical—it could, for example, be on the time available for pairing, or on the spatial distribution of homologoues at the time when pairing is initiated—but the ultimate result is to stabilize homologous pairing, and in doing so to convert a potentially unstable polyploid into one with regular segregation and a high degree of fertility. The evolutionary advantage of such a system is obviously great, and of high selective value to any polyploid organism.

That polyploidy is of significance in the evolution of plants is suggested by its prevalence and distribution. Although known only in the yeasts among the fungi, polyploidy is common in all other groups. The remnants of once prevalent floras, such as our present-day equisetums and lycopodiums, are characterized by very high polyploidy, suggesting that this step has a high survival value. Polyploidy is less common among the gymnosperms, but it is rampant among the ferns and angiosperms. 30% to 35% are polyploid, with the grasses having about 75% of their species exhibiting it to one degree or another. Genetic segregation among such forms will, of course, be curtailed when a given gene is present more than twice, but this evolutionary disadvantage is apparently outweighed by the relative vigor of a select number of polyploids and their ability to permit the species to exploit new and diverse ecological habitats, the latter activity arising out of the new gene combinations brought into being by the mixture of two previously separated species. This does not mean that all polyploids stemming from interspecific crosses are necessarily vigorous. Breeding experiments have shown that many are genetic misfits that cannot compete with their diploid ancestors; this is particularly true in a stable environment where no unoccupied territory is available for exploitation and where the diploid forms are well adapted. It is probable, therefore, that many more polyploids are formed than eventually become established, but their success would be favored in times of environmental change that lead to the formation of new habitats and foster the migration of floras. Many polyploid complexes, indeed, can be traced back to the Pleistocene and Pliocene Epochs, the former a time of oscillating ice sheets in the Northern Hemisphere and the latter a period of continuous and extensive mountain building.

The existence of polyploids which have had a long evolutionary history raises the question of whether, and in what manner, their genomes have undergone change since the time of their origin, having in mind here that many of the genetic loci will be represented four or more times in each nucleus and hence might be more readily subjected to alteration with time. This is not easy to answer in terms of structural loci coding for proteins, but the relation of repetitive DNA such as rDNA to total DNA can readily be determined through hybridization studies. This kind of DNA is particularly subject to gains or losses, reflecting apparently the needs of the cell for such genic or regulatory activity. For example, at the time of origin of either an autoploid or an alloploid the amount of rDNA to total DNA should be twice the average of the diploid parents. Is all of the rDNA needed to maintain a functional polyploid? If not, can adjustments be made? The answer, based on studies done on the genus *Nicotiana,* seems to be that the proportion of rDNA to total DNA is lower in polyploids than in related diploids. Thus in *N. tabacum* $(2n = 48)$, there are 4,500 cistrons coding for 70S cytoplasmic ribosomal nucleic acid as compared to 3,500 and 4,100 in *N. paniculata* and *N. sylvestris,* respectively, both of which are diploid $(2n = 24)$. An even wider discrepancy is evident for the cistrons coding for chloroplastic ribosomal nuclei acid, with *N. tabacum* possessing 4,600, *paniculata* 7,600 and *sylvestris* 11,300 (the polyploid nature of chloroplasts raises some question as to the significance of these latter figures). From these kinds of information it can be assumed that once a polyploid, regardless of its origin, gains a foothold in an environment, its genome also undergoes an appropriate adjustment during the course of subsequent adaptation.

Polyploidy in Animals. The relative scarcity of polyploidy among animal species presents a sharp contrast to its widespread prevalence among the higher plants. It has been proposed that polyploidy would be quite limited among animals because the sharp separation of the two sexes rests upon a sensitively balanced chromosomal mechanism. This is undoubtedly true in such instances as *Drosophila* where sex determination rests upon a balance between the X chromosomes and the autosomes, and intersexes would be formed if this balance were disturbed, but in *Melandrium* where a well-defined sex-determining mechanism exists and where the Y chromosome is male-determining, triploid and tetraploid individuals with at least one Y chromosome are fully fertile males, and intersexes are generally absent. Similarly constituted karyotypes in humans would also be male, but live-born triploids have been very rare, and survival has been brief. However, triploidy and tetraploidy among abortuses is not rare—triploidy is the second most common chromosomal anomaly among abortuses, trisomy being the most frequent—but the developmental disturbance must be severe since usually only fetal fragments are found at the time of abortion.

Naturally occurring polyploid populations of hermaphroditic earthworms and planarians exist, all of which indicates that the paucity of polyploids among animals cannot be ascribed wholly to sex chromosome difficulties. Much more likely is the fact that polyploidy arises in single individuals, and since cross breeding is a rule, even among hermaphroditic species, the likelihood of a polyploid finding a suitable mate is highly uncertain. Where polyploidy has been, or is, present, it is much more the fact that it is intimately bound up with a parthenogenetic mode of development, thus escaping the meiotic difficulties of pairing and segregation (see Chapter 8).

In addition to the $2n - 4n$ populations of African frogs previously mentioned (p. 433), the hymenopteran sawflies seem to possess an authentic case of polyploidy unassociated with parthenogenesis. Most of the species of *Neodiprion* and *Diprion* have a haploid number of 7 chromosomes in the males and a diploid number of 14 in the females, the males developing parthenogenetically as is usual in this group. *D. simile,* however, has 28 in the females and 14 in its male members. Meiosis in the males is similar to that found in other hymenopteran males in that the first division is missing and the second is equational, whereas in the females bivalents but not quadrivalents are formed. On this basis, *D. simile* can be considered to be an allotetraploid, but it may well be that the small size of the chromosomes, and their low chiasma frequency, precludes quadrivalent formation, thus obscuring what may be a case of autotetraploidy.

BIBLIOGRAPHY

Avery, A. G., et al., *The Genus Datura.* New York: Ronald Press, 1959.

Barclay, I. R., "High Frequencies of Haploid Production in Wheat (*Triticum aestivum*) by Chromosome Elimination," *Nature 256* (1975), 410-411.

Beatty, R. A., *Parthenogenesis and Polyploidy in Mammalian Development.* New York: Cambridge University Press, 1957.

Bogart, J. P., and M. Tandy, "Polyploidy Amphibians: Three More Diploid-Tetraploid Cryptic Species of Frogs," *Science, 193* (1977), 334-35.

Bungenberg de Jong, C., "Polyploidy in Animals," *Bibliog. Genetics, 17* (1957), 111-228.

Carson, H. L., "Chromosome Tracers of the Origin of Species," *Science, 168* (1970), 1414-18.

Christensen, B., "Studies on Cyto-taxonomy and Reproduction in the Enchytracidae. With Notes on Parthenogenesis and Polyploidy in the Animal Kingdom," *Hereditas, 47* (1961), 387-450.

Ford, C. E., et al., "Chromosoma Polymorphism in the Common Shrew," *Nature, 180* (1957), 392-93.

Hamerton, J. L., et al., "The Somatic Chromosomes of the Hominoidea," *Cytogenetics, 2* (1963), 240-63.

Hurley, J. E., and S. Pathak, "Elimination of Nucleolus Organizers in a Case of 13/14 Robertsonian Translocation," *Hum. Genetics, 35* (1977), 169-73.

John, B., and K. R. Lewis, "Chromosome Variability and Geographic Distribution in Insects," *Science, 162* (1966), 711-21.

Kasha, K. J., and K. N. Kao, "High Frequency Haploid Production in Barley *(Hordeum vulgare L.)," Nature, 225* (1970), 874-875.

Ledoux, L. (ed.), *Genetic Manipulation with Plant Material.* New York: Plenum, 1974.

Levan, A., "Meiosis of *Allium porrum,* a Tetraploid Species with Chiasma Localization," *Hereditas, 26* (1940), 454-62.

Lewis, H., "Speciation in Flowering Plants," *Science, 152* (1966), 167-72.

Nicklas, B. R., "The Chromosome Cycle of a Primitive Cecidomyiid—*Mycophila speyeri," Chromosoma, 11* (1960), 402-18.

Nitsch, C., "Single Cell Culture of a Haploid Cell: The Microspore," in *Genetic Manipulation with Plant Material,* L. Ledoux, ed. New York: Plenum, 1974.

Ohno, S., et al., "Post-Zygotic Chromosomal Rearrangements in Rainbow Trout (*Salmo irideus* Gibbons)," *Chromosoma, 4* (1965), 117-29.

Sparrow, A. H., and A. F. Nauman, "Evolution of Genome Size by DNA Doublings," *Science, 192* (1976), 524-29.

441

Staiger, H., "Genetical and Morphological Variation in *Purpura lapillus* with Respect to Local and Regional Differentiation of Population Groups," *Col. Intern. Biol. Marine Sta. Biol. de Roscoff* (June 27-July 4, 1956).

————, "Der chromosomendimorphismus bein prosobranchier *Purpura lapillus* in Beziehung zur Okologie der Art," *Chromosoma, 6* (1954), 419-78.

Stebbins, G. L., "Chromosomal Variation and Evolution," *Science, 152* (1966), 1463-69.

————, "The Significance of Hybridization for Plant Taxonomy and Evolution," *Taxon, 18* (1969), 26-35.

Suomalainen, E., "On Polypoidy in Animals," *Proc. Finn. Acad. Sci. Letters,* (1958), 1-15.

Van Bruekelen, E. M. W., et. al., "Parthenogenetic Monohaploids $(2n = X = 12)$ from *Solanum tuberosum* L. and *S. verrucosum* Schlechtd., and the production of homozygous potato diploids." *Euphytica 26* (1977), 263-271.

White, M. J. D., "Principles of Karyotype Evolution in Animals," *Proc. XI Intern. Cong. Genetics,* (1963), 391-97.

————, "Speciation in Animals," *Australian J. Science, 22* (1959), 32-39.

————, "Models of Speciation," *Science, 159* (1968), 1065-70.

Wiens, D., and B. A. Barlow, "Permanent Translocation Heterozygosity and Sex Determination in East African Mistletoes," *Science, 187* (1975), 1208-09.

8

Variation: Sources and Consequences Involving Variant Chromosomal Systems

The avenues of evolutionary change that can be successfully explored are limited by the physical and biological environment in which the organisms live and with which they must come to terms; the variation of which hereditary materials are capable, and which must be present at an appropriate time and place if they are to persist and contribute to the next generation; and the mechanisms and rates by which these hereditary materials, old as well as new, are transmitted to the next generation of offspring. Sexual reproduction is, of course, an integral cog in the machinery of evolution; through recombination and segregation it generates a number of genetically dissimilar gametes, and through fertilization it brings these gametes together and presents to the environment individuals of diverse genotypes which may be selected for or against. Because of this diversity, sexual reproduction is not without its disadvantages, for not all genotypes and karyotypes are balanced, viable, or of immediate adaptability, and most species have some means, of varying effectiveness, for ridding the population of these. The fetal wastage in humans is an example of embryonic screening, while the gametophytic generation of higher plants removes those genotypes and karyotypes which, temporarily at least, are inadequate. When the varied mechanisms are coupled with differential rates of reproduction among the individuals in a population, evolutionary change or natural selection is therefore inevitable, even in the absence of obvious environmental pressures.

A number of variant systems in sexually reproducing species has already been discussed in other contexts. Thus, in the case of *Drosophila pseudoobscura,* both inversion heterozygosity and inversion frequencies indicate that a delicate adjustment exists between genotype (karyotype) and environment, with temperature, season, altitude, and/or latitude responsible for shifts in one direction or another, while the translocation complexes of *Oenothera* contain integrated segregation and recognition mechanisms which foster a genetically and cytologically heterozygous status quo. Other examples of meiotic drive similarly push populations of organisms in nonrandom directions. On the other hand, many species have taken a quite distinctly different route to continued survival: In whole or in part, they have abandoned a sexual for an asexual means of reproduction, leading as a result to altered patterns of inheritance.

Table 8.1. Plant genera exhibiting various forms of asexual reproduction; not all species or even all individuals of a species are necessarily asexual

---

A. *Vegetative Reproduction:*
    1. Nonfloral parts: *Opuntia* (stem joints); *Fragaria* (runners); *Bryophyllum* (leaves);
        Lilium tigrium (bulbils); potato (tuber); tulips, narcissi, lilies (bulbs); gladioli (corms)
    2. Floral parts: *Agave, Allium, Cardamine, Citrus, Deschampsia, Festuca, Mangifera,*
        Nigritella, Poa, Polygonum, Saxifraga; Xanthoxylum
B. *Adventitious Embryony**
    Citrus, Mangifera, Nigritella, Opuntia, Xanthoxylum
C. *Gametophytic Apomixis*
    1. Diplospory: *Antennaria, Taraxacum, Ixeria, Hieracium, Potentilla, Eupatorium,*
        Erigeron, Burmannia, Balanophora, Elatostoma, Calamagrostis
    2. Apospory: *Hieracium, Malus, Hypericum, Ranunculus, Crepis, Potentilla, Atraphaxis,*
        Rubus, and in the ferns† *Dryopteris, Cyrtomium, Pteris, Asplenium, Pellaea*
    3. Diploid parthenogenesis: *Malus, Sorbus, Ranunculus, Poa, Calamagrostis, Potentilla,*
        Hypericum, Antennaria, Crepis, Taraxacum, Hieracium
    4. Apogamety: *Alchemilla, Taraxacum, Alnus, Hieracium, Allium, Burmannia,* and in the
        ferns *Adiantum, Asplenium, Dryopteris, Polystichum, Pteris*

---

*May need stimulus from growing pollen tube to initiate development of egg.
†Nonsporogenous cells give rise to gametophytes.

Source: Compiled from Grant, 1971; Fryxell, 1957; Manton, 1952; Stebbins, 1950; Nygren, 1954; and Gustafson, 1946–47.

Since a large proportion of these species owe their origin to interspecific hybridization, coupled generally with varying degrees of polyploidy, the asexual route ensures immediate even if short-term reproductive success for those individuals that are environmentally adapted by circumventing the difficulties that would arise meiotically as a result of poor synapsis and irregular segregation. Asexual reproduction also stabilizes hybridity and perpetuates any hybrid vigor that might have arisen from the union of dissimilar gametes. The genetic plasticity upon which subsequent evolution will depend would seem to be seriously diminished by the establishment of asexual mechanisms, but the number of obligate asexual forms is small compared to those species exhibiting a combination of sexual

Fig. 8.1. Modes of sexual and asexual reproduction in plants and animals.

A. Normal sexual life cycle

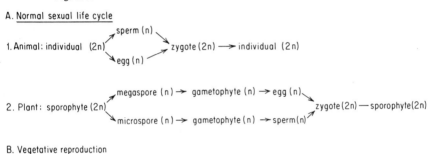

B. Vegetative reproduction

1. Animal: budding, fragmentation of body or parts of body
2. Plant: runners, stolons, rhizomes, corms, tubers, bulbs, bulbils, stem joints and other propagules

C. Adventitious embryony

Sporophyte (2n) ⟶ somatic cell of ovule (2n) ⟶ seed with embryo (2n) ⟶ sporophyte (2n)

D. Gametophytic apomixis

E. Animal parthenogenesis

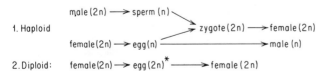

*egg may be 2n as a result of no meioiis, or as the result of the fusion of two haploid cells.

Table 8.2. Animal species exhibiting various forms of diploid parthenogenesis.

Species	Ploidy	Comments
Diptera		
Cnephia mutata (simuliid)	3*n*, alloploid (?)	2*n* sexual form cytologically homozygous; 3*n* females contain at least 70 different inversions, giving over 700 different karyotypes
Drosophila mangabeirai	2*n*	Obligate; 40% egg mortality; fusion of nonsister nuclei after second meiotic division
Lonchoptera dubia	2*n*	Four biotypes known, each with its own inversion heterozygosity and in different ecological niches; fusion of two haploid nuclei in egg
Orthoptera		
Pycnoscelus surinamensis	2*n* = 34 to 74	Found worldwide; two meiotic divisions, but each mitotic in nature
Brunneria borealis	2*n*	Found in Mexico and N. America, but related 2*n* sexual form in S. America
Sago pedo	4*n*	Localized colonies in Europe, bisexual relatives in Israel and Turkey; meiotic divisions, but no synapsis and no reduction of chromosome number
Moraba virgo	2*n*	2*n* = 15, related bisexual form 2*n* = 18; isolated colonies on acacia in Australia; shows translocation and inversion heterozygosity; pre-mieotic doubling of chromosome number followed by regular meiotic divisions
Lepidoptera		
Solenobia triquetrella	2*n*, 4*n*	2*n* forms co-exist in Alps, 4*n* widely distributed in Europe; fusion of two central nuclei in egg maintains heterozygosity
Apterna helix	2*n*	Both sexual and parthenogenetic forms 2*n* = 62; sexual form locally near Munich; parthenogenetic form throughout central Europe; fusion of second division spindles at metaphase restores number to 62, followed by a mitotic separation at telophase
Crustacea		
Artemia salina	2*n*, 3*n*, 4*n*	Sexual and parthenogenetic forms worldwide; first division anaphase nuclei fuse to give a diploid second division, one nucleus of which is expelled as a polar body, the other remaining as the egg nucleus
Isopoda		
Trichoniscus elizabethae var. *coelebs*	3*n*	2*n* sexual form in dry, unglaciated regions of southern France, 3*n* ranges north to Sweden and Finland; some 3*n* males formed from 3*n* females, so some fertilization may take place

Table 8.2. (Continued)

Species	Ploidy	Comments
Vertebrata		
Poecilia (Mollienisia) formosa	2n	Probably an interspecific hybrid in Texas and Mexico; eggs stimulated to develop by being penetrated by sperm from related sexual species
Ambystoma jeffersonianum group	3n	Several 2n sexual species and several 3n parthenogenetic species; pre-meiotic doubling of chromosomes; sperm stimulates development of egg
Cnemidophorus uniparens	3n, 4n	Parthenogenetic females are 3n and *XXX*, 4n = *XXXX* or *XXXY*, the latter males, indicating sexual reproduction with 2n species; premeiotic chromosome doubling

Source: Compiled from White, 1973; Suomalainen, 1962; Cuellar, 1977.

and asexual reproduction. The latter, of course, can thus partake of the advantages of both kinds of reproduction.

The several kinds of variant reproductive systems are depicted in Fig. 8.1, with representative members listed in Tables 8.1 and 8.2. In normal sexual reproduction, or *amphimixis,* a reduction in chromosome numbers is brought about by meiosis and is compensated for by the process of fertilization and the union of gametic nuclei. Thus, a haploid-diploid alternation of stages is a regular and recurring aspect of all sexual species. A functional asexual cycle consequently must have a suitable substitute for both meiosis and fertilization, two processes which are quite different from, and independent of, each other, and which in most species of higher plants are separated in time by the gametophytic generation rather than following each other sequentially as in animals. The various types of asexual reproduction can be conveniently grouped under the term *apomixis. Vegetative reproduction* is not a form of apomixis in a strict sense, but it is included in Table 8.1 and Fig. 8.1 for comparative convenience.

Still other species have retained a sexual mode of reproduction, but they have added various features, some of them sufficiently complex and ingenious to strain credulity. The genetic basis and origin of most deviant mechanisms are poorly understood, but each of them brings about an alteration in the cytogenetic structure of a species and, obviously, in the transmission of genes from one generation to the next.

Asexual Reproduction in Plants

Vegetative reproduction occurs in a variety of species. Sometimes it is the only means of perpetuation of the species, but usually it is a supplement to the sexual mode. Any part of the plant may be involved: underground stolons, rhizomes, bulbs, tubers and corms, surface runners, adventitious buds on stem joints or leaves, and sprouts from cut tree stumps. The commercial strawberry, for example, can reproduce itself effectively by means of runners which root at their tips or at leaf nodes; meiosis, fertilization, and seed formation are normal even though its ancestry includes interspecific hybridization between *Fragaria virginiana* of North America and *F. chiloensis* of South America. An arctic species of *Stellaria* (Caryophylaceae) and several species of *Potentilla* (Rosaceae) also reproduce by runners exclusively, but these are polyploid, seedless forms, with close sexual relatives having lower chromosome numbers. A group of *Opuntia* species of the American Southwest spread themselves over wide areas by stem joints which break off and root, but these are also flowering species which hybridize readily among themselves. The reproductive behavior of the common pond weed, *Elodea canadensis,* is determined by latitude and temperature. In northern waters it reproduces almost exclusively by fragmentation of floating stems or winter buds; in warmer waters to the south it reproduces by normal seed production as well as vegetatively. *Lilium tigrinum* is unusual among members of this large genus by producing *bulbils* in the axils of its upper leaves; it is also triploid, although what this might have to do with bulbil formation is not clear or certain.

Many other species have their flowering parts, and hence seed formation, replaced by structures called *propagules* which drop off and produce offspring directly, a process known as *vivipary.* It is a common device among a number of species of grasses and *Allium* (onion), virtually all of them aggressive weedy forms. For example, *A. carinatum* and *A. oleraceum* are two viviparous European species. The former is diploid in the Alps and $3n$ in northern Europe; the latter, which is more widespread, consists of $4n$ and $5n$ individuals. A cross with *A. carinatum* as the pollen parent and *A. pulchellum,* a related sexual diploid species, indicates that a single dominant gene governs the expression of vivipary, but in most other instances in which similar crosses can be made, sexuality appears to be dominant to asexuality.

Vegetative reproduction is, of course, a cloning phenomenon; all of the offspring are genetically identical to the plant producing them. Most such species, however, are not restricted to vegetative reproduction, and a constant supply of variations is injected into the population by the seeds produced sexually, a process fostered by the fact that outcrossing is the rule in such species.

Adventitious embryony is comparable to vegetative reproduction in that maternal inheritance, or cloning, is maintained, but it differs in that seed production occurs as part of the life cycle. An unreduced cell in some part of the ovule other than sporogenous tissue—i.e., nucellus, chalaza, or integument—gives rise directly to an embryo encased in maternally derived seed coats. It is a form of reproduction characteristic of many diploid *Citrus* species, but much more commonly it is associated with polyploidy at the tetraploid or higher level, as in the grass *Bouteloua* and in some sections of the blackberry genus, *Rubus* (Table 8.1). In virtually all species that have been thoroughly investigated, interspecific hybridization and outcrossing are involved. The genomes must have been sufficiently similar to permit crossing to take place but dissimilar enough to trigger a set of physiological changes which eliminated the gametophytic stage of the life cycle and to replace the events of meiosis and fertilization with the mechanism of adventitious embryony.

Most apomictic species have retained a semblance of the sporophyte-gametophyte-sporophyte alternation of generations. Referred to as *gametophytic apomixis* because a gametophyte is formed, the process utilizes either somatic or archesporial cells, the latter being cells that, under normal circumstances, undergo meiosis and a reduction division to form eventually a haploid embryo sac. If somatic cells are involved (*apospory*), these are usually of a nucellar or integumental origin, and a diploid embryo sac is formed directly by a series of cell divisions. If the cells are of archesporial or, as in ferns, sporogenous, origin (*diplospory*), the meiotic processes are missing or abortive, but a gametophyte of unreduced character results. Interestingly enough, these meiotic cells are scrubbed clean of ribosomes and mRNAs, as if preparing for the expression of the haploid genome which does not form. The nature of the stimulus inducing the appearance of either apospory or diplospory is unknown, as is its relation to unreduced tissues, but once formed, the unreduced gametophyte can take one of two courses of action if viable seeds are to be formed: (1) An egg can form and develop directly into a sporophyte (*diploid parthenogenesis*) or (2) some somatic cell of the gametophyte, usually a synergid, can undergo a series of mitosis to form the sporophyte (*apogamety*). It is possible that the gametophytes (embryo sacs), whether formed by apospory or diplospory, can produce relatively normal eggs and sperm and that these can unite during fertilization, but it is unlikely since there would be a jump in ploidy with each generation, and this seems not to have occurred as a regular phenomenon. If apospory or diplospory occurs in the sporophyte, diploid parthenogenesis or apogamety usually follows at the gametophyte stage, with apogamety being common among the aposporous ferns and parthenogenesis more so among the flowering plants (Fig. 8.2). None of the apomictic features are to be found among the gymnosperms.

Although interspecific hybridization and polyploidy are closely associated with apomixis, both seem to favor apomixis without being its cause directly. Where breeding experiments have been possible between facultative and sexual forms, a genetic basis for apomixis is indicated. As pointed out earlier, vivipary in *Allium carinatum* seems to be due to a single dominant gene, but this is perhaps an unusual situation. Studies carried out on guayule (*Parthenium argentatum*) illustrate

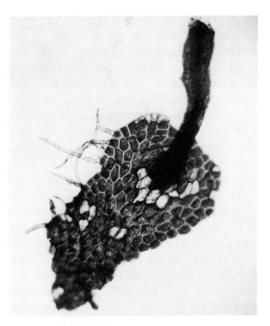

Fig. 8.2. An emerging sporophyte from a somatic cell of an unreduced gametophyte of the grammatid fern, *Xiphopteris sp.*, of French Guyana. The gametophyte contains no antheridia or archegonia (courtesy of Dr. D. Bierhorst).

this point more fully. Most of the apomictic forms of guayule are polyploid and self-incompatible, while the diploid individuals are sexual and self-compatible; occasional polyploids, on the other hand, may also be sexual. The results of many breeding experiments suggest that three pairs of genes could explain the behavior of individual plants and also indicate the sequence of events by which apomixis might have arisen. Assume that gene *a* in a homozygous condition leads to the formation of unreduced eggs (apospory or diplospory), gene *b* prevents fertilization, and gene *c* causes the eggs to develop parthenogenetically. Plants with the genotype *aaBBCC* would form unreduced eggs, but these could not develop apomicitcally. They might, however, be fertilized, in which case triploids would result. Plants with the genotype *AAbbCC* would produce reduced eggs, but embryos would not form because fertilization would be prevented. Those with an *AABBcc* genotype would have normal sexual behavior; *cc* would have no effect in the presence of *A* and *B* because the eggs would be reduced and fertilization would normally take place. Only those plants with an *aabbcc* constitution would be apomictic. In a population of mixed genotypes, with sexuality and self-compatibility the rule, segregation would occur, and fully sexual plants could give rise to apomictic offspring. Polyploidy in increasing degrees would serve to reinforce the genetic basis determining apomixis.

Hybridization appears to favor apomixis in two ways. First, it is the most appropriate method for bringing together diverse genomes that by chance have the proper combination of factors to promote apomixis. This must be a chance formation, because crosses of species closely related to other apomictic forms do not necessarily yield hybrids that are apomictic, thus leaving little doubt that hybridization is only a means to an end rather than direct cause of apomixis. Second, the heterotic effects often found in hybrids allow for a greater expansion in the range of ecological habits over that found in the parent species, and apomixis would tend to preserve such adapted genotypes through maternal inheritance if they should occur. Facultative apomixis is consequently a device that, on one hand, permits the mass production of seeds of like genotype, and on the other, preserves a store of potential variability that may be later released through occasional sexual seed production.

From a taxonomic point of view, apomixis creates a complex and difficult problem, because hybridization, polyploidy, and apomixis tend to confuse clear-cut differences between species. This is particularly true because most apomicts are not strict obligates as to mode of reproduction, exhibiting occasional, but successful, sexual reproduction. In such genera as *Rubus, Crepis, Hieracium, Crataegus, Bouteloua,* and *Poa,* thousands of "species" have been described, named, and perpetuated in the literature, but once their source is known it is clear that a species designation is taxonomically inappropriate. Each variant, however, possesses the potentiality for preservation and further proliferation, leading to what have been termed *agamic complexes.*

The structure and origin of an agamic complex can be visualized by considering the species of *Crepis* in western North America. Seven diploid sexual species have a base number of 11 chromosomes ($2n = 22$, a number which may itself be of an ancient tetraploid character), but through hybridization they gave given rise to a polyploid, apomictic series having somatic numbers of 33, 44, 55, up to 88. The sexual species are much more restricted in area as compared to the polyploid derivatives (Fig. 8.3) and are morphologically and ecologically

Fig. 8.3. Distribution of the diploid and polyploid forms of *Crepis* in the western United States. Many of the polyploid forms are apomictic as well. E. B. Babcock and G. L. Stebbins Jr., *Carnegie Inst. Wash. Publ.,* 504 (1938).

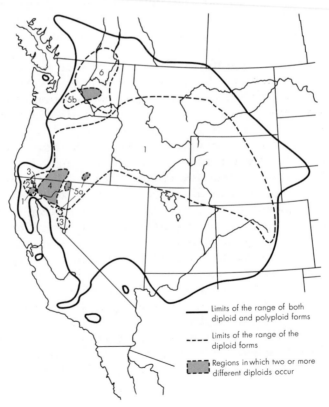

Limits of the range of both diploid and polyploid forms

Limits of the range of the diploid forms

Regions in which two or more different diploids occur

distinct from each other. The apomicts, however, show all manner of gradations of morphology and ecological preference, but these possess no new characters to distinguish them sharply from their diploid progenitors. Some apomicts show combinations of characters which suggest that more than two diploids must have been involved in their formation in some stepwise fashion. They are, therefore, variations on a theme rather than new and distinctly different members of the genus, and as a result of their polyploidy they are most unlikely to contribute to further speciation. Some of the plants in the complex produce occasional offspring through sexual reproduction, thus contributing further to the segregation, recombination, and shuffling of generic characters. The polymorphism of the complex is thereby extended, and apomixis ensures the perpetuation of any strain that shows ecological fitness.

Parthenogenesis in Animals

Reproduction by detached groups of somatic cells, or by the fragmentation of individuals, comparable to vegetative reproduction in plants, occurs only among the more primitive invertebrates such as sponges and some worms, but among the normally sexual groups parthenogenesis is widespread in the animal kingdom (Table 8.2). It can also be induced artifically by stimulation of the egg, even in a vertebrate such as the rabbit, so that its natural occurrence is somehow related to overcoming the developmental barrier that is normally breached by fertilization. *Haploid parthenogenesis* is both a form of reproduction and a means of sex determination. Fertilized eggs develop into diploid females, unfertilized parthenogenetic ones develop into haploid males. As far as is known, haploid parthenogenesis has evolved independently six or seven times in the animal kingdom: in the Rotifera, Acarina (mites), and in four orders of insects: in all members of the Hymenoptera, a few species of coccids and whiteflies in the Homoptera, a single species, *Micromalthus debilis,* in the Coleoptera, and occasionally in the Thysanoptera (thrips). Oogenesis in these forms is essentially normal. However, because the males are haploid, spermatogenesis cannot follow a normal course of events. The first meiotic division is generally abortive. A monopolar spindle develops, but the chromosomes do not divide so that a reduction in chromosome number does not take place. Instead, a non-nucleated bud of cytoplasm may be pinched at the polar end of the cell, all of which suggests that this form of chromosome and cell behavior was derived from diploid forms by suppression of meiosis and by making unnecessary the stimulation of fertilization for development. In some forms the bud is pinched off at the end of an abortive second division. Meiosis can be considered to have taken place in the mites and whiteflies of the Homoptera since sperm are formed, but the divisions are essentially mitotic in nature. This is true also in male Hymenoptera, that is, the second division is normal and mitotic in character, and two instead of four spermatids result. That this type of spermatogenesis is genetically fixed is indicated by the fact that it occurs also in the rare diploid males of *Habrobracon,* where homologous chromosomes capable of pairing in meiosis are available.

In an evolutionary sense, male haploidy means that every mutation is immediately exposed to the selective action of the environment. The same, of course, is true for any haploid organism. A greater degree of homozygosity and far fewer lethals are, therefore, to be expected in such populations as compared to those of normal diploids where a higher level of recessive heterozygosity can exist. In this sense, the entire haploid set of chromosomes is comparable to the X chromosome of those species, including the human, which is exposed in an unprotected state in males. For example, where comparative data have been obtained from natural populations, *Drosophila* X chromosomes do not contain the frequency of lethal and sublethal genes found in the autosomes, the latter being always paired in males and females.

Diploid parthenogenesis produces only females from unfertilized eggs. If obligatory, the population consists only of females (occasional males have been reported, but their source is not known and they seem not to be necessary for the continuation of the species). *Facultative parthenogenesis* is rare in animals as compared to its frequency among higher plants, perhaps because of chromosomally based sex-determining mechanisms, and is used to describe those instances of normally bisexual species giving rise to egg development without fertilization. Thus, in the phasmids (walking sticks), one species had 44% of its unfertilized eggs hatch and another species had 66% hatch. To what extent this would occur in a naturally breeding population is uncertain, but by selection the frequency of parthenogenesis in *Drosophila mercatorum* can be increased from 1% to about 6%. The females obtained in this manner seemed to be homozygous for all loci, indicating that meiosis took place and was followed by a doubling of the chromosomes in the haploid egg. *Cyclical parthenogenesis* is also known, and in such groups as the aphids sexual reproduction alternates with parthenogenesis in a systematic manner.

Obligate parthenogenesis can be a closed genetic system, but it need not be; it all depends upon whether in the formation of an egg meiosis and the recombination of genes have been abolished. A group of obligate organisms can be considered a clonal line of relatively unvarying genotype, more or less subject to the vicissitudes of a changing environment, although random mutations and chromosomal changes can accumulate slowly and be a source of variation. It also appears that parthenogenetic individuals cannot compete successfully with their sexual progenitors and, consequently, do not, as a rule, co-exist with them, but the advantage of the parthenogenetic systems is that the rate of increase can be doubled since only females are produced and a colony can be established by a single functional female. The fact that they do not compete well with the sexual forms suggests that their best chance for establishment is at the periphery of the range of the species or where new territory is available for invasion. Moreover, most obligatory groups are polyploid and are thereby provided with a buffered genotype that can exist and reproduce in more diverse and more extreme ecological situations than can their related diploid ancestors.

Some obligatory forms undergo a regular meiosis, with the somatic number of chromosomes being restored by fusion of the egg nucleus with one of the polar bodies, by fusion of the two second-division metaphases followed by a normal second-division anaphase and telophase, or by a fusion of two of the cleavage nuclei, introducing thereby a wider range of possibilities for genotypic and phenotypic variation. Even so, the entire range of variability is limited by the degree of heterozygosity of only the single parent, and once homozygosity is established for any given gene it must remain so except for mutations and chromosomal alterations. It is not surprising, therefore, that obligatory diploid parthenogenesis does not characterize an entire assemblage of species and that its sporadic occurrence in the animal kingdom is indicative of repeated independent origins. It is possible, even likely, that the impetus to move in a parthenogenetic direction is brought on by some meiotic disturbance that interferes partially with normal sexual reproduction. Subsequent mutations could reinforce such circumstances and lead to the establishment of an obligatory system.

Cyclical parthenogenesis, on the other hand, utilizes the advantages of both parthenogenetic and sexual reproduction. Such behavior characterizes many of the aphids, and *Tetraneura ulmi* can serve as an example. The fertilized eggs winter over, and in the spring the female nymph forms a leaf gall on the European elm. Within the gall, development of a wingless adult form is attained, followed by the parthenogenetic production of about 40 winged offspring, which migrate to a summer food plant (in the case of *T. ulmi,* grasses are the summer host). A succession of generations is produced through parthenogenesis (the exules), the last of which, coinciding with the approach of cold weather, includes winged sexual males and parthenogenetic females. These return to the elm, their winter host, where winged sexual females are produced, and together with the winged males of the previous generation will produce the fertilized eggs for wintering-over.

Chromosome Diminution and Elimination

In presenting the basic facts of cytology and genetics it is usual to state that the chromosomal compositions of the germinal cells and of the cells of the soma are equivalent. In plants this is particularly true, because the same meristematic cells that give rise to the vegetative organs at one stage of life cycle can give rise, at a later stage, to the sporogenous tissues in which meiosis occurs. Vegetative propagation, including vivipary and adventitious embryony, is practical proof of this underlying assumption, for the offspring are clonal derivatives of the parent plant. In many animals, however, the germ line is defined early in the embryonic development, and although these cells are not isolated from the soma in a physiological sense, they can be considered to be so genetically. The soma of an animal represents an evolutionary dead end, lost from a population with the death of the individual, and any changes occurring in it affect only that individual. It has now become quite clear that the somatic cells of many species may undergo cytological changes of various sorts—some random, some peculiar to various organs—even while the germ cells or their progenitors may preserve the original chromosomal constitution. The constancy of the species is dependent upon the constancy of behavior of the cells of the germ line.

It was Boveri who first pointed out that in the horse nematode, *Parascaris equorum* (*Ascaris megalocephala*) (Fig. 8.4) the chromosomal compositions of the soma and the germ line were not identical, and similar circumstances have since been found to be a regular occurrence in a number of species, particularly in certain groups of insects. The variations are, however, not confined to the soma; equally dramatic changes are known to occur in the cells of the germ line. A good example is the genus *Sciara* (fungus gnat), although the dog rose, *R. canina,* the gall midges, and a number of coccid (Homoptera) species also display interesting forms of chromosomal elimination and/or unusual behavior.

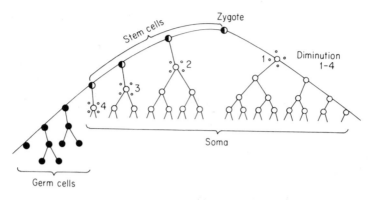

Fig. 8.4. Diagram of the process of chromosomal diminution as it occurs in the early cleavage divisions of *Parascaris equorum*. The compound, or possibly polycentric, chromosomes lose their heterochromatic ends as the germ line segregates from the soma, and the euchromatic portion of the chromosomes break up into many smaller ones which, subsequently, go through regular mitoses to form the multicellular soma.

Sciara displays a unique set of chromosomal mechanisms. All species behave in much the same manner, and the description to be given of *S. coprophila* is fairly representative. In most organisms there is at least one member of each pair of chromosomes in every cell of the body. In the germ line of *Sciara,* however, there are present, in addition to the basic set of four pairs of chromosomes, a variable number of anomalous chromosomes, distinctive in appearance and behavior. These are the *limited chromosomes,* so called because of their partial restriction to the germ line. They are apparently devoid of mendelizing genes so far as any genetic evidence can reveal, and they can be recognized by their heteropycnotic appearance during mitosis. During the early cleavage stages of the embryo, they are eliminated from the somatic nuclei and pass into the cytoplasm, where they disintegrate.

Oogenesis in *Sciara* is normal in all respects, but spermatogenesis exhibits peculiarities of segregation and behavior. The unequal contributions to the zygote by the egg and the sperm are a reflection of these peculiarities. The egg pronucleus, following meiosis, contains a haploid set of chromosomes consisting of an X, three autosomes, and one or more limited chromosomes. These are of paternal or maternal derivation, segregation in oogenesis apparently being random. The sperm, however, in addition to one or more limited chromosomes, contributes two sister X chromosomes and three autosomes, all of maternal origin.

Figure 8.5 indicates the course of events as they occur in the soma and germ line of both sexes. The first spermatocyte division is a monocentric mitosis, the spindle being monopolar, and the maternal homologues are separated from those of paternal origin.

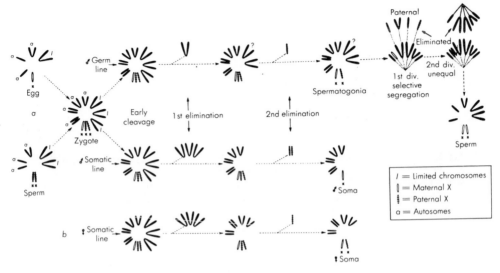

Fig. 8.5. Chromosome history during somatic development and gametogenesis in the male *Sciara coprophila* and somatic development in the female. Oogenesis follows a regular course of events. (Crouse, 1943).

The maternal chromosomes move to the single pole while the paternal ones back away from the pole, even though their centromeres are directed poleward. This is true for the X chromosome and the autosomes only; the limited chromosomes, whatever their origin, pass to the single pole along with the maternal X and autosomes and are incorporated in the secondary spermatocyte. The paternal chromosomes collect in a tiny bud that is pinched off and which later disintegrates. Males of *Sciara,* therefore, never transmit traits of paternal origin.

The division in the single remaining secondary spermatocyte is also anomalous and unequal. The spindle, unlike that of the first division, is bipolar, and the chromosomes reach the metaphase plate in normal fashion. One chromosome, however, is precocious in its behavior, and it passes to one pole before anaphase movement has commenced for the remaining chromosomes. Because it is longitudinally divided, both chromatids pass to the same nucleus. Anaphase movement of the remaining chromosomes then takes place, giving two nuclei of unequal constitution. The nucleus bearing the two halves of the precocious chromosome becomes incorporated into a functional sperm; the other nucleus degenerates. By means of translocations between the X chromosome and the autosomes, it has been demonstrated that the precocious chromosome is the X chromosome and that its centromere is responsible for its early anaphase movement and eventual passage into the functional sperm nucleus.

Elimination of chromosomes from both male and female soma follows a constant and unique pattern. As stated earlier, the zygote contains three X chromosomes (two of paternal origin), three pairs of autosomes, and one or more limited chromosomes (three are indicated in Fig. 8.5). This complement is maintained through the first few cleavage divisions only. At the fifth or sixth division, the limited chromosomes are eliminated. They reach the metaphase plate in normal fashion along with the other chromosomes, their centromeres divide, and anaphase movement begins, but owing to anaphase lag they are left in the middle of the spindle and are not included in the daughter nuclei. It would be of considerable interest to know whether the lag is due to a delayed or interrupted pattern of replication which would interfere with anaphase movement, and if so, how the delay is controlled.

At the seventh or eighth cleavage division, an additional elimination of chromosomes from the soma occurs. In this instance it is the paternal X chromosome that is involved, and the mode of elimination is similar to that which removes the limited chromosomes. The circumstances differ, however, depending upon whether the soma is to become male or female. If male, the elimination of two paternal X chromosomes takes place; if female, only one X chromosome is deleted from the complement, and this again is one of those of paternal origin. The male soma consequently is XO, the female XX.

After the germ cells, which possess the zygotic complement of chromosomes, have been set aside from the soma, there is a third and a fourth elimination. There must first be an elimination of one or more of the limited chromosomes in the germ line as well, although this has not been cytologically described. A paternal X chromosome is then lost. The method of elimination in this case differs, however, from those previously described, because elimination appears to take place from resting nuclei, with the X chromosome moving directly through the nuclear envelope into the cytoplasm, where it disintegrates. The male and female germ cells consequently are identical in chromosomal constitution, and it must be assumed that the nature of the ovarian or testicular germinal tissue is determined by the constitution of the soma.

The mechanism by which zygotes identical in chromosomal constitution develop into males or females is not well understood. From the genetic studies that have been done, it appears that the male is without influence in the determination of sex. *S. coprophila* generally produces unisexual broods, and it has been suggested that the male producers differ from the female producers in the type of X chromosomes in the mother and that the female producers are heterozygous (XX′), the male producers homozygous (XX). Since no difference has been detected cytologically between the two types of X chromosomes, it can be assumed that the genetical composition of the mother determines the type of somatic elimination that will occur, and that in turn will determine the sex.

The dog rose of Europe, *Rosa canina,* and its widespread relatives are tetraploid and pentaploid species of hybrid origin. No certain ancestors are known, their self-compatibility permits the retention of hybridity and polyploidy despite their highly unusual meiotic divisions, but enough outcrossing has taken place to give rise to a large array of microspecies which, in turn, also breed true. Their somatic number of chromosomes is 28 or 35 (the basic number for the genus *Rosa* is 7), and in both micro- and macrosporogenesis 7 pairs of homologous chromosomes in the form of bivalents and either 14 or 21 univalents arrange themselves on the metaphase plate in meiosis. Since all species behave similarly, the pentaploid *R. canina* will serve as an example. The 7 bivalents, which form chiasmata and presumably undergo crossing over and gene recombination, segregate regularly to give each reduced pollen grain and embryo sac a set of these chromosomes. The 21 univalents behave quite differently, depending upon the type of cell. In the microsporocytes the bivalents divide first and pass to the poles, to be followed by the univalents which have divided equationally. An occasional univalent fails to be included in the telophase nucleus because of segregational irregularity and is lost. The process in the second meiotic division is similar: The 7 normal chromosomes segregate regularly while the univalents either are distributed to one pole or the other or are lost in the cytoplasm. It has been stated that the univalents divide at both meiotic divisions, but this seems most unlikely since it presupposes either a polynemic chromosome in the pre-meiotic cells or another round of replication between meiotic divisions, neither of which has any experimental support. In any event, a high degree of pollen sterility results. Most of the pollen grains that can germinate and undergo division to form the tube and generative nuclei contain 7, 8, or 9 chromosomes, with the 7-chromosome ones having the best chance of reaching the egg. Only occasionally are aneuploid plants encountered.

In megasporogenesis the 7 bivalents are likewise formed, but the univalents, instead of segregating irregularly as in microsporogenesis, are precociously grouped at the micropylar pole of the spindle. Here they are joined by the 7 segregating chromosomes, with all being included in the telophase nucleus. The second division proceeds regularly to give a linear tetrad of two large megaspores in the micropylar position, each with 28 chromosomes, and two small chalazal megaspores, each containing but 7 chromosomes. One of the large megaspores becomes the functional embryo sac, and on fertilization by a 7-chromosome sperm the somatic number of 35 is restored. Somatic numbers higher or lower than 35 are also found, but the 7-chromosome pollen tubes are the most successful in competition.

Obviously, the recombination and segregation of genes in *R. canina* can occur only for those located in one of the 7 pairs of homologues which form bivalents. Whatever genetic influence the 21 univalents might have is transmitted maternally as a block, but that each of them carries a functioning set of genes is indicated by the fact that monosomics, that is, those with 34 or fewer chromosomes, have reduced vigor and fertility. From the meiotic picture obtained from *R. canina,* the karyotype could be represented as *AABCD,* with *A* representing the 7 basic chromosomes which can pair with their homologues, and with *B, C,* and *D* the groups of univalents which seem to show no homology among themselves or with the basic seven. When a cross is made between *R. canina* and *R. rubiginosa,* the F_1 hybrids show 7 bivalents regularly forming in meiosis, although some multivalents are found occasionally. However, a cross between *R. canina* and the diploid *R. rugosa* yields hybrids in which 11 to 14 bivalents are found; the only explanation that fits the facts would be that there is "internal segmental polyploidy" and that some of the univalents can pair among themselves. The karyotype of *R. canina* can therefore most accurately be represented as $A_1A_1A_2CD,$ and the *canina-rugosa* hybrid as $A_1A_2CCD.$ Since the roses of this group can be freely crossed, the several sets of 7 chromosomes share homologies with each other, but only in the hybrids do these partial homologies reveal themselves. Consequently, the system is analogous to the translocation complexes of *Oenothera* in that heterozygosity is preserved at the same time that a minimum of recombination is permitted. In *Oenothera,* however, the difference between complexes lies in the different gene content of the internal regions of the chromosomes which do not pair and hence do not undergo recombination; in the dog roses, however, each complex has its own gene content, and since hybridization occurs rather freely, microspecies which breed true are continuously being formed.

The role of the limited chromosomes has remained obscured,

but recent work on gall midges sheds some light on the problem. These insects are taxonomically related to the sciarid group and share some of their chromosomal behavior. In the cecidomyiid midge, *Wachtiella persicariae*, the 2n number is 40, but the sperm contains only 8 chromosomes, the reduced egg 32. The extra 24 in the egg are E, or *elimination, chromosomes* and appear comparable to the limited chromosomes of *Sciara*. In spermatocytes, the 24 are eliminated, and only 8 pairs show regular pairing and segregation; in the egg, the 8 plus the 24 E chromosomes pass into the pronucleus. The 24 E chromosomes are eliminated from the soma of both sexes in the early cleavage divisions. However, by appropriate centrifugation of early embryos, it is possible to delay migration of the germ cells to their gonadal site. The E chromosomes are then eliminated from all nuclei. Migration of the germ cells then takes place, and development of the embryo proceeds. The embryos and larvae have a normal appearance and development, but the pupae are abnormal in that eggs are arrested in formation and sperm are absent. It would appear, therefore, that the E chromosomes are necessary for the development of functional gonads and/or gametes, but they seem to serve no other purpose.

A circumstance similar to the selective elimination or retention of particular members of the karyotype has previously been described in Chapter 3, in this instance the lyonization of one of the X chromosomes in the soma of the mammalian female. As was pointed out, the human X chromosome to be shunted aside genetically is randomly selected by some as yet undetermined mechanism unless it has an aberrant morphology, in which case it is the aberrant X that is lyonized. In mules it is the donkey X, as compared to the horse X, that is heterochromatinized about 90% of the time. It was also pointed out that in some of the marsupials the selected X is eliminated from the somatic nuclei rather than being simply heterochromatinized and replicated at each cell division. Whether such an elimination would be possible in human females is uncertain. XX and XO karyotypes do not induce the same somatic phenotype, but whether the difference is due to the presence or absence of a lyonized X in somatic cells, or to the presence or absence of a second functional X in the germ line, where it might exert an influence hormonally, is not known. In any event, the E chromosomes of the gall midges and the mammalian X chromosome to be lyonized share similar functions in that, in the presence of the remainder of the genome, they are necessary for normal gonadal and embryonic development, but are dispensable for the remainder of somatic growth and differentiation.

A final example of selective chromosome elimination, in this case for purposes of sex determination and through the mechanisms of nondisjunction, is provided by the vole, *Microtus oregoni*. The male soma is XY, with $2n=18$, but the male germ-line cells contain only 17 chromosomes with the X chromosome missing. Consequently, these cells are OY as a result of nondisjunction occurring early in the formation of the germ-line, with OY, but not XY, cells being capable of developing into testicular tissue with spermatogonia and spermatocytes. The male reduced gametes, therefore, are either O $(n=8)$ or Y $(n=9)$. The female soma, however, is XO $(2n=17)$, but the oogonia are XX by virtue again of nondisjunction early in the development of the female germ-line, and with both O and XX cells being produced. Only the latter develop into functional oogonia. Female offspring result, therefore, from the fertilization of an X egg by an O sperm and males when a Y sperm unites with an X egg.

Cytology of the Coccids

The cytological literature contains numerous other examples of the selective behavior of the karyotype, in whole or in part, and of the exquisite controls built into the genotype to guide or direct this behavior, but no group approaches, in diversity or uniqueness, the chromosome cycles found in the scale insects (*Coccoidea* of the order Homoptera). Although an XX-XO sex-determining is probably primitive for the group, various species may exhibit male haploidy; parthenogenesis; heterochromatinization of the paternal set of chromosomes; elimination of paternal chromosomes either as a set or randomly as individual chromosomes; and involvement of the polar bodies in embryogeny.

Fig. 3.13 depicts the typical somatic chromosomes of the coccids. The centromeres are diffuse, so the chromosomes lie parallel to the metaphase plate with a sheet of microtubules extending from their surfaces to the poles. Each chromosome at metaphase, however, is made up of four half-chromatids, each of which is fully compacted and distinctly separate from its sister half-chromatid. The half-chromatids are clearly visible in anaphase as well, but they apparently undergo reassociation with their sister half-chromatids prior to the next prophase. It is not clear how the existence of half-chromatids, which must result from a replicative process occurring a whole cell cycle in advance of when separation takes place, squares with the general view that a chromatid consists only of a single double helix of DNA; no studies with labelled isotopes have yet been performed to explore this matter.

I. NAUTOCOCCUS

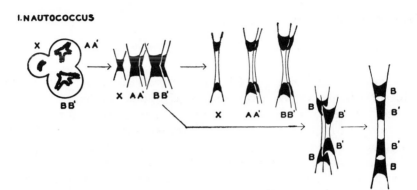

2. NAUTOCOCCUS ASYNAPTIC A A'

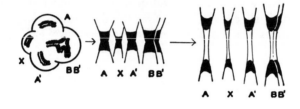

3. PROTORTONIA

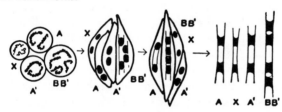

Fig. 8.6. Meiosis in two genera of coccids. 1. The spermatocyte nucleus of *Nautococcus* is partially vesiculated, synapsis of the autosomal homologues is regular, terminalization is complete by metaphase, flaring spindle elements are formed by each chromosome as the chromatids orient on the metaphase plate, and anaphase disjunction is equational even when the *B* and *B'* find themselves in the same spindle; 2. the same as above except that the *A* pair of homologues are asynaptic; 3. vesiculation is complete as is asynapsis, the vesicles become biacuminate and fuse to form a composite spindle, and with the first division being equational (Hughes-Schrader, 1948).

A further deviation during cell division is indicated in Fig. 8.6. Each meiotic chromosome or bivalent may be enclosed in a separate vesicle instead of in a single nucleus. A beginning of this may be seen in the genus *Nautococcus* where a tendency toward vesiculation of the nucleus is evident. When asynapsis of the A pair of homologues occurs, each chromosome is found in a separate lobe of the nucleus. Vesiculation is complete in the genus *Protortonia,* as is the failure of the A pair of homologues to synapsis.

As the meiotic cycle approaches metaphase, the vesicles become transformed into elements of the spindle. The spindle, therefore, is of composite origin. The chromatids of each chromosome or bivalent orient in single-file fashion within the tubular spindle elements, fibers flare out from the vesicles toward the poles, and anaphase movement is accomplished by a constriction of the tubular spindle elements and a shortening of the fibers. The first meiotic division of all paired and unpaired chromosomes is equational since sister chromatids segregate from each other; this is made certain even for the paired homologues (BB') by the linear insertion of the two chromatids of one homologue between the chromatids of the other homologue (Fig. 8.6). In the second division of meiosis, the autonomy of the X chromosome is revealed by the fact that it will pass to one pole by becoming inserted between the chromatids of one of the segregating autosomes (Fig. 8.7). If the autosomes behaved as univalents during the first division, their nonsister, but homologous, chromatids will reassociate during the second meiotic prophase, and then segregate from each other at the second anaphase.

The lecanoid family of the coccids exhibits many of the features of the entire group discussed above while introducing new ones reminiscent of the elimination events which take place in *Sciara* and the gall midges. As Fig. 8.8 indicates, four general types of meiotic behavior can be recognized. The genus *Puto* is believed to be a primitive member of the group. It has an XX–XO sex-determining system, and the meiotic events are depicted in Fig. 8.9. The remaining genera of the lecanoid assemblage, however, exhibit a variety of unusual behaviors. In the somatic cells of the male *Phenacoccus acericola* the paternal set of chromosomes becomes heterochromatinized at the blastula stage of development, and remains so for the duration of the life cycle. The males, therefore, are physiologically haploid, but karyologically diploid. That it is the paternal set of haploid chromosomes that becomes altered has been demonstrated experimentally: if fathers are x-rayed, the fragmented chromosomes are found only in the heteropycnotic set of the sons; if the mothers are x-rayed, the breaks are only in the euchromatic set. Some sterility in the sons is evident when high doses of x-rays are used, indicating that the paternal chromosomes, like the E chromosomes in the cecidomyiid midges, play a developmental role of some sort in the germ-line. As might be expected, it is also the paternal set which replicates later in the S period.

Meiosis in *P. acericola* is asynaptic. The paternal set of chromosomes is heteropycnotic in prophase, and is located as a clump at one side of the nucleus. By metaphase all of the chromosomes are equally

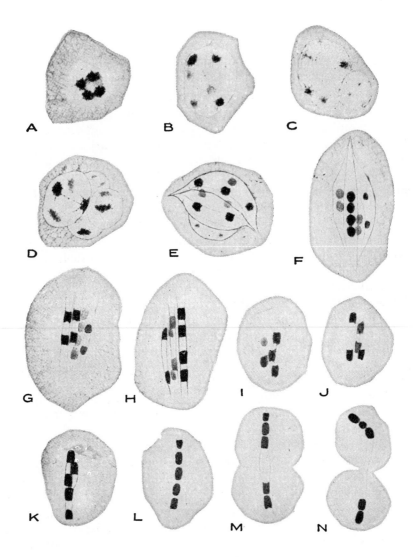

Fig. 8.7. Meiosis in *Protortonia primitiva*. A - D, meiotic prophase, illustrating the vesiculation of the nucleus; E and F, the bipolarity of the nucleus established by the stretching and fusion of the vesicles to give a composite spindle; G and H, anaphase of the first division; I - L, fusion of the spindle elements in the secondary spermatocytes to give a linear aggregate of five chromosomes; L, the aggregate consists of the undivided X and the separate chromatids of the two autosomes; M and N, anaphase II with the X terminal in the upper group in M, and central in the upper group in N. The chromatids of the two autosomes must alternate in the linear aggregate (as indicated in Fig. 8.6) in order that each spermatid receives its proper chromosomal complement; K illustrates how the X can be inserted into the linear aggregate (Hughes-Schrader, 1948).

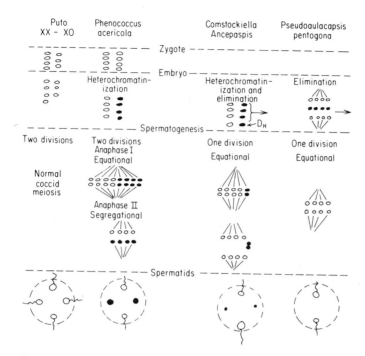

Fig. 8.8. Some of the systems of meiotic behavior, heterochromatinization and chromosome elimination in the coccids (modified from Brown, 1963). See text for explanation.

condensed, each orients individually on the metaphase plate, and segregation is equational. The second division finds the paternal set returning to a heteropycnotic state. A unipolar spindle forms at each pole and separates them from the more diffusely staining maternal set, leading to the formation of a quadrinuclear cell, two nuclei of which degenerate (these contain the paternal chromosomes) while the other two form functional sperm. Oogenesis seems to be normal in all respects.

In *Comstockiella sabalis,* one set of chromosomes, presumably paternal in origin, is heterochromatinized as in *Phenacoccus,* but a difference appears at the time of spermatogenesis. The heterochromatic set, with the exception of one particular chromosome (referred to as D_H to distinguish it from its euchromatic homologue D_E), degenerates and is lost prior to meiosis. A single division takes place, with the euchromatic chromosomes, including the D_E member, dividing equationally, while the D_H either lags at anaphase and is lost, or is ejected from the telophase nucleus. Only two functional sperm develop. A further variation of the system is found in another

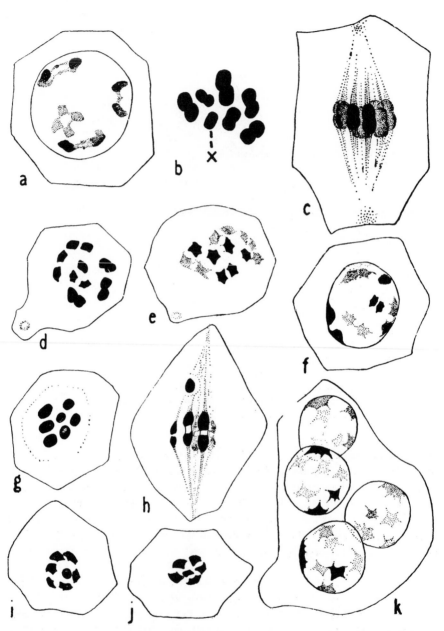

Fig. 8.9. Meiosis in male *Puto*. a, diakinesis; b and c, polar and side views of metaphase I (there is no vesiculation of the nucleus in this genus); d and e, secondary spermatocytes following equational division in anaphase I; f, reassociation of homologous chromatids in a meiosis II nucleus leading up to metaphase II (g); h, early anaphase II with the X chromatid passing first to one of the poles; i and j, two spermatid nuclei in telophase; k, the quadrinucleate spermatid from which four sperm will be formed (Hughes-Schrader, 1948).

member of the *Comstockiella* group, *Ancepaspis tridentata* ($2n=6$). As in *C. sabalis,* a single heterochromatic chromosome enters meiotic prophase, but it may be any one of the three haploid chromosomes (they are distinguishable from each other on the basis of size), with inclusion being a matter of random choice (Fig. 8.10). There is some suggestion that the single D_H chromosome in *C. sabalis* (it is always the same one) might be an X chromosome playing some essential role in spermatogenesis, but it is difficult to see how this suggestion can be applied to the situation in *A. tridentata* unless one postulates a moveable "sex realizer."

The fact that it is the paternal set of chromosomes which becomes heterochromatinized would suggest that these chromosomes are somehow imprinted by passage through the father. This may well be so in some, but not in all, instances. The rare males of the lecanoid species, *Pulvinaria hydrangeae,* are diploid, exhibit one set of heterochromatinized chromosomes, but have both sets derived from the mother through a parthenogenetic process which involves the

Fig. 8.10. Meiosis in the scale insect *Ancepaspis tridentata.* The testis consists of a series of cysts; in each cyst the spermatocyces are all alike in that they contain a haploid set of three chromosomes plus a homologue for one of the three, the D_H member; which of the three chromosomes is in a diploid state is a matter of chance in contrast to the situation obtaining in *Comstockiella sabalis* where the extra chromosome (D_H) is always the same one. It is not known when the homologues for the other two chromosomes in *A. tridentata* were lost. The two chromatids of the D_H chromosome do not enter a nucleus, but form pycnotic bodies in a binucleate spermatid; the size of the pycnotic bodies reveals which of the three chromosomes contributed the D_H member (modified from Brown and Nur, 1964 and Kitchin, 1970).

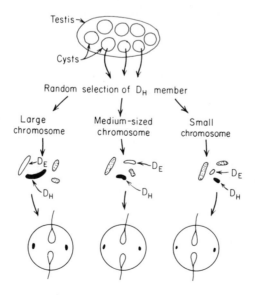

fusion of two nuclei, both of which are derived from the egg pronucleus. Cytoplasmic factors or position may have been involved, but at least no male influence entered as a determining feature.

The sequence of events is carried still another step along a given path of evolution in the diaspidid, or armored, scales. The males of *Pseudoaulacapsis pentagona* arise from fertilized eggs, but the paternal chromosomes are eliminated in the late blastula stage; the males consequently are true haploids, and their meiotic processes are reduced to a single mitotic-like division to produce to two sperm. In *Icerya purchasi,* the cottony-cushion scale, fertilization has been dispensed with, haploid parthenogenesis occurs, and a haplo-diploid sex-determining system similar to that in the Hymenoptera is in force. On the other hand, the *I. purchasi* male is a rarity even though the "females" arise from fertilized eggs. Rather than being true females, they are, in fact, self-fertilizing hermaphrodites. The hermaphroditic gonad consists of a center of haploid cells. When the reduction to a haploid state occurred is not known, but the hollowing out of the gonad results from a number of single-division meioses which produce bundles of sperm from the gonadal cysts. The oogonia, with their nurse cells, develop in the ovarian wall of the gonad, and as they mature and undergo meiosis, they are fertilized by the sperm which are freed into the central cavity.

The sexual flexibility of the coccids is made more evident by the existence of diploid as well as haploid parthenogenesis. Males are relatively rare, and are sometimes absent, in a number of lecanoid species. Some seem obligatorily parthenogenetic since no males are known, e.g., *Aspidiotis hederae* and *Marchalina hellenica*, while others such as *Lecanium hesperdium* are sexual or parthenogenetic depending upon the presence and frequency of available males.

A last feature of coccid chromosomal behavior that might be mentioned is the use of polar bodies in development. Ordinarily polar bodies play no further role after their formation, although a well-known exception is that in the silkworm where fertilized polar bodies often sink back into the egg to contribute to somatic development and a recognizable mosaicism. In the coccids, the polar bodies are made use of in another way. *Pseudococcus citri* has a diploid number of 10. A polar nucleus containing 15 chromosomes is formed by the fusion of the first and second polar bodies: the first with 10 chromosomes resulting from the equational first division, and the second from the segregational (reductional) second division. This triploid nucleus divides several times, after which the cells fuse in pairs or with some of the early cleavage nuclei to give cells with 25, 30, or 35 chromosomes each. Endomitosis may double the chromosome number to form giant cells which then take in a fungal symbiont to become *mycetocytes*. The fungus is passed from one generation to the next through the egg. The mycetocytes, by division, become the source of the *mycetome organ* of the adult. Presumably the mutual advantage must be similar to that found in a lichen.

The cytological literature contains numerous other examples of selective behavior of karyotypes or of individual chromosomes, but those discussed are sufficient to indicate the control which the genotype has acquired to govern its own behavior and transmission. We have, however, little genuine understanding of how these systems have come into existence, although their confinement largely to the insect groups would suggest that short life cycles are conducive of a good deal of evolutionary experimentation that would be out of the question with other forms.

BIBLIOGRAPHY

Babcock, E. B., *The Genus Crepis*, Vols. I and II. Berkeley: University of California Press, 1947. (See especially Chapter 5 in Vol. II.)

Beatty, R. A., *Parthenogenesis and Polyploidy in Mammalian Development*. New York: Cambridge University Press, 1957.

Brown, S. W., "The *Comstockiella* System of Chromosome Behavior in the Armored Scale Insects (Coccoidea: Diaspididae)," *Chromosoma, 14* (1963), 360–406.

———, "Automatic Frequency Response in the Evolution of Male Haploidy and Other Coccid Chromosome Systems," *Genetics, 49* (1964), 797–817.

———, "Chromosome Systems of the Ericoccoidae (Coccoidea-Homoptera)," *Chromosoma, 22* (1967), 126–150.

———, and H. S. Chandra, "Chromosome Imprinting and the Differential Regulation of Homologous Chromosomes," in *Cell Biology: A Comprehensive Treatise*, Vol. 1 (L. Goldstein and D. M. Prescott, eds.), New York: Academic Press, 1978.

———, and U. Nur, "Heterochromatic Chromosomes in the Coccids," *Science, 145* (1964), 130–136.

Clausen, J., "Partial Apomixis as an Equilibrium System in Evolution," *Caryologia* (suppl.) (1954), 469–479.

———, *Stages in the Evolution of Plant Species*. Ithaca: Cornell University Press, 1951.

Crouse, H. V., "Translocations in Sciara: Their Bearing on Chromosome Behavior and Sex Determination," *Univ. Missouri Agr. Exp. Sta. Res. Bull., 379* (1943), 1-75.

————, "The Controlling Element in Sex Chromosome Behavior in Sciara," *Genetics, 45* (1960), 1429-1443.

Cuellar, O., "Animal Parthenogenesis," *Science, 197* (1977), 837-843.

Darlington, C. D., *Evolution of Genetic Systems.* New York: Basic Books, 1958.

Grant, V., *Plant Speciation.* New York: Columbia University Press, 1971.

Hughes-Schrader, S., "Cytology of Coccids (Coccoidea-Homoptera)," *Adv. Genetics, 2* (1948), 127-203.

Kitchin, R. M., "A Radiation Analysis of a *Comstockiella* Chromosome System: Destruction of Heterochromatic Chromosomes during Spermatogenesis in *Parlatoria olae* (Coccoidea-Homoptera)," *Chromosoma, 31* (1970), 165-197.

Manton, I., *Problems of Cytology and Evolution in the Pteridophyta.* New York: Cambridge University Press, 1952.

Nur, U., "Meiotic Parthenogenesis and Heterochromatinization in a Soft Scale, *Pulvinaria hydrangeae*, (Coccoidea-Homoptera)," *Chromosoma, 14* (1963), 123-139.

Rieffel, S. M., and H. V. Crouse, "The Elimination and Differentiation of Chromosomes in the Germ-Line of Sciara," *Chromosoma, 19* (1966), 231-276.

Stebbins, G. L., "Chromosomal Variation and Evolution," *Science, 152* (1966), 1463-1469.

————, "The Significance of Hybridization for Plant Taxonomy and Evolution," *Taxon, 18* (1969), 26-35.

————, *Chromosomal Evolution in Higher Plants.* Reading: Addison-Wesley, 1971.

Suomalainen, E., "On Polyploidy in Animals," *Proc Finn. Acad. Sci. Letters* (1958), 1-15.

————, "Significance of Parthenogenesis in the Evolution of Insects," *Ann. Rev. Entomology, 7* (1962), 349-366.

White, M. J. D., "Cytological Studies on Gall Midges (Cecidomyidae)," *Univ. Texas Publ., 5007* (1950), 1-80.

————, *Animal Cytology and Evolution* (3rd ed.). New York: Cambridge University Press, 1973.

9

Evolution of the Karyotype

T he preceding chapters have been largely concerned with cytogenetic phenomena as isolated problems, leaving in abeyance an
extensive consideration of their evolutionary origin or potential. But
just as a species has a history, so too does a chromosome and a
karyotype, and the changes, internal and external, which, in many
instances lead to reproductive isolation and speciation, are intimately
associated with karyological changes. Indeed, White has stated that
"evolution is essentially a cytogenetic process." It may be difficult
to accept so unequivocal a statement without qualification, but the
relevance of White's position rests upon the observation that even
closely related species are often separated from each other karyologically by chromosomal alterations of many kinds: some obvious,
others detectable only with difficulty, some understandable, and
others obscure as to origin and significance. To be sure, there are
species such as *Platanus orientalis* and *P. occidentalis,* separated from

each other by a hundred million years or more, which exhibit no obvious chromosomal differences between them, and which, when crossed, produce F_1 hybrids with seemingly complete meiotic pairing, high pollen fertility, and good seed set, but a great many other species, when compared with closely related forms, reveal numerous chromosomal changes: some small, some large, some common to large arrays of species, and some peculiar to one group or another. In the genus *Drosophila,* for example, the sibling species of *melanogaster* and *simulans,* quite similar to each other morphologically, yet reproductively isolated, differ by one large inversion in chromosome 3 and several smaller alterations in other parts of the genome; staining for C-bands reveals still further changes not immediately evident by other techniques (Fig. 9.1), but whose significance remains to be assessed. Their reproductive isolation, therefore, would appear to rest primarily on genic (molecular?) rather than obvious chromosomal changes, and with the genic alterations having a significant effect reproductively but only a minor one morphologically. *D. pseudoobscura* and *D. simulans,* two species somewhat more distantly related to each other, differ by chromosomal changes that would require more than 50 break points, followed by extensive rearrangement of chromatin; these changes are largely within rather than being between chromosome arms. *D. melanogaster* and *D. pseudoobscura,* still more distantly related, have had their chromosomes, on a com-

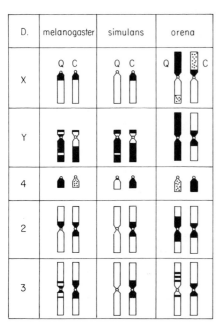

Fig. 9.1. Differences brought out by quinacrine mustard staining Q- and C-banding in the somatic metaphase chromosomes of *D. melanogaster, D. simulans* and *D. orena,* all members of the *melanogaster* subgroup. The karyotypes of *melanogaster* and *simulans* are generally alike in size and morphology, but differ in banding, while the X and Y of *orena,* an African species, are larger, contain more heterochromatin, and differ in position of centromere. Similar changes in the Y chromosome have also taken place in the *D. hydei* subgroup, but not to the same extent. The increased heterochromatin in *orena* is reflected in the fact that it has 30% satellite DNA as compared to only 6% in *simulans.* ☐ = no staining; ▨ = light staining; ■ = intense staining (redrawn from Lemeunier, et. al., 1978).

parative basis, so extensively repatterned that areas of homology, on a band-for-band basis, can be established only in a tenative way, even though it is more than likely that they share many genes in common. In species lacking polytene chromosomes, the difficulties of karyo-type comparisons, and hence of determining patterns of karyotype evolution, become compounded, and resort must be made to other techniques and other approaches if we are to understand the relation between chromosomal evolution and speciation in general. In a number of angiosperm genera such chromosomal changes have been identified; they are associated with self-compatibility and the occupation of marginal habitats, and, in concert, are responsible for the erection of reproductive barriers in a comparatively brief span of time. At the molecular level there is a growing realization that both single-copy and repetitive DNA undergo nucleotide changes with a finite frequency—a molecular clock, as it were—thus providing a means other than chromosomal to be made of the time span that related genomes have been separated from each other.

The chromosome, therefore, is like any other biological, time-sifted organelle. It changes with time, and its repatterning is subject to constraints imposed on it both internally by the organism and externally by the environment; that is, it must fulfill its physiological functions, whatever they might be, and it must be transmitted in division if it is to be kept within the species. Neither the individual chromosome, be it prokaryotic or eukaryotic, nor the multichromo-somal karyotype of a higher organism is a random affair; whatever their nature, all must be channeled by selection if they are to con-tribute to, and participate in, the existence of a species. Undoubt-edly, thousands of chromosomal changes arise and disappear without effect on the species, or even in the individual in which they occur; White, for example, estimates that in a particular Australian grass-hopper, a new translocation appears in one out of every 1,000 gametes, while virtually every individual in natural populations of the diploid species of *Tradescantia* possesses one or more inversions, although their rate of occurrence is not known. But the enormous diversity of cytogenetic changes, some of which have been described earlier, indicates that some pass through the mitotic, meiotic, and developmental barriers that any species gradually erects as it achieves specific uniqueness, and each change that is established, whether it replaces the old or attains a balanced polymorphism with it, provides a new genomic environment within which the next succeeding change must contend.

We have indicated that there are barriers which karyological changes must pass through, or surmount, if they are to be successful.

In mitosis the changes must be such that gains, losses, or rearrangements of chromatin, resulting from the breakage and reunion that takes place constantly, segregate properly at anaphase in addition to being physiologically viable. Ring, or circular, chromosomes may be characteristic of prokaryotes, but since in eukaryotes they frequently form interlocked or double-sized rings as a result of replication difficulties (Fig. 6.32), they cannot pass muster for any protracted period (the dinoflagellate chromosomes may provide a clear exception), and they seem to have had no significant place in eukaryotic history. In meiosis, pairing and crossing over (chiasma formation) are preludes to proper segregation and the production of balanced gametes or spores. Both may be dispensed with, but only if another mechanism—such as touch-and-go pairing, or the precocious movement of the X chromosome to one pole in XO males of grasshopper species—is substituted to ensure regular segregation. The most difficult barrier, however, to incorporated chromosomal change is a developmental one. As White has pointed out, a parthenogenetic species, including those among the higher plants which bypass the gametophytic generation, should be able to accumulate any number of chromosomal changes which can pass the mitotic or meiotic hurdle (most parthenogenetic species are facultative and possess some semblance of a meiotic process) and should therefore show a far higher degree of chromosomal heterozygosity than the usual sexually breeding forms, but the fact that it does not can only be interpreted to mean that the changes that lead to a repatterning of the chromosomes are more often than not developmentally detrimental, and hence are not established in the genome. Chromosomes and karyotypes differ, of course, in the degree to which they can be associated with or withstand rearrangements, gains, or losses. X chromosome-autosome translocations in *D. melanogaster* are generally lethal in a dominant sense; autosome-autosome translocations are far less likely to be so. A loss of more than 50 bands is lethal, even in a heterozygous state. In humans, on the other hand, a translocation between chromosomes 13 and 14 has been described that led to the loss of the short arm of each, and hence of the nucleolar organizer regions in those short arms, but no phenotypic effect was detectable. Loss of most of the short arm of chromosome 17 has been shown in at least one individual to be similarly tolerated without effect, but the disastrous *cri du chat* syndrome is also due to chromatin loss, in this instance, in the short arm of chromosome 5 (5*p* 14-5). The evolution of the karyotype, therefore, can involve a variety of changes with a variety of consequences, some of which we can explore here.

It needs to be recognized here that no attempt is being made to identify, or define, a species in terms of karyotypic structure. Rearrangements of the karyotype can contribute to speciation, but at best they are but one facet of a much larger and more complicated problem. The species concept, as Levin points out, is a human construct, a taxonomic tool for dealing with organic diversity in a manageable way. Although similarities and dissimilarities judged to be of importance in the separation of species are generally more easily handled and more readily defensible in animal than in plant groups, the degree to which any species designation approaches reality is a continuously debatable subject. The same can be said of karyotypic changes even though it is quite obvious, in some instances, that the rearrangements of chromatin are related to the establishment and maintenance of the diversity upon which evolution (and the taxonomist) feed. What is not always evident is the degree to which all of these factors mesh in the act of speciation. The inversion heterozygosity of *D. pseudoobscura,* for example, has been correlated with ecological preference and, possibly, with incipient speciation, but the same conclusions cannot be applied to other species of *Drosophila* which may be even more diverse karyologically. When, therefore, we identify karyotypic diversity with particular species, we do so with the knowledge that it is but one aspect of evolution, albeit an important one.

From the Prokaryotype to the Eukaryotype?

Here we shall avoid any speculations concerned with the origin of the first chromosomes from nonchromosomal beginnings since no cytogenetic purpose would be served. Instead, we shall base our discussion of karyotype evolution on the chromosomes as we now know them. Further, we shall not include the viruses in this discussion since it seems reasonably clear from the nucleotide homologies that they share with their host—that is, the *Rous sarcoma* virus shows definite homologies with the genome of the chicken as does the *lambda* phage with *E. coli*—that they are derivative rather than transitional forms and that they have played no direct role in the evolution of higher organisms. Indirectly, of course, their effect on individuals and even on species can, indeed, be significant.

The origin of the prokaryotic and eukaryotic genomes, and their relation to each other, is by no means as clear as was once assumed. It is a customary textbook posture, arrived at more by faith that present-day complexity arose out of primitive simplicity than by sound fossil evidence of a sequential nature, that the prokaryotes (including both the archaebacteria and eubacteria) are ancestral to the eukaryotes, with the mesokaryotes a questionable intermediate on the basis of a mixture of traits, including their low histone content, and their peculiar manner of cell division and of packaging DNA. There can be no question, of course, that all living species once shared a common ancestor some time in the past; the universality of the genetic code, and the basic aspects of the protein manufacturing mechanisms, including the structure and roles of ribosomes and tRNAs, and the several types of DNA-protein and RNA-protein interactions involved in transcription and translation, permit no other interpretation.

Recent evidence, however, casts doubt on the view that the eukaryotes arose out of a prokaryotic group less than 10^9 years ago. Several features of the archaebacteria, the absence of basic proteins associated with the chromosomes of such primitive water molds as *Phycomyces* and *Allomyces* and the variable histone status of other fungi, some of the red algae, and the dinoflagellates lead to the not unreasonable suggestion that these species may represent transitional stages of evolutionary change out of which the eukaryotes emerged. On this basis, one can view the increasing structural and/or behavioral complexity of the eukaryotes as a gradual change requiring genetic controls of an increasingly sophisticated nature as the cellular organelles increased in number and acquired specialized functions and as multicellularity and cell differentiation made their appearance, and that this progressive increase in complexity was made possible not only by sequestering increased amounts of DNA and numbers of chromosomes within a membrane-bound nucleus but also by the insertion within and between genes of noncoding stretches of DNA to modulate transcriptional and translational activity, and hence gene expression. The difficulty with accepting this point of view, despite the convincing data in Fig. 1.1, is that, apart from the genetic code and its mode of operation, there are within the eukaryotic nuclear genome few "fossil" vestiges, genetic or molecular, of an earlier prokaryotic ancestry. It is perhaps more reasonable to accept the views of Woese and Fox that the archaebacteria, eubacteria (including the mitochondrial and chloroplastic genomes) and ur-karyotes (or eukaryotes) are equally ancient, and, in addition, to view the dinoflagellates as having split off from the eukaryotic line at a later date, possibly from a unicellular green algal stock. As Darnell and others have stressed, the large amounts of both coding and noncoding DNA and the interrupted gene sequences of the eukaryotes can equally well be viewed as a more primitive state, reflecting a period in evolutionary history before the processes of transcription, translation, and splicing were firmly fixed, that the interrupted sequences were retained since they "would allow cells to readily try out various combinations because of recombination of the pieces into one transcription unit" (Darnell), and that out of this transitional state both bacterial groups and the eukaryotes had their origin, the time of separation being considerably earlier than the 10^9 years ago that has been postulated. When and how the accumulation of DNA, and its organization into interrupted coding sequences, occurred remains uncertain, but it is assumed that out of this transitional state the bacterial genome emerged as a streamlined, tightly knit, "operon" governed, specialized product of evolutionary change, which adapted to an equally specialized existence. They rid themselves of the intervening, noncoding sequences—"frozen remnants of history," to quote Gilbert—as they tightened up their genome, to become characterized by rapid genetic and metabolic response to changing conditions, large population turnover, and efficiency in exploiting appropriate habitats. This ability to explore the environment, and to bring the genome quickly to a different adaptive state, has been demonstrated recently in the soil bacterium, *Bacillus subtilis*. Two distinctly different strains, variously labelled with antibiotic markers for resistance or sensitivity, and with transformation the only means of recombination, were incubated together in a sterile peat soil. After eight days of growth, 80% of the population had a newly evolved genotype which departed significantly from those of either of the original strains, but one more appropriate to its new environment. The prokaryote is, therefore, a specialized, not a primitive, genetic system, still retaining its evolutionary flexibility by means of transformation and/or episomal and plasmid vectors and of responding as the environmental pressures dictate and as random nucleotide changes alter the genome. The eukaryotic line, also emerging from this state, retained much of its DNA and its interrupted coding sequences along with a potential for "trial-and-error" experimentation, with novel proteins, generating new functions and new phenotypes, the result of imprecise transcription and splicing producing a variety of mRNAs from a single gene. Gilbert has argued that this kind of an interrupted gene structure would set the stage for, and maintain on a continuing basis, a rapid proliferation of eukaryotic types with diversity of function and structure, and hence of evolution. As he has pointed out, single base changes are, of course, the mutational events which lead to altered protein sequences, but if such base substitutions occur at the boundaries of internal coding (*exons*) and noncoding (*introns*) sequences, errors in splicing are a likely occurrence, adding or deleting variable numbers of codons at one step, and thus producing significant changes in amino acid sequence. As a result, it is proposed that a single, appropriately placed mutation in an interrupted gene can have far-reaching and more significant consequences by promoting rapid evolutionary change through new proteins than can a comparable mutation in an uninterrupted prokaryotic gene. In addition, the process makes superfluous the former assumption that a duplication of an essential gene is a necessary first or only step in the evolution of a new gene with a new function. In Gilbert's words, the circumstance of interrupting gene sequences by introns enables evolution to "seek new solutions without destroying the old."

Crick, on the other hand, seeking to understand the evolution of the RNA splicing phenomenon, suggests that interrupted gene sequences may not have been primitive, but rather may have been constructed from several distinct exons which, with their flanking introns, were brought together by a random shuffling of the genome by aberrations. The new sequence, coding for a larger protein, might combine the features and functions of the previously separate, but smaller, exons. Thus, in the vertebrate globin gene, which consists on three coding units separated by noncoding introns, the middle exon codes for that portion of the polypeptide which interacts with the heme group. Part of the globin gene, therefore, might well have been evolved from a gene which coded for a heme-related protein, and which had been in existence much earlier.

The validity of the phylogenetic scheme just discussed will, of course, be strengthened or weakened by future studies, but a more detailed examination of the currently available aspects of gene structure and function makes quite clear the sharp differences separating the prokaryotic and eukaryotic groups, differences which suggest a common ancestry from which each emerged rather than that one group is a direct derivative of the other. Some of these aspects have been mentioned earlier for other reasons, but they bear reiteration here in the present comparative context. It would be expected that the greatest degree of similarity between the two groups would be found in the elements of the coding machinery and that the differences would widen as other, less central features are examined. Since the code is universal, the tRNAs should reflect this, and this indeed is found (Fig. 9.2); the anticodes conform to expectations and are flanked by A and U in a modified or unmodified state, the amino acid-accepting arm always terminates in a –CCA–OH sequence, and the loops are similar in shape and contain similar numbers of nucleotides. These features, then, are unvarying and are presumed to be maintained by rigorous selection pressure, but when comparisons

Fig. 9.2. The nucleotide sequences of two tRNAs—tRNA^{serI} above the line and tRNA^{phe} below—each being derived from three different organisms. The spaces do not indicate interruptions in the linear sequence of the molecules, but are there merely to delimit the acceptor stems and the several loops, and to make comparisons more readily. Although there are obvious similarities within each of the two kinds of tRNAs, it is equally obvious that in both instances the yeast tRNAs show a far greater degree of correspondence with those of the rat and wheat than they do with those of E. coli. Also, the extra loop in E. coli tRNA^{serI} has an extra nucleotide as compared to the others, and there is an unpaired C between the extra loop and the anticodon loop that is missing in the yeast and rat tRNA^{serI}s. The nucleotides other than A, U, G, and C are as follows: T = ribothymidylic acid; ψ = pseudouridylic acid; D = 5,6 dihydrouridylic acid; I = inosinic acid; X = unknown nucleotide; Y = a fluorescent nucleotide. The sequences were taken from Madison (1976).

Fig. 9.3. The nucleotide sequences of the 5S rRNAs of E. coli (above) and Xenopus laevis (below). Each is 120 nucleotides long, but as one moves from the 3′ end (left) to the 5′ end of the molecule, and makes a nucleotide-by-nucleotide comparison, there are only 31 nucleotides that can be matched; these are indicated by the dots above the E. coli sequence. Attempts to shift the sequences in relation to each other does not seem to improve the register. However, each molecule has complementary sequences at the two ends which, when paired, can impart a secondary structure; the E. coli molecule also has three additional complementary sequences internally, while the eukaryotic 5S rRNA generally has only two.

UGCCUGGCGGCCGUAGCGCGGUGGUCCCACCUGACCCCAUGCCGAACUCAGAAGUGAAACGCCGUAGCGCCGAUGGUAGUGUGGGGUCUCCCCAUGCGAGAGUAGGGAACUGCCAGGCAU

CGGAUGCGGUGUGGGUGGGACUUUCAGUCAGACUAGAGACUAGAGCUCGCUUAGGUCCAUAUGUUCCGGACCAAUCAUGGUGGUCCAUAUGUUCCGGACCUUUAUGUCCGGCGAACCCUGGCGGCCGCCUUGGGCAGCAGUAGCCAGGCAUCCGGAA

481

are made of two tRNAs—tRNA^{serl} of *E. coli,* rat, and yeast and tRNA^{phe} of *E. coli,* yeast, and wheat—the internal nucleotides show considerable variation, without, however, altering the secondary and tertiary shapes of the molecules. Those of the rat and yeast and those of yeast and wheat are obviously more nearly alike than either is to that of *E. coli.* In addition and what is also not evident from Fig. 9.2 is the processing which transforms a precursor eukaryotic tRNA into a finished product by the removal of an insertion (intron) of about 15 nucleotides in length, and the splicing together of the two ends. An examination of the 5S rRNAs, another part of the coding machinery at the ribosomal level, reveals a similar conservancy as well as divergence. The rRNAs of the archaebacteria are as different in nucleotide sequence from those of the eubacteria as they are from those of the eukaryotes. If we examine the 5S rRNAs, the nucleotide sequences in both *E. coli* and *Xenopus laevis* are similar in length (120 nucleotides) but not in serial order (Fig. 9.3); a further point of dissimilarity is that the *Xenopus* 5S DNAs are flanked on either side by noncoding sequences of unknown, but presumably necessary, function. Related to these differences, but a step further along the line of protein involvement, is the fact that enzymes performing similar task in the prokaryotes and eukaryotes do not necessarily possess similar amino acid sequences, indicating that the genes coding for these enzymes have undergone considerable nucleotide change since the time of divergence (assuming, of course, that the genes existed prior to divergence).

The most striking differences that distinguish the genomes of prokaryotes from those of the eukaryotes is in gene arrangement, structure, and regulation. Genes engaged in the synthesis and management of an essential molecule—histidine or lactose, for example—are clustered into an operon in the prokaryotes, and governed by means of protein regulatory agents, while their eukaryotic counterparts, concerned with similar metabolic pathways, are scattered throughout the genome. Diffusable proteins do not seem to govern their activity, although, as shall be pointed out, the RNAs of the nucleus are undoubtedly involved. Specific regulatory elements, however, have been uncovered in eukaryotic genomes. One has been identified lying outside of, but adjacent to, the *rosy* locus in *D. melanogaster,* another is closely associated with the expression and mutability of the *white* locus, while the complex *bithorax* locus in the same species contains within it two sites which seem to have a regulatory function not unlike that of a prokaryotic operon. These regulatory units have been identified only with difficulty; their very elusiveness suggests that there are but few of them, that they are minute in size, or that present techniques are not sufficiently discriminating. Comparable units, however, have also been described in several fungal species, but when located it is apparent that such regulatory units must have a *cis,* and not a *trans,* arrangement if they are to be effective. Because of this it seems unlikely that diffusable proteins play a regulatory role, but it does suggest that every eukaryotic gene has its own regulatory sequence adjacent to it, and points to a role for the repetitive DNA which is interspersed among the structural genes. These sequences reanneal at slow or intermediate rates rather than at the very fast rate characteristic of constitutive heterochromatin. In the genome of the sea urchin, for example, there are several hundred thousand of these elements, making up about two or more thousand distinct families, with each family, or subset, represented by 30 to 200 copies. A few families, however, are represented by as many as several thousand copies per genome. Arguing that any system of gene regulation would seem to require a battery of genes functioning in coordinated, but overlapping, patterns, Davidson and Britten suggest that each subset of repetitive DNA copies occupies preferred locations adjacent to those genes that come into activity during a particular stage of development, or in a particular kind of terminally differentiated cell. If these genes are scattered throughout the genome, so, too, apparently are the members of a subset, and there would be no hindrance to the expression of a set of genes in an organized fashion provided that the subset functions or responds in a similarly organized fashion. Although the mRNAs are to be translated in the cytoplasm, it appears that the regulation of gene expression takes place in the nucleus, but not at the level of the chromosome itself.

The hypothesis of Davidson and Britten, based on data obtained largely from sea urchins but supported also by information from a wide variety of sources, proposes that the structural genes of an organism produce three groups of cytoplasmic messages (mRNAs), the distinction between them being made on quantitative grounds. Those with the fewest number of copies per cell (one to several) constitute the *complex class,* their structural diversity being sufficient to code for about 10^4 or more different proteins. *Moderately prevalent* mRNAs, having a sequence diversity that can account for only about one-tenth as many proteins, occur in numbers ranging from 15 to 300 copies per cell, while the *superprevalent* mRNAs, represented by over 10^4 copies per cell, are found only in differentiated cells producing a particular protein in massive amounts; e.g., those in the oviduct of the chick producing ovalbumin, or the reticulocytes of the mouse or chick which produce globin. The number of genes giving rise to the superprevalent mRNAs are relatively few, the largest number being those which produce the complex class. When specific tissues of an organism are compared in terms of shared or unique mRNAs, it is among the complex class that the differences are most evident. For example, only about 20% of the complex mRNAs in the sea urchin are found in all cells, and a comparison of complex mRNAs from a variety embryo and adult tissues reveals a diversity of sequences that can be equated to several thousand diverse structural genes. It is postulated, then, that differentiated structure and function at the cellular level is attained through regulation of the complex mRNAs, their number running into the thousands. The other mRNAs are also regulated, but it is possible that the genes giving rise to the superprevalent mRNAs may have a regulatory mechanism uniquely different in order to produce such a large amount of mRNAs. Although not a part of the Davidson-Britten hypothesis, is it possible that the moderately prevalent mRNAs are the products of those genes necessary for the functioning of any metabolically active cell?

A portion of the RNAs found in the nucleus (the *nRNAs*) is representative of *all* of the structural genes contained in a genome. In the sea urchin this amounts to about 25% of the nRNAs; it may be as low as the 4 to 6% found in cultured cells of *Drosophila*. A given mRNA may be absent from, or present to a very minor extent in, the cytoplasm, but its precursor molecule, not yet processed into a functional mRNA, can still be strongly represented among the RNAs of the nucleus. This suggests that all of the cells of an organism contain all of the structural genes in functional order, and that all of them are being transcribed; in addition, since the average piece of nRMA has a half-life of about 20 minutes, these genes must be transcribing at approximately the same rates and on a more-or-less continuous basis. Such a postulate contradicts, of course, the long held view that differentiation results from the "turning on or off" of selected structural genes, and points instead to a mechanism which exercises a control over the rate at which the various mRNAs are processed and moved across the nuclear envelope and into the cytoplasm for translation into proteins.

The remainder of the nRNA consists of sequences which are complementary to the repetitive DNA which exists as families of repeated units in the genomes of eukaryotes. Each differentiated cell, or stage of development, possesses an RNA containing certain repetitive sequence transcripts in high concentration while other repetitive transcripts are at a low level or are absent. Thus, in the gastrula stage of the sea urchin, the most highly represented set of repetitive transcripts differs from that found in adult intestinal tissue. In addition, the complementary sequence transcripts of each repetitive family are found in nRNA, thus setting the stage for the formation of intranuclear RNA–RNA duplexes which are thought to play a specific role in the processing of mRNAs, and therefore in the regulation of gene expression.

The gist of the Davidson–Britten hypothesis of gene regulation is illustrated in Fig. 9.4. It is assumed that the genome consists of two kinds of transcribing units. The *constitutive transcription units* (CTUs), each of which includes a structural gene plus any flanking repetitive sequences and a site for transcription initiation, are transcribed continuously into *constitutive transcripts* (CTs) which form a portion of the single copy sequences present in the nRNAs of all cells. The *integrating regulatory transcription units* (IRTUs), constituting a major portion of the genome, are made up of interspersed repetitive and single copy sequences (not those of structural genes), or groups of repetitive sequences, both complementary strands of which are transcribed in a cell-specific manner to form *integrating regulatory transcripts* (IRTs). The term "integrating" is used in order to imply that a single IRTU, because it may contain several different repetitive sequences, can be involved in the control of the coordinated expression of a number of structural genes in the same cell.

The CTs and IRTs together form the bulk of the nRNAs. Since both will contain similar repetitive sequences, with the IRTs in greater concentration than the CTs because of their repeated nature and because both complementary strands are transcribed, circumstances are present for the reassociation of complementary RNA strands to give RNA–RNA duplexes. If the nature of the duplexes is determined by their concentration in the nucleus rather than by some unknown selective factor, some of them will be IRT–IRT duplexes, but if they have no attached message they will eventually be degraded. Others will be CT-IRT duplexes, and it from these that the functional mRNAs will be processed. It is suggested that the CTs that are the precursors of the moderately prevalent mRNAs are regulated by repetitive sequences that are derived from larger families of repeats than those that are related to the complex mRNAs. If this is so, then the concentration of their IRTs will be higher, their rate of duplex formation will be more rapid, and the moderately prevalent mRNAs, as compared to the complex mRNAs, will show up more frequently in polysome preparations.

It is the double-stranded nature of the duplexes that will prevent or delay the degradation of the CTs by nucleases, and enable

(a)   CTU (structural gene region)

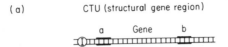

(b)   Constitutively transcribed to yield CT:
                 a' mRNA precursor b'

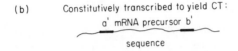

                 sequence

(c)   Three possible examples of IRTU

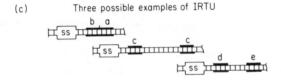

(d) Transcribed under control of sensor structure (ss) to yield IRT:

(e) Intranuclear reassociation to yield IRT·CT duplexes:

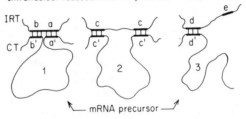

Fig. 9.4. A diagrammatic representation of the Davidson-Britten model of gene regulation. All events described are assumed to take place within the nucleus. Lower case letters denote repetitive sequences; all other regions are single-copy in nature. A, a transcribing region or a constitutive transcription unit (CTU), which includes a structural gene with its interspersed sequences (if present), flanking repetitive sequences and a transcription initiation site, I. B, *n*RNA or constitutive transcript (CT) derived from the transcribing region shown in A; the CT is also a precursor *m*RNA. C, several possible integrating regulatory transcription units (IRTU), with sensors (SS) which control the transcription of IRTUs in response to external signals; it is through the regulation of transcription of the IRTUs that control of gene expression is initially expressed. D, the transcribed products, integrating regulatory transcripts (IRT), derived from the IRTUs. E, a group of possible RNA–RNA duplexes resulting from complementary base pairing between the CTs and IRTs, a step which is required for the processing of the precursor *m*RNAs into functional *m*RNAs for transport to the cytoplasm for translation. It is possible, indeed probable, that IRT-IRT duplexes will also be formed, but if so they would not be processed (Fig. 1 from Davidson and Britten, 1979, copyright 1979 by the American Association for the Advancement of Science.)

the appropriate endonucleases to transform the precursor molecules into functional mRNAs. How the processing is accomplished remains to be fully determined, but it must include the removal of intragenic noncoding sequences when they are present, the splicing of the several RNA segments to fashion an uninterrupted message, the capping of the 5′ end of the message with 7-methylguanosine, and the attachment of a Poly(A) tail to the 3′ end. Only the histone mRNAs seem to be processed and transported to the cytoplasm without the added post-transcriptional details.

While we are still far from a complete understanding of gene regulation during growth and development in eukaryotic systems, the Davidson–Britten hypothesis gives greater credence to the view that the relation of the prokaryotes to the eukaryotes is far more remote than was earlier believed, and that they have pursued very different evolutionary pathways since diverging from some putative, common ancestor. Stated in what is probably a much over-simplified manner, prokaryotic gene regulation seems to function exclusively at the level of transcription, and by means of regulatory protein vectors interacting with the genes; eukaryotic gene regulation, on the other hand, is a post-transcriptional phenomenon of ribonucleic acid interactions taking place in a membrane-bound nucleus, although cell-specific patterns of repetitive transcripts suggest that some kind and degree of regulation must also operate at the level of transcription as well. The eukaryotes and prokaryotes seem also to share other similarities of gene regulation even though it is premature to state if these are fundamentally alike or merely superficially so. For example, the movable insertion elements that have been identified with the mutable *white* and *rosy* loci in *D. melanogaster,* and the similarly mobile controlling elements found in maize (p. 490), bear some resemblance to the insertion elements of bacteria (Fig. 5.30) in that they contribute added diversity of expression to the species through mutation, and by transposing genes in, out, or about the genome. Other patterns of eukaryotic regulation will be dealt with in the next section.

Selective Regulation of the Karyotype

The overriding importance of sexual reproduction in biparental inheritance, and the necessity of maintaining two distinct phenotypes within each species, has been such as to subordinate one or more chromosomes, in whole or in part, to subserve these ends. No other single feature of diploid existence in either plant or animal groups has exerted a comparable and as concentrated an influence on the karyotype, with the result that genes governing non-sexual characters are scattered, in apparently random fashion, throughout the eukaryotic genome, while those necessary for maintaining the separateness of the sexes tend to be grouped into one or more specific chromosomes. This does not mean regulation of other parts of the genome is not as closely determined as are the regulatory processes governing the expression of sexual phenotypes, but it has been done by means other than that of organizing the karyotype in an obviously specific manner.

Genetic regulation is a matter of timing, of bringing the various parts of the karyotype into transcribing activity, or shutting those parts off, at appropriate times. Parts of chromosomes and even entire chromosomes may be removed from the nucleus, but this is not a customary mode of genetic regulation. A differentiated cell is a regulated one capable of expressing only a part of its genome. Thus a *beta* cell in the pancreas of a mammal produces insulin, not hemoglobin or hydrochloric acid, although it contains the genes for doing so; the latter two substances can only be produced, respectively, by the *erythroblasts* of the hemopoietic tissue and by the *parietal cells* of the stomach. As will be pointed out later, the mass of DNA may govern, in a quantitative manner, the rate processes of a cell, but the mechanism of qualitative orchestration remains unclear.

A number of observations bear on this topic. It has been shown in *E. coli* that about 95% or more of the genome is transcribed, but of this amount approximately one-third of the genome is transcribed only once in every 300 to 1,000 cells, a very low level of activity. It has been proposed that these genes are of the inducible variety which are called into action when environmental circumstances change and a new phenotype is appropriate. A parallel situation has been demonstrated in amphibia. Gurdon has transferred nuclei from various somatic cells to a fertilized, but enucleated, egg to determine to what degree these nuclei would support development. The nuclei from young cells of the blastula show totipotency, that is, they can fully supplant the removed nucleus, and they give rise to embryos showing normal patterns of growth and development. As the nuclei from cells in progressively more differentiated tissues are transplanted into the enucleated egg, the frequency of normal embryos decreases until only abnormal ones or none at all are produced. Based on these and other observations, it appears that a greater and greater portion of the genome is deprived of transcriptional ability and that after a given time this ability cannot be recovered. The end result, if not the actual mechanism itself, is similar to the heterochromatinization of euchromatic elements, but it is uncertain how this explanation would fit in with the Davidson–Britten hypothesis of gene regulation (Fig. 9.4).

A more subtle form of intrachromosomal regulation and response involves the relation of adjacent genes to each other. As pointed out before (p. 367), the *Bar* mutation in *D. melanogaster,* which alters the number of facets in the compound eye, is a duplication of the 16A region of the X chromosome (Fig. 6.9), but the facet number is affected both by the position of 16A regions as well as their number. Thus, +/+ females and *B/Y* males have the same number of 16A regions, but the males have a reduced facet number while females are normal. The *cis* arrangement (16A, 16A/Y in males) has a different and more drastic effect from the *trans* arrangement in females (16A/16A). That this is not a phenomenon related to sexual differences is made clear by the fact that the facet number in females is more reduced in *ultra-Bar* (16A, 16A, 16A/16A) individuals than in *Bar* females whose arrangement is 16A, 16A/16A, 16A. A 3+1 arrangement, therefore, has a greater phenotypic effect than 2+2.

The *Bar* and *ultra-Bar* phenotypes are unvarying and predictable, and they provide an example of a *stable, or S-type, position effect* operating in euchromatin. The *variegated, or V-type, position effect* is related to the influence of heterochromatin on euchromatic genes. If by means of an aberration, the normal allele of *white* (w^+) on the X chromosome is removed from its euchromatic milieu and placed adjacent to constitutive heterochromatin, a heterozygous female (w^+w) will have eyes with patches of red and white instead of the normal solid red color. The production of red eye pigment, governed by the w^+ gene, is obviously suppressed in some regions of the eye but not in others. In addition, a spreading effect is also evident. Thus, if the fly is also heterozygous for *roughest* (rst^+/rst), which lies 1.5 map units proximal to *w,* the eye will similarly show patches of smooth and rough areas. The areas of roughness will invariably coincide with the *w* patches, but they may or may not do so with the red patches.

Heterochromatin can, therefore, exert an inhibitory effect on the functioning of normal euchromatic genes, with the effect diminishing as the distance from the heterochromatic juncture increases. The influence may, in some instances, spread as far as 50 bands of salivary chromosome length; in the *w-rst* case, *rst* would be closer to heterochromatin than *w;* hence all *w* areas would be *rst,* but all *rst* areas need not be *w.* The addition of extra heterochromatin, an added Y, for example, would tend to diminish the magnitude of the *V-type position effect* but have no effect on the *S-type.* The mechanism of such inhibition, as well as the basis of the *V-type position effect,* remains unknown.

A similar inhibitory and spreading action can be caused by facultative heterochromatin in mice. The translocation of autosomal genes to an X chromosome that is being inactivated will cause the autosomal genes to exhibit variegation.

The *S-type position effect* is rare, possibly nonexistent, in plants, but a number of situations have been described in maize which resemble the *V-types* of *Drosophila* in that the activity of a gene, presumably its transcribing ability, is altered by the presence of a controlling element which come to lie adjacent to, or is integrated into, the gene in question (Table 9.1). The suspicion that these elements are heterochromatic stems from their possible relation to the knobs, and in particular to the terminal knob, of chromosome 9, as well as to the relation of heterochromatin and variation in other organisms, but their ability to move within the genome from site to site suggests an even more striking similarity to the transposable nature of the episomes and insertion elements of *E. coli*. Like the insertion elements, those in maize seem also to have preferred sites, but the degree of homology, if any, with those in *E. coli* is debatable.

The existence of the maize regulatory elements remained unrecognized until they were uncovered by the events of the breakage-fusion-bridge cycle which brought on unexpected kinds of mutable changes. A two-element system consisting of a regulatory element and a receptor element is typical, and the *Dissociation-Activator* (*Ds-Ac*) system described by McClintock is representative of this group. *Ds* has a more or less standard location on chromosome 9, and is a receptor unit governed by a regulatory unit, *Ac*. The latter moves too frequently to make mapping feasible. *Ds* is so named because, in the presence of *Ac*, it sometimes causes chromosomal instability; that is, when *Ds* is excised from the chromosome and

Table 9.1. Regulator, receptor, and mutability circumstances in maize; receptors respond only to their own regulators

Regulator	Receptor	Comments
Ac (Mp)	Ds	Mutations are evident when Ds moves to another locus. The A_1, A_2, Bz, C_1, and Wx loci are most commonly involved, but they include the P and R loci if *modulator (Mp)* is identical to Ac.
Spm (En)	A_1, A_2, B^t, C_1, C_2, I, P, Pr, R, and Wx	*Spm (Suppressor-mutator)* and *En (Enhancer)* can regulate gene activity in the absence of mutation events, and mutation effect can be lost while suppressor action is retained. A modifying gene *(Mod)* can restore full activity to a weakly active *Spm*.
Dt	a^{m-1}, a,	Acts only on a^{m-1} and a, causing them to mutate to A_1.

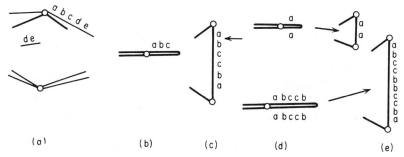

(a)   (b)   (c)   (d)   (e)

Fig. 9.5. Diagram of the chromatid type of breakage-fusion-bridge cycle as it occurs during gametogenesis in maize following the formation of a dicentric chromatid after breakage at the *Ds* locus (Fig. 9.6). The altered chromatid is indicated by the heavy lines in *A. B,* the chromatid following replication, which occurred between the end of the second meiotic division and the first microspore division. C, anaphase of the first microspore division, with the dicentric chromatid being broken at the region between *a* and *b* (arrow). D, replication and fusion to continue the cycle, resulting in two very differently sized chromatids at the anaphase of the second microspore division (E). The cycle would continue in the endosperm as replication and fusion precedes each division, but would cease when a gamete containing a broken chromosome enters into fertilization to form a zygote. The broken chromosome would then heal, and fusion would cease. (after McClintock, 1941).

moves to a new location, the chromosome is broken at that point, with the result that a fusion of broken ends of sister chromatids occurs, and a breakage-fusion-bridge cycle is initiated (Fig. 9.5). Also, when *Ds* moves, those genes at its previous position are no longer under the control of the *Ds-Ac* system; they may return to their former phenotypic expression, or they may have acquired a new allelic status with a new phenotypic expression. If *Ds* occupies a new position, the genes adjacent to it will now show varying degrees of instability (Fig. 9.6). *Ds*, therefore, not only alters the expression of genes in its immediate neighborhood; it is also a source of new mutations and hence of considerably variability.

Fig. 9.6. Top: a portion of chromosome 9 of maize showing the standard position of *Ds* between the centromere and *Wx*, and shifts in the location of *Ds* that have been determined. When shift #1 is made, the effect is on the Sh_1 gene, while shift #2 affects the *C* locus. Both shifts allow the *Wx* locus to assume its normal behavior while Sh_1 or *C* would come under the mutable influence of *Ds*. Bottom; diagram of chromosome 9 in pachytene, with a break at the standard position caused by *Ds* moving to another location (after McClintock, 1951).

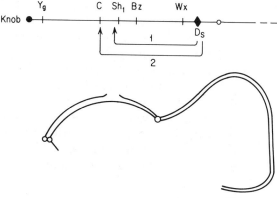

491

Ds and *Ac* generally act together. They are, in fact, rather similar to each other, and may have had a common origin sometime in the past, but *Ds* alone is ineffective, and its presence cannot be detected in the absence of an active *Ac*. *Ac*, however, when inserted adjacent to or within the locus of a gene, can act as a one-element system to bring that gene under its control. When *Ac* moves to another location, the gene in question may return to its previous state, or it may show an altered allelic form of a stable or mutable character. Both *Ds* and *Ac* are themselves mutable, suggesting that these elements have gene-like qualities. The different states of *Ac* are detectable by changes in the manner of its control of *Ds*, and in the type of mutability made manifest, while changes in the state of *Ds* are ascertained by the relative frequency and kind of events—breakage or gene instability—taking place at the site where *Ds* is located. Various combination of *Ds* and *Ac* states can produce a broad spectrum of events ranging from many chromosomal breaks and few mutable loci, to no breaks and many mutations. The phenotypic changes range from those that are scarcely detectable to those indistinguishable from the dominant wild-type expressions.

The mutations induced by the presence of *Ds* are generally simulated in the sense that the removal of *Ds* returns the gene to its normal state. In this sense *Ds*, although a receptor of the influence emanating from *Ac*, is a regulator of the gene associated with it just as *Ac* is the distant regulator of *Ds*. Thus, when *Ds* is in its normal chromosome 9 position (Fig. 9.6), it lies adjacent to, or within, the locus of *Waxy* (*Wx*), and *Wx* behaves as though it had mutated to some form of a recessive state, or was absent. The generally unchanged condition of *Wx*, however, is indicated by its reversion to normal function with the removal of *Ds* to a different location. The genes affected by *Ds* can "mutate" to higher or lower degree of expressivity, or toward dominance or recessiveness, while the action of *Ac*, which may be represented from 0 to 3 times within the endosperm, affects the timing but not the frequency of action of *Ds*, that is, the larger the number of *Ac* elements present, the later in development is the action of *Ds*, and hence the smaller the amount of tissue showing a phenotypic change.

The first *Dotted* (Dt_1) locus to be discovered is, like *Ds*, located in chromosome 9, but its action is more like *Ac* in that it acts at a distance on a specific locus, A_1, on chromosome 3. Thus the action of *Dt* is to cause a_1 or a^{m-1}, two colorless mutants, to mutate to *A*, with the presence of anthocyanin in either leaves or the aleurone layer of the endosperm as an indicator of the presence of *A*. For example, a plant homozygous or a_1/a_1 in the presence of *Dt*, will

Table 9.2. Effect of varying numbers of *Dt* loci on an endosperm containing varying numbers of a_1 alleles. Gene a^p, which is unaffected by the presence of *Dt*, produces a pale-colored aleurone against which the mutation of $a_1 \longrightarrow A_1$ shows quite clearly. Only the figures in the first column are observed (Rhoades, 1941); the $a_1 a_1 a^p$ and $a_1 a_1 a_1$ columns are estimates but are based on the knowledge that the increase per a_1 is linear, in contrast to the exponential increase brought on by increases in the number of *Dt* loci.

	$a_1 a^p a^p$	$a_1 a_1 a^p$	$a_1 a_1 a_1$
Dtdtdt	7.2	14.4	21.6
DtDtdt	22.2	44.4	66.6
DtDtDt	121.9	243.8	365.7

show the appearance of purple stripes on brownish leaves, or purple spots on a colorless kernel. Unlike *Ac*, one dose of which causes a maximum frequency of change when *Ds* is present, *Dt* in the endosperm causes the frequency of *A* spots to rise disproportionately fast as the number of *Dt* elements goes from 1 to 3 (Table 9.2), but if the number of *Dt* elements is kept constant, and the number of a_1 is varied, the rise is more or less arithmetic. The size of the spots induced by *Dt* is always the same, however, indicating that *Dt* acts at a specific time during development of the endosperm. As with *Ds* and *Ac*, no normal allele of *Dt* is known, although McClintock has uncovered a *Dt* element in normal strains of maize by manipulation of the breakage-fusion-bridge cycle. Whether *Dt* acts directly on the structural gene to cause it to "mutate," or whether a *Ds*-like element, lying adjacent to a_1, responds to the action of *Dt*, is not known.

Two other *Dt* loci are also known: Dt_2 from a Brazilian strain of maize and located in the long arm of chromosome 6, and Dt_3 from Peru, located toward the end of the long arm of chromosome 7. All three have a similar effect on a_1 and a^{m-1}, the mutation frequencies are comparable when the genetic background is held constant, and an exponential increase is found when the dosage of each *Dt* element goes from 1 to 3. The fact that more than three *Dt* loci can be obtained in individual endosperm nuclei permits a further extension of the data in Table 9.2; the frequency of dotting increases with increasing numbers of *Dt* elements, but the character of the mutant areas, i.e., their size, shape and intensity of color, remains unchanged.

The response of a_1 and a^{m-1} to the several *Dt* elements differs considerably; the mutability of a^{m-1} is very much higher than that of a_1 to a given dose of *Dt*, and the mutation events are not restricted in time. Thus, a cross of $a_1 a_1 dtdt \times a^{m-1} a_1 DtDt$ will produce two kinds of seeds with their respective endosperms being $a_1 a_1 a_1 Dtdtdt$ and $a^{m-1} a_1 a_1 Dtdtdt$. The latter show about 400 times more A spots than the former, and while the $a_1 \xrightarrow{\hspace{1cm}} A$ spots average about 78 cells in extent, those from an $a^{m-1} \xrightarrow{\hspace{1cm}} A$ event may involve only a single cell, or it may embrace the entire seed.

Amount of DNA per Genome

Tables 9.3 and 9.4 indicate the rather extraordinary range of DNA values found among eukaryotic species, stressing in yet another way the separateness of the eukaryotes from the prokaryotes. Overlaps do occur in a limited range of values (Fig. 1.1), suggesting a possible phylogenetic relationship, but a detailed examination reveals the relation to be fortuitous only. Thus, *Bacillus megatherium* with 0.70 picograms (pg) of DNA per haploid genome, and *Clostridium welchii* with 0.24 pg, possess more DNA than the eukaryotic *Neurospora crassa* with only 0.017 pg. However, 20% of the latter consists of noncoding repetitive sequences while the two bacterial species consist almost entirely of unique stretches of codons, making rather meaningless any comparison of prokaryotes and eukaryotes based only on the amounts of DNA. Only *Methanobacterium thermautotrophicum* seems to have any significant (6%) amount of rapidly reannealing DNA, suggesting a degree of repetitiveness not found in most prokaryotes.

The very large amounts of DNA in eukaryotic species have proved to be a cytogenetical dilemma. The role of this DNA comes into question, since it seems obvious that the amount of DNA is not an accurate indicator of the number of genes needed for the delineation and maintenance of the species. Part of this dilemma is resolved by the discovery of the interrupted nature of eukaryotic genes. Gilbert has estimated that the introns range in length from 10 to 10,000 bases and that it is likely that they will prove to contain five to ten times the amount of DNA found in exons. Such speculation, which remains to be substantiated, helps to make understandable the situation exposed by Judd and others in their study of the amount of DNA vs. gene content in the bands of *Drosophila* polytene chromosomes; that is, as they point out, the genes are small in number compared to the amount of DNA per genome, and only a small proportion of the heterogeneous nuclear RNA transcribed becomes the coding core of a functional mRNA.

Table 9.3. Amounts of DNA/haploid in picograms ($\times$ 10^{-12}) in representative bacteria and higher plants

	Picograms	Common Names
Bacteria		
Escherichia coli	0.009	Colon bacterium
Salmonella typhimurium	0.011	—
Clostridium welchii	0.024	—
Bacillus megatherium	0.070	—
Fungi		
Neurospora crassa	0.017	Pink bread mold
Aspergillus nidulans	0.044	Black bread mold
Dictyostelium sp.	0.100	Slime mold
Gymnosperms		
Ephedra fragilis	8.4	—
Picea glauca	60.0	Blue spruce
Pinus rigidia	100.0	Pitch pine
Pinus resinosa	140.0	Red pine
Angiosperms		
Monocots		
Avena sativa	4.30	Oats
Hordeum vulgare	5.40	Barley
Triticum aestivum (6x)	5.60 (per genome)	Wheat
Briza minor	7.30	s-c; grass, annual*
Briza madia	8.50	s-i; grass, fac. per.
Secale cereale	8.80	Rye
Narcissus tazetta	9.50	Paperwhite, bulb
Briza maxima	10.80	s-c, grass, annual
Allium cepa	16.25	Onion, bulb
Tradescantia paludosa	18.00	Spiderwort, fleshy per.
Trillium luteum	20.00	Yellow wakerobin, rhiz.
Scilla sibirica	22.30	Squill, bulb
Lilium longiflorum	35.00	Easter lily, bulb
Haemanthus katherinae	120.00	Blood lily, bulb
Dicots		
Arabidopsis thaliana	0.26	Ephemeral, annual
Vigna radiata	0.50	Mung bean, annual
Linum usitatissimum	0.70	Flax, annual
Capsella bursa-pastoris	0.81	Shepherd's purse, annual
Glycine max	1.10	Soybean, annual
Crepis capillaris	1.20	Composite, annual
Raphanus sativa	1.50	Radish, annual
Senecio vulgaris	1.74	Composite, winter-blooming
Haplopappus gracilis	1.80	Composite, annual
Antirrhimum majus	1.83	Snapdragon, fac. per.
Vicia sativa	2.75	Bean, annual
Lactuca sativa	3.10	Lettuce, biennial
Pisum sativum	4.80	Garden pea, annual
Lathyrus maritimus	7.80	Pea, per.
Ranunculus repens	11.60	Buttercup, per.
Hepatica americana	13.40	Liverleaf, per.
Vicia faba	13.00	Broad bean, annual

*s-c = self-compatible; s-i = self-incompatible; per. = perennial; fac. per. = facultative perennial; rhiz. = rhizomatous root.

Table 9.4. Amounts of DNA per haploid genome in picograms $(\times\ 10^{-12})$ in representative animals

	Picograms	Common names
Invertebrates		
Tube sponge	0.05	
D. melanogaster	0.12	Fruit fly
Chironomus sp.	0.20	Black midge
Apis mellifera	0.35	Honeybee
Crassostrea virginica	0.69	Oyster
Aurelia sp.	0.73	Jellyfish
Strongylocentrotus sp.	0.89	Sea urchin
Musca domestica	0.89	Housefly
Spisula solidissimum	1.20	Surf clam
Balanus sp.	1.40	Barnacle
Cancer anthoai	1.70	Crab
Callinectes sapidus	2.00	Blue crab
Uca pugilator	2.20	Fiddler crab
Limulus polyphemus	2.80	Horseshoe crab
Loligo sp.	4.50	Squid
Gryllus domestica	6.00	Hearth cricket
Vertebrates		
Fishes (*Pisces*)		
Spheroides maculatus	0.25	Puffer
Xystreurys liolepis	0.40	Sole
Xyphophorus helleri	0.40	Swordtail
Esox lucias	0.85	Pike
Fundulus heteroclitus	1.50	Grunnion
Salmo irideus	2.50	Rainbow trout
Torpedo ocellata	2.75	Shark
Lepidosiren paradox	55.00	African lungfish
Amphibia		
Xenopus laevis	2.89	African clawed toad
Rana temporaria	4.10–5.00	Leopard frog
Bufo bufo	6.00	Toad
Plethodon cinerius	20.00	Eastern small salamander
Triturus cristatus	22.50	Crested salamander
Triturus viridescens	36.00	Salamander
Plethodon vehiculum	69.30	Western salamander
Amphiuma means	85.00	Congo eel
Necturus maculosa	85.00	Mudpuppy
Reptilia		
Coluber constrictor	1.40	Black racer snake
Chelonia sp.	2.45	Turtle
Alligator mississipiensis	2.50	American alligator
Aves		
Passer domestica	0.95	House sparrow
Columbia livia	1.00	Pigeon
Gallus sp.	1.15	Chicken
Anser sp.	1.45	Goose

Table 9.4. (Continued)

	Picograms	*Common names*
Mammalia		
Microtus oregoni	2.70	Creeping vole
Canis familiaris	2.80	Dog
Homo sapiens	3.00	Man
Rattus rattus	3.05	Rat
Bos taurus	3.20	Bull
Mus musculus	3.25	Mouse
Felis cattus	3.55	Cat

Much of the dilemma remains, however. Does the noncoding DNA, both that outside of as well as that interspersed within coding segments, have regulatory function as proposed by Davidson and Britten, and thereby justify its retention, or is the metabolic cost of retaining this much noncoding DNA balanced against the evolutionary flexibility its presence is purported to provide to the species? Are there other balance-sheet relations which might account for the sometimes massive amounts? Viewed in another way, what is the relation of Gilbert's introns to the repetitive sequences that have been uncovered in every eukaryote so investigated. Table 9.5 provides a sampling of species, with percentages and, where known, characteristics of their repetitive sequences. The highly repetitive, short, and very rapidly reannealing sequences, represented by 10^5 to 10^6 or more copies per genome, are apparently unrelated to the introns and are found not within genes but almost exclusively are related to constitutive heterochromatin. Their function remains unknown, but as the DNA per genome increases, so too does their number seem to increase. The slow and intermediate frequencies of repeated sequences, which are generally of larger size and fewer in number, undoubtedly include those related to the formation of the rRNAs, tRNAs, and histones, each with its own set of possible introns. Are the introns from one gene to another, or even within a single gene, repeats of each other? What is the relation of the introns to the many families of repeats found in *Xenopus laevis,* to those which Bonner has reported to be widely dispersed in the genome of the rat, or to the 300 to 400 and 1,000 nucleotide long repeats found by Thompson in enormous numbers in the pea genome? How is the concept of the intron to be integrated into the 1 band: 1 transcribing gene concept in *Drosophila?*

Table 9.5. Composition of various representative genomes as to unique, or single-copy, and repetitive sequences, and as to the nature of the sequences

Species	Picograms per Haploid Genome	Unique DNA, % of Genome	Repetitive DNA, % of Genome	Comments
Apis mellifera (honeybee)	0.35	90%	10%	Unique and repetitive DNA in sequences longer than 2,000 nucleotides
Musca domestica (housefly)	0.89	15%	28% 57%	Repeats 1,500 nucleotides long Repeats 300 nucleotides long
Gallus sp. (chicken)	1.15	70%	24% 3% 3%	120 copies, long repeats 330,000 copies, shorter repeats 1.1×10^6 copies, short, fast reannealing repeats
Xenopus laevis (Afr. clawed toad)	2.70	76%	6% 18%	Multigenic, transcribing repeats, including 5S, 18S, and 28S rRNAs, tRNAs, and histone loci; very little variation among families Short 300 nucleotide repeats, about 800 distinct families with about 2,000 copies per family
Mus musculus (mouse)	3.25	49%	13% 18% 13%	120 copies, slow reannealing 7,500 copies, intermediate reannealing rate 10^6 copies, very rapid reannealing rate
Secale cereale (rye)	9.00	26%	30.5% 33% 10%	850 copies, slow reannealing 37,000 copies, interm. reannealing Between 6×10^5 and 8×10^6 copies, rapid reannealing
Pisum sativum (garden pea)	13.00	30%	70%	10^7 different repetitive units in a continuous frequency range from 100 to 10,000
Necturus maculosus (mudpuppy)	85.00	12%	20% 54% 14%	300 long repeats 50,000 intermediate repeats 6.7×10^6 short repeats

Source: Various sources, but see Straus, 1975.

Most questions of the kind posed cannot yet be answered, but it seems quite clear that the large amounts of eukaryotic DNA are not fully explained either by invoking the existence of introns (their presence in higher plants remains be to demonstrated), by viewing the noncoding DNA as regulatory, or by equating increasing amounts with increasing complexity and its increased need for regulation. It is generally assumed that the amount and informational content of the DNA characteristic of a species must be consistent with all of its requirements for growth, development, and maintenance; the reasonableness of this position is borne out by the relationships shown in Fig. 1.1 and which are related to the minimal amounts of DNA found in the major groups of organisms, but the meaning of amounts of DNA beyond the minimal levels continues to elude the cytogeneticist.

The increases in the number of repeated sequences can be explained more easily than can their evolutionary significance. Once a repeated sequence has arisen, perhaps by accident of duplication or transposition, it is reasonable to invoke the phenomenon of unequal crossing over, and the faster, in the absence of contrary pressure, will be the rate of accumulation. By translocation and inversion, these could become dispersed throughout the genome even if their origin was initially a localized one (Fig. 9.7), and mutation, coupled with amplification, could lead to the formation of families of sequences which diverge from each other. When in the course of evolutionary history these changes began, and whether coding sequences could become shifted to a noncoding status are questions still to be answered, but when a comparison is made between repeated sequences of related species, as has been done in a number of cereal genomes, there is ample evidence to indicate that all sequences, both single-copy as well as repetitive, are undergoing rapid changes, thus altering sequences shared by all species, and bringing new sequences into being that are species-specific.

Most structural genes, and possibly some of their regulatory units as well, are composed of unique, single-copy sequences of DNA. However, not all single-copy sequences are of a structural or regulatory nature; it has, in fact, been estimated that probably no more than 10%, and probably a good deal less, of the single-copy sequences in vertebrate species perform essential structural or regulatory roles. The percentage may be less in higher plant species. From his analysis of the garden pea, Thompson has stated that even if it is assumed that the coding genes were to be found among the sequences longer than 1,000 nucleotides, this would constitute only 2% of the total DNA of the genome, although the amount would still be sufficient to account for about 70,000 genes of average size.

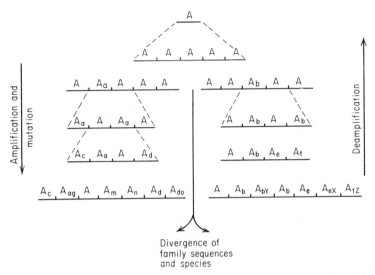

Fig. 9.7. A possible scheme for the increase and evolution of sequences of repetitive families of DNA. It is assumed that amplification and mutation would not only cause the divergence of repeated families, but could also be associated with species divergence as well. These processes would account for repetitive and related families both within a species and among groups of related species. With the passage of time, and as the genome is rearranged as a result of deletions, inversions, translocations, and the like, small sequences could lose their identity as family members to become part of larger sequences; by this process the *Xenopus* pattern of short sequences could possibly evolve into the *Drosophila* pattern of longer sequences. Further, if amplification occurs through unequal crossing over among similar sequences and adjacent, or by mistakes in replication, deamplification could also take place, and advanced or specialized species, with their lower DNA values, could arise by a shedding of nonessential DNA (see Mills, et. al., 1973, and Flavell, et. al., 1977).

A very goodly number of single-copy sequences must, therefore, be derived from repetitive sequences through the accumulation of random mutations. Do such mutations interfere with the regulatory functions of the repetitive sequences? The remainder of the single-copy sequences in the garden pea consists of approximately 3×10^6 copies in the 300 to 400 nucleotide class of fragments. The 70% repetitive DNA has been estimated to consist of 10^7 units of short length distributed into frequency classes having a continuous range of 100 to 10,000 copies per class. Since the repetitive units are three times more numerous than the single-copy elements, a ratio consistent with the Davidson–Britten hypothesis, the pea genome must consist of repetitive units variously interspersed with other repetitive units as well as repetitive units alternating with single-copy sequences. The task of understanding the origin, structure, behavior, and evolutionary history of even the smallest eukaryotic chromosome or genome is a formidable one.

500

By way of contrast, the mung bean, *Vigna radiata,* with its DNA value of 0.5 pg per haploid genome, has its genome differently constructed. Davidson and his colleagues have referred to the garden pea genome construction, with its short repeated sequences, as the *Xenopus* pattern (after the genus in which it was first described). It would appear that all metazoans, and possibly all higher plants as well, having DNA values above 0.7 pg per genome are similarly constructed, although among the higher plants those having 5 or more pg/haploid genome average around 80% repetitive DNA, while those having 4.0 pg or less average only 62%. The mung bean, however, falls into the *Drosophila* pattern, along with such other species as *D. melanogaster* (0.12 pg), *Chironomus* (0.2 pg), *Dictyostelium* (0.1 pg), and *Apis mellifera* (0.35 pg), which means that a large part of its genome consists of unique, single-copy sequences and that these as well as the repeated sequences are of lengths averaging over several thousand nucleotides. Thus, in *Apis,* and on the basis of rates of reassociation of fragments of 330 and 2,200 nucleotides in length, 90% of the genome consists of single-copy sequences, with few or no repetitive sequences interspersed within distances less than 2,000 nucleotides in length. On the other hand, the housefly, an insect of a totally different order, fits the *Xenopus* pattern. With a DNA value of 0.89 pg, less than 15% of its genome consists of single-copy sequences. The remaining 85% is of repetitive DNA, one-third of which averages 1,500 nucleotides in length and the remaining two-thirds about 330 nucleotides. Since the lower DNA values are found in advanced or specialized species, it would appear that the evolutionary patterns of advancement or specialization have been to remove, in many instances, the short repetitive sequences, possibly by reversing the process of amplification, and to conserve, or build up, the longer sequences as the remaining elements of the genome (Fig. 9.7). It remains to be determined whether the *Xenopus* and *Drosophila* patterns of genome construction are as distinctively different as they seem to be, and whether they are to be distinguished on the basis of DNA values as well as their patterns of interspersion.

Quite apart from questions relating to the nature and genetic function of the large amounts of eukaryotic DNA, increasing amounts of DNA generate relationships stemming from quantitative rather than qualitative considerations. There is no question about the genetic control of cell and developmental parameters, but it is equally clear that the amount of DNA per genome, regardless of its informational content, imposes its own kind of control on a number of size and timing phenomena: nuclear, nucleolar, and cell size, duration of mitosis and meiosis, duration of the S period, and rates of cell metabolism, particularly that of an oxidative nature. It is even possible, in birds and mammals, to show a negative correlation between DNA values and both heartbeat rate and respiratory rates. Among the higher plants, there is a strong relation between the amount of DNA and minimum generation time, that is, the length of time from germination to the production of mature seed, as well as with annual, biennial, and perennial habit.

Table 9.6 and Figs. 9.8 and 9.9 document some of these relationships in higher plants. The addition of a picogram of DNA to the genome lengthens the mitotic cycle by nearly 4 min and the meiotic cycle by 2 hr. The initial effect of the DNA is, of course, at the cellular level—chromosome, nuclear, nucleolar, and cell size, as well as the duration of various processes—but eventually such effects are expressed phenotypically by the whole individual. Many of the quantitative features of an organism's phenotype are, therefore, not simply the product of an interaction between the genotype and the environment, but they are also due, as Bennett has emphasized, to the *nucleotype* as well, that is, the amount of DNA expressed as a noninformational quantity.

The effect of the nucleotype, at both cellular and whole organism levels, is on rate phenomena, putting thereby a limit on the length of the life cycle and, as a result, a limit on cell, organ, and organism size. Plants with low DNA values—*Arabidopsis thaliana,*

Fig. 9.8. Duration of the mitotic cell cycle as a function of the DNA content of the haploid genome. Perennials are indicated by circled dots, annuals by dots (see also Table 9.6).

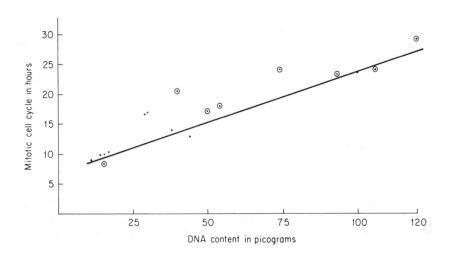

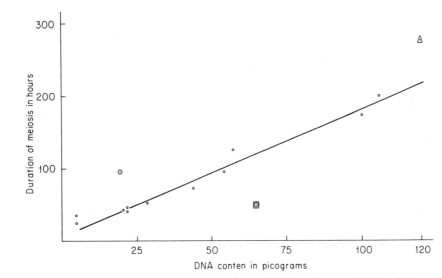

Fig. 9.9. Duration of the meiotic cell cycle in higher plants as a function of the DNA content of the haploid genome. ⊙ = *Ornithogalum virens;* ⊡ = *Endymion nonscriptus;* △ = *Trillium erectum* (see also Table 9.6).

Table 9.6. Duration of mitosis and meiosis in a number of plant species, together with their DNA values in picograms and their annual or perennial habit; where the DNA values are different, both are given.

Species	Picograms per Haploid Genome	Mitosis in Hours	Meiosis in Hours	Plant Habit
Crepis capillaris	1.20	10.8	—	Annual
Haplopappus gracilis	1.85	10.5	36.0	Annual
Pisum sativum	3.9, 4.8	10.8	—	Annual
Ornithogalum virens	6.43	—	96.0	Perennial
Secale cereale	8.8, 9.6	12.8	51.2	Annual
Vicia faba	13.0, 14.8	13.0	72.0	Annual
Allium cepa	14.8, 16.25	17.4	72.0	Perennial
Tradescantia paludosa	18.0	18.0	126.0	Perennial
Endymion nonscriptus	21.8	—	48.0	Perennial
Tulipa kaufmanniana	31.2	23.0	—	Perennial
Lilium longiflorum	35.3	24.0	192.0	Perennial
Trillium erectum	40.0	29.0	274.0	Perennial

Source: Van't Hof, 1965, and Bennett, 1972.

Capsella bursa-pastoris, and *Veronica persica,* for example—also have the smallest reproductive and vegetative parts: fruits, flowers, seeds, and plant habit. If we confine our attention to diploid forms only among the higher plants (Table 9.3), we find that herbaceous species with high DNA values are obligate perennials—those that do not normally go from germination to seed within a single growing season; those with the lowest values are annuals, although the facultative perennials—which can germinate and set seed within a single season, as does an annual—fall within the range of the annuals. A further distinction may be made among the annuals; the shortest-lived species have the lowest values, these including the ephemeral desert species as well as those of northern or alpine habitats where the growing season is short. Included among the latter, as Bennett has shown, are those that come into flower and set seed during the English winter months. However, facultative perennials such as the snapdragon, *Antirrhinum majus,* have as low a DNA value (1.83 pg per haploid genome) as the ephemeral annual, *Haplopappus gracilis.* A small amount of DNA can, therefore, determine by genetic control the length of its life cycle, but at high DNA values the very mass of DNA intrudes to exercise an independent control over rate processes which would prevent, for example, a species with high DNA values from having a brief life history, or vice versa, an ephemeral species having high DNA values.

One of the few instances known where this relationship does not hold is in several annual species of the genus *Pyrrhopappus* of the family Cichoriaceae (Compositae). As Price and Bachmann have shown, their DNA values (13.3 pg per nucleus) and the duration times of their mitotic cycles are in accord with the linear relations depicted in Figure 9.8, but while other annual species of the same family having much lower DNA values (2.4–3.2 pg) require a minimum of 11 weeks to complete their life cycle, the *Pyrrhopappus* species do so within a span of only 7 weeks. A genetic rather than a strictly nucleotypic control must be in force in this instance. Further, this kind of correlation cannot be extended indiscriminately when polyploid forms are involved. As Bennett and Smith have shown with wheat, rye, and the hybrid *Triticale,* the time span of meiosis on through pollen maturation decreases with increasing polyploidy. Thus diploid wheat completes meiosis in 42.0 hr, the tetraploid in 30.0 hr, and the hexaploid in 24.0 hr. Diploid and tetraploid rye completed meiosis in 51.2 hr and 38.0 hr, respectively, while hexaploid and octoploid *Triticale* showed times of 35.0 hr and 22.0 hr.

The positive correlation of rate phenomena such as cell cycles and generation times with DNA values raises the question of adaptation. Do environmental pressures for shorter life cycles put a premium on lower DNA values, or are species with lower DNA values more opportunistic and more successful competitors in situations where, for example, a biological "hit-and-run" tactic is appropriate, that is, desert annuals taking advantage of a brief wet period, mosquitoes breeding in a rain puddle, population explosions of aphids in favorable circumstances? This cannot be the only causal explanation for low DNA values, however, or no overlap between facultative perennials and annuals would be possible. Commoner has proposed that increasing amounts of DNA lead to increasing sequestration of nucleotides from the cellular pool, placing greater demands on the cellular machinery to maintain the supply, and thereby slowing down the metabolic processes which are governed by the concentration of free nucleotides. A shedding of DNA, particularly that of a repetitious nature, a process that would have an effect on rate processes, would result in smaller, possibly fewer, chromosomes with lower overall DNA values. Many of the species with low DNA values are obviously specialized forms, and it would be tempting to make such a correlation (Tables 9.3 and 9.4). Thus in addition to the species already mentioned, a number of other instances can be cited. Virtually all teleost fishes that are advanced and specialized, with specialization taking many forms of expression, have lower DNA values than those having the typical fusiform shape. Fishes in deep water have more DNA than those in shallow waters. In the orthopteran genus, *Nemobius,* the American species having smaller and fewer chromosomes are correlated with advanced morphology, more complex mating "calls," and more restricted habitats. Yet, on the other hand, as Stebbins points out, *Trillium, Lilium,* and the mistletoe genus, *Phoradendron,* have both extremely high DNA values at the same time that they are end points in their particular trends of evolution; however, only the mistletoe could be considered specialized, at least in terms of its nutritional habits. Undoubtedly, the same can be said for the amphibians *Amphiuma* and *Necturus;* they are, without question, terminally derivative and not ancestral species.

A rather unique experiment carried out by Mills and his colleagues bears on the question of low DNA values and survival under rigorous circumstances, and it indicates how evolutionary and environmental pressures can operate effectively at the molecular level. The example is from a prokaryotic form, but applicability to the eukaryotes may be more direct than is immediately obvious. Use was made of an RNA molecule derived from a *QB* virus that was capable of replicating *in vitro* in the presence of a *QB* replicase. The RNA molecule contained coding sequences for the replicase and coat proteins, together with those sequences needed by the replicase for recognition and replication. Since the replicase was supplied *in vitro,* and further since only replication and not encapsulation of the genome within a protein coat was involved, the experiment was designed to test whether the coding sequences for the replicase and coat protein were dispensable if the criterion for selection was simply rapid completion of RNA synthesis. By means of serial transfer the

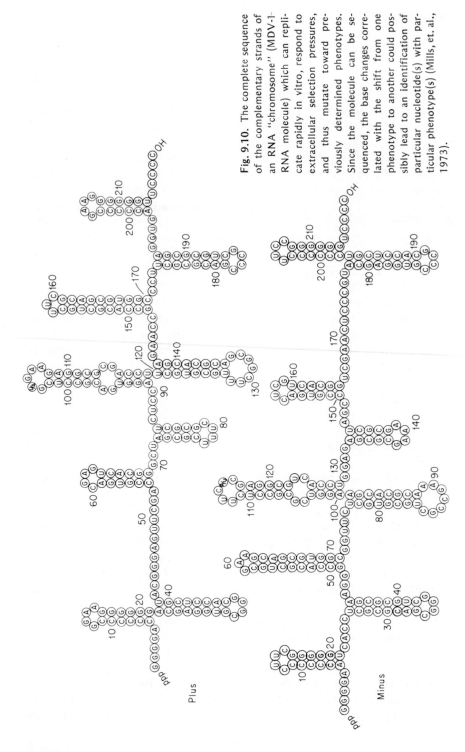

Fig. 9.10. The complete sequence of the complementary strands of an RNA "chromosome" (MDV-1 RNA molecule) which can replicate rapidly in vitro, respond to extracellular selection pressures, and thus mutate toward previously determined phenotypes. Since the molecule can be sequenced, the base changes correlated with the shift from one phenotype to another could possibly lead to an identification of particular nucleotide(s) with particular phenotype(s) (Mills, et. al., 1973).

time of replication was progressively decreased; the shorter molecules as well as those with faster growth rates, were those best adapted to the selection pressures imposed, with the result that by the 74th transfer, the minimum size and sequence consistent with *in vitro* replication were reached. The coding sequences for replicase and coat protein were lost in the process, and only 12%, or 550 nucleotides, remained of the original *QB* strands. Other kinds of selection pressures, similarly applied, provided other variant *QB* molecules, indicating that a diversity of genomic molecules, capable of expressing a variety of phenotypes, can be produced; indeed, a wide variety of genomes expressing an equally diverse set of predetermined phenotypes can readily be obtained by such imposed selection pressures.

One such replicating molecule, which can respond to selection pressures in the manner of the *QB* RNA virus, is illustrated in Fig. 9.10. The G–C content of the 218 nucleotide "chromosome" is nearly 70%, and the strands of the replicative (*RF*) form are obviously complementary to each other, but what is particularly striking and is reminiscent of the secondary structure of tRNA molecules, and of the coding sequence of the coat protein gene of the MS2 virus (Fig. 1.3), is the large amount of intrastrand, reverse repeats that give rise to hairpins and loops. These can scarcely be random sequences even though their significance remains to be explained; molecular stability and enzyme resistance, due to both intra- and interstrand pairing, may have been selected for, but it would appear that selective pressures can be brought to bear not only on the linear sequences themselves (the molecular genotype) but also on the two- and three-dimensional structures of the secondary and tertiary configurations (the molecular phenotypes). As has been suggested, new types of selection pressures became available through such mechanisms; although more difficult to design, experiments leading to increases as well as decreases of DNA could be achieved, as occurs at the *bb* locus in *D. melanogaster*. Thus, Wallace and Kass have proposed that the estimated 3,000 to 4,000 hairpins and loops in *Drosophila* play a role as sensors, which, when activated, bring adjacent genes into transcriptional activity. At the same time the tandem, but inverted, nature of the hairpins allows for exchange between pedestals, providing an opportunity for excising or adding segments of DNA, or for shifting the position of the loops (Fig. 9.11). Some 40,000 inverted repeats have been detected in the main band DNA of the mouse, with the pedestals, 60% of which also have loops at their tops, averaging 1000 base pairs each. If these are regions governing the activity of genes adjacent to them, they represent a fluid portion

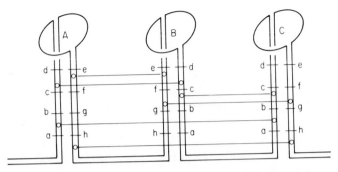

Fig. 9.11. Six repeated duplications, any adjacent pair of which constitutes a reverse repeat, while successive pairs constitute a tandem duplication. Each adjacent pair can undergo intra-strand pairing to form a pedestal, topped by an unpaired loop; the configuration of successive loops could act as sensors to govern gene activity. Intra-strand crossing over can occur between adjacent pedestals, or between alternate pedestals, with varying results. The first recombination, #1, brings about no change; #2 causes sensors A and B to exchange place in sequence; #3 leaves the sensors unchanged; while #4 causes B and C to exchange place. Any sensor can, therefore, move about in the control region, and without loss of DNA. Numbers 5 and 6 are between alternate pedestals, and their occurrence leads to a loss of both pedestals and sensors; #5 removes A and B together with their pedestals, while #6 excises A and C. Nonessential DNA can be shed in this manner, but if essential, such exchanges could constitute lethal mutations. Presumably adjacent exchanges would be more frequent than alternate ones, but the excision of a sensor with its pedestal causes adjacent exchanges to have more serious consequences, thereby hastening the alteration of a genome (Wallace and Kass, 1974).

of the genome, capable of responding to a variety of environmental pressures.

Flavell and Smith have demonstrated that cultivated varieties of wheat and rye differ in the number of rRNA genes present in the genome, a situation also known in hyacinth and maize as well as in *D. melanogaster*. If these repeated sequences can vary in number, it would seem quite possible for other sequences to vary in number if a change in number is the means whereby a genome can respond to environmental pressures. The addition of an inhibiting drug in ever-increasing amounts to cells in culture is one method for the testing of such an hypothesis, particularly since the occurrence of resistance in bacteria and insects to antibiotics and pesticides is well known and has a genetic basis. With this in mind, Schimke and his colleagues have shown that the structural genes coding for a specific enzyme can respond in this manner. Certain strains of cultured mouse cells are resistant to 3000 times as much methotrexate as that needed to kill an unselected and sensitive cell. Methotrexate, a 4-amino analog of folic acid, inhibits dihydrofolate reductase, an enzyme involved in the pathways of synthesis of glycine, purine and thymidylate. The resistant cells contain 250 times more dihydrofolate reductase than the sensitive ones, and, judging from DNA/mRNA hydridization studies supported by cytological examination, they also have 250 copies of the responsible structure genes as compared to one in the sensitive clone. It is assumed that the initial step toward resistance was the fortuitous appearance of a duplication of the reductase locus; once arisen, the duplication became the focus by which, through step-wise addition of more copies of the gene by means of unequal crossing-over, the genome could respond to the pressure brought on by increased concentrations of the drug. Relaxing the pressure caused some of the isolates to drop back to intermediate levels of enzyme activity and, presumably, fewer copies of the gene, but other strains retained their high levels, suggesting that the genomes can not only respond through gene amplification, but can also, by some unknown mechanism, preserve its high adaptiveness in the face of relaxed pressure.

Evolution of Individual Chromosomes

Evolution can be selective in its action of chromosomes, affecting the karyotype as a unit, a single chromosome to the exclusion of others, or parts of individual chromosomes. Centromeres, telomeres, and nucleolar organizers are products of evolution even though we are incapable of reconstructing their past histories. The presence of constitutive heterochromatin, and of euchromatin with its coding and noncoding sequences, provide further evidence of change which evolution has fixed in the eukaryotic chromosome, and which differentiates it from that characterizing the prokaryotes.

There is no concrete evidence that any heterochromatin, constitutive or facultative, can be returned to a transcribing, euchromatic state; selection pressure on cultured cells to favor an allele on an inactive X is ineffective, even if cell death results from such inaction. Also, the influence of viruses, hormones and chemical treatments is insufficient to alter the phenotype brought on by inactivation. From what we now know, therefore, heterochromatin seems to constitute a stable, irreversible, genetically inert state of DNA, and euchromatin can be heterochromatized and made genetically inactive when specific purposes are to be served: for example, dosage compensation through X chromosome inactivation in mammals, some kind of limited inheritance as in the dog roses, gall midges and coccids, or the gradual maturation of polymorphonuclear leukocytes from relatively undifferentiated myelocytes (Fig. 3.42). Heterochromatin, however, is not without influence. In addition to contributing to the mass of DNA and thereby influencing rate processes, constitutive heterochromatin has several evolutionary roles it can play: when located at the centromere it both reduces crossing over in its neighborhood, and because of its higher breakage rate and its tendency to pair with other segments of heterochromatin in a nonhomologous manner, it makes centric fusions and translocation of the *Oenothera* type readily possible without loss of essential chromatin. It can affect the function of euchromatic genes when they are placed, by rearrangement, adjacent to heterochromatin, although this role must be accidental rather than selected for in any positive sense. It may also serve in a buffering capacity, minimizing insults to the more critical euchromatin of the genome. The last suggestion, called the *bodyguard hypothesis,* stems from the observation that heterochromatin, by virtue of its location, is more frequently involved in breakage and rearrangement, whether this be spontaneous or induced by chemicals or radiation; it is distributed in interphase at the periphery of the nucleus (Fig. 3.37), the more so in differentiated cells than in embryonic ones, and hence is more immediately accessible to incoming influences than the more internally distributed euchromatin. Dispersal of the heterochromatin can be brought about by growing cultured cells in hypotonic solutions— this has been done with cells of the mouse, *Peromyscus eremicus*—and if mitomycin C, a potent chromosome-breaking agent and mutagen, is then added, a shift in breakage frequency is from 99% in heterochromatin to about 36%, the latter figure being more nearly that of the percentage of heterochromatin in the total genome.

The evolution of one chromosome, or a pair of homologues, differentiated wholly or in part for a specific purpose and thus set apart from the remainder of the genome, is exhibited in those chromosomes concerned with sex determination. In some instances, the differentiation is functional only, with no morphological alterations in the chromosomes; in others, structural modifications have occurred as well, with the members of a pair of homologues diverging from each other to the extent that they become heteromorphic; one sex, therefore, becomes heterogametic, the other homogametic. Most species with recognizable sex chromosomes are XX, XY (or some modification thereof), with the male heterogametic; the birds, snakes, and *Lepidoptera* are WW, WZ with the female heterogametic.

The origin of sex chromosomes, which has undoubtedly occurred a number of different times in multicellular organisms, postdates the establishment of an alternation of generations, and the accompanying features of fertilization and meiosis. When the mating system is based on a simple allelic pair of genes, no chromosomal differentiation is needed. In *Paramecium* and *Neurospora,* with *A* and *a* mating types, "sex" is expressed in the haploid state only, and on the basis of an acceptance or rejection system of mating, so the selective pressure to develop elaborately differentiated chromosomes does not exist. It is more than likely that the differences between the sexes in many species where sex chromosomes are not recognizably different from other chromosomes of the genome are similarly governed by allelic pairs of genes, with sex being more of a physiological than a morphological expression of the genotype. This category would include many of the hermaphroditic lower invertebrates and the higher plants with their several forms of self-incompatibility systems. In diploid species where the differences between the sexes are expressed morphologically and/or behaviorally, sex chromosomes are not always evident, but differentiation of the sex chromosomes occurs when reinforcement of the sexual differences comes into play, and when one of the chromosomes becomes confined to a single sex, thus isolating it in a genetic sense. This kind of genetic isolation and the random nature of mutations would set the two members of a pair of homologues off on independent paths of evolution, with their differences of gene content accentuated progressively until their homologies would be lost except for that required for pairing and segregational purposes. The independent paths of evolution leading to the formation of heteromorphic homologues, together with the confinement of one chromosome (the Y or the Z) to the heterogametic sex, would inevitably be associated with differences in gene dosage in that X or W chromosome genes would be represented twice in one sex, and only once in the other. Since not all genes in the X are concerned with the expression of sex, but govern as well a spectrum of morphological and biochemical traits, mechanisms would be required to correct the genetic imbalance that would increase as the difference and isolation between the homologues was accentuated with time and by the accumulation of mutations. The origin of heteromorphic sex chromosomes is, therefore, intimately related to gene dosage compensation, but the question is whether dosage compensation arose in response to the different gene content of the heteromorphic sex chromosomes or whether it was the driving force responsible for the evolution of sex chromosomes.

Gene dosage compensation is evident in *Drosophila,* in amphibia, and in mammals, but with different manifestations. In *Drosophila,* the activity of a single X in an XY male is genetically equivalent to the two Xs in a female, as measured by such varied phenomena as pigment production in the eye, enzyme levels in a variety of cells, and RNA synthesis in salivary gland cells. Only the genes *eosin* and *facet* in the X chromosome of *D. melanogaster* appear to be uncompensated. It would appear that compensation is achieved by varying the rate of transcription, and since both female XXs are active, as judged by the simultaneous presence of two enzyme variants in equal quantities, the single male X in an XY individual must be transcribing at double the rate of a given X chromosome in an XX female. The level of X chromosome activity in both males and females is geared to the number of sets of autosomal genes (Fig. 9.12), and the fact that the X in an XY, 3A male can be pushed to an activity well above its normal level in an XY individual would suggest that the compensation is positively regulated rather than being negatively depressed.

The phenomenon can also be expressed at the chromosomal level, with the single X of the male having twice the diameter of the Xs in the female. This is particularly evident in hybrid individuals between *Drosophila insularis* and *D. tropicalis* (Fig. 9.13); synapsis has not occurred except in the heterochromatic chromocenter, and the increased diameter of the X in the male is readily apparent. Interestingly enough, dosage compensation is not a result of an increased level of polyteny; instead, it is due to an increased rate of RNA synthesis. The rate of synthesis of the male is, therefore, increased rather than that of the female being depressed, and the compensatory effect is regulated at the level of transcription.

A somewhat similar situation has been described in *Xenopus laevis*. An individual heterozygous for the anuclealate mutant produces as much rRNA as does a homozygote with two nucleolar organizers. In mammals, however, dosage compensation of X chromosome genes is brought about by an inactivation mechanism, the genetic activity of only one X being expressed in somatic cells. All others are heterochromatized and appear at Barr bodies in interphase cells. In eutherian (placental) mammals, heterochromatinization of one or the other X in the females is a random process, while inactivation in marsupials selectively removes the paternally derived X from genetic participation in somatic cells.

Lucchesi has envisioned the evolution of heteromorphic sex chromosomes taking place by mutations first accumulating in the vicinity of that allele which will be confined to the heterogametic sex, that is, *A* is in an *Aa,* as opposed to an *aa*, individual. The closer these mutations will be to the allele, the less likely will they be removed by recombination or lost by elimination. Also, since the mutations will generally be toward dysfunction, or absence of function, of the genes involved, they could be heterochromatized or lost without serious effect; the effect would spread along the chromo-

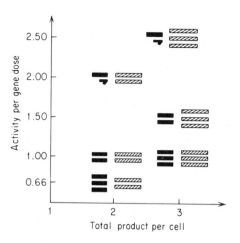

Fig. 9.12. The activity per gene dose as a function of the total product per cell. For a given number of autosomal sets, the total product per cell is the same whether there is one, two or three X chromosomes; the X chromosomes are therefore regulated as to the degree to which they respond to cellular needs. X chromosomes are solid lines; autosome sets crosshatched lines; and Y chromosomes J-shaped (Lucchesi, 1978, copyright 1978, by the American Association for the Advancement of Science.)

some. The separation of the sexes would be reinforced by the progressive isolation of the affected chromosome, but a necessary concomitant would be the creation of a dosage compensation system to offset the lost genetic activity in the heterogametic sex. How this is achieved is not certain, but a regulatory process in *Drosophila* is evident. X chromosome genes translocated to autosomes retain their capacity to be regulated, while autosomal genes do not gain this capacity by being translocated to an X chromosome; the capacity to be regulated is, therefore, genic and not chromosomal.

A not too dissimilar mechanism can explain the origin of the heteromorphic sex chromosomes of the mammals, but obviously the regulatory mechanism of inactivation must be distinctly different.

Fig. 9.13. The salivary gland chromosomes of female (left) and male (right) hybrid offspring from a cross of *Drosophila insularis* with *D. tropicalis*. Synapsis is absent except at the chromocenter. The single X in the male has a much greater thickness than those in the female. Measurements of DNA values indicated that this is not due to an increased degree of polyteny in the single male X, but probably to increased levels of RNA and/or protein (Dobzhansky, 1957).

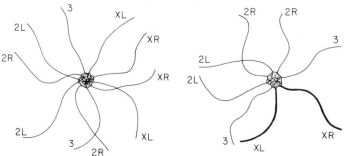

The genetic situation in maize can, perhaps, illustrate the manner by which X and Y chromosomes could come into existence. Maize is a monoecious species with separate staminate and pistillate structures. A recessive gene *ts* (= *tassel seed*) converts the terminal male tassel into a pistillate structure when homozygous, while another recessive gene *sk* (= *silkless*) causes ovule abortion. *Ts* is epistatic to *sk* when both are recessive so that the double homozygous recessive, *ts ts sk sk,* has normal lateral female inflorescences while the tassel has been transformed into a pistillate structure; *ts ts sk sk* plants are, therefore, female. The male, totally staminate, plants are *Ts ts sk sk.* The two genes are unlinked so that a cross between the two genotypes (*Ts ts sk sk* X *ts ts sk sk*) produces only male and female plants. The chromosomes bearing *ts* and *Ts* become the equivalent of primitive X and Y chromosomes, respectively. Modifying genes could aid in reinforcing the expression of "maleness" and "femaleness" in the genome, and some 40 or more are known in maize which affect the staminate and pistillate structure, but if the new Y chromosome is to achieve a heteromorphic status, *Ts* has to become the focus around which dysfunctional mutations accumulate, setting the stage for subsequent heterochromatinization and further structural change.

Since sex-determining mechanisms have arisen independently a number of times in both plant and animal kingdoms, it is to be expected that they vary in detail even though the end result, cytogenetically, is a heteromorphic pair of chromosomes. In *Drosophila* the X chromosomes are female-determining, but with no specific genes determining femaleness; the autosomes are male-determining, and the two sexes represent a balance between the X and the autosomes (Tables 7.3 and 7.4). The *Drosophila* Y is concerned with male fertility, not with determination of other features of the male phenotype. The human Y is male-determining only, the *H-Y* locus having a testes-determining effect along with the TDF locus in the short arm (Fig. 4.46). No mammalian species, in fact, lacks a Y chromosome, although XX males are known. Although sterile, they show male development due to the presence of an autosomal dominant gene (*Sxr=sex-reversed*) which behaves similarly to the H-Y locus. It seems that three genes are necessary for the development of the mammalian testis. There is some question as to the location of the female-determining factors (or influence) in the human genome, and the lethality of unbalanced genomes makes determination difficult and uncertain. In the plant *Melandrium,* the Y is male-determining, and the X exerts a female influence, with specific regions of the chromosomes controlling specific aspects of sex expression (Fig. 7.5).

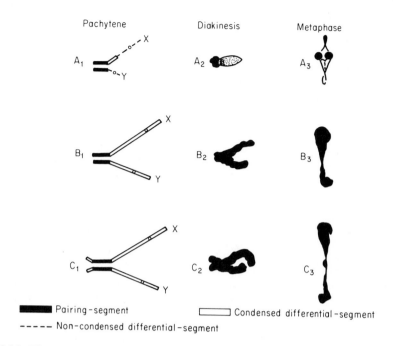

Pachytene Diakinesis Metaphase

Pairing-segment
Non-condensed differential-segment
Condensed differential-segment

Fig. 9.14. Diagrammatic representation of the structure of the sex chromosomes in five genera of marsupials, and at several stages of meiosis. Both chromosomes have undergone considerable change in the A group of genera, less so in the others. A group, structures typical of the X and Y in *Dasyurus, Sarcophilus* and *Phascolarctus;* B, *Trichosurus;* C, *Pseudochirus* (Koller, 1936).

As Lucchesi has emphasized, it is the heteromorphic and partially isolated chromosome that undergoes the greatest change since it is no longer subject to recombination, and the phemenon of dose compensation lessens its genetic influence. This is the Y or the Z chromosome, and in advanced sex-determining systems these chromosomes are heterochromatic to varying degrees (Fig. 9.14), and, because of their relative genetic inertness, they are more subject to loss. The widespread prevalence of XX-XO sex-determining systems in insect groups is indicative of this loss, with whatever residual genetic influence the Y may have had transferred presumably to the autosomes, and with the single X fully regulated by some mechanism of dosage compensation. Such XX-XO species, however, can give rise to XX-XY or $X_1 X_1 X_2 X_2 - X_1 X_2 Y$ forms through translocations, the particular change dependent upon whether the chromosomes involved are acrocentric or metacentric (Fig. 9.15), while an XY system can, again by translocation, be converted into an $XY_1 Y_2$ condition.

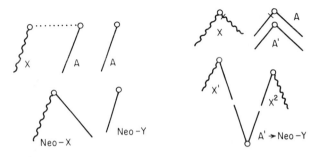

Fig. 9.15. The transformation of an original XO type of male into a neo-XY by the centric fusion of an acrocentric X with an acrocentric autosome, and with the remaining autosome recoming a neo-Y chromosome, confined thenceforth to the male lineage (left). A similar transformation of an XO male into an X_1X_2Y male by a reciprocal translocation between a metacentric X and a metacentric autosome, and with the remaining autosome becoming the neo-Y (right).

Conversion of an XX–XO species into a neo–XX–XY system has occurred a number of times in the Orthoptera. The species *Hesperotettix viridis* is of interest in this respect. In the subspecies *H.v. brevipennis* an XO condition prevails, while an apparently recently derived neo-XY is present in *H.v. protensis.* The subspecies *H.v. viridis* has both conditions, the Y and the arm of the X with which it is homologous being euchromatic and capable of synapsis. It would appear that the neo-Y is of such recent origin that it has neither become fixed in the population nor has it acquired a heterochromatic state. A neo-XY derived from an XO state has also been observed in a phasmid (walking stick) and in a beetle, *Tribolium confusum,* but in both instances sufficiently long ago in the past for the neo-Y to become heterochromatinized.

A neo-X_1X_2Y derivation from an original XO condition has been found among the mantids. The females would be $X_1X_1X_2X_2$. There is some suggestion that the change occurred once, and the Y, on becoming heterochromatinized, has also undergone diminution with time (Fig. 9.16). Assuming that the two arms of the original neo-Y were equal in length to those of the two Xs with which it pairs, a condition found in *Choeradodis, Stagmatoptera,* and *Tauromantis,* an extreme in size is reached in the genus *Melliera* in which the size ratio of Y to X is about 0.07. In this instance, it would appear that the neo-Y serves little function other than that of a pairing device for purposes of regular segregation.

The transformation of an XY system into an $X_1Y_1Y_2$ condition, and then back to an XY, with loss of one of the heterochromatic Ys, has been thought to have occurred in some species of *Drosophila,* and it has been reasonably well documented in the buprestid beetle, *Agrilus anxius.* Two strains exist, one feeding on

516

poplar and the other on birch, with the former (strain A) having a $2n$ formula of $18A + XY$ and the Y being a large chromosome. Strain B has a $2n$ formula of $20A + Xy$ (with y being quite small). It would appear that strain A was derived from strain B by translocation between an acrocentric X and an acrocentric autosome, and with the subsequent loss of the small y. Thus, the origin is presumed to be $10AA + Xy \rightarrow 9AA + A\ \widehat{AX}\ y \rightarrow 9AA + \widehat{AX}\ A$, with the solitary A becoming the neo-Y of strain A.

In the XO state, the X chromosome has no pairing partner, and it must reach a pole by means other than segregation from its partner. It does so by passing to one pole or the other before the other bivalents segregate, but the means are difficult to determine. In one instance, at least, such a lone X has also been shown to influence the segregation of another, presumably unrelated, pair of autosomes. The mole cricket, *Gryllotalpa dactyla,* possesses a pair of readily identifiable heteromorphic, autosomal homologues, one being twice the size of the other. The large homologue always passes to the same pole as the X although there is no obvious connection between them. Micromanipulation experiments reveal that the X exerts a determining influence; even if the X is moved in the cell, the heteromorphic pair adjusts itself to the X so that the X and the large autosome reach the same pole. Two kinds of sperm are obviously

Fig. 9.16. Sex trivalents of 10 species of X^1X^2Y mantids at metaphase I, showing the variable size of the Y chromosome (located at the bottom of each trivalent). Some variation in the pairing and the lengths of the free arms of the X chromosomes is also evident, and can be presumed to be the result of evolutionary modification, since the X^1X^2Y mantids appear to have a monophyletic origin (Hughes-Schrader, 1958).

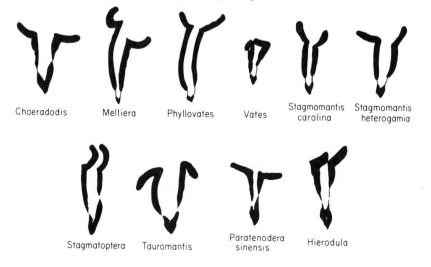

| Choeradodis | Melliera | Phyllovates | Vates | Stagmomantis carolina | Stagmomantis heterogamia |

Stagmatoptera   Tauromantis   Paratenodera sinensis   Hierodula

produced: one with an X, the large homologue, and the rest of the autosomes which segregate normally and the other with no X, the small homologue, and a set of the autosomes. This is basically an $X_1 X_2 Y$ system in terms of distribution, if not in pairing relationships. The female is assumed to be $X_1 X_1 X_2 X_2$ since the males always carry the heteromorphic pair of homologues, and the small homologue is assumed to be the equivalent of a Y chromosome.

Karyotype Changes Within Taxa

During past geologic periods plant and animal taxa of various kinds have made their appearance, and to judge from those extant today, their acquisition of taxonomic identity has often been paralleled by karyotypic uniqueness as well. Among generic groups such as *Drosophila,* maize, *Tradescantia, Trillium, Triturus, Mus,* and the human, no perceptive cytogeneticist would mistake any of their karyotypes for other than what it is. The degrees of difference within taxa will vary, however, and patterns of change used for taxonomic purposes may, or may not, be accompanied by detectable modifications of the karyotype. Even more difficult would be the determination of the adaptive significance of chromosomal change, although it is generally accepted that a reordering of the genome brings about new gene groups, possibly with unanticipated but significant position effects, and new linkage groups result in different patterns of recombination, and hence different gametic diversities.

Unlike gene mutations which can occur repeatedly and often at predictable rates, chromosomal rearrangements are unique events. Each occurs within a single individual, and it is unlikely that any two are exactly alike. When two related species or taxa are separated from each other by extensive chromosomal alterations, it would appear that a whole series of independent events, occurring one at a time, have accumulated and become focused in one species or the other to bring about their karyological separation.

Speciation can occur without detectable chromosomal change. The *Platanus* situation has been previously mentioned. In *Drosophila,* three closely related species—*mulleri, aldrichi,* and *wheeleri*—show no rearrangement of chromatin that is discernible even at the level of polytene banding patterns; a similar situation has been observed among a group of *Drosophila* species endemic to Hawaii. Speciation in these instances, with whatever reproductive barriers have been established between these sibling species, appears to be a matter of relatively recent genic change rather than of chromosomal factors.

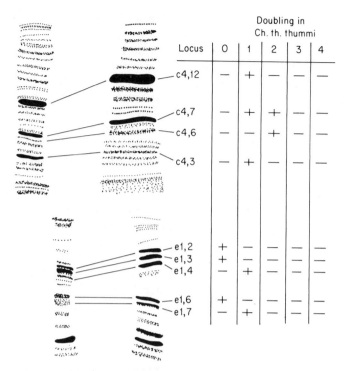

	Locus	\multicolumn{5}{c}{Doubling in Ch. th. thummi}
		0	1	2	3	4
	c4,12	−	+	−	−	−
	c4,7	−	+	+	−	−
	c4,6	−	−	+	−	−
	c4,3	−	+	−	−	−
	e1,2	+	−	−	−	−
	e1,3	+	−	−	−	−
	e1,4	−	+	−	−	−
	e1,6	+	−	−	−	−
	e1,7	−	+	−	−	−

Fig. 9.17. Bands in the polytene chromosomes of *Chironomus thummi thummi* and *Ch. th. piger.* In the two figures, the chromosomes of *piger* are on the left. Top, bands from a portion of the left arm of chromosome II; bottom, bands from a portion of the left arm of chromosome I. The chart to the right indicates the degree of doubling that has taken place in *thummi* as compared to *piger* (Keyl, 1965).

A limited step in the direction of chromosomal repatterning can be seen in subspecies of *Chironomus: C. thummi thummi* and *C. thummi piger* (Fig. 9.17). The former has 1.27 times more DNA than the latter, and it is due largely to an increased DNA amount found in specific bands variously located in the chromosomes. The differences in amounts are not the same in all bands, and they vary by factors of 2, 4, 8, and 16. These are not instances of lateral redundancy but most probably represent examples of linear duplication and reduplication confined to a localized region of one chromosome in the genome and brought into existence by unequal crossing over or mistakes of replication (Fig. 3.30). Genic evidence of duplications in *Chironomus* is lacking, and *piger* may be derived from *thummi* by a reduction of DNA, but as Ohno has extensively documented, information derived from other species points to duplications as at least one source of new genes. The four hemoglobin genes in the human and particularly the adjacent *beta* and *delta* loci, code for polypeptides too similar in amino acid content and sequence to have evolved independently; it is further believed the myoglobin gene is ancestral to those now coding for hemoglobin. Evolution of the loci governing the structure of the isozymic subunits of the tetrameric lactate dehydrogenase can be similarly interpreted in terms of duplication, followed by divergence through mutation. Humans and other mammals have A, B, and C loci, as do birds, but whereas some fishes have but a single locus for the enzyme, others such as the *Salmo* species may have as many as eight.

A number of pseudoallelic series in *Drosophila* can be similarly viewed as duplicate loci, for example, the *vermilion, white, dumpy, Lozenge,* and *bithorax* loci. The repeated clusters of related units have been identified and separated by mutant, linkage, and complementation criteria, and they have been characterized cytologically by association with doublet bands (Figs. 6.10 and 6.11). These pieces of data would imply that duplications, followed by divergence through mutation, represent the only source of new loci with new function, but the discovery of interrupted eukaryotic genes and of RNA·RNA splicing as a post-transcriptional event makes possible, theoretically, the transformation of pieces of scattered noncoding DNA into actively coding DNA, thereby creating new genes from present DNA by the manipulation of the mRNAs. This concept will be as difficult to prove as has been the idea of new genes from duplicated old genes, but duplications, coupled with the imprecision of RNA·RNA splicing, could provide an added impetus to evolutionary rates.

The North American diploid species of *Tradescantia* present another facet of the problem of chromosomal repatterning. The karyotypes of the several species are quite uniform within the group, interspecific hybrids are readily formed, and pollen viability and seed set are high, but no individual plant, whether of a pure species or of hybrid origin, seems to be free of paracentric inversions. Their random location within arms and their frequent occurrence would suggest that, as in *Drosophila willistoni,* they are without physiological effect or adaptive significance, and hence that they differ from those in *D. pseudoobscura* which exhibit varying degrees of environmental superiority, particularly when heterozygous. Since these are perennial species, however, the idea is not an easily testable one, and it may well be, as in *Clarkia,* a genus related to the *Oenotheras,* that chromosomal and genetic heterozygosity is closely correlated with adaptive superiority.

Under ordinary circumstances it would seem most unlikely, given the random nature of mutations and the uniqueness of chromosomal aberrations, that the acquisition of both chromosomal and genetic heterozygosity could be attained in any other than a gradual process. The situation in *Clarkia,* however, suggests that the circumstances may well be otherwise. *C. speciosa,* an annual endemic to western North America, is normally outcrossing, with selfing leading to reduced vigor, fertility, and chiasma formation. A derivative form, *C. speciosa polyantha,* differs from its parent by a minimum of ten different chromosomal arrangements based on five reciprocal translocations to give karyotypes which include rings of 4, 6, 8, and 10 chromosomes. When *polyantha* is selfed, the structural heterozygotes are twice as frequent as the homozygotes, and they seem to be better able to survive. Presumably, the superiority of genic and/or structural heterozygosity is involved. Lewis, in fact, states that in this genus, at least, "the essence of speciation is chromosomal reorganization in an ecologically marginal site." Figure 9.18 indicates the relationships among a number of annual species of *Clarkia. C. franciscana* is separated from the related *C. rubicunda* by three translocations and four inversions, all of which seem to have been acquired within a relatively brief span of time. Hybridization experiments show that *C. rubicunda* is clearly derived from *C. amoena.* The five species in the *C. unguiculata* complex differ from one another by three to six translocations; they are similar to each morphologically, but the hybrids are highly sterile.

The coincident occurrence of a number of unlikely circumstances surround the speciation situation in *Clarkia.* Since all are annuals, there is a rapid overturn of individuals. Lewis believes that some drastic ecological change such as a drought reduces the normally outcrossing species to a few individuals. Forced inbreeding accompanies (or is responsible for?) extensive chromosomal breakage and rearrangement, leading to structural heterozygosity and low fertility, and out of which, by chance and continued inbreeding,

Fig. 9.18. Species relationships of two complexes in the genus Clarkia, with the parent species at the base, and the neospecies arising from it. The haploid chromosome numbers are indicated numerically, and the self-pollinating species are underlined. All other species are normally outcrossed (from Lewis, 1973).

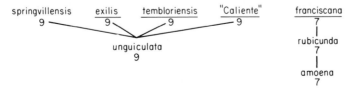

emerge genic and structural homozygotes having gene combinations that are adaptable. These homozygotes find an initial home in ecologically marginal areas, but with time they can compete with the parental species when they come in contact. The neo-species are not better adapted than the parental species; they simply can gain a foothold in marginal, unoccupied sites. Any hybrids formed between parents and neo-species are structurally heterozygous and of low fertility and cannot compete unless the population is once again reduced to low numbers or if marginal areas are available for occupancy. It may well be, therefore, that a short life cycle, the availability of unoccupied habitats, and species hybridization, self-compatability, and an unstable karyotype are the prerequisites to speciation of the sort found in *Clarkia*. The situation seems not to be unique, being found among some of the annual species of the Compositae, but it is unlikely that it would be characteristic of woody or perennial species.

Chromosome Size and Number. The experimental studies carried out on the *QB* RNA virus indicate that a genome can rid itself of unnecessary genomic elements if the selective pressures are sufficiently severe and constantly applied; granting that the *in vitro* conditions of the *QB* experiments were highly artificial, the results nonetheless suggest that, given time and the right conditions, a karyotype can respond to external conditions and that changes in chromatin content, leading to variations in chromosome size and number, reflect adaptive phenomena operating at the chromosomal level. The significance of the *QB* studies is made more pertinent by the parallelism encountered in nature and among eukaryotic species: within related groups of species, where specialization can be ascertained with some degree of assurance, DNA values, chromosome size, and, more often than not, chromosome number as well are correlated with the degree to which a species is or is not specialized. The trends are such that the karyotypes of specialized species, including those that occupy varied as opposed to stable environments, consist generally of smaller and fewer chromosomes than those of the less specialized relatives. There is, in addition, a comparable correlation when rates of metabolism are involved.

Unfortunately, no comparably obtained experimental evidence is available which proves the obverse of the concept, namely, that a lowered selection pressure, or less specialization, or a more stable environment will lead to an increase in DNA values and larger chromosome size. Circumstantial evidence, however, is most supportive of such a correlation. It has already been pointed out that the *bobbed* phenotype in *D. melanogaster* makes its appearance when the number of rRNA coding sequences drops below 130 and that a stock of *bb* flies will, over a period of time, revert to a wild-type phenotype. The nature of the selection pressure remains unknown, but presumably it depends upon a metabolic need for a specific level of rRNA molecules to meet the demands of the life cycle of the species. As such, it is a process akin to gene amplification and to gene dosage compensation in that the genome possesses a variety of mechanisms for responding to pressures by manipulating its chromosomes to meet its own metabolic means.

The correlation of animal karyotypes with the stability of the environment and with rates of metabolism follows definite trends. *Amphiuma* and *Necturus,* both having massive amounts of DNA, have a notably slower metabolic rate and a more sluggish behavior than other amphibians, and they exist in a very stable environment. As pointed out earlier, deep-sea fishes generally have higher DNA values than those in open, more turbulent waters, while the teleost fishes that depart from a fusiform shape and acquire specialized parts or shapes have lower DNA values than their more primitive fusiform ancestors with many bony parts. In terms of metabolic rates—based either on oxygen consumption or heart rate—the birds have far lower DNA values and higher metabolic rates than do the reptiles from which they arose, while the mammals are generally intermediate. Also, within amphibian, avian, and mammalian groups or subgroups, a similar relation holds generally: the lower the DNA values, the more rapid the metabolic rate. It appears that those forms with high DNA values are simply incapable of having a short life history or of existing in variable environments and of carrying on an existence that puts a strain on metabolic rates or energy transfer systems.

If we turn now to the plant world, the monocots among the angiosperms generally have larger chromosomes than do the dicots. There is considerable overlap, but within each of the two major groups, the more primitive families have larger ones than the more advanced groups. Thus the Liliales have larger chromosomes than do the Orchidales, while the Ranales, the most primitive of the dicots, have larger ones than do the Campanulales or Compositae. A comparable situation occurs in the ferns as well. The primitive Osmundaceae and Hymenophyllaceae have considerably larger chromosomes than do the more specialized Salviniaceae, while the Polypodiaceae are intermediate, both taxonomically and chromosomally. On the other hand, the evolutionary sequence of life forms is assumed to be from woody to perennial to annual habit, but the woody species generally have smaller and more numerous chromosomes than do the herbaceous perennials presumably derived from them; the herbaceous annuals return to a smaller size and fewer chromosomes. The fact should be emphasized that there is no basis for assuming that large amounts of DNA have the same significance and play the same role in all groups.

The genus *Crepis* provides illustrative material to demonstrate the relation of chromosome size to both specialization and habit (Fig. 9.19). Four of the species—*kashmirica, sibirica, conyzaefolia,* and *mungieri*—are perennials with rhizomatous root systems; *leontodentoides* is perennial or biennial; *capillaris* is generally an annual, but it may, under some circumstances, behave as a biennial; *suffrenicana* and *fuliginosa* are strict annuals. In terms of size, *kashmirica* chromosomes are five times larger than those of *fuliginosa*, with the others of intermediate values, and this relationship is paralleled by comparable reductions in the size of florets and achenes. A similar trend, in this instance in the genera *Allium* and *Vicia*, can be demonstrated between DNA values and seed mass, while Bennett has also shown that a similar correlation is found with the time for fusion of egg and sperm nuclei in the embryo sac after the penetration of the pollen tube.

What has been described above for chromosome size, which is a reasonable criterion of DNA amount, applies also to chromosome number, although in less exact fashion (*C. capillaris* would be an exception since its chromosomes are fewer but larger than *mungieri* or *leontodontoides*). Among genera having numerous species or different structure and habitat, the more advanced and specialized members have lower haploid numbers than do their more primitive relatives, and they are generally outcrossing species capable of rapid colonization of new habitats. Members of the family Compositae can again illustrate the point. The basic number of the family is probably $n = 9$, a number encountered in the woody and/or herbaceous perennial species of such genera as *Haplopappus, Solidago* (goldenrod), *Aster, Chrysopsis* (golden aster), and *Helianthus* (sunflower). Semi-shrubby or herbaceous perennial species of *Haplopappus* have haploid numbers of 6 and 5, while the annual *H. gracilis* has only 2, 3, or 4. *Chrysopsis pilosa* ($n = 4$) is also an annual species, as is the related monotypic genus *Bradburia* ($n = 3$). The same size reduction seen in *Crepis* (Fig. 9.18) is also evident in these annual forms. There is no compelling evidence to suggest that the 9-chromosome forms are polyploid derivatives of some lower numbered species, and when strictly woody families are examined, they are generally characterized by a single base number ranging between 11 and 19. It is the

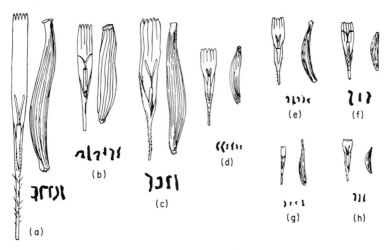

Fig. 9.19. Species of the genus Crepis showing the size relations of chromosomes, florets (lacking ovaries), and achenes, all drawn to the same magnification so that the sizes are relative to each other. A, *C. sibirica;* B, *C. kashmirica;* C, *C. conyzaefolia;* D, *C. mungierii;* E, *C. leontodontoides;* F, *C. capillaris;* G, *C. suffreniana;* H, *C. fuliginosa* (rearranged from Babcock, 1947). See text for further explanation.

evolution of herbaceous types out of a woody primitive group that carries the base number to lower values; exceptions are known, but they are rare.

Among both higher plants and animals, therefore, a causal relation exists between a reduction in DNA values, as evidenced by decreased chromosome size and numbers, and a specialized way of life. Among the higher plants, the trends of specialization may take a number of forms: a reduction in size or specialization of fruits and flowers (Fig. 9.19); from homospory to heterospory in the ferns; from a woody to an herbaceous habit; from a perennial to a biennial to an annual life cycle; from mesophytic to xerophytic conditions; from long growing seasons to shorter ones. A possible reversal of the latter trend, that is, an increase in chromosome size of plants in temperate or boreal regions as compared to more tropical climates is seen in the Graminaceae, but the relationship is not a certain one if climatic adaptation is taken into account. *Trillium,* for example, is not a primitive member of the Liliales despite its high DNA values, but its slow-paced rates of both mitotic and meiotic cell division reflect its pattern of growing through division at low temperatures and with slow metabolic rates; each stage of division is stretched out proportionally rather than establishing a delay at some particular step in the divisional cycle. One must assume that its DNA values, as well as those of *Amphiuma* and *Necturus,* are as rigorously selected

for as are the reduced karyotypes of the annual species of *Crepis* or
Haplopappus. Both American and Asiatic species of *Trillium* have
probably maintained their karyotypes for many millions of years,
and it is hardly likely that they would have done so without good
adaptive cause. The gymnosperms, with their high DNA values (Fig.
1.1), must have similar good cause, for they have persisted for even
longer periods of time.

The mechanism(s) of reduction in chromosome numbers is
known with more certainty than is a reduction in size and has been
demonstrated in several instances. Among the higher plants in which
a derivative species can be hybridized interspecifically with its putative parental form, it has been shown that unequal translocations,
accompanied by losses of centromeres and centric heterochromatin,

Fig. 9.20. The mitotic and meiotic chromosomes of *Crepis neglecta* and *C. fuliginosa*. Top
row, left, haploid complement of *neglecta*; right, haploid complement of *fuliginosa*. Middle
row, left, the diploid complement of *neglecta*, showing the large amount of centric heterochromatin, and the largely heterochromatic C_N chromosomes; right, the somatic complement of the hybrid. Bottom row, the meiotic metaphase I configurations in the hybrid,
indicating that there has been a translocation between the A and D chromosomes to give a
chain of four, $A_F A_N D_F D_N$: A translocation between the B and C chromosomes produces,
in the hybrid, a $B_N B_F C_N$ trivalent, and with the C_F chromosome lost. Univalents are
produced when pairing fails (Tobgy, 1943).

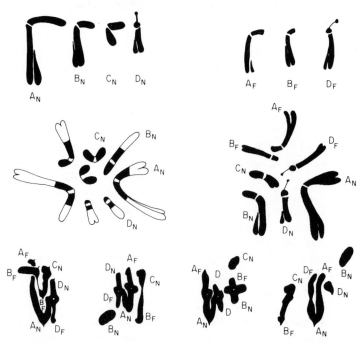

are involved. Thus, as Fig. 9.20 illustrates, *Crepis fuliginosa* has been derived from *C. neglecta;* it is also more reduced in growth habit and more restricted in distribution than *C. neglecta,* suggesting a recent origin. *C. kotschyana* (n = 4) has been similarly derived from a 5-chromosome species such as *C. foetida,* and *Haplopappus gracilis* (n = 2) has been derived from *H. ravenii* (n = 4). Of the latter two species, *H. gracilis* occupies more xeric habitats than does *H. ravenii,* and when conditions are favorable, it can complete its life cycle with dispatch.

In contrast to the perennial *Oenotheras* in which the transloca- tion complexes maintain structural heterozygosity and competitive capability in many habitats, without reduction in chromosome size or number, the complex of features characterizing the *Crepis-Haplopappus* annual species—that is, reduced chromosome size and number and an abbreviated life cycle—must be selected for, and related to, reproductive success in pioneer habitats. This seems best achieved by genotypes of limited variability but of high survival, and once again the pattern of change is consistent with the life style. Linkages are greatly tightened by fewer and smaller chromosomes, localized chiasma formation, and self-compatibility, and as genetic homozygosity is increased, so too would be the preservation of adaptive gene complexes grouped meaningfully in a limited number of chromosome arms. Not all aggressive species are weedy annuals, but a high proportion are self-compatible, short-lived seed producers fitting the above description. The perennials, if weedy, more often than not do so by apomictic or vegetative means and so may cir- cumvent the chromosomal mechanisms which reduce the DNA content.

The conspicuous types of karyotype repatterning in the animal kingdom differs from that in the plant world, but the correlation of DNA amount with growth and development is similar. A comparison of the metamorphic frog with the neotenous *Necturus* has already been mentioned. The incompletely metamorphic Orthoptera—for example, the hearth cricket, *Gryllus domesticus*—has nearly two orders of magnitude more DNA per haploid genome than do the fully metamorphic and more short-lived *Drosophila* species (1.1 × 10^{10} vs. 1.7 × 10^8 nucleotides), while the more primitive goose has about 50% more DNA per cell than does the more advanced sparrow (2.9 × 10^9 vs. 1.9 × 10^8), although both have the same number of chromosomes (Table 9.7). It is more than likely that the rapidly metabolizing hummingbird among the avian group and the shrew (*Sorex cinerius*) among the mammals would have the lowest DNA values in the vertebrates.

Table 9.7. Diploid chromosome numbers and chromosome characteristics in the Aves

Species	2n	Acrocentrics	Meta- or Submetacentrics	Z*	W*
			Chromosome Characteristics		
Anas acuta (pintail duck)	82	78–76	4–6	s-m	a
Lophortyx gambelli (Gambell's quail)	80	64	16	s-m	a
Calypte anna (Anna's hummingbird)	74	50	24	s-m	a
Bubo virginianus (Gr. horned owl)	82	72	10	m	m
Anser c. caerulesceus (blue goose)	82	74	8	s-m	s-m
Zonotrichia albicollis (wh.-th. sparrow)	82	60	22	s-t	a
Corvus corax (raven)	78	64	14	a	s-t

*Designations under the Z and W chromosomes: a = acrocentric; s-m = submetracentric; s-t = subtelocentric; m = metacentric.

Source: M. L. Becak et al., *Chromosome Atlas: Fish, Amphibians, Reptiles and Birds*, Vols. 2 and 3 (New York: Springer-Verlag, 1973, 1975).

The most obvious route of chromosomal change in the vertebrates has been by way of Robertsonian fusions, previously discussed in Chapter 7 (Figs. 7.6 through 7.9). Such changes have also been described in the cricket *Nemobius* and in several groups of lizards (Figs. 9.21 and 9.22). Since the total number of chromosome arms cannot be accounted for solely on the basis of centric fusions, some centromere loss or gain has been involved as well. In the Acrididae, one of the large assemblages of XX-XO grasshoppers, the primitive male karyotype, found in 90% of the species, is $2n = 23$, with all chromosomes being acrocentric. Autosome-autosome and X-autosome fusions have led to a variety of diploid numbers, with $2n = 11$ the lowest encountered. One autosome-autosome fusion would lessen the diploid number to 21, two fusions to 19, and six fusions to 11, while an X-autosome fusion would reduce the number by one, at the same time converting the XO to an XY pattern (Fig. 7.9). A Y-autosome fusion, in XX-XY species, would also reduce the $2n$ number by one and convert an XY to an $X_1 X_2 Y$ system, a circumstance encountered in the coccinellid beetle, *Chilocorus stigma*.

Similar shifts from acrocentrics to metacentrics, without loss of arm number, can be found as well among the mammals. Thus in the genus *Ovis*, the goat ($2n = 60$) has all acrocentrics while the domestic

Fig. 9.21. Karyotypes of the haploid chromosomes of various species of the cricket genus, Nemobius. The X chromosome is to the left in all cases (from White, 1973, after Baumgartner and Ohmachi).

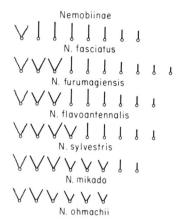

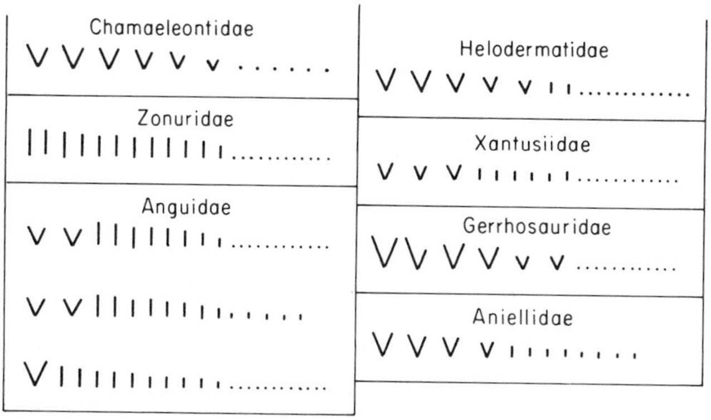

Nemobiinae

N. fasciatus

N. furumagiensis

N. flavoantennalis

N. sylvestris

N. mikado

N. ohmachii

Fig. 9.22. Karyotypes of lizards from seven families, indicating that centric fusions together with pericentric inversions have refashioned the karyotypes. It is uncertain as to the role played by the tiny dot chromosomes (White, 1973, after Matthey).

| Chamaeleontidae |
| V V V V V v . . . . . . |

| Zonuridae |
| \| \| \| \| \| \| \| \| \| \| \| \| ı . . . . |

| Anguidae |
| V V \| \| \| \| \| \| \| ı . . . . |

| V V \| \| \| \| \| \| \| ı ı . . . . . |

| V \| \| \| \| \| ı ı ı . . . . |

| Helodermatidae |
| V V V V v ı ı . . . . |

| Xantusiidae |
| V v V \| ı ı ı ı . . . . |

| Gerrhosauridae |
| VV V V v v . . . . |

| Aniellidae |
| V V V v ı ı ı ı ı ı . . |

sheep ($2n = 54$) has three pairs of metacentrics. Table 9.8 lists the chromosome numbers and types found among the primates. With the exception of the tarsier, which seems to exhibit little cytogenetic relationship either within its own prosimian group or with the anthropoids, there is a general tendency for the more primitive, or generalized, members to have high diploid numbers and more acrocentrics than metacentrics. Figure 9.23 shows this for two species of tree shrews; the number of arms (74) remains constant, while the diploid numbers are strikingly different. The possible phylogeny and chromosomal evolution in the Lemuroidea is given in Fig. 9.24. Again, it would seem that centric fusion is responsible for most of the karyological changes, although pericentric inversions, transforming an acrocentric to a metacentric, can complicate the picture.

Table 9.8. Chromosomes in representative primates (see Table 7.5 for the Anthropoidea)

Species	Common Name	2n	Acrocentrics	Two-armed Chromosomes	X	Y
Tupaiiformes						
Tupaia montana	Tree shrew	68	60	6	M	A
T. palawanensis	Tree shrew	52	30	20	M	A
Urogale everetti	Tree shrew	44	8	36	—	—
Tarsiiformes						
Tarsius bancanus	Tarsier	80	66	14	—	—
Lemuriformes						
Microcebus murinus	Miller's mouse lemur	66	64	2	—	—
L. mongoz	Mongoose lemur	60	54	4	A	A
L. fulvus fulvus	Brown lemur	48	30	16	A	A
L. macaco	Black lemur	44	22	20	A	A
Lorisiformes						
Perodicticus potto	Potto	62	36	24	S	A
Galago crassicaudatus	Bush baby	62	54	6	S	A
G. senegalensis	Bush baby	38	6	30	S	A
Cebidae (New World Monkeys)						
Cebus capucinus	Capuchin ringtail	54	26	26	A	A
Alouatta seneculus	Red howler monkey	44	30	12	A	M
Saimiri sciurea	Squirrel monkey	44	12	30	S	A
Ateles belzebuth	Golden spider monkey	34	2	30	M	A
Cercopithecidae						
Cercopithecus l horsti	l'Horst's monkey	72	18	52	M	M
Erythrocebus patas	Patas monkey	54	10	42	S	M
Colobus polykomos	Colobus monkey	44	0	42	M	A
Macaca mulatta	Rhesus monkey	42	0	40	M	M
Papio papio	Olive baboon	42	0	40	M	M

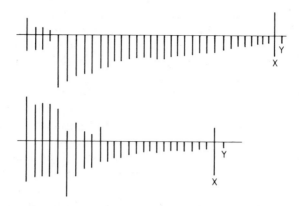

Fig. 9.23. The haploid karyotype of two species of the tree shrews (Tupaiidae). Both possess 74 chromosome arms, but *Tupaia montana* has an *n* number of 34 while that of *Tupaia palawanensis* is 26, with the difference related to the number of acrocentrics vs metocentrics. In both species, the X is approximately metacentric (more so in *montana* than in *palawanensis*) while the Y is a small acrocentric (redrawn from Arrighi, et. al., 1969).

Fig. 9.24. A possible family tree of the lemurs, assuming that higher chromosome numbers and acrocentrics are primitive, with centric fusions reducing the haploid numbers. Solid lines are judged to be correct, dashed lines tentative (after Buettner-Janusch, 1966, with data from Chu and Bender).

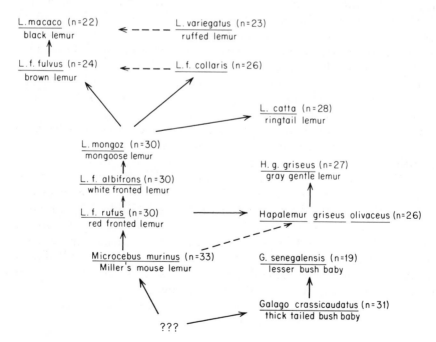

In the South American Ceboidea, the capuchin monkey, *Cebus capucinus,* is less specialized than the spider monkeys, *Ateles sp.,* while in the Old World Cercopithecoidea, *Cercopithecus* is less specialized than the olive baboon, *Papio,* or the colobus monkey. Among the anthropoids, the human would seem to stand intermediate between the highly specialized 44-chromosome gibbons and the 48-chromosome chimpanzee and gorilla, and on the basis of polypeptide comparisons the human and the chimpanzee share about a 99% correspondence, suggesting that biochemical evolution has been far more conservative than has the behavioral or anatomical. An obvious chromosomal difference between the human and the chimpanzee is the difference in acrocentrics vs. metacentrics, and the human chromosome 2 is derived by centric fusion of two acrocentrics in the latter species. However, the rearrangements are far more extensive when viewed closely. Lejeune has shown that 6 pericentric inversions, 1 telocentric fusion, 4 intercalary deletions or insertions, and 16 deletions or additions of terminal Q-band material in addition to much heterochromatic variation separate the two species.

Robertsonian fusions, while prevalent among animal groups, are not the only type of change. Males of the acrididian species, *Dactylotum bicolor,* have a diploid number of 17, all of which are acrocentric, indicating a loss of at least six centromeres plus an unspecified amount of chromatin. In at least one instance, again in the Acrididae, an increase in the diploid number has occurred to give a $2n = 25$ karyotype. The postulated mechanism involves a pericentric inversion to transform an acrocentric to a metacentric, followed by a dissociation of this same chromosome to give two acrocentrics or telocentrics. The evolutionary significance of dissociations to bring about karyotypic changes is suggested by the situation in the morabine grasshoppers of Australia. The basic $2n = 17$, but with a range from 13 to 21, with dissociations of autosomal metacentrics about one-half as frequent as fusions. No dissociations have been detected in either X or Y chromosomes.

The circumstances in avian species is also of interest. The avian karyotype contains many small chromosomes, the majority of them acrocentric, and the basic number of $2n = 82$ (Table 9.7). The number of metacentrics and subtelocentrics seem to be independent, to a degree, of the diploid number, suggesting that centromere shifts by way of pericentric inversions are common. Again, it is clear that the more specialized or advanced species, for example, Anna's hummingbird, has the lowest number of chromosomes.

The significance of Robertsonian fusion to adaptiveness is

debatable. Simple centric fusions, with little loss of heterochromatin, would seem, at first glance, to have little influence on linkage relations since the arms of metacentric chromosomes, to judge from *Drosophila* data, are independent of each other as regards recombination. Significant loss of heterochromatin, on the other hand, such as seem to have occurred in some species such as *Thais* or *Chilocorus*, could have an important effect on metabolic rates and introduce into a species a kind of metabolic as well as chromosomal polymorphism. This could lead to the ecological preferences found in *Thais* and be paralleled by the ecological preferences brought on by the paracentric inversions in *D. pseudoobscura*.

Robertsonian fusions may have been given undue prominence as a means of karyotypic change because of ease of detection. Paracentric inversions can be detected only as a result of crossing over within the inversion loop, and the resultant anaphase I bridge and fragment (Fig. 6.13), or in polytene chromosomes, and in some groups—*Drosophila, Tradescantia,* and *Paris,* but not in grasshoppers —these outstrip all other rearrangements in frequency. White, in fact, has estimated that as many as 6,000 may have become established in one or another of the 300 species of *Drosophila* that have been well studied. Pericentric inversions (32 in number) and heterochromatic "additions" (38) have been almost as frequent as centric fusions (58) in *Drosophila,* but even these figures may be on the low side, as the evidence from *D. miranda* would suggest. This species has evolved from a *D. pseudoobscura-persimilis* stock (Fig. 6.19). Hybrids between *pseudoobscura* and *miranda* are possible under laboratory conditions; pairing on a band-for-band basis in polytene chromosomes is much disrupted, and up to 100 breaks, followed by extensive rearrangements, would be needed to derive the *miranda* karyotype from its ancestral stock. Many of the changes were minute in character and of uncertain origin and could only have been detected in polytene chromosomes.

The karyotype of *D. virilis,* consisting of five acrocentrics and a dot chromosome, is judged to be the primitive karyotype of the genus. Similar karyotypes are found in *palustris, guttifera, tripunctata,* and possibly *funebris,* while *repleta* has a very much reduced Y chromosome, and *hydeioides* has both a reduced Y and the absence of the dot chromosome (Fig. 9.25). In other *virilis* groups, and using *virilis* as a starting point, a more detailed examination has revealed a pericentric inversion in Primitive III, *ezoana* and *littoralis,* a 3–4 fusion in *littoralis,* a 2–3 fusion in *texana* and *americana,* and an X–4 fusion in *americana* to create an $X_1 Y_1 Y_2$ system,

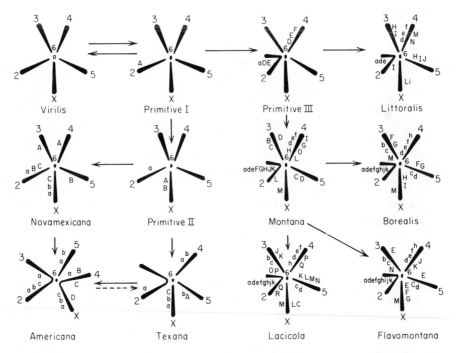

Fig. 9.25. Chromosome phylogeny in the *D. virilis* species group. Chromosomes are identified by numbers except for the X; inversions are identified by capital letters in the species in which they first occurred, by a small letter thereafter; centric fusions (translocations) can be recognized by the union of two chromosomes (e.g., the 3-4 fusion in *littoralis*). A pericentric inversion in chromosome 2 occurs in the six species to the right of the diagram (Patterson and Stone, 1952).

with Y_2 being an earlier acrocentric 4. The paracentric inversions are also indicated, both as to the species in which they first occurred and as to those species in which the rearrangements are fixed.

Three species of *Drosophila* have a haploid number of 7, indicating that dissociations have occurred, although very infrequently. The origin of the additional centromere in these species remains obscure. The trend in the genus, therefore, has been a reduction in number through fusion, an occasional loss of the dot chromosome, and an extensive rearrangement of chromatin by inversions. Reciprocal translocations have been relatively rare, but small, internal changes of uncertain origin have been numerous, and they seem to have brought about a more genetic reshuffling than have the larger and more obvious rearrangements.

Karyotype Trends

The basic nature of the eukaryotypic chromosome imposes its own restrictions on the variety of rearrangements of chromatin that are both possible and viable, but even within these strictures, and excluding polyploidy, the karyotypes extant reveal that many avenues of change have been exploited by one group of organisms or another. The pattern of rearrangements in the species of *Oenothera* as contrasted with those in *Crepis* or *Haplopappus,* on the one hand, or in *Trillium,* on the other, or those in the orthopteran Acrididea as contrasted with those in the lepidopteran genus, *Lysandra,* or in *Drosophila pseudoobscura,* make it obvious, however, that whatever the frequency and nature of initial change, the selection processes have moved the karyotypes along specific pathways. Certain karyotypic changes seem to be preserved while others are either not formed or are somehow prevented from becoming established. Karyotypic trends, therefore, are not random phenomena, and it is more than likely, although evidence is scanty, that patterns of rearrangements and of selection pressures are equally nonrandom.

Karyotype changes stem from chromosome breaks followed by the reunion of broken ends, with breakage and repair systems determining the eventual form of change. A study carried out by Giles on pollen and roottip mitoses in *Tradescantia* indicates the nonrandomness of spontaneous chromatin breaks or rearrangements. A number of points were established: post-replicative chromatid aberrations were far more frequent than pre-replicative changes which appear as chromosome breaks (i.e., those involving both chromatids at the same locus as opposed to a chromatid break which involves only a single chromatid); breaks of all kinds were nonrandomly distributed along the chromosome arms, being far more frequent in the vicinity of the centromere (heterochromatin?) than elsewhere; the frequency of changes detected was far too high to be accounted for by the level of background radiation even though the pattern of break distribution was similar to that induced by X-radiation; from one individual plant to another the frequency of spontaneous aberrations varied significantly, indicating a genetic control of break frequency and/or repair efficiency; and, the frequency of aberrations was three times higher in F_1 and F_2 interspecific hybrids than in the parental species, pointing to hybridization as a means of hastening karyotypic evolution, particularly if hybridization occurred in an environment diverse enough to provide an opportunity for the testing of a variety of new genotypes and karyotypes.

A comparison of many of the illustrations in this volume, or in any other such as that of White on animal karyotypes, will reveal that there are many distinctive karyotypes characterizing both large and small taxonomic groups. These are so different from each other, and so unique to their own taxon, that they could not have been attained by other than change operating along determined pathways. White has coined the term *karyotypic orthoselection* to describe this phenomenon, thus placing emphasis on the mechanisms of establishment instead of on patterns of initial change. Massive screening of populations, including human populations, for karyotypic changes have revealed that spontaneously occurring aberrations are not uncommon and that there is probably always a pool of change from which karyotypic selection can be made. What is not clear is why certain pathways of change are preferable to others which seem equally accessible for exploitation, although it seems not unreasonable to assume that trends in so basic a feature as the karyotype must have an adaptive significance that is being exploited.

Table 9.9 lists a number of examples of karyotypic orthoselection that relate to specific kinds of rearrangements; it is not possible to define the changes that give an entire taxon its karyotypic character. Some of the changes have been discussed previously, and some are worthy of additional comment. The point was made earlier that in both plant and animal groups, and particularly within large genera possessing both primitive and advanced species, it is the more advanced, or perhaps more accurately the specialized, forms that show reduced chromosome numbers, accompanied in some instances by lowered DNA values and a reduction in chromosome size. In many animal groups, this has occurred by the centric fusion of acrocentric chromosomes, thus moving the karyotype along the pathway of chromosomal symmetry; that is, the chromosomes, most of which are of relatively similar size, become two-armed and leave medially placed centromeres. Other than the suggestion that such centric fusions may be associated with ecologically disturbed areas, as in the marine snail *Thais,* there is little evidence to point to the evolutionary importance of symmetry; how much centric chromatin is lost in such fusions is not always obvious, and it is possible to have such fusions occur without loss of centromeres, and probably with little loss of chromatin. In the Compositae, however, the reduction in chromosome numbers has been through unequal, but reciprocal, translocations, with loss of centromeres and of chromatin. The parallel reduction in chromosome size (Fig. 9.19) undoubtedly means a lowering of DNA values per haploid complement; it would be of interest to know just what kind of chromatin was lost in the process of reduction: heterochromatin, Gilbertian exons or introns, spacer DNA, or repetitive DNA of varying lengths and complexity? In any event, the result is a trend toward asymmetric chromosomes of different sizes, a circumstance clearly associated in plants with the annual habit. It remains to be seen just what karyotypic changes are associated with the lower DNA values in the specialized fishes.

The *Oenothera* complex, seen as well in other genera of the Onagraceae and in *Rhoeo discolor,* is also a translocation phenomenon, but one involving reciprocal and equal exchanges between metacentric chromosomes and facilitated by the presence of abundant centric heterochromatin within which breakage and exchange would be more frequent and more viable than exchanges involving other parts of the karyotype. Since the breakpoints seem to be confined to constitutive heterochromatin, no critical linkages are involved or disturbed, as they might well be in the Compositae. In the latter, homozygosity, attained through tightened linkages and self-compatibility, is an accompaniment to the annual habit and an efficient seed production, with all features combining to produce a highly competitive system in appropriate peripheral habitats. The *Oenothera* system of translocations complexes, on the other hand, is based on a maximized heterozygosity which provides hybrid vigor; it, too, is reinforced by self-compatibility, but for the purpose of maintaining the complexes instead of preventing them from being broken up through outcrossing. A degree of homozygosity results at the ends of the pairing regions and ensures a regular segregation, but heterozygosity is achieved through a system of balanced lethals coupled with a chemical recognition pattern that brings together dissimilar gametes. *Paeonia californica, P. brownii,* and the several annual species of *Clarkia* seem to be moving along a similar, but not yet fully established, pathway, as are roach populations of *Periplanata americana* and *Balberus discoidalis* and similar populations of the Brazilian scorpion, *Tityus bahiensis.* In these species, the number of translocations is not fixed, and there is no evidence that balanced lethals, selective fertilization, or adaptive preferences are involved.

The presence of centric heterochromatin is probably a necessary karyotypic feature for the frequent occurrence of centric fusions and the less frequency reciprocal exchanges of the *Oenothera* type. Even the unequal exchanges leading to a reduction in chromosome number and size seem to involve such constitutive heterochromatin, although euchromatic breaks are part of the pattern as well. Constitutive heterochromatin by itself, however, is not a sufficient cause for these exchanges. Many other species in both the plant and animal kingdoms have ample amounts of heterochromatin, but they have not

Table 9.9. Examples of karyotype orthoselection in both plants and animals

Type of Orthoselection	Organism		Comments
	Major Group	Species	
Paracentric inversions	Diptera	Many, but not all, species of *Drosophila*	Some show ecological adaptation and heterozygous superiority; *pseudo-obscura-persimilis* has them concentrated in chr. 3; *repleta* and *melanica* groups in chr. 2; *grimshawi* group in chr. 4, thus permitting phylogenetic families to be constructed
		Simulium, Chironomus, Sciara, Anopheles, Tenipes	Presence in Diptera may be due to ease of detection, while absence in other forms may be due to restriction of chiasma formation to noninverted areas of chromosomes
	Angiosperms	*Tradescantia, Paris, Paeonia, Polygonatum*	Evolutionary significance unknown, but inversions seem to be tolerated without detriment
Pericentric inversions	Orthoptera	North American species of *Trimopteris, Cicrotettix* and *Aerochoreutes*; Australian species of *Keyacris* (*Moraba*), *Austroicetes,* and *Warramunga*	Sporadic in dipteran species; tolerated if chiasmata confined to ends of chromosomes
	Aves	See Table 9.7	Little change in chromosome number, but shift from acrocentrics to meta- or submetacentrics with phylogenetic advancement

538

Category	Taxon	Genera	Comments
	Mammalia	*Peromyscus, Rattus, Mastomys*	Some apparent pericentrics, based on centromere positions, may well be due to variations in heterochromatin amount
	Angiosperms	*Agave, Yucca*	Asymmetrical karyotypes of these two genera may be due in part to derivation from more symmetrical karyotypes by way of pericentric inversions
Centric fusions (generally leads to symmetrical karyotypes)	Orthoptera Coleoptera Mollusca	*Keyacris, Nemobius, Chilocorus Thais*	Evolutionary significance suggested on basis of persistence and distribution; see Fig. 7.7
	Reptilia	Many lizards	See Figs. 9.21 and 9.22
	Mammalia	Many primates	See Table 9.8; much of the reduction in chromosome numbers can be attributed to centric fusions
Reciprocal translocations (equal and unequal)	Orthoptera	*Periplaneta, Blaberus*	Considerable polymorphism in a state of flux
	Scorpions	*Tityus, Isometrus*	Similar state of polymorphism
	Angiosperms	*Oenothera* (including subgenera and species) *Rhoeo, Datura*	Involve metacentric chromosomes with breaks in centric heterochromatin, hence no change in symmetry
		Paeonia	Acrocentric and metacentric chromosomes involved, with breaks variously placed in the karyotype
		Aster, Crepis, Haplopappus, Delphinium, Aconitum, Aloe, Gasteria, Haworthia	Reduced chromosome numbers and asymmetrical karyotypes due to unequal reciprocal translocations and loss of centromeres and heterochromatin

Table 9.9. (Continued)

Type of Orthoselection	Organism		Comments
	Major Group	Species	
Asymmetrical karyotypes	Reptilia and Aves	Lizards, snakes, and all birds	See Fig. 9.26 and Table 9.7; karyotypes are made up of macrochromosomes and microchromosomes, with the karyotype being carried from the reptiles to the birds as the latter evolved.
	Angiosperms	*Yucca, Agave*	See Fig. 9.26; these karyotypes undoubtedly evolved from the more symmetrical karyotypes of the lily group
Dissociations	Lepidoptera	*Lysandra* sp.	From $n = 24$ to $n = $ ca. 220
		Agrodiaetus sp.	From $n = 10$ to $n = $ ca. 80
		Erebia sp.	From $n = 7$ to $n = 51$
		Actinote sp.	From $n = 14$ to $n = 150$
		Leptidea sp.	From $n = 54$ to $n = 104$
			There is a question of polyploidy being involved, although this is generally discounted, and there is the further possibility of diffuse centromeres, with fragmentation high and all fragments surviving

Source: White, 1964, 1973; Stebbins, 1971; John and Lewis, 1966.

undergone changes of this type. The selective pressures that initiate and sustain such trends remain undetermined, but once initiated it seems to continue, and as it continues the prospects for future evolutionary change seem to become increasingly narrow.

Orthoselection also seems to be operative on inversions as well. As regards those of the paracentric variety, the liliaceous genus *Trillium* is relatively free of them while in the closely related genus *Paris* no individual seems free of them. *D. willstoni* and *D. subobscura* possess more than 40 different inversions scattered throughout their genomes, and with no obvious ecological relations evident, although most of the inversions, as would be expected, are more concentrated in the center of their range, and isolated inversions are found at the periphery. In *D. simulans, D. virilis,* and *D. novamericana* none have been found. In the *D. pseudoobscura-persimilis* complex (Fig. 6.19), the many inversions are concentrated in the left arm of chromosome 3; in species of the *D. melanica* group, chromosome 2 is a favored location. As an explanation for clustered inversions, it has been suggested that inversions occur frequently and that the appearance of one with a favored effect would lead to the retention of others with similar influence and in the same neighborhood; while that explanation might satisfy the *pseudoobscura* circumstances, it cannot serve equally to cover those in *willstoni*.

Pericentric inversions have played a less prominent role in orthoselection, very likely because of the sterility induced when crossing over occurs. Some 30 or more have been established among the investigated species of *Drosophila,* transforming acrocentrics into metacentrics or submetacentrics. In the deer mouse, *Peromyscus,* in which the number of acrocentrics can range from 30 to as low as 12 or 14, and without accompanying change in chromosome number, pericentrics seem numerous. Their prevalence in some grasshopper groups is also well established. The genera *Trimerotropis* and *Circotettix* of North America and several of the morabine species of Australia display a significant frequency of these rearrangements. On occasion, the reverse will occur, a metacentric being converted by dissociation to two acrocentrics. In the morabine grasshoppers, two pairs of homologues exhibit heterozygosity for pericentric inversions; the presence of one pair of heterozygous inversions confers superiority; the presence of two pairs is detrimental.

A final example of orthoselection, if indeed that is what is really involved, is that relating to chromosome size. Ordinarily, the chromosomes of a genome are of a similar size, the largest being no more than two or three times as long as the shortest (B chromosomes are an exception). In a few species, however, chromosome size is such that a distinct bimodality is evident to such an extent that one can speak of macrochromosomes and microchromosomes. Among the angiosperms, the genera *Yucca* and *Agave* of the Liliales each have 5 large and 25 small chromosomes in the haploid set (Fig. 9.26). During division, either mitotic or meiotic, the large chromosomes arrange themselves around the periphery of the spindle while the small chromosomes orient themselves within the body of the spindle. A similar distribution of size, although of a more graded sort, occurs among the avian and lizard groups (Fig. 9.22). Whether this is an accident of evolutionary history or an example of a genome selected for unknown reasons is not clear, but as is true for most karyotypic circumstances, it is easier to describe than it is to explain in satisfying terms.

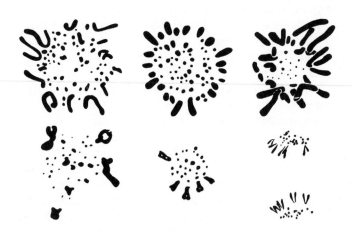

Fig. 9.26. Macro- and micro-chromosomes of bird and plant species at metaphase I of meiosis, illustrating both the size differences within a genome and the manner by which the micro-chromosomes tend to congregate in the center of the spindle. Top left, *Oceanodroma leucorroa* (Procellariidae); top middle, *Melopsittacus undulatus* (Psittacidae); top right, *Cuculus canorus* (Cuculidae); bottom left, *Anas platyrhyncha* (Anatidae); bottom middle and right, metaphase I and anaphase I in the plant *Yucca flaccida* (Liliaceae) (Matthey, 1949 and O'Mara, 1932).

BIBLIOGRAPHY

Angerer, R. C., et al., "Single Copy DNA and Structural Gene Sequence Relationships Among Four Sea Urchin Species," *Chromosoma, 56* (1976), 213–22.

Arrighi, F. E., et al., "Chromosomes of the Tree Shrews (Tupaiidae)," *Cytogenetics, 8* (1969), 199–208.

Babcock, E. B., *The Genus Crepis,* Vols. I and II: Berkeley: University of California Press, 1947.

Bachmann, K., et al., "Genome Size in Birds," *Chromosoma, 37* (1972), 405–16.

————, "Nuclear DNA Amounts in Vertebrates," in *Evolution of Genetic Systems,* H. H. Smith, ed. New York: Plenum, 1972.

Bachmann, K., and E. L. Rheinsmith, "Nuclear DNA Amounts in Pacific Crustacea," *Chromosoma, 43* (1973), 225–36.

Bainton, D. F., and M. G. Farquhar, "Origin of Granules in Polymorphonuclear Leukocytes," *Jour. Cell Biol., 28* (1966), 277–301.

Becak, M. L., et al., *Chromosome Atlas: Fish, Amphibians, Reptiles and Birds,* Vols. 2 and 3. New York: Springer-Verlag, N.Y., 1973, 1975.

Bender, M. A., and L. E. Mettler, "Chromosome Studies of Primates," *Science, 128* (1958), 186–90.

Bennett, M. D., "Nuclear DNA Content and Minimum Generation Time in Herbaceous Plants," *Proc. Roy. Soc. Lond., B 181* (1972), 109–35.

————, and J. B. Smith, "The Effects of Polyploidy on Meiotic Duration and Pollen Development in Cereal Anthers," *Proc. Roy. Soc. Lond., B 181* (1972), 81–107.

Bloom, W. L., "Origin of Reciprocal Translocations and Their Effect in *Clarkia speciosa,*" *Chromosoma, 49* (1974), 61–76.

————, "Translocation Heterozygosity, Genetic Heterozygosity and Inbreeding in *Clarkia speciosa,*" *Evolution, 31* (1977), 256–64.

Britton, R. J., and E. H. Davidson, "Gene Regulation for Higher Cells: A Theory," *Science, 165* (1969), 349–57.

Buettner-Janusch, J., *Origins of Man.* New York: John Wiley, 1966.

Camenzind, R., and R. B. Micklas, "The Non-random Chromosome Segregation in Spermatocytes of *Gryllotalpa hexadactyla,*" *Chromosoma, 24* (1968), 324–35.

Carlson, E. A., "Comparative Genetics of Complex Loci," *Quart. Rev. Biol., 34* (1959), 33–67.

Cech, T. R., and J. E. Hearst, "An Electron Microscope Study of Mouse Foldback DNA," *Cell, 5* (1975), 429–46.

Commoner, B., "Roles of Deoxyribonucleic Acid in Inheritance," *Nature, 202* (1964), 960-68.

Crain, W. R., et al., "Contrasting Patterns of DNA Sequence Arrangement in *Apis mellifera* (Honeybee) and *Musca domestica* (Housefly)," *Chromosoma, 59* (1976), 1 12

Crick, F., "General Model for Chromosomes of Higher Organisms," *Nature, 234* (1971), 25-27.

————, "Split Genes and RNA Splicing," *Science, 204* (1979), 264-71.

Darnell, J. E., Jr., "Implications of RNA—RNA Splicing in Evolution of Eukaryotic Cells," *Science, 202* (1978), 1257-60.

Davidson, E. H., and R. J. Britten, "Regulation of Gene Expression: Possible Role of Repetitive Sequences," *Science, 204* (1979), 1052-59.

Davidson, E. H., et. al., "Comparative Aspects of DNA Organization in Metazoa," *Chromosoma, 51* (1975), 253-59.

Dobzhansky, Th., "The X-chromosome in the Larval Salivary Glands of Hybrids *Drosophila insularis* X *Drosophila tropicalis,*" *Chromosoma, 8* (1957), 691-98.

Duffy, P. A., "Chromosome Variation in *Peromyscus:* A New Mechanism," *Science, 176* (1972), 1333-34.

Fincham, J. R. S., and G. R. K. Sastry, "Controlling Elements in Maize," *Ann. Rev. Genetics, 8* (1974), 15-50.

Flamm, W. G., "Highly Repetitive Sequences of DNA in Chromosomes," *Int. Rev. Cytology, 32* (1972), 1-52.

Flavell, R. B., et al., "Repeated Sequence DNA Relationships in Four Cereal Genomes," *Chromosoma, 63* (1977), 205-22.

Gilbert, W., "Why Genes in Pieces?" *Nature, 271* (1978), 501.

Gillespie, D., and R. C. Gallo, "RNA Processing and RNA Tumor Virus Origin and Evolution," *Science, 188* (1975), 802-11.

Graham, D. E., et. al., "Interspersion of Repetitive and Nonrepetitive DNA Sequences in the Sea Urchin Genome," *Cell, 1* (1974), 127-37.

Graham, J. B., and C. A. Istock, "Gene Exchange and Natural Selection Cause *Bacillus subtilis* to Evolve in Soil Culture," *Science, 204* (1979), 637-38.

Grant, V., "Chromosome Repatterning and Adaptation," *Adv. in Genetics, 8* (1956), 89-107.

Green, M. M., "The Case for DNA Insertion Mutations in *Drosophila,*" in *DNA: Insertion Elements, Plasmids and Episomes.* Cold Spring Harbor, N.Y.: Cold Spring Harbor Laboratory, 1977.

————, Insertion Mutants and the Control of Gene Expression in *Drosophila melanogaster,*" in *The Clonal Basis of Development* (S. Subtelny and I. M. Sussex, eds.), New York: Academic Press, 1978.

Hahn, W. E., et. al., "One Strand Equivalent of the *Escherischia coli* is Transcribed: Complexity and Abundance Classes of mRNA," *Science, 197* (1977), 582-5.

Hinegardner, R., "Evolution of Cellular DNA Content in Teleost Fishes," *Amer. Nat., 102* (1968), 517-23.

Hsu, T. C., "A Possible Function of Constitutive Heterochromatin: The Bodyguard Hypothesis," *Genetics, 79 (Suppl.)* (1975), 137-50.

Hughes-Schrader, S., "The Chromosomes of Mantids (Orthoptera, Manteidae) in Relation to Taxonomy," *Chromosoma, 4* (1958), 1-55.

John B., and K. R. Lewis, "Chromosome Variability and Geographic Distribution in Insects," *Science, 152* (1966), 711-21.

Keyl, H.-G., Duplikationen von Untereinheiten der chromosomalen DNA während der Evolution von *Chironomus thummi*," *Chromosoma, 17* (1965), 138-40.

King, M.-G., and A. C. Wilson, "Evolution at Two Levels in Humans and Chimpanzees," *Science, 188* (1975), 107-16.

Koller, P. C., "The Genetical and Mechanical Properties of the Sex Chromosomes," *J. Genetics, 32* (1936), 451-72.

Lejeune, J., et al., "Comparison de la structure fine des chromatides d'*Homo sapiens* et de *Pan troglodytes*," *Chromosoma, 43* (1973), 423-44.

Lemeunier, F., et al., "Relationships Within the *Melanogaster* Subgroup Species of the Genus *Drosophila* (*Sophophora*). III. The Mitotic Chromosome and Quinacrine Fluorescent Patterns of the Polytene Chromosomes," *Chromosoma, 69* (1978), 349-62.

Levin, D. A., "The Nature of Plant Species," *Science, 204* (1979), 381-4.

Lewis, H. "The Origin of Diploid Neospecies of *Clarkia*," *Amer. Nat., 107* (1973), 161-70.

Lewis, H., and P. H. Raven, "Rapid Evolution in *Clarkia*," *Evolution, 12* (1958), 319-66.

Lucchesi, J. C., "Gene Dosage Compensation and the Evolution of Sex Chromosomes," *Science, 202* (1978), 711-16.

Madison, J. T., "Transfer RNAs," in *Handbook of Genetics*, Vol. 5, R. C. King, ed. New York: Plenum, 1975.

Magenis, R. E., et al., "Giemsa-11 Staining of Chromosome 1: A Newly Described Heteromorphism," *Science, 202* (1978), 64-65.

Markert, C. L., et al., "Evolution of a Gene," *Science, 189* (1975), 102-14.

McCarron, M., et. al., "Organization of the *rosy* Locus in *Drosophila melanogaster:* Further Evidence in Support of a *cis*-Acting Control Element Adjacent to the Xanthine Dehydrogenase Structural Element," *Genetics, 91* (1979), 275-93.

McClintock, B., "Spontaneous Alterations in Chromosome Size and Form in *Zea Mays*," *Cold Spring Harbor Symp. Quant. Biol., 9* (1941), 72-81.

McClintock, B., "Genetic Systems Regulating Gene Expression During Development," In *Development Biology* (suppl.), (1967), 84-112.

———, "Chromosome Organization and Gene Expression," *Cold Spring Harbor Symp. Quant. Biol., 16* (1951), 13-47.

———. "The Control of Gene Action in Maize," *Brookhaven Symp. Biol., 18* (1965), 162-84.

Miller, L., and **D. D. Brown,** "Variation in the Activity of Nucleolar Organizers and Their Ribosomal Gene Content," *Chromosoma, 28* (1969), 430-44.

Mills, D. R., et al., "Complete Nucleotide Sequence of a Replicating Molecule," *Science, 180* (1973), 916-27.

Mizuno, S., and **H. C. Macgregor,** "Chromosomes, DNA Sequences, and Evolution in Salamanders of the Genus *Plethodon*," *Chromosoma, 48* (1974), 239-96.

Murray, B. G., "The Cytology of the Genus *Briza L.* (Gramineae)." I. Chromosome Numbers, Karyotypes and Nuclear DNA Variation," *Chromosoma, 49* (1975), 299-308.

Murray, M. G., and **W. F. Thompson,** "Sequence Repetition and Interspersion in Pea DNA," Carnegie Inst. Yearbook 77, (1978), 316-23.

Nuffer, M. G., "Dosage Effect of Multiple *Dt* Loci in Mutation of *a* in Maize Endosperm," *Science, 121* (1955), 399-400.

———, "Mutation Studies at the A_1 Locus in Maize," I. A Mutable Allele Controlled by *Dt. Genetics 46* (1961), 624-40.

Ohno, S., *Evolution by Gene Duplication.* New York: Springer-Verlag, N.Y., 1970.

O'Mara, J., "Chromosome Pairing in *Yucca flaccida*," *Cytologia, 3* (1932), 66-76.

Patterson, J. T., and **W. S. Stone,** *Evolution in the Genus Drosophila.* New York: Macmillan, 1952.

Peterson, P. A., "The Position Hypothesis for Controlling Elements in Maize," in *DNA: Insertion Elements, Plasmids and Episomes,* Cold Spring Harbor, N.Y.: Cold Spring Harbor Laboratory, 1977.

Price, H. J., and **K. Bachmann,** "Mitotic Cycle Time and DNA Content in Annual and Perennial Microseridinae (Compositae, Cichoriaceae)," *Plant Sys. Evol., 126* (1976), 323-30.

Rees, H., "DNA in Higher Plants," in *Evolution of Genetic Systems,* H. H. Smith, ed. New York: Plenum, 1972.

———, and **R. N. Jones,** "The Origin of the Wide Species Variation in Nuclear DNA Content," *Int. Rev. Cytology, 32* (1972), 53-92.

Rhoades, M. M., "The Genetic Control of Mutability in Maize," *Coldspring Harbor Symp. Quant. Biol., 9* (1941), 138-44.

Rigby, P. W. J., et al., "Gene Duplication in Experimental Enzyme Evolution," *Nature, 25* (1974), 200-04.

Rizzo, P. J., "Basic Chromosomal Proteins in Lower Eukaryotes: Relevance to the Evolution and Function of Histones," *J. Mol. Evol., 8* (1976), 79-94.

Rothfels, K., et al., "Chromosome Size and DNA Content of Species of *Anemone L.* and Related Genera (Ranunculaceae)," *Chromosoma, 20* (1966), 54-74.

Schmid, C. W., et al., "Inverted Repeat Sequences in the *Drosophila* Genome," *Cell, 5* (1975), 159-72.

Schimke, R. T., et. al., "Gene Amplification and Drug Resistance in Cultured Murine Cells," *Science, 202* (1978), 1051-5.

Smith, M., "Recombination and the Rate of Evolution," *Genetics, 78* (1974), 290-304.

Sparrow, A. H., et al., "A Survey of DNA Content per Cell and per Chromosome of Prokaryotic and Eukaryotic Organisms: Some Evolutionary Considerations," in *Evolution of Genetic Systems,* H. H. Smith, ed. New York: Plenum, 1972.

Spofford, J. B., "Heterosis and the Evolution of Duplications," *Amer. Nat., 103* (1969), 407-32.

Stebbins, G. L., "Chromosomal Variation and Evolution," *Science, 152* (1966), 1463-69.

―――, et al., "Chromsomes and Phylogeny in the Compositae, Tribe Cichoricae," *Univ. Calif. Pub. (Botany), 26* (1952-53), 401-29.

Straus, N. A., "Repeated DNA in Eukaryotes," in *Handbook of Genetics,* Vol. 5, R. C. King, ed. New York: Plenum, 1976.

Thompson, W. F., "Perspectives in the Evolution of Plant DNA," Carnegie Inst. Yearbook 77 (1978), 310-11.

Tobgy, H. A., "A Cytological Study of *Crepis fuliginosa, C. neglecta,* and Their F_1 Hybrid, and Its Bearing on the Mechanism of Phylogenetic Reduction in Chromosome Number," *J. Genetics, 45* (1943), 67-111.

Van't Hof, J., "Relationships Between Mitotic Cycle Duration, S Period Duration and the Average Rate of DNA Synthesis in the Root Meristem Cells of Several Plants," *Exptl. Cell Res., 39* (1965), 48-58.

―――, and C. A. Bjerknes, "18 μm Replication Units of Chromosomal DNA Fibers of Differentiated Cells of Pea (*Pisum sativum*)," *Chromosoma, 64* (1978), 287-94.

―――, and C. A. Bjerknes, "Chromosomal DNA Replication in Higher Plants," *BioScience, 29* (1979), 18-22.

————, et. al., "Replicon Properties of Chromosomal DNA Fibers and the Duration of DNA Synthesis of Sunflower Root Tip Meristem Cells of Different Temperatures," *Chromosoma, 66* (1978), 161–71.

————, et. al., "The Size and Number of Replicon Families of Chromosomal DNA of *Arabidopsis thaliana*," *Chromosoma, 68* (1978), 269–85.

Wachtel, S. S., "H-Y Antigen and the Genetics of Sex Determination," *Science, 198* (1977), 797–99.

Wallace, B., and T. L. Kass, "The Structure of Gene Control Regions," *Genetics, 77* (1974), 544–58.

Wellauer, P. K., et al., "X and Y Chromosomal Ribosomal DNA of *Drosophila:* Comparison of Spacers and Insertions," *Cell, 14* (1978), 269–78.

Westergaard, M., "The Mechanisms of Sex Determination in Dioecious Flowering Plants," *Adv. in Genetics, 9* (1958), 217–81.

White, M. J. D., "Principles of Karyotype Evolution in Animals," *Genetics Today, 1* (1963), 391–97.

————, "Models of Speciation," *Science, 159* (1968), 1065–70.

Young, M. W., and B. H. Judd, "Nonessential Sequences, Genes, and the Polytene Chromosome Bands of *Drosophila Melanogaster*," *Genetics, 88* (1978), 723–42.

Appendix

Journals

American Journal of Human Genetics
American Naturalist
American Scientist
Biochemical Genetics
Canadian Journal of Genetics and Cytology
Caryologica
Cell
Chromosoma
Cytogenetics & Cell Biology (formerly Cytogenetics)
Cytologia
Experimental Cell Research
Evolution
Genetical Research
Genetics
Genetika
Hereditas
Heredity
Journal de Génétique Humaine
Journal of Cell Biology
Journal of Genetics
Journal of Heredity
Journal of Medical Genetics
Journal of Molecular Biology
Molecular & General Genetics
Mutation Research
Nature
Proceedings of the National Academy of Sciences
Proceedings of the Royal Society of London, Series B
Science
Scientific American
Somatic Cell Genetics

Reviews

Advances in Genetics
Annual Review of Genetics
Cold Spring Harbor Symposia on Quantitative Biology
International Review of Cytology
The Cell, Volumes 1-6.
Symposia of the Society of Experimental Biology

Subject Index

Species Index

565

Author Index

Abelson, J., 353
Achtman, M., 59
Adelberg, E., 19
Albert, B. N., 292, 294, 353
Allen, J. R., 59
Allfrey, V. G., 19
Amati, P., 270, 279
Ames, B., 313, 353
Anderson, R. P., 403
Angerer, R. C., 543
Arber, W., 19
Arrighi, F. E., 181, 531, 543
Auerbach, C., 18
Avery, A. G., 412, 441
Avery, O. T., 18

Babcock, E. B., 88, 177, 452, 471, 525, 543
Bachmann, K., 504, 543, 544
Baglioni, C., 373
Bailey, W. T., 18
Bainton, D. F., 115, 543
Baker, B. S., 279

Balbiani, E. G., 162
Balkeslee, A. F., 18, 412
Baltimore, D., 61
Baramki, T. H., 19
Barclay, I. R., 441
Barlow, B. A., 405, 442
Barr, M., 114–116
Barry, E. G., 281
Barski, G., 19
Bartalos, M., 19
Basile, R., 163, 181
Bassett, E. G., 31, 60, 329, 353
Bateson, W., 17
Beadle, G. W., 18, 405
Beatty, R. A., 441, 471
Becak, M. L., 528, 543
Bedbrook, J. R., 53, 59
Beerman, W., 168, 169, 177, 178
Beers, R. E., 31, 60, 329, 353
Belford, H. S., 134, 137, 178
Bender, M. A., 531, 543
Bennett, M. D., 134, 137, 178, 210, 232, 281, 502–504, 524, 543

573